Student Solutions Manual for Hirsch/Goodman's

Understanding Intermediate Algebra

A Course for College Students

Fifth Edition

Cheryl Cantwell
Seminole Community College

Australia • Canada • Mexico • Singapore • Spain • United Kingdom • United States

Assistant Editor: *Julie Foster*
Marketing Manager: *Leah Thomson*
Marketing Communications: *Samantha Cabaluna*
Marketing Assistant: *Maria Salinas*
Production Coordinator: *Dorothy Bell*
Cover Design: *Andrew Ogus Book Design*
Cover Illustration: *Judith Harkness*
Print Buyer: *Christopher Burnham*
Printing and Binding: *Globus Printing*

For more information about this or any other Brooks/Cole product, contact:
BROOKS/COLE
511 Forest Lodge Road
Pacific Grove, CA 93950 USA
www.brookscole.com
1-800-423-0563 (Thomson Learning Academic Resource Center)

Printed in the United States of America

10 9 8 7 6 5 4 3 2

ISBN 0-534-38341-6

This book is dedicated to the loving memory of my mother, Nancy Varner, who was my typist for 15 years. She put a lot of hard work and long hours into this manual.

TABLE OF CONTENTS

CHAPTER 1

Exercises 1.1

1. $\{3, 4, 5, 6, 7, 8, 9, 10, 11\}$

3. $\emptyset$

5. $\{41, 43, 47\}$

7. $\{0, 7, 14, 21, 28, \dots\}$

9. $\{1, 2, 3, 6, 9, 18, 27, 54\}$

11. $A \cap B = \{0, 3, 6\}$

13. $A = \{0, 1, 2, 3, 4, 5, 6\}$
 $B = \{3, 6, 9, 12, 15, 18, 21, 24, 27, 30, 33\}$

 $A \cup B = \{0, 1, 2, 3, 4, 5, 6, 9, 12, 15, 18, 21, 24, 27, 30, 33\}$

15. $A = \{0, 1, 2, 3, 4, 5, 6\}$
 $D = \{7, 8, 9, 10, 11, 12, 13\}$

 $A \cup D = \{0, 1, 2, 3, 4, 5, 6, 7, 8, 9, 10, 11, 12, 13\}$

17. $66 = 2 \cdot 33$
 $= 2 \cdot 3 \cdot 11$

19. $128 = 2 \cdot 64$
 $= 2 \cdot 2 \cdot 32$
 $= 2 \cdot 2 \cdot 2 \cdot 16$
 $= 2 \cdot 2 \cdot 2 \cdot 2 \cdot 8$
 $= 2 \cdot 2 \cdot 2 \cdot 2 \cdot 2 \cdot 4$
 $= 2 \cdot 2 \cdot 2 \cdot 2 \cdot 2 \cdot 2 \cdot 2$

21. Prime

23. $91 = 7 \cdot 13$

25. True, -8 is an integer.

27. True

29. True, 1.8 is a rational number.

31. False, $\sqrt{19}$ is irrational.

33. False, $0 \in W$, but $0 \notin N$.

35. True, the rational numbers are a subset of the real numbers.

37. $6 \cdot 2$ $\quad$ $6 + 2$
 12 $\quad$ 8
 $>, \geq, \neq$

39. $3 \cdot 0$ $\quad$ $3 + 0$
 0 $\quad$ 3
 $<, \leq, \neq$

41. $\{x \mid x < 4\}$

43. $\{a \mid a \leq -3\}$

45. $\{y \mid y \geq -5\}$

47. $\{y \mid -2 < y < 5, y \in Z\}$

49. $\{r \mid 5 \leq r \leq 10\}$

51. $\{z \mid -3 < z \leq 0, z \in Z\}$

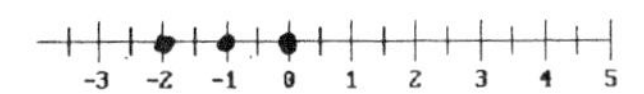

53. No solution; There are no numbers greater than −4 and less than or equal to −7.

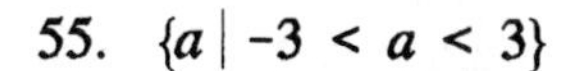

55. $\{a \mid -3 < a < 3\}$

57. $C = \{-4, -3, -2, -1, 0, 1, 2, 3, 4, 5, 6\}$
$D = \{1, 2, 3, 4, 5, 6, 7, 8\}$

$C \cup D = \{-4, -3, -2, -1, 0, 1, 2, 3, 4, 5, 6, 7, 8\}$
$= \{x \mid -4 \le x < 9, x \in Z\}$

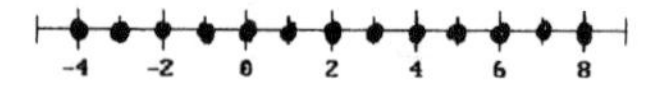

59. $A = \{-3, -2, -1, 0, 1, \ldots\}$
$D = \{1, 2, 3, 4, 5, 6, 7, 8\}$

$A \cap D = \{1, 2, 3, 4, 5, 6, 7, 8\} = D$

61. $A \cup B = \{x \mid x \ge -3 \text{ or } x < -3\} = R$

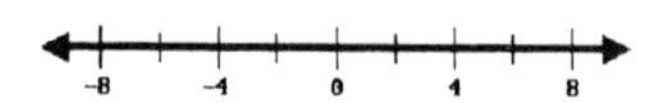

63. $C = \{x \mid -4 \le x \le 6\}$

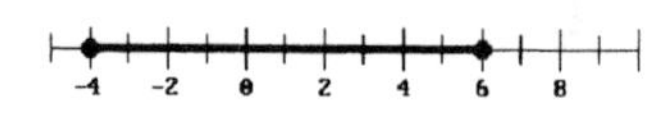

$D = \{x \mid 1 \le x < 9\}$

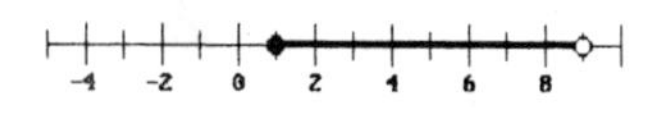

$C \cap D$ is where the two graphs overlap.

$C \cap D = \{x \mid 1 \le x \le 6\}$

65. $A \cup C = \{x \mid x \ge -3 \text{ or } -4 \le x \le 6\}$
$= \{x \mid x \ge -4\}$

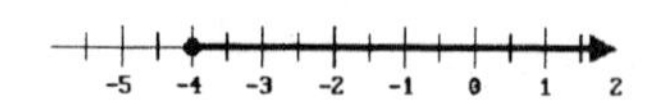

67. True, associative property for addition

69. True, commutative property for addition

71. True, distributive property

73. False

75. True, distributive property

77. False

79. True, associative property for multiplication

81. True, multiplicative inverse

83. True, additive inverse property

85. True, commutative property of addition

87. True, distributive property

89. True, closure for multiplication

91. False

93. Let $x = 0.674674\overline{674}$
$1000x = 674.674\overline{674}$

$$\begin{array}{r} 1000x = 674.674\overline{674} \\ -\quad x = 0.674\overline{674} \\ \hline 999x = 674 \end{array}$$

$$x = \frac{674}{999}$$

Let $x = 0.9292\overline{92}$

$100x = 92.92\overline{92}$

$$\begin{aligned} 100x &= 92.92\overline{92} \\ -\quad x &= 0.92\overline{92} \\ \hline 99x &= 92 \end{aligned}$$

$$x = \frac{92}{99}$$

95. No, subtraction is not commutative. For example, $6 - 3 \neq 3 - 6$.

No, subtraction is not associative. For example, $6 - (4 - 2) \neq (6 - 4) - 2$.

97. 1. Additive identity property
2. Distributive property
4. Additive inverse property
5. Associative property for addition
6. Additive inverse property
7. Additive identity property

Exercises 1.2

1. $-3 + 8 = +(8 - 3) = 5$

3. $-3 - 8 = -3 + (-8)$
$= -11$

5. $-3(-8) = 24$

7. $1.692 - 3.965 + 8.754$
$= 1.692 + (-3.965) + 8.754$
$= -2.273 + 8.754$
$= 6.481$

9. $-3 - 4 - 5 = -3 + (-4) + (-5)$
$= -7 + (-5)$
$= -12$

11. $-3 - 4(5) = -3 - 20$
$= -3 + (-20)$
$= -23$

13. $-3(-4)(-5) = 12(-5)$
$= -60$

15. $3(-4 - 5) = 3(-9)$
$= -27$

17. $8 - 4 \cdot 3 - 7 = 8 - 12 - 7$
$= -4 - 7$
$= -11$

19. $8 - (4 \cdot 3 - 7) = 8 - (12 - 7)$
$= 8 - 5$
$= 3$

21. $(8 - 4)(3 - 7) = (4)(-4)$
$= -16$

23. $(2.6)^2 - |7.8 - 13.69| = 6.76 - |-5.89|$
$= 6.76 - 5.89$
$= 0.87$

25. $\dfrac{-20}{-5} = 4$

27. $\dfrac{-5 - 11}{-9 + 4} = \dfrac{-16}{-5}$
$= \dfrac{16}{5}$

29. $\dfrac{-10 - 2 - 4}{-2} = \dfrac{-12 - 4}{-2}$
$= \dfrac{-16}{-2}$
$= 8$

31. $\dfrac{-10 - (2 - 4)}{-2} = \dfrac{-10 - (-2)}{-2}$
$= \dfrac{-10 + 2}{-2}$
$= \dfrac{-8}{-2}$
$= 4$

33. $\dfrac{-10 - 2(-4)}{-2} = \dfrac{-10 - (-8)}{-2}$
$= \dfrac{-10 + 8}{-2}$
$= \dfrac{-2}{-2}$
$= 1$

35. $\frac{-4(-3)(-6)}{-4(-3) - 6} = \frac{12(-6)}{12 - 6}$

$= \frac{-72}{6}$

$= -12$

37. $8 - 3(5 - 1) = 8 - 3(4)$
$= 8 - 12$
$= -4$

39. $-7 - 2(4 - 6) = -7 - 2(-2)$
$= -7 - (-4)$
$= -7 + 4$
$= -3$

41. $7 + 2[4 + 3(4 + 1)] = 7 + 2[4 + 3(5)]$
$= 7 + 2(4 + 15)$
$= 7 + 2(19)$
$= 7 + 38$
$= 45$

43. $7 - 2[4 - 3(4 - 1)] = 7 - 2[4 - 3(3)]$
$= 7 - 2(4 - 9)$
$= 7 - 2(-5)$
$= 7 - (-10)$
$= 7 + 10$
$= 17$

45. $8 - \left(\frac{10}{-5}\right) = 8 - (-2)$
$= 8 + 2$
$= 10$

47. $8\left(\frac{10}{-5}\right) = 8(-2)$
$= -16$

49. $\frac{12}{-4} - \left(\frac{10}{-2}\right) = -3 - (-5)$
$= -3 + 5$
$= 2$

51. $\frac{-8 + 2}{4 - 6} - \frac{6 - 11}{-3 - 2} = \frac{-6}{-2} - \frac{-5}{-5}$

$= 3 - 1$
$= 2$

53. $-12 - \frac{6 - 2(-3)}{-3} = -12 - \frac{6 + 6}{-3}$

$= -12 - \frac{12}{-3}$

$= -12 - (-4)$
$= -12 + 4$
$= -8$

55. $(-6)^2 = 36$

57. $2^2 + 3^2 + 4^2 = 4 + 9 + 16$
$= 13 + 16$
$= 29$

59. $2(5)^2 = 2(25)$
$= 50$

61. $-3^4 = -81$

63. $-8 - 2(-4)^2 = -8 - 2(16)$
$= -8 - 32$
$= -40$

65. $2(-5)(-6)^2 = 2(-5)(36)$
$= (-10)(36)$
$= -360$

67. $2(-5) - 6^2 = 2(-5) - 36$
$= -10 - 36$
$= -46$

69. $\frac{5[-8 - 3(-2)^2]}{-6 - 6 - 2} = \frac{5[-8 - 3(4)]}{-12 - 2}$

$= \frac{5(-8 - 12)}{-14}$

$= \frac{5(-20)}{-14}$

$= \frac{-100}{-14}$

$= \frac{50}{7}$

71. $-3 - 2[-4 - 3(-2 - 1)]$
$= -3 - 2[-4 - 3(-3)]$
$= -3 - 2(-4 + 9)$
$= -3 - 2(5)$
$= -3 - 10$
$= -13$

73. $-3\{5 - 3[2 - 6(3 - 5)]\}$
$= -3\{5 - 3[2 - 6(-2)]\}$
$= -3[5 - 3(2 + 12)]$
$= -3[5 - 3(14)]$
$= -3(5 - 42)$
$= -3(-37)$
$= 111$

75. $|3 - 8| - |3| - |-8|$
$= |-5| - |3| - |-8|$
$= 5 - 3 - 8$
$= 2 - 8$
$= -6$

77. $|-3 - 2 - 4| - (2 - 3)^3$
$= |-5 - 4| - (-1)^3$
$= |-9| - (-1)$
$= 9 + 1$
$= 10$

79. $(-3 - 2)^2 - (-3 + 2)^2$
$= (-5)^2 - (-1)^2$
$= 25 - 1$
$= 24$

81. $$\frac{42.2}{1.63 - (2.1)(5.8)} = \frac{42.2}{1.63 - 12.18} = \frac{42.2}{-10.55} = -4$$

83. Beginning balance: \$1542.75

Balance on 8/12: \$1542.75 - \$234.25 - \$134.82 - \$68.05 = \$1105.63

Balance on 8/14: \$1105.63 + \$352 = \$1457.63

Balance on 8/15: \$1457.63 - \$274.15 = \$1183.48

Balance on 8/16: \$1183.48 - \$392.24 = \$791.24
Since the balance dropped below \$1000, deduct \$10.
\$791.24 - \$10.00 = \$781.24

Balance on 8/21: \$781.24 + \$1241.00 = \$2022.24

Balance on 8/24: \$2022.24 - \$382.15 - \$112.12 - \$142.55 - \$155.00 = \$1230.42

85. $29{,}035 - (-1300) = 30{,}335$ ft

87. (a) $$6\left(\frac{3}{7}\right)^2 + 4\left(\frac{3}{7}\right) - 3 = 6\left(\frac{9}{49}\right) + 4\left(\frac{3}{7}\right) - 3$$

$$= \frac{54}{49} + \frac{12}{7} - 3$$

$$= \frac{54}{49} + \frac{84}{49} - \frac{147}{49}$$

$$= -\frac{9}{49}$$

(b) -0.18

89. $$\begin{aligned} x + y + z &= -2 + (-3) + 5 \\ &= -5 + 5 \\ &= 0 \end{aligned}$$

91. $$\begin{aligned} xyz &= (-2)(-3)(5) \\ &= 6(5) \\ &= 30 \end{aligned}$$

93. $$\begin{aligned} -x^2 - 4x + 2 &= -(-2)^2 - 4(-2) + 2 \\ &= -4 - 4(-2) + 2 \\ &= -4 + 8 + 2 \\ &= 4 + 2 \\ &= 6 \end{aligned}$$

95. $$\begin{aligned} |xy - z| &= |(-2)(-3) - 5| \\ &= |6 - 5| \\ &= |1| \\ &= 1 \end{aligned}$$

97. $$\frac{3x^2y - x^3y^2}{3x - 2y} = \frac{3(-2)^2(-3) - (-2)^3(-3)^2}{3(-2) - 2(-3)}$$

$$= \frac{3(4)(-3) - (-8)(9)}{-6 + 6}$$

$$= \frac{12(-3) - (-72)}{0}$$

$$= \frac{-36 + 72}{0}$$

$$= \frac{36}{0}$$

$= \text{undefined}$

99. $$\begin{aligned} s_e &= s_y\sqrt{1 - r^2} \\ &= 2.3\sqrt{1 - (0.74)^2} \\ &= 2.3\sqrt{1 - 0.5476} \\ &= 2.3\sqrt{0.4524} \\ &= 2.3(0.6726) \\ &= 1.55 \end{aligned}$$

101. (a) $$\sigma_r = \sqrt{\frac{1 - \rho^2}{n - 1}}$$

$$= \sqrt{\frac{1 - (0.72)^2}{100 - 1}}$$

$= 0.070$

(b) $$\sigma_r = \sqrt{\frac{1 - \rho^2}{n - 1}}$$

$$= \sqrt{\frac{1 - (0.64)^2}{50 - 1}}$$

$= 0.110$

103. $P = 1.92364t - 3588$

(a) $P = 1.92364(1990) - 3588 = 240.044$ million or 240,044,000

(b) $P = 249 - 240.044 = 8.956$ million

105. $E = 16.6333(t - 1980) + 97.1$

(a) 1994: $E = 16.6333(1994 - 1980) + 97.1$
$= \$329.9662$ billion

1995: $E = 16.6333(1995 - 1980) + 97.1$
$= \$346.5995$ billion

1996: $E = 16.6333(1996 - 1980) + 97.1$
$= \$363.2328$ billion

1997: $E = 16.6333(1997 - 1980) + 97.1$
$= \$379.8661$ billion

(b) $1995 - 1994 = 346.5995 - 329.9662$
$= \$16.6333$ billion

$1996 - 1995 = 363.2328 - 346.5995$
$= \$16.6333$ billion

$1997 - 1996 = 379.8661 - 363.2328$
$= \$16.6333$ billion

Percent increase for 1994 to 1995: $\dfrac{16.6333 \text{ billion}}{329.9662 \text{ billion}} = 0.0504$
$= 5.04\%$

1995 to 1996: $\dfrac{16.6333 \text{ billion}}{346.5995 \text{ billion}} = 0.0480$
$= 4.80\%$

1996 to 1997: $\dfrac{16.6333 \text{ billion}}{363.2328 \text{ billion}} = 0.0458$
$= 4.58\%$

107. $(-3)(4) = [(-1)(3)](4)$ Theorem $(-1)x = -x$

$= (-1)[(3)(4)]$ associative property of multiplication

$= (-1)[12]$ multiplication

$= -12$ Theorem $(-1)x = -x$

Exercises 1.3

1. $6x + 2x = (6 + 2)x$
$= 8x$

3. $6x(2x) = (6)(2)x \cdot x$
$= 12x^2$

5. $2x - 6x = (2 - 6)x$
$= -4x$

7. $2x(-6x) = (2)(-6)x \cdot x$
$= -12x^2$

9. $3m - 4m - 5m = (3 - 4 - 5)m$
$= -6m$

11. $3m(-4m)(-5m) = (3)(-4)(-5)m \cdot m \cdot m$
$= 60m^3$

13. $-2t^2 - 3t^2 - 4t^2 = (-2 - 3 - 4)t^2$
$= -9t^2$

15. $-2t^2(-3t^2)(-4t^2) = (-2)(-3)(-4)t^2t^2t^2$
$= -24t^6$

17. $2x + 3y + 5z$

19. $2x(3y)(5z) = (2)(3)(5)xyz$
$= 30xyz$

21. $x^3 + x^2 + 2x$

23. $x^3(x^2)(2x) = 2x^3x^2x$
$= 2x^6$

25. $-5x(3xy) - 2x^2y = (-5)(3)x \cdot x \cdot y - 2x^2y$
$= -15x^2y - 2x^2y$
$= (-15 - 2)x^2y$
$= -17x^2y$

27. $-5x(3xy)(-2x^2y)$
$= (-5)(3)(-2)x \cdot x \cdot x^2y \cdot y$
$= 30x^4y^2$

29. $2x^2 + 3x - 5 - x^2 - x - 1$
$= (2 - 1)x^2 + (3 - 1)x + (-5 - 1)$
$= 1x^2 + 2x - 6$
$= x^2 + 2x - 6$

31. $10x^2y - 6xy^2 + x^2y - xy^2$
$= (10 + 1)x^2y + (-6 - 1)xy^2$
$= 11x^2y - 7xy^2$

33. $3(m + 3n) + 3(2m + n)$
$= 3m + 9n + 6m + 3n$
$= (3 + 6)m + (9 + 3)n$
$= 9m + 12n$

35. $6(a - 2b) - 4(a + b)$
$= 6a - 12b - 4a - 4b$
$= (6 - 4)a + (-12 - 4)b$
$= 2a - 16b$

37. $8(2c - d) - (10c + 8d)$
$= 16c - 8d - 10c - 8d$
$= (16 - 10)c + (-8 - 8)d$
$= 6c - 16d$

39. $x(x - y) + y(y - x) = x^2 - xy + y^2 - xy$
$= x^2 - 2xy + y^2$

41. $a^2(a + 3b) - a(a^2 + 3ab)$
$= a^3 + 3a^2b - a^3 - 3a^2b$
$= (1 - 1)a^3 + (3 - 3)a^2b$
$= 0a^3 + 0a^2b$
$= 0 + 0$
$= 0$

43. $5a^2bc(-2ab^2)(-4bc^2)$
$= (5)(-2)(-4)a^2abb^2bcc^2$
$= 40a^3b^4c^3$

45. $(2x)^3(3x)^2$
$= (2x)(2x)(2x)(3x)(3x)$
$= (2 \cdot 2 \cdot 2 \cdot 3 \cdot 3)(x \cdot x \cdot x \cdot x \cdot x)$
$= 72x^5$

47. $2x^3(3x)^2 = 2x^3 \cdot 3x \cdot 3x$
$= 2 \cdot 3 \cdot 3 \cdot x^3 \cdot x \cdot x$
$= 18x^5$

49. $(-2x)^5(x^6)$
$= (-2x)(-2x)(-2x)(-2x)(-2x)(x^6)$
$= (-2)(-2)(-2)(-2)(-2)(x \cdot x \cdot x \cdot x \cdot x)(x^6)$
$= -32x^{11}$

51. $(-2x)^4 - (2x)^4$
$= (-2x)(-2x)(-2x)(-2x) - (2x)(2x)(2x)(2x)$
$= (-2)(-2)(-2)(-2)(x \cdot x \cdot x \cdot x)$
$- (2)(2)(2)(2)(x \cdot x \cdot x \cdot x)$
$= 16x^4 - 16x^4$
$= (16 - 16)x^4$
$= 0$

53. $(-2x)^3 - (2x)^3$
$= (-2x)(-2x)(-2x) - (2x)(2x)(2x)$
$= (-2)(-2)(-2)(x \cdot x \cdot x) - (2)(2)(2)(x \cdot x \cdot x)$
$= -8x^3 - 8x^3$
$= (-8 - 8)x^3$
$= -16x^3$

55. $4b - 5(b - 2) = 4b - 5b + 10$
$= (4 - 5)b + 10$
$= -1b + 10$
$= -b + 10$

57. $8t - 3[t - 4(t + 1)]$
$= 8t - 3(t - 4t - 4)$
$= 8t - 3(-3t - 4)$
$= 8t + 9t + 12$
$= 17t + 12$

59. $a - 4[a - 4(a - 4)]$
$= a - 4(a - 4a + 16)$
$= a - 4(-3a + 16)$
$= a + 12a - 64$
$= 13a - 64$

61. $x + x[x + 3(x - 3)] = x + x(x + 3x - 9)$
$= x + x(4x - 9)$
$= x + 4x^2 - 9x$
$= 4x^2 - 8x$

63. $x - \{y - 3[x - 2(y - x)]\}$
$= x - [y - 3(x - 2y + 2x)]$
$= x - [y - 3(3x - 2y)]$
$= x - (y - 9x + 6y)$
$= x - (-9x + 7y)$
$= x + 9x - 7y$
$= 10x - 7y$

65. $3x + 2y[x + y(x - 3y) - y^2]$
$= 3x + 2y(x + xy - 3y^2 - y^2)$
$= 3x + 2y(x + xy - 4y^2)$
$= 3x + 2xy + 2xy^2 - 8y^3$
$= -8y^3 + 2xy^2 + 2xy + 3x$

67. $6s^2 - [st - s(t + 5s) - s^2]$
$= 6s^2 - (st - st - 5s^2 - s^2)$
$= 6s^2 - (-6s^2)$
$= 6s^2 + 6s^2$
$= 12s^2$

69. (a) $3x - 15y - 11x$
$= 3(2.82) - 15(7.25) - 11(2.82)$
$= 8.46 - 108.75 - 31.02$
$= -100.29 - 31.02$
$= -131.31$

(b) $3x - 15y - 11x$
$= -8x - 15y$
$= -8(2.82) - 15(7.25)$
$= -22.56 - 108.75$
$= -131.31$

71. (a) $2(3x - 4) - 5(3x + 8)$
$= 2[3(2.82) - 4] - 5[3(2.82) + 8]$
$= 2(8.46 - 4) - 5(8.46 + 8)$
$= 2(4.46) - 5(16.46)$
$= 8.92 - 82.3$
$= -73.38$

(b) $2(3x - 4) - 5(3x + 8)$
$= 6x - 8 - 15x - 40$
$= -9x - 48$
$= -9(2.82) - 48$
$= -25.38 - 48$
$= -73.38$

73. (a) $P = 16.75(t - 1980) + 463.5$
$P = 16.75t - 33165 + 463.5$
$P = 16.75t - 32701.5$

(b) $P = 16.75(1998 - 1980) + 463.5 = 765$ thousand physicians
$P = 16.75(1998) - 32701.5 = 765$ thousand physicians

75. $A = (12)(8) - (3)(x)$
$A = 96 - 3x$

77. $A = x(x + 5) - (2)(x)$
$A = x^2 + 5x - 2x$
$A = x^2 + 3x$

1.4 Exercises

1. number: n

8	more than	a number
8	+	n

$8 + n$

3. number: n

3 less than twice a number

$2n - 3$

5. number: n

4 more than 3 times a number is 7 less than the number.

$4 + 3n = n - 7$

$4 + 3n = n - 7$

7. 1st number: n
2nd number: m

sum of two numbers is 1 more than their product

$n + m = 1 + nm$

$n + m = 1 + nm$

9. smaller number: n
larger number: 5 more than twice the smaller

$5 + 2n$

$5 + 2n$

11. smallest number: n
middle number: 3 times the smallest

$3n$

largest number: 12 more than the middle number

$12 + 3n$

$12 + 3n$

13. 1st integer: n
2nd consecutive integer: $n + 1$

15. 1st even integer: n
2nd consecutive even integer: $n + 2$
3rd consecutive even integer : $n + 4$

17. 1st integer: n
2nd integer: $n + 1$
cube of 1st integer: n^3
cube of 2nd integer: $(n + 1)^3$
sum of the cubes: $n^3 + (n + 1)^3$

19. 1st number: n
2nd number: $40 - n$

21. 1st number: n
2nd number: $2n$
3rd number: $100 - (n + 2n) = 100 - 3n$

23. width: n
length: $3n$

$A = \text{width} \cdot \text{length}$
$A = n \cdot 3n$
$A = 3n^2$

$P = 2 \cdot \text{width} + 2 \cdot \text{length}$
$P = 2(n) + 2(3n)$
$P = 2n + 6n$
$P = 8n$

25. 2nd side's length: n
1st side's length: $2n$
3rd side's length: $4 + n$

P = 1st side's length + 2nd side's length + 3rd side's length
$P = 2n + n + 4 + n$
$P = 4n + 4$

27. (a) $12 + 9 + 10 = 31$ coins

(b) value of nickels: $12(0.05) = \$0.60$
value of dimes: $9(0.10) = \$0.90$
value of quarters: $10(0.25) = \$2.50$
Total value = \$0.60 + \$0.90 + \$2.50 = \$4.00

29. (a) $(n + d + q)$ coins

(b) value of nickels: $5n$ cents
value of dimes: $10d$ cents
value of quarters: $25q$ cents
Total value = $5n + 10d + 25q$ cents

31. width: w
length: $3w$

(a) regular fence $= 2 \cdot \text{width} = 2w$ meters
(b) cost of regular fence = ($2w$ meters)(\$2 per meter)
= $4w$ dollars

(c) heavy-duty fence $= 2 \cdot \text{length} = 2(3w) = 6w$ meters
(d) cost of heavy-duty fence = ($6w$ meters)(\$5 per meter)
= $30w$ dollars

(e) Total cost = cost of regular fence + cost of heavy-duty fence
$= 4w + 30w$
$= 34w$ dollars

33. number of nickels: n cents
number of dimes: $20 - n$ cents
value of nickels: $5n$ cents
value of dimes: $10(20 - n) = 200 - 10n$ cents
Total value = value of nickels + value of dimes
$= 5n + 200 - 10n$ cents
$= 200 - 5n$ cents

35. (a) $A = 1500(1.06) = \$1590$
(b) $A = 1500(1.06)^2 = \$1685.40$
(c) $A = 1500(1.06)^n$
(d) $A = 1500(1.06)^{15} = \$3594.84$

37. (a) For 2 hours = 120 minutes per month:

Standard Plan: $0.10(120) = \$12.00$

Saver Plan: $0.06(120) + 9 = \$16.20$

For 1 hour = 60 minutes per month:

Standard Plan: $0.10(60) = \$6.00$

Saver Plan: $0.06(60) + 9 = \$12.60$

(b) Standard Plan: $0.10t$

Saver Plan: $0.06t + 9$

(c) For 225 minutes both plans cost the same. The Standard Plan is less expensive if your monthly long distance usage is under 225 minutes, while the Saver Plan is less expensive for over 225 minutes.

39. (a) Original price = \$1250
25% discount = $0.25(1250) = \$312.50$

Sale price = $1250 - 312.50 = \$937.50$
Tax = $0.05(937.50) = \$46.88$

Final price = $937.50 + 46.88 = \$984.38$

(b) Original price = x
25% discount = $0.25x$

Sale price = $x - 0.25x = 0.75x$
Tax = $0.05(0.75x) = 0.0375x$

Final price = $0.75x + 0.0375x = 0.7875x$

41. (a) Oct. 28: \$2800

Oct. 29: $2800 - 0.50(2800) = \$1400$

Oct. 30: $1400 + 0.50(1400) = 2100$

(b) Oct. 28: x
Oct. 29: $x - 0.50x = 0.5x$

Oct. 30: $0.5x + 0.50(0.5x)$
$= 0.5x + 0.25x$
$= 0.75x$

Exercises 1.5

1. $5(x - 3) - (x + 2) = -5$

Test $x = 0$:
$5(0 - 3) - (0 + 2) = -5$
$5(-3) - 2 = -5$
$-15 - 2 = -5$
$-17 = -5$
No

Test $x = 3$:
$5(3 - 3) - (3 + 2) = -5$
$5(0) - 5 = -5$
$0 - 5 = -5$
$-5 = -5$
Yes

Test $x = 5$:
$5(5 - 3) - (5 + 2) = -5$
$5(2) - 7 = -5$
$10 - 7 = -5$
$3 = -5$
No

3. $3(a - 5) + 2(1 - a) = -11$

Test $a = -3$:
$3(-3 - 5) + 2[1 - (-3)] = -11$
$3(-8) + 2(4) = -11$
$-24 + 8 = -11$
$-16 = -11$
No

Test $a = 0$:
$3(0 - 5) + 2(1 - 0) = -11$
$3(-5) + 2(1) = -11$
$-15 + 2 = -11$
$-13 = -11$
No

Test $a = 2$:
$3(2 - 5) + 2(1 - 2) = -11$
$3(-3) + 2(-1) = -11$
$-9 - 2 = -11$
$-11 = -11$
Yes

5. $x^2 - 4x = 5$

Test $x = -1$:

$$\begin{aligned}(-1)^2 - 4(-1) &= 5\\ 1 + 4 &= 5\\ 5 &= 5\end{aligned}$$

Yes

Test $x = 5$:

$$\begin{aligned}5^2 - 4(5) &= 5\\ 25 - 20 &= 5\\ 5 &= 5\end{aligned}$$

Yes

Test $x = 2$:

$$\begin{aligned}2^2 - 4(2) &= 5\\ 4 - 8 &= 5\\ -4 &= 5\end{aligned}$$

No

7. $a(a + 8) = a + 8$

Test $a = -1$:

$$\begin{aligned}-1(-1 + 8) &= -1 + 8\\ -1(7) &= 7\\ -7 &= 7\end{aligned}$$

No

Test $a = 1$:

$$\begin{aligned}1(1 + 8) &= 1 + 8\\ 1(9) &= 9\\ 9 &= 9\end{aligned}$$

Yes

Test $a = 3$:

$$\begin{aligned}2(3 + 8) &= 3 + 8\\ 3(11) &= 11\\ 33 &= 11\end{aligned}$$

No

9.

$$\begin{aligned}2x - 7 &= 5\\ 2x - 7 + 7 &= 5 + 7\\ 2x &= 12\\ \frac{2x}{2} &= \frac{12}{2}\\ x &= 6\end{aligned}$$

11.

$$\begin{aligned}4y + 2 &= -1\\ 4y + 2 - 2 &= -1 - 2\\ 4y &= -3\\ \frac{4y}{4} &= \frac{-3}{4}\\ y &= -\frac{3}{4}\end{aligned}$$

13.

$$\begin{aligned}m + 3 &= 3 - m\\ m + 3 + m &= 3 - m + m\\ 2m + 3 &= 3\\ 2m + 3 - 3 &= 3 - 3\\ 2m &= 0\\ \frac{2m}{2} &= \frac{0}{2}\\ m &= 0\end{aligned}$$

15.

$$\begin{aligned}3t - 5 &= 5t - 13\\ 3t - 5 - 3t &= 5t - 13 - 3t\\ -5 &= 2t - 13\\ -5 + 13 &= 2t - 13 + 13\\ 8 &= 2t\\ \frac{8}{2} &= \frac{2t}{2}\\ 4 &= t\end{aligned}$$

17.

$$\begin{aligned}11 - 3y &= 38\\ 11 - 3y - 11 &= 38 - 11\\ -3y &= 27\\ \frac{-3y}{-3} &= \frac{27}{-3}\\ y &= -9\end{aligned}$$

19.

$$\begin{aligned}5s + 2 &= 3s - 7\\ 5s + 2 - 3s &= 3s - 7 - 3s\\ 2s + 2 &= -7\\ 2s + 2 - 2 &= -7 - 2\\ 2s &= -9\\ \frac{2s}{2} &= \frac{-9}{2}\\ s &= -\frac{9}{2}\end{aligned}$$

21.

$$\begin{aligned}3.24x - 5.2 &= 7.74x + 0.3\\ 3.24x - 5.2 - 3.24x &= 7.74x + 0.3 - 3.24x\\ -5.2 &= 4.5x + 0.3\\ -5.2 - 0.3 &= 4.5x + 0.3 - 0.3\\ -5.5 &= 4.5x\\ \frac{-5.5}{4.5} &= \frac{4.5x}{4.5}\\ -\frac{11}{9} &= x\\ x &\approx -1.22\end{aligned}$$

23.
$$\begin{aligned} 5 - 21x &= 3x - 16 \\ 5 - 21x + 21x &= 3x - 16 + 21x \\ 5 &= 24x - 16 \\ 5 + 16 &= 24x - 16 + 16 \\ 21 &= 24x \\ \frac{21}{24} &= \frac{24x}{24} \\ \frac{7}{8} &= x \end{aligned}$$

25.
$$\begin{aligned} 3x - 12 &= 12 - 3x \\ 3x - 12 + 3x &= 12 - 3x + 3x \\ 6x - 12 &= 12 \\ 6x - 12 + 12 &= 12 + 12 \\ 6x &= 24 \\ \frac{6x}{6} &= \frac{24}{6} \\ x &= 4 \end{aligned}$$

27.
$$\begin{aligned} 3x - 12 &= -12 - 3x \\ 3x - 12 + 3x &= -12 - 3x + 3x \\ 6x - 12 &= -12 \\ 6x - 12 + 12 &= -12 + 12 \\ 6x &= 0 \\ \frac{6x}{6} &= \frac{0}{6} \\ x &= 0 \end{aligned}$$

29.
$$\begin{aligned} 2t + 1 &= 7 - t \\ 2t + 1 + t &= 7 - t + t \\ 3t + 1 &= 7 \\ 3t + 1 - 1 &= 7 - 1 \\ 3t &= 6 \\ \frac{3t}{3} &= \frac{6}{3} \\ t &= 2 \end{aligned}$$

31.
$$\begin{aligned} 3(x - 1) &= 2(x + 1) \\ 3x - 3 &= 2x + 2 \\ 3x - 3 - 2x &= 2x + 2 - 2x \\ x - 3 &= 2 \\ x - 3 + 3 &= 2 + 3 \\ x &= 5 \end{aligned}$$

33.
$$\begin{aligned} 3(x + 1) + x &= 2(x + 3) \\ 3x + 3 + x &= 2x + 6 \\ 4x + 3 &= 2x + 6 \\ 4x + 3 - 2x &= 2x + 6 - 2x \\ 2x + 3 &= 6 \\ 2x + 3 - 3 &= 6 - 3 \\ 2x &= 3 \\ \frac{2x}{2} &= \frac{3}{2} \\ x &= \frac{3}{2} \end{aligned}$$

35.
$$\begin{aligned} 0.06x + 0.0725(22.500 - x) &= 1{,}500 \\ 0.06x + 1631.25 - 0.0725x &= 1{,}500 \\ -0.0125x + 1631.25 &= 1{,}500 \\ -0.0125x + 1631.25 - 1631.25 &= 1{,}500 - 1631.25 \\ -0.0125x &= -131.25 \\ \frac{-0.0125x}{-0.0125} &= \frac{-131.25}{-0.0125} \\ x &= 10{,}500 \end{aligned}$$

37.
$$\begin{aligned} \frac{x}{2} - 1 &= 5 \\ 2\left(\frac{x}{2} - 1\right) &= 2(5) \\ \frac{2}{1} \cdot \frac{x}{2} - 2(1) &= 2(5) \\ x - 2 &= 10 \\ x - 2 + 2 &= 10 + 2 \\ x &= 12 \end{aligned}$$

39.
$$\begin{aligned} \frac{2x}{3} + 2 &= \frac{x}{2} \\ 6\left(\frac{2x}{3} + 2\right) &= 6\left(\frac{x}{2}\right) \\ \frac{6}{1} \cdot \frac{2x}{3} + 6(2) &= \frac{6}{1} \cdot \frac{x}{2} \\ 4x + 12 &= 3x \\ 4x + 12 - 4x &= 3x - 4x \\ 12 &= -x \\ \frac{12}{-1} &= \frac{-x}{-1} \\ -12 &= x \end{aligned}$$

41. $$5x - \frac{2}{3} = \frac{x}{4}$$

$$12\left(5x - \frac{2}{3}\right) = 12\left(\frac{x}{4}\right)$$

$$12(5x) - \frac{12}{1} \cdot \frac{2}{3} = \frac{12}{1} \cdot \frac{x}{4}$$

$$60x - 8 = 3x$$
$$60x - 8 - 60x = 3x - 60x$$
$$-8 = -57x$$

$$\frac{-8}{-57} = \frac{-57x}{-57}$$
$$\frac{8}{57} = x$$

43. $$6x - \frac{2}{5} = \frac{3x}{4}$$

$$20\left(6x - \frac{2}{5}\right) = 20\left(\frac{3x}{4}\right)$$

$$20(6x) - \frac{20}{1} \cdot \frac{2}{5} = \frac{20}{1} \cdot \frac{3x}{4}$$

$$120x - 8 = 15x$$
$$120x - 8 - 120x = 15x - 120x$$
$$-8 = -105x$$

$$\frac{-8}{-105} = \frac{-105x}{-105}$$
$$\frac{8}{105} = x$$

45. $$\frac{3}{5}x - \frac{2}{3} = 5x$$

$$15\left(\frac{3}{5}x - \frac{2}{3}\right) = 15(5x)$$

$$\frac{15}{1} \cdot \frac{3}{5}x - \frac{15}{1} \cdot \frac{2}{3} = 15(5x)$$

$$9x - 10 = 75x$$
$$9x - 10 - 9x = 75x - 9x$$
$$-10 = 66x$$

$$\frac{-10}{66} = \frac{66x}{66}$$
$$-\frac{5}{33} = x$$

47. $$\frac{3x - 2}{4} = 5$$

$$\frac{4}{1} \cdot \left(\frac{3x - 2}{4}\right) = 4(5)$$

$$3x - 2 = 20$$
$$3x - 2 + 2 = 20 + 2$$
$$3x = 22$$

$$\frac{3x}{3} = \frac{22}{3}$$

$$x = \frac{22}{3}$$

49. $$\frac{7x - 1}{2} = \frac{2}{3}$$

$$\frac{6}{1} \cdot \left(\frac{7x - 1}{2}\right) = \frac{6}{1} \cdot \frac{2}{3}$$

$$3(7x - 1) = 4$$
$$21x - 3 = 4$$
$$21x - 3 + 3 = 4 + 3$$
$$21x = 7$$

$$\frac{21x}{21} = \frac{7}{21}$$

$$x = \frac{1}{3}$$

51. $C = 4.695(t - 1991) + 117.37$

(a) $C = 4.695(1996 - 1991) + 117.37$
$= \$140.845$ thousand $\approx \$141{,}000$

(b) $\$140{,}000 = 140$ thousand
$C = 140$

$$140 = 4.695(t - 1991) + 117.37$$
$$140 = 4.695t - 9347.745 + 117.37$$
$$140 = 4.695t - 9230.375$$
$$140 + 9230.375 = 4.695t$$
$$9370.375 = 4.695t$$

$$\frac{9370.375}{4.695} = \frac{4.695t}{4.695}$$

$$1995.8 = t$$

The year was 1995.

53.
$$\begin{aligned} 5x + 7y &= 4 \\ 5x + 7y - 7y &= 4 - 7y \\ 5x &= 4 - 7y \\ \frac{5x}{5} &= \frac{4 - 7y}{5} \\ x &= \frac{4 - 7y}{5} \end{aligned}$$

55.
$$\begin{aligned} 2x - 9y &= 11 \\ 2x - 9y - 2x &= 11 - 2x \\ -9y &= 11 - 2x \\ \frac{-9y}{-9} &= \frac{11 - 2x}{-9} \end{aligned}$$

or

$$y = \frac{2x - 11}{9}$$

57.
$$\begin{aligned} 2(x - y) &= 3x + 4 \\ 2x - 2y &= 3x + 4 \\ 2x - 2y - 2x &= 3x + 4 - 2x \\ -2y &= x + 4 \\ -2y - 4 &= x + 4 - 4 \\ -2y - 4 &= x \end{aligned}$$

59.
$$\begin{aligned} 2x - 5 &< 7x - 2 \\ 2x - 5 - 2x &< 7x - 2 - 2x \\ -5 &< 5x - 2 \\ -5 + 2 &< 5x - 2 + 2 \\ -3 &< 5x \\ \frac{-3}{5} &< \frac{5x}{5} \\ -\frac{3}{5} &< x \end{aligned}$$

or

$$x > -\frac{3}{5}$$

61.
$$\begin{aligned} 4x - 25 &< x - 8 \\ 4x - 25 - x &< x - 8 - x \\ 3x - 25 &< -8 \\ 3x - 25 + 25 &< -8 + 25 \\ 3x &< 17 \\ \frac{3x}{3} &< \frac{17}{3} \\ x &< \frac{17}{3} \end{aligned}$$

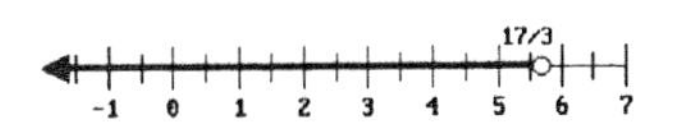

63.
$$\begin{aligned} 3 + (2x - 1) &< 4x - 6 \\ 2x + 2 &< 4x - 6 \\ 2x + 2 - 4x &< 4x - 6 - 4x \\ -2x + 2 &< -6 \\ -2x + 2 - 2 &< -6 - 2 \\ -2x &< -8 \\ \frac{-2x}{-2} &> \frac{-8}{-2} \\ x &> 4 \end{aligned}$$

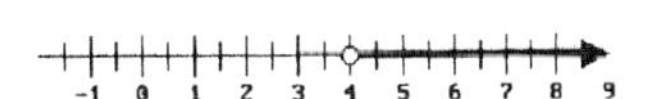

65.
$$\begin{aligned} 3x - (2x - 7) &\le 4x + 6 \\ 3x - 2x + 7 &\le 4x + 6 \\ x + 7 &\le 4x + 6 \\ x + 7 - x &\le 4x + 6 - x \\ 7 &\le 3x + 6 \\ 7 - 6 &\le 3x + 6 - 6 \\ 1 &\le 3x \\ \frac{1}{3} &\le \frac{3x}{3} \\ \frac{1}{3} &\le x \end{aligned}$$

or

$$x \ge \frac{1}{3}$$

67. $6x - (3x + 1) \geq 2x + (2x - 3)$
$$6x - 3x - 1 \geq 4x - 3$$
$$3x - 1 \geq 4x - 3$$
$$3x - 1 - 4x \geq 4x - 3 - 4x$$
$$-x - 1 \geq -3$$
$$-x - 1 + 1 \geq -3 + 1$$
$$-x \geq -2$$
$$\frac{-x}{-1} \leq \frac{-2}{-1}$$
$$x \leq 2$$

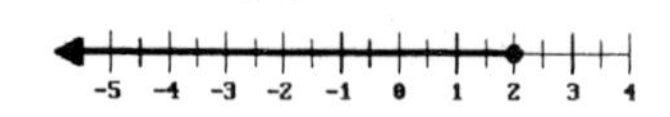

69. $3.4 - (2x - 5.8) \leq 2.6x + 4$
$$3.4 - 2x + 5.8 \leq 2.6x + 4$$
$$-2x + 9.2 \leq 2.6x + 4$$
$$-2x + 9.2 + 2x \leq 2.6x + 4 + 2x$$
$$9.2 \leq 4.6x + 4$$
$$9.2 - 4 \leq 4.6x + 4 - 4$$
$$5.2 \leq 4.6x$$
$$\frac{5.2}{4.6} \leq \frac{4.6x}{4.6}$$
$$\frac{26}{23} \leq x$$

or

$$x \geq \frac{26}{23}$$

CHAPTER 1 REVIEW EXERCISES

1. $A = \{1, 2, 3, 4\}$

3. $C \cap D = \{b\}$

5. $A = \{1, 2, 3, 4\}$
$B = \{6, 7, 8, 9, \ldots\}$

$A \cup B = \{1, 2, 3, 4, 6, 7, \ldots\}$

7. $A = \{1, 2, 3, 4, 6, 12\}$

9. $A = \{1, 2, 3, 4, 6, 12\}$
$B = \{0, 12, 24, 36, \ldots\}$

$A \cap B = \{12\}$

11. $B = \{0, 12, 24, 36, \ldots\}$
$C = \{0, 6, 12, 18, 24, \ldots\}$

$B \cap C = \{0, 12, 24, 36, \ldots\}$

13. $A = \{-1, 0, 1, 2, 3, 4\}$
$B = \{3, 4, 5, 6, 7, 8, 9, 10, 11, 12\}$

$A \cap B = \{3, 4\} = \{x \mid 3 \leq x \leq 4, x \in Z\}$

15. $\{x \mid x \leq 4\}$

17. $\{b \mid -8 \leq b \leq 5\}$

19. $\{a \mid -2 < a \leq 4\}$

21. False, $\frac{1}{2} \in Q$.

23. True, it is a repeating decimal.

25. False, $\pi \in I$.

27. True

29. True, commutative property of addition.

31. True, distributive property

33. True, multiplicative inverse property

35. False

37. $(-2) + (-3) - (-4) + (-5) = -5 + 4 + (-5)$
$= -1 + (-5)$
$= -6$

39. $6 - 2 + 5 - 8 - 9 = 4 + 5 - 8 - 9$
$= 9 - 8 - 9$
$= 1 - 9$
$= -8$

41. $(-2)(-3)(-5) = 6(-5)$
$= -30$

43. $(-2)^6 = (-2)(-2)(-2)(-2)(-2)(-2)$
$= 64$

45. $(-2) - (-3)^2 = (-2) - 9$
$= -11$

47. $(-6 - 3)(-2 - 5) = (-9)(-7)$
$= 63$

49. $5 - 3[2 - (4 - 8) + 7]$
$= 5 - 3[2 - (-4) + 7]$
$= 5 - 3(6 + 7)$
$= 5 - 3(13)$
$= 5 - 39$
$= -34$

51. $5 - \{2 + 3[6 - 4(5 - 9)] - 2\}$
$= 5 - \{2 + 3[6 - 4(-4)] - 2\}$
$= 5 - [2 + 3(6 + 16) - 2]$
$= 5 - [2 + 3(22) - 2]$
$= 5 - (2 + 66 - 2)$
$= 5 - (66)$
$= -61$

53. $\dfrac{4[5 - 3(8 - 12)]}{-6 - 2(5 - 6)} = \dfrac{4[5 - 3(-4)]}{-6 - 2(-1)}$

$= \dfrac{4(5 + 12)}{-6 + 2}$

$= \dfrac{4(17)}{-4}$

$= \dfrac{68}{-4}$

$= -17$

55. $x^2 - 2xy + y^2 = (-2)^2 - 2(-2)(-1) + (-1)^2$
$= 4 - 4 + 1$
$= 1$

57. $|x - y| - (|x| - |y|)$
$= |-2 - (-1)| - (|-2| - |-1|)$
$= |-1| - (2 - 1)$
$= 1 - 1$
$= 0$

59. $\dfrac{2x^2y^3 + 3y^2}{zx^2y} = \dfrac{2(-2)^2(-1)^3 + 3(-1)^2}{0(-2)^2(-1)}$

$= \dfrac{2(-2)^2(-1) + 3(1)}{0(4)(-1)}$

$= \dfrac{-8 + 3}{0}$

$= -\dfrac{5}{0}$

undefined

61. $t = \dfrac{\overline{X} - a}{\dfrac{Sx}{\sqrt{n}}}$

$= \dfrac{100 - 95}{\dfrac{7.1}{\sqrt{30}}}$

$= 3.86$

63. $(2x + y)(-3x^2y) = -6x^3y - 3x^2y^2$

65. $(2xy^2)^2(-3x)^2$
$= (2xy^2)(2xy^2)(-3x)(-3x)$
$= (2)(2)(-3)(-3)x \cdot x \cdot x \cdot x \cdot y^2 \cdot y^2$
$= 36x^4y^4$

67. $3x - 2y - 4x + 5y - 3x$
$= (3 - 4 - 3)x + (-2 + 5)y$
$= -4x + 3y$

69. $-2r^2s + 5rs^2 - 3sr^2 - 4s^2r$
$= (-2 - 3)r^2s + (5 - 4)rs^2$
$= -5r^2s + 1rs^2$
$= -5r^2s + rs^2$

71. $(y - 5) - (y - 4) = y - 5 - y + 4$
$= -1$

73. $5rs(2r + 3s) = 10r^2s + 15rs^2$

75. $7y - 9(2x + 1) = 7y - 18x - 9$

77. $-2r + 3[s - 2(s - 6)]$
$= -2r + 3(s - 2s + 12)$
$= -2r + 3(-s + 12)$
$= -2r - 3s + 36$

79. $7 - 3\{y - 2[y - 4(y - 1)]\}$
$= 7 - 3[y - 2(y - 4y + 4)]$
$= 7 - 3[y - 2(-3y + 4)]$
$= 7 - 3(y + 6y - 8)$
$= 7 - 3(7y - 8)$
$= 7 - 21y + 24$
$= -21y + 31$

81. 1st number: n
2nd number: m

Five less than the product of two numbers is 3 more than their sum.
$nm - 5 = 3 + n + m$

$nm - 5 = 3 + n + m$

83. 1st integer: n
2nd consecutive odd integer: $n + 2$
3rd consecutive odd integer: $n + 4$

sum of first two is 5 less than the third.
$n + n + 2 = n + 4 - 5$

$n + n + 2 = n + 4 - 5$

85. width: n
length: $4n - 5$

$A = \text{width} \cdot \text{length}$
$A = n(4n - 5)$
$A = 4n^2 - 5n$

$P = 2 \cdot \text{width} + 2 \cdot \text{length}$
$P = 2(n) + 2(4n - 5)$
$P = 2n + 8n - 10$
$P = 10n - 10$

87. 1st number: n
2nd number: m

sum of the squares is 8 more than the product of the number
$n^2 + m^2 = 8 + nm$

$n^2 + m^2 = 8 + nm$

89. number of dimes: n
number of nickels: $40 - n$
value of dimes: $10n$
value of nickels: $5(40 - n)$

Total value = value of dimes + value of nickels
Total value $= 10n + 5(40 - n)$
$= 10n + 200 - 5n$
$= 5n + 200$ cents

91.
$$\begin{aligned} 5x - 2 &= -2 \\ 5x - 2 + 2 &= -2 + 2 \\ 5x &= 0 \\ \frac{5x}{5} &= \frac{0}{5} \\ x &= 0 \end{aligned}$$

93.
$$\begin{aligned} 3x - 5 &= 2x + 6 \\ 3x - 5 - 2x &= 2x + 6 - 2x \\ x - 5 &= 6 \\ x - 5 + 5 &= 6 + 5 \\ x &= 11 \end{aligned}$$

95.
$$\begin{aligned} 11x + 2 &= 6x - 3 \\ 11x + 2 - 6x &= 6x - 3 - 6x \\ 5x + 2 &= -3 \\ 5x + 2 - 2 &= -3 - 2 \\ 5x &= -5 \\ \frac{5x}{5} &= \frac{-5}{5} \\ x &= -1 \end{aligned}$$

97.
$$\begin{aligned} 5(a - 3) &= 2(a - 4) \\ 5a - 15 &= 2a - 8 \\ 5a - 15 - 2a &= 2a - 8 - 2a \\ 3a - 15 &= -8 \\ 3a - 15 + 15 &= -8 + 15 \\ 3a &= 7 \\ \frac{3a}{3} &= \frac{7}{3} \\ a &= \frac{7}{3} \end{aligned}$$

99.
$$\begin{aligned} 6(q - 4) + 2(q + 5) &= 8q - 19 \\ 6q - 24 + 2q + 10 &= 8q - 19 \\ 8q - 14 &= 8q - 19 \\ 8q - 14 - 8q &= 8q - 19 - 8q \\ -14 &= -19 \end{aligned}$$

No solution

101.
$$\begin{aligned} 3x - 5x &= 7x - 4x \\ -2x &= 3x \\ -2x + 2x &= 3x + 2x \\ 0 &= 5x \\ \frac{0}{5} &= \frac{5x}{5} \\ 0 &= x \end{aligned}$$

103.
$$\begin{aligned} x - \frac{2}{3} &= 2x + 4 \\ 3\left(x - \frac{2}{3}\right) &= 3(2x + 4) \\ 3x - \frac{3}{1} \cdot \frac{2}{3} &= 6x + 12 \\ 3x - 2 &= 6x + 12 \\ 3x - 2 - 3x &= 6x + 12 - 3x \\ -2 &= 3x + 12 \\ -2 - 12 &= 3x + 12 - 12 \\ -14 &= 3x \\ \frac{-14}{3} &= \frac{3x}{3} \\ -\frac{14}{3} &= x \end{aligned}$$

105.
$$\begin{aligned} \frac{x - 3}{5} &= x + 1 \\ \frac{5}{1} \cdot \frac{x - 3}{5} &= 5(x + 1) \\ x - 3 &= 5x + 5 \\ x - 3 - x &= 5x + 5 - x \\ -3 &= 4x + 5 \\ -3 - 5 &= 4x + 5 - 5 \\ -8 &= 4x \\ \frac{-8}{4} &= \frac{4x}{4} \\ -2 &= x \end{aligned}$$

107.
$$\begin{aligned} 3.2x + 0.14 &= 1.4x - 21.46 \\ 3.2x + 0.14 - 0.14 &= 1.4x - 21.46 - 0.14 \\ 3.2x &= 1.4x - 21.6 \\ 3.2x - 1.4x &= 1.4x - 21.6 - 1.4x \\ 1.8x &= -21.6 \\ \frac{1.8x}{1.8} &= \frac{-21.6}{1.8} \\ x &= -12 \end{aligned}$$

109.

$$\begin{aligned} 7x + 8y &= 22 \\ 7x + 8y - 7x &= 22 - 7x \\ 8y &= 22 - 7x \\ \frac{8y}{8} &= \frac{22 - 7x}{8} \\ y &= \frac{22 - 7x}{8} \end{aligned}$$

111.

$$\begin{aligned} 3x - 6y &= 2 - 4x + 2y \\ 3x - 6y - 2y &= 2 - 4x + 2y - 2y \\ 3x - 8y &= 2 - 4x \\ 3x - 8y - 3x &= 2 - 4x - 3x \\ -8y &= 2 - 7x \\ \frac{-8y}{-8} &= \frac{2 - 7x}{-8} \\ y &= \frac{2 - 7x}{-8} \text{ or } y = \frac{7x - 2}{8} \end{aligned}$$

113.

$$\begin{aligned} 3x + 12 &< 2x - 9 \\ 3x + 12 - 2x &< 2x - 9 - 2x \\ x + 12 &< -9 \\ x + 12 - 12 &< -9 - 12 \\ x &< -21 \end{aligned}$$

115.

$$\begin{aligned} 5 - (2x - 7) &> 3x - 15 \\ 5 - 2x + 7 &> 3x - 15 \\ 12 - 2x &> 3x - 15 \\ 12 - 2x + 2x &> 3x - 15 + 2x \\ 12 &> 5x - 15 \\ 12 + 15 &> 5x - 15 + 15 \\ 27 &> 5x \\ \frac{27}{5} &> \frac{5x}{5} \\ \frac{27}{5} &> x \text{ or } x < \frac{27}{5} \end{aligned}$$

117.

$$\begin{aligned} 3.2 - (5.3x - 1.2) &\le 3.4x - 1.8 \\ 3.2 - 5.3x + 1.2 &\le 3.4x - 1.8 \\ 4.4 - 5.3x &\le 3.4x - 1.8 \\ 4.4 - 5.3x + 5.3x &\le 3.4x - 1.8 + 5.3x \\ 4.4 &\le 8.7x - 1.8 \\ 4.4 + 1.8 &\le 8.7x - 1.8 + 1.8 \\ 6.2 &\le 8.7x \\ \frac{6.2}{8.7} &\le \frac{8.7x}{8.7} \\ 0.713 &\le x \end{aligned}$$

or

$$x \ge 0.713$$

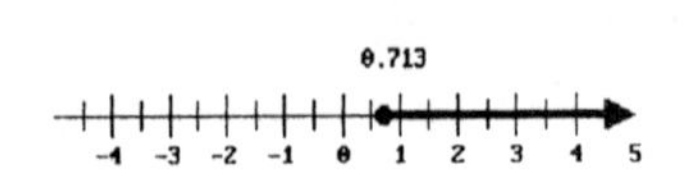

CHAPTER 1 PRACTICE TEST

1. $A = \{2, 3, 5, 7, 11, 13, 17, 19, 23\}$
   $B = \{2, 3, 5, 7\}$

   (a) $A \cap B = \{2, 3, 5, 7\}$

   (b) $A \cup B = \{2, 3, 5, 7, 11, 13, 17, 19, 23\} = A$

3. (a) $\{a \mid a > 4\}$

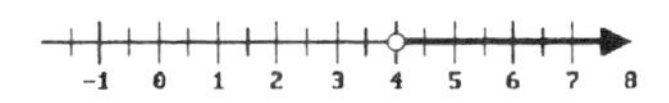

   (b) $\{x \mid -3 \le x < 10\}$

5. (a) $-3 - (-6) + (-4) - (-9)$
   $= -3 + 6 + (-4) + 9$
   $= 3 + (-4) + 9$
   $= -1 + 9$
   $= 8$

   (b) $(-7)^2 - (-6)(-2)(-3)$
   $= 49 - (-6)(-2)(-3)$
   $= 49 - (-36)$
   $= 85$

   (c) $|3 - 8| - |5 - 9| = |-5| - |-4|$
   $= 5 - 4$
   $= 1$

   (d) $6 - 5[-2 - (7 - 9)]$
   $= 6 - 5[-2 - (-2)]$
   $= 6 - 5(0)$
   $= 6 - 0$
   $= 6$

7. (a) $(5x^3y^2)(-2x^2y)(-xy^2)$
   $= (5)(-2)(-1)x^3x^2xy^2yy^2$
   $= 10x^6y^5$

   (b) $3rs^2 - 5r^2s - 4rs^2 - 7rs$
   $= (3 - 4)rs^2 - 5r^2s - 7rs$
   $= -rs^2 - 5r^2s - 7rs$

   (c) $2a - 3(a - 2) - (6 - a)$
   $= 2a - 3a + 6 - 6 + a$
   $= (2 - 3 + 1)a + (6 - 6)$
   $= 0$

   (d) $7r - \{3 + 2[s - (r - 2s)]\}$
   $= 7r - [3 + 2(s - r + 2s)]$
   $= 7r - [3 + 2(3s - r)]$
   $= 7r - (3 + 6s - 2r)$
   $= 7r - 3 - 6s + 2r$
   $= 9r - 6s - 3$

9. number of dimes: x
   number of nickels: $34 - x$
   value of dimes: $10x$
   value of nickels: $5(34 - x)$

Total value = value of dimes + value of nickels
Total value $= 10x + 5(34 - x)$
$= 10x + 170 - 5x$
$= 5x + 170$ cents

11.
$$3s - 5t = 2t - 4$$
$$3s - 5t + 5t = 2t - 4 + 5t$$
$$3s = 7t - 4$$
$$3s + 4 = 7t - 4 + 4$$
$$3s + 4 = 7t$$
$$\frac{3s + 4}{7} = \frac{7t}{7}$$
$$\frac{3s + 4}{7} = t$$

CHAPTER 2

Exercises 2.1

1. Let G = course grade,
   f = final exam grade

$G = 0.75(82) + 0.25f$
$G = 61.5 + 0.25f$

f	G
73	79.75
74	80
75	80.25
100	86.5
110	89
115	90.25

(a) She would need a 74 on her final exam.

(b) She would need between 110 and 115. If the final exam is based on 100 points, this would be impossible.

3. Let G = course grade,
   f = final exam grade

$G = 0.75(78) + 0.25f$
$G = 58.5 + 0.25f$

f	G
85	79.75
86	80
87	80.25

She would need an 86 on her final exam.

5. Let C = weekly commission,
   G = gross sales

$C = 0.09G$

G	C
6650	598.50
6660	599.40
6670	600.30

He would need weekly gross sales between \$6660 and \$6670.

7. number of Euros: x
   number of U. S. dollars: y

$$x = \frac{y}{0.8702}$$

y	x
260	298.78
261	299.93
262	301.08

300 Euros is worth between 261 and 262 U. S. dollars.

9. number of British Pounds: x
   number of U. S. dollars: y

$$x = \frac{y}{1.4388}$$

y	x
500	347.51
525	364.89
550	382.26
575	399.64

400 British Pounds are worth approximately 575 U. S. dollars.

11. Let P = weekly pay,
    G = gross sales

strictly commission plan:

$P = 0.09G$

base salary + commission plan:

$P = 120 + 0.06G$

	$P = 0.09G$	$P = 120 + 0.06G$
3800	342	348
3900	351	354
4000	360	360
4100	369	366
4200	378	372

For gross sales greater than \$4000, the strictly commission plan is better. For gross sales less than \$4000, the salary plus commission plan is better.

13. Let P = perimeter,
w = width

then $3w + 1$ = length

$P = 2 \cdot \text{width} + 2 \cdot \text{length}$
$P = 2(w) + 2(3w + 1)$
$P = 2w + 6w + 2$
$P = 8w + 2$

W	P
9	74
10	82
11	90

The width must be between 9 and 10 in. and the length between 28 and 31 in.

15. Let I = interest earned,
x = amount invested

$I = 0.083x$

x	I
12000	996
12050	1000.20
12100	1004.30

She must invest approximately $12,050.

17. Let I = interest earned,
x = amount invested at 8%
then $12500 - x$ = amount invested at 5%

$I = 0.08(x) + 0.05(12500 - x)$
$I = 0.08x + 625 - 0.05x$
$I = 0.03x + 625$

x	I
0	625
2000	685
5000	775
7000	835
10000	925
12500	1000

The minimum interest, $625, is earned when $0 is invested at 8% and $12,500 is invested at 5%

More interest is earned as more money is invested at 8%.

The maximum interest, $1000 is earned when all $12,500 is invested at 8%.

19. Let I = interest earned,
x = amount invested at 4.2%
then $24{,}000 - x$ = amount invested at 7.8%

$I = 0.042(x) + 0.078(24{,}000 - x)$
$I = 0.042x + 1872 - 0.078x$
$I = 1872 - 0.036x$

x	I
0	1872
10,000	1512
15.000	1332
18,000	1224
19,000	1188
24,000	1008

The maximum interest is $1872 when all of the money is invested at 7.8%. The minimum interest is $1008 when all of the money is invested at 4.2%. To earn $1200 in interest approximately $18,000 is invested at 4.2% and $6000 is invested at 7.8%.

21. Let n = total number of copies printed,
t = number of minutes the faster copier works,
$t - 5$ = number of minutes the slower copier works

$n = 9t + 4(t - 5)$
$n = 9t + 4t - 20$
$n = 13t - 20$

t	n
10	110
11	123
12	136
13	149
14	162

It will take between 12 and 13 minutes to print 142 pages.

23. Let x = number of \$850 computers sold,
$58 - x$ = number of \$600 computers sold
C = total money collected

$$C = 850x + 600(58 - x)$$
$$C = 850x + 34800 - 600x$$
$$C = 250x + 34800$$

x	C
20	39800
21	40050
22	40300
23	40550
24	40800

22 \$850 computers were sold.

25. value of stock yesterday: x
value of stock today: y

$$y = \frac{47}{61}x$$

x	y
3220	2480.98
3230	2488.69
3240	2496.39
3250	2504.10

The value of the stock yesterday was between \$3240 and \$3250.

27. Let h = number of hours Lewis works,
$h - ½$ = number of hours Arthur works,
n = total number of forms they can process

$$n = 200h + 300(h - ½)$$
$$n = 200h + 300h - 150$$
$$n = 500h - 150$$

h	n
5	2350
6	2850
7	3350
8	3850

It will take between 6 and 7 hours.

29. Let x = number of AM radios,
$24 - x$ = number of AM/FM radios,
C = total cost

$$C = 35x + 50(24 - x) + 70$$
$$C = 35x + 1200 - 50x + 70$$
$$C = 1270 - 15x$$

x	C
16	1030
17	1015
18	1000
19	985

18 AM radios, 6 AM/FM radios

Exercises 2.2

1. $x - (x - 3) = 4$
$x - x + 3 = 4$
$3 = 4$
Contradiction

3. $4(w - 2) = 4w - 8$
$4w - 8 = 4w - 8$
Identity

5. $2(y + 1) - (y - 5) = 4(y - 2) - 3(y - 5)$
$2y + 2 - y + 5 = 4y - 8 - 3y + 15$
$y + 7 = y + 7$
Identity

7. $6(2y + 1) - 4(3y - 1) = y - (y + 10)$
$12y + 6 - 12y + 4 = y - y - 10$
$10 = -10$
Contradiction

9. $5(x - 3) - (x + 2) = 8 - x$

$x = -0.08$: $5(-0.08 - 3) - (-0.08 + 2) = 8 - (-0.08)$
$5(-3.08) - 1.92 = 8.08$
$-15.4 - 1.92 = 8.08$
$-17.32 = 8.08$
does not satisfy

$x = 0$: $5(0 - 3) - (0 + 2) = 8 - 0$
$5(-3) - 2 = 8$
$-15 - 2 = 8$
$-17 = 8$
does not satisfy

$x = 5$: $5(5 - 3) - (5 + 2) = 8 - 5$
$5(2) - 7 = 3$
$10 - 7 = 3$
$3 = 3$

satisfies

11. $x^2 - 5x = 5 - x$

$x = -5$: $(-5)^2 - 5(-5) = 5 - (-5)$
$25 + 25 = 10$
$50 = 10$

does not satisfy

$x = -\frac{1}{2}$: $\left(-\frac{1}{2}\right)^2 - 5\left(-\frac{1}{2}\right) = 5 - \left(-\frac{1}{2}\right)$

$\frac{1}{4} + \frac{5}{2} = \frac{11}{2}$

$\frac{11}{4} = \frac{11}{2}$

does not satisfy

$x = 5$: $5^2 - 5(5) = 5 - 5$
$25 - 25 = 0$
$0 = 0$

satisfies

13. $3x - 12 = 12 - 3x$
$6x - 12 = 12$
$6x = 24$
$x = 4$

15. $3x - 12 = -12 + 3x$
$-12 = -12$
Identity, true for all real numbers

17. $2t + 1 = 7 + t$
$t + 1 = 7$
$t = 6$

19. $2(x - 1) = 2(x + 1)$
$2x - 2 = 2x + 2$
$-2 = 2$
Contradiction, no solution

21. $3(x + 1) - x = 2(x + 3)$
$3x + 3 - x = 2x + 6$
$2x + 3 = 2x + 6$
$3 = 6$
Contradiction, no solution

23. $5(2t - 1) - 3t = 5 - 7t$
$10t - 5 - 3t = 5 - 7t$
$7t - 5 = 5 - 7t$
$14t - 5 = 5$
$14t = 10$
$t = \frac{5}{7}$

25. $6(3 - z) + 2(4z - 5) = z - (3 - z)$
$18 - 6z + 8z - 10 = z - 3 + z$
$2z + 8 = 2z - 3$
$8 = -3$
Contradiction, no solution

27. $4(3x - 1) - 5(3x - 2) = 2(x + 3) - 5x$
$12x - 4 - 15x + 10 = 2x + 6 - 5x$
$-3x + 6 = -3x + 6$
Identity, true for all numbers

29. $x(x - 2) - 15 = x(x + 5) - 3(x + 5)$
$x^2 - 2x - 15 = x^2 + 5x - 3x - 15$
$x^2 - 2x - 15 = x^2 + 2x - 15$
$-2x - 15 = 2x - 15$
$-4x - 15 = -15$
$-4x = 0$
$x = 0$

31. $6x - [2 - (x - 1)] = x - 5(x + 1)$
$6x - (2 - x + 1) = x - 5x - 5$
$6x - (3 - x) = -4x - 5$
$6x - 3 + x = -4x - 5$
$7x - 3 = -4x - 5$
$11x - 3 = -5$
$11x = -2$
$x = -\frac{2}{11}$

33. $3 - [4t + 5(2t - 1) - 3t] = 3(4 - t)$
$3 - (4t + 10t - 5 - 3t) = 12 - 3t$
$3 - (11t - 5) = 12 - 3t$
$3 - 11t + 5 = 12 - 3t$
$-11t + 8 = 12 - 3t$
$8 = 12 + 8t$
$-4 = 8t$
$-\frac{1}{2} = t$

35.
$$\frac{x}{2} - \frac{3x}{5} = 4$$
$$10\left(\frac{x}{2} - \frac{3x}{5}\right) = 10(4)$$
$$\frac{10}{1} \cdot \frac{x}{2} - \frac{10}{1} \cdot \frac{3x}{5} = 40$$
$$\begin{aligned} 5x - 6x &= 40 \\ -x &= 40 \\ x &= -40 \end{aligned}$$

37.
$$\frac{y}{2} - \frac{4 - y}{3} = 4$$
$$6\left(\frac{y}{2} - \frac{4 - y}{3}\right) = 6(4)$$
$$\frac{6}{1} \cdot \frac{y}{2} - \frac{6}{1} \cdot \frac{4 - y}{3} = 24$$
$$\begin{aligned} 3y - 2(4 - y) &= 24 \\ 3y - 8 + 2y &= 24 \\ 5y - 8 &= 24 \\ 5y &= 32 \\ y &= \frac{32}{5} \end{aligned}$$

39.
$$\frac{2y - 1}{4} = y - 3$$
$$4\left(\frac{2y - 1}{4}\right) = 4(y - 3)$$
$$\begin{aligned} 2y - 1 &= 4y - 12 \\ -1 &= 2y - 12 \\ 11 &= 2y \\ \frac{11}{2} &= y \end{aligned}$$

41.
$$\frac{3a}{2} - \frac{1}{3} = \frac{a + 5}{4}$$
$$12\left(\frac{3a}{2} - \frac{1}{3}\right) = 12\left(\frac{a + 5}{4}\right)$$
$$\frac{12}{1} \cdot \frac{3a}{2} - \frac{12}{1} \cdot \frac{1}{3} = 3(a + 5)$$
$$\begin{aligned} 18a - 4 &= 3a + 15 \\ 15a - 4 &= 15 \\ 15a &= 19 \\ a &= \frac{19}{15} \end{aligned}$$

43. 1st number: x
2nd number: $5 + 2x$

$$\begin{aligned} x + 5 + 2x &= 23 \\ 3x + 5 &= 23 \\ 3x &= 18 \\ x &= 6 \\ 5 + 2x &= 5 + 2(6) = 17 \end{aligned}$$

The numbers are 6 and 17.

45. 1st integer: x
2nd integer: $x + 1$
3rd integer: $x + 2$
4th integer: $x + 3$

$$\begin{aligned} x + x + 1 + x + 2 + x + 3 &= 1 + (x + 2) \\ 4x + 6 &= x + 3 \\ 3x + 6 &= 3 \\ 3x &= -3 \\ x &= -1 \\ x + 1 &= -1 + 1 = 0 \\ x + 2 &= -1 + 2 = 1 \\ x + 3 &= -1 + 3 = 2 \end{aligned}$$

The integers are −1, 0, 1, and 2.

47. value in U. S. dollars: x

$$\begin{aligned} x &= 500(0.8640) \\ x &= 432 \end{aligned}$$

The U. S. value is $432.

49. value on Jan. 4: x
value on Jan. 5: $x - 0.25x = 0.75x$
value on Jan. 6: $0.75x + 0.25(0.75x)$

$$\begin{aligned} 0.75x + 0.25(0.75x) &= 2500 \\ 0.75x + 0.1875x &= 2500 \\ 0.9375x &= 2500 \\ x &= 2666.67 \end{aligned}$$

It was worth $2666.67 on Jan. 4.

51. width: x
length: $2x$

$$\begin{aligned} P &= 2w + 2l \\ 42 &= 2x + 2(2x) \\ 42 &= 2x + 4x \\ 42 &= 6x \\ 7 &= x \\ 2x &= 2(7) = 14 \end{aligned}$$

The garden is 7 meters by 14 meters.

53. 1st side: $x - 5$
2nd side: x
3rd side: $2(x - 5)$

$$\begin{aligned} P &= \text{sum of three side lengths} \\ 33 &= x - 5 + x + 2(x - 5) \\ 33 &= x - 5 + x + 2x - 10 \\ 33 &= 4x - 15 \\ 48 &= 4x \\ 12 &= x \\ x - 5 &= 12 - 5 = 7 \\ 2(x - 5) &= 2(12 - 5) = 14 \end{aligned}$$

The sides have lengths 7 cm, 12 cm, and 14 cm.

55. original rectangle
width: x
length: $1 + 3x$

new rectangle
width: $x + 2$
length: $2(1 + 3x)$

$$\begin{aligned} \text{original perimeter} &= 2(x) + 2(1 + 3x) \\ &= 2x + 2 + 6x \\ &= 8x + 2 \end{aligned}$$

$$\begin{aligned} \text{new perimeter} &= 2(x + 2) + 2[2(1 + 3x)] \\ &= 2x + 4 + 4 + 12x \\ &= 14x + 8 \end{aligned}$$

The new perimeter is 3 less than 5 times the original length.

$$\begin{aligned} 14x + 8 &= 5(1 + 3x) - 3 \\ 14x + 8 &= 5 + 15x - 3 \\ 14x + 8 &= 15x + 2 \\ 8 &= x + 2 \\ 6 &= x \\ 1 + 3x &= 1 + 3(6) = 19 \end{aligned}$$

The original dimensions were 6 by 19.

57. gross sales: x

$$\begin{aligned} 0.08x &= 600 \\ x &= 7500 \end{aligned}$$

The gross sales need to be $7500.

59. final exam score: x

$$\begin{aligned} 0.45(x) + 0.55(72) &= 80 \\ 0.45x + 39.6 &= 80 \\ 0.45x &= 40.4 \\ x &= 89.8 \end{aligned}$$

The minimum score is 89.8.

61. number of $5 bills: x
number of $10 bills: $x + 1$
number of $1 bills: $25 - (x + x + 1) = 24 - 2x$

$$\begin{aligned} 5(x) + 10(x + 1) + 1(24 - 2x) &= 164 \\ 5x + 10x + 10 + 24 - 2x &= 164 \\ 13x + 34 &= 164 \\ 13x &= 130 \\ x &= 10 \\ x + 1 &= 10 + 1 = 11 \\ 24 - 2x &= 24 - 2(10) = 4 \end{aligned}$$

He has 10 $5-bills, 11 $10-bills, and 4 $1-bills.

63. number of 20-lb packages: x
number of 25-lb packages: $50 - x$

$$\begin{aligned} 20(x) + 25(50 - x) &= 1075 \\ 20x + 1250 - 25x &= 1075 \\ -5x + 1250 &= 1075 \\ -5x &= -175 \\ x &= 35 \\ 50 - x &= 50 - 35 = 15 \end{aligned}$$

There are 35 20-lb packages and 15 25-lb packages.

65. pairs of shoes: x
pairs of boots: $70 - x$

$$\begin{aligned} 368 &= 200 + 3(70 - x) + 2(x) \\ 368 &= 200 + 210 - 3x + 2x \\ 368 &= 410 - x \\ -42 &= -x \\ 42 &= x \end{aligned}$$

She sold 42 pairs of shoes.

67. lbs of $2/lb coffee: x

$$\begin{aligned} 2(x) + 3(30) &= 2.60(x + 30) \\ 2x + 90 &= 2.60x + 78 \\ 90 &= 0.60x + 78 \\ 12 &= 0.60x \\ 20 &= x \end{aligned}$$

20 lb of the $2-per-pound coffee should be used.

69. number of orchestra seats: x
number of balcony seats: $56 - x$

$$\begin{aligned} 96(x) + 76(56 - x) &= 5016 \\ 96x + 4256 - 76x &= 5016 \\ 20x + 4256 &= 5016 \\ 20x &= 760 \\ x &= 38 \end{aligned}$$

38 orchestra seats were purchased.

71. plumber's hours: x
assistant's hours: $x + 2$

$$\begin{aligned} 32x + 16(x + 2) &= 320 \\ 32x + 16x + 32 &= 320 \\ 48x + 32 &= 320 \\ 48x &= 288 \\ x &= 6 \end{aligned}$$

The plumber worked 6 hours.

73. 1st car's hours: x
2nd car's hours: x

$$\begin{aligned} 55x + 60x &= 345 \\ 115x &= 345 \\ x &= 3 \end{aligned}$$

They will meet in 3 hours
which will be 6:00 P.M.

75. 1st car's hours: x
2nd car's hours: x

$$\begin{aligned} 35x + 50x &= 595 \\ 85x &= 595 \\ x &= 7 \end{aligned}$$

It will take 7 hours.

77. 1st person's hours: x
2nd person's hours: $x + 3$

$$\begin{aligned} 17x &= 7(x + 3) \\ 17x &= 7x + 21 \\ 10x &= 21 \\ x &= 2.1 \end{aligned}$$

It would take 2.1 hours.

79. number of hours going: x
number of hours returning: $17 - x$

$$\begin{aligned} 48x &= 54(17 - x) \\ 48x &= 918 - 54x \\ 102x &= 918 \\ x &= 9 \end{aligned}$$

Distance $= 48(9) = 432$

The convention is 432 km away.

81. 1 hr 50 min = 110 min

number of minutes for old copier: x
number of minutes for new copier: $110 - x$

$$\begin{aligned} 35x + 50(110 - x) &= 5125 \\ 35x + 5500 - 50x &= 5125 \\ -15x + 5500 &= 5125 \\ -15x &= -375 \\ x &= 25 \end{aligned}$$

old copier: $35(25) = 875$

The older copier made 875 copies.

83. experienced worker's hours: x
new worker's hours: x

$$\begin{aligned} 60x + 30x &= 6750 \\ 90x &= 6750 \\ x &= 75 \end{aligned}$$

It will take 75 hours.

85. new worker's hours: x
experienced worker's hours: $x + 3$

$$\begin{aligned} 60(x + 3) + 30x &= 6750 \\ 60x + 180 + 30x &= 6750 \\ 90x + 180 &= 6750 \\ 90x &= 6570 \\ x &= 73 \\ x + 3 &= 73 + 3 = 76 \end{aligned}$$

It will take the new worker 73 hours
and the experienced worker 76 hours.

87. experienced worker's hours: x
trainee's hours: $x + 3$

$$\begin{aligned} 80x + 48(x + 3) &= 656 \\ 80x + 48x + 144 &= 656 \\ 128x + 144 &= 656 \\ 128x &= 512 \\ x &= 4 \\ x + 3 &= 4 + 3 = 7 \end{aligned}$$

The trainee works 7 hours, hence
they will finish the job 7 hours
after 10:00 a.m. which is 5:00 p.m.

Exercises 2.3

1. $x + 1 > x - 1$
$1 > -1$
Identity

3. $2x - (x - 3) \geq x + 5$
$$2x - x + 3 \geq x + 5$$
$$x + 3 \geq x + 5$$
$$3 \geq 5$$
Contradiction

5. $-6 \leq r - (r + 6) < 3$
$$-6 \leq r - r - 6 < 3$$
$$-6 \leq -6 < 3$$
Identity

7. $0 < x - (x - 3) \leq 5$
$$0 < x - x + 3 \leq 5$$
$$0 < 3 \leq 5$$
Identity

9. $4 + 3u > 10$

$u = -3.4$: $4 + 3(-3.4) > 10$
$$-6.2 > 10$$
does not satisfy

$u = 2$: $4 + 3(2) > 10$
$$10 > 10$$
does not satisfy

11. $6 - 4y \leq 8$

$y = -2.5$: $6 - 4(-2.5) \leq 8$
$$16 \leq 8$$
does not satisfy

$y = -1$: $6 - 4(-1) \leq 8$
$$10 \leq 8$$
does not satisfy

13. $-3 \leq 3 - (4 - z) < 3$
$$-3 \leq 3 - 4 + z < 3$$
$$-3 \leq z - 1 < 3$$

$z = -2$: $-3 \leq -2 - 1 < 3$
$$-3 \leq -3 < 3$$
satisfies

$z = 4.6$: $-3 \leq 4.6 - 1 < 3$
$$-3 \leq 3.6 < 3$$
does not satisfy

15. $5 < x < 8$
makes sense

17. $-6 > w \geq -8$
makes sense
$-8 \leq w < -6$

19. $-3 < a < -2$
makes sense

21. $2 > x > -3$
makes sense

$-3 < x < 2$

23. $5y \geq 2 - 3y$
$$8y \geq 2$$
$$y \geq \frac{1}{4}$$

25. $\frac{2}{3}x - 4 \leq 3x + \frac{1}{2}$

$$6\left(\frac{2}{3}x - 4\right) \leq 6\left(3x + \frac{1}{2}\right)$$

$$\frac{6}{1} \cdot \frac{2}{3}x - 6(4) \leq 6(3x) + \frac{6}{1} \cdot \frac{1}{2}$$

$$4x - 24 \leq 18x + 3$$
$$-24 \leq 14x + 3$$
$$-27 \leq 14x$$
$$\frac{-27}{14} \leq x$$

or

$$x \geq -\frac{27}{14}$$

27. $3x + 4 \geq 2x - 5$
$$x + 4 \geq -5$$
$$x \geq -9$$

29. $11 - 3y < 38$
$$-3y < 27$$
$$y > -9$$

31. $3a - 8 < 8 - 3a$
$$6a - 8 < 8$$
$$6a < 16$$
$$a < \frac{8}{3}$$

-6 -4 -2 0 2 4 6 8

33. $3t - 5 > 5t - 13$
$-2t - 5 > -13$
$-2t > -8$
$t < 4$

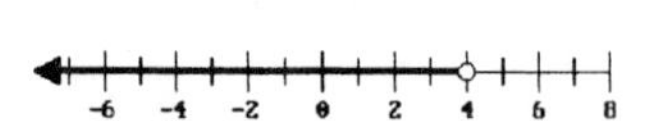

35. $2y - 3 \le 5y + 7$
$-3y - 3 \le 7$
$-3y \le 10$
$y \ge -\frac{10}{3}$

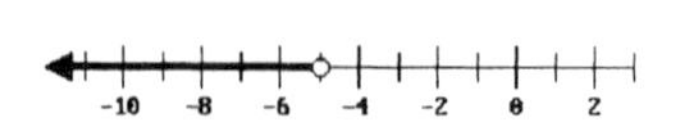

37. $5(a - 2) - 7a > 2a + 10$
$5a - 10 - 7a > 2a + 10$
$-2a - 10 > 2a + 10$
$-4a - 10 > 10$
$-4a > 20$
$a < -5$

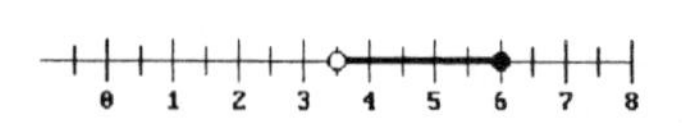

39. $\frac{3x + 2}{5} < x + 4$

$5\left(\frac{3x + 2}{5}\right) < 5(x + 4)$

$3x + 2 < 5x + 20$
$-2x + 2 < 20$
$-2x < 18$
$x > -9$

41. $7 - 5(t - 2) \le 2$
$7 - 5t + 10 \le 2$
$-5t + 17 \le 2$
$-5t \le -15$
$t \ge 3$

43. $2(y + 3) + 3(y - 4) < 3(y + 2) + 2(y + 1)$
$2y + 6 + 3y - 12 < 3y + 6 + 2y + 2$
$5y - 6 < 5y + 8$
$-6 < 8$
Identity, true for all real numbers

45. $1 \le c + 3 < 5$
$-2 \le c < 2$

47. $6 < 2k - 1 \le 11$
$7 < 2k \le 12$
$\frac{7}{2} < k \le 6$

49. $-1 < 4 - t < 3$
$-5 < -t < -1$
$5 > t > 1$
or $1 < t < 5$

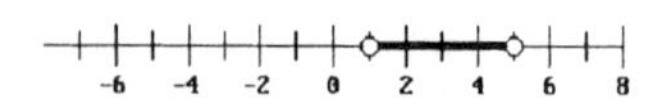

51. $1 \le 8 - 3t \le 12$
$-7 \le -3t \le 4$
$\frac{7}{3} \ge t \ge -\frac{4}{3}$
or $-\frac{4}{3} \le t \le \frac{7}{3}$

53. $2 < 5 - 3(x + 1) < 17$
$2 < 5 - 3x - 3 < 17$
$2 < 2 - 3x < 17$
$0 < -3x < 15$
$0 > x > -5$
or $-5 < x < 0$

55. $-2 < x - 5 \le 1$
$3 < x \le 6$

57. $-2 \le 5 - x < 1$
$-7 \le -x < -4$
$7 \ge x > 4$
or $4 < x \le 7$

59. $1 \le 1 - (5z - 2) \le 11$
$1 \le 1 - 5z + 2 \le 11$
$1 \le 3 - 5z \le 11$
$-2 \le -5z \le 8$
$\frac{2}{5} \ge z \ge -\frac{8}{5}$
or $-\frac{8}{5} \le z \le \frac{2}{5}$

61. $0.39 \le 0.72x - 1.5 < 8.1$
$1.89 \le 0.72x < 9.6$
$2.625 \le x < 13.\overline{3}$

63. number: x

$3x - 4 < 17$
$3x < 21$
$x < 7$

All numbers less than 7.

65. number: x

$6x + 12 > 3x$
$12 > -3x$
$-4 < x$

The numbers must be greater than -4.

67. side length: x

$P \le 72$
$4x \le 72$
$x \le 18$

The maximum length is 18 feet.

69. length: x

$P \ge 80$
$2w + 2l \ge 80$
$2(8) + 2(x) \ge 80$
$16 + 2x \ge 80$
$2x \ge 64$
$x \ge 32$

The length must be at least 32 cm.

71. width: x

$50 \le P \le 70$
$50 \le 2w + 2l \le 70$
$50 \le 2(x) + 2(18) \le 70$
$50 \le 2x + 36 \le 70$
$14 \le 2x \le 34$
$7 \le x \le 17$

The width has a range 7 in. to 17 in., inclusively.

73. width: x
length: $3x + 5$

$P = 2w + 2l$
$P = 2(x) + 2(3x + 5)$
$P = 2x + 6x + 10$
$P = 8x + 10$
$P - 10 = 8x$
$\frac{P - 10}{8} = x$

$12 \le x \le 16$
$12 \le \frac{P - 10}{8} \le 16$

$$8(12) \le 8\left(\frac{P - 10}{8}\right) \le 8(16)$$

$$96 \le P - 10 \le 128$$
$$106 \le P \le 138$$

The range of values for the perimeter is 106 in. to 138 in.

75. shortest side: x
medium side: $x + 2$
longest side: $2x$

$$30 \le P \le 50$$
$$30 \le x + x + 2 + 2x \le 50$$
$$30 \le 4x + 2 \le 50$$
$$28 \le 4x \le 48$$
$$7 \le x \le 12$$

The shortest side has a range of values, 7 cm to 12 cm.

77. reserved ticket price: x
at the door ticket price: $x - 2$

$$C \ge 3750$$
$$300x + 150(x - 2) \ge 3750$$
$$300x + 150x - 300 \ge 3750$$
$$450x - 300 \ge 3750$$
$$450x \ge 4050$$
$$x \ge 9$$

The minimum price is \$9.

79. number of dimes: x
number of nickels: $40 - x$

$$10(x) + 5(40 - x) \le 285$$
$$10x + 200 - 5x \le 285$$
$$5x + 200 \le 285$$
$$5x \le 85$$
$$x \le 17$$

He can have at most 17 dimes.

81. number of elephants: x
number of pandas: $24 - x$

$$20(24 - x) + 25(x) \ge 575$$
$$480 - 20x + 25x \ge 575$$
$$480 + 5x \ge 575$$
$$5x \ge 95$$
$$x \ge 19$$

The minimum number of elephants is 19.

83. hours tutoring: x
hours at restaurant: $30 - x$

$$15x + 10(30 - x) \ge 475$$
$$15x + 300 - 10x \ge 475$$
$$5x + 300 \ge 475$$
$$5x \ge 175$$
$$x \ge 35$$

It is impossible to make \$475 with just 30 hours.

85. number of super-dogs: x
number of regular-dogs: $100 - x$

$$25(100 - x) + 45(x) \ge 3860$$
$$2500 - 25x + 45x \ge 3860$$
$$20x \ge 1360$$
$$x \ge 68$$

He must sell at least 68 super-dogs.

87. number of shares of Stock A: $1000 - x$
number of shares of Stock B: x

$$2(1000 - x) + 3(x) \ge 2400$$
$$2000 - 2x + 3x \ge 2400$$
$$x + 2000 \ge 2400$$
$$x \ge 400$$

The minimum number of shares of Stock B is 400.

Exercises 2.4

1. $|x| = 4$
$x = 4$ or $x = -4$

3. $|x| < 4$
$-4 < x < 4$

5. $|x| > 4$
$x > 4$ or $x < -4$

7. $|x| \le 4$
$-4 \le x \le 4$

9. $|x| \ge 4$
$x \ge 4$ or $x \le -4$

11. No solution, absolute value is nonnegative.

13. $|x| > -4$
True for all real numbers, since absolute value is nonnegative. It is greater than -4.

15. $|x| \le -4$
No solution, absolute value is nonnegative.

17. $|t - 3| = 2$
$t - 3 = 2$ or $t - 3 = -2$
$t = 5$ or $t = 1$

19. $|t| - 3 = 2$
$|t| = 5$
$t = 5$ or $t = -5$

21. $|5 - n| = 1$
$5 - n = 1$ or $5 - n = -1$
$-n = -4$ $\quad$ $-n = -6$
$n = 4$ or $n = 6$

23. $|a - 5| < 3$
$-3 < a - 5 < 3$
$2 < a < 8$

25. $|a - 1| \ge 2$
$a - 1 \ge 2$ or $a - 1 \le -2$
$a \ge 3$ or $a \le -1$

27. $|2a - 5| < -1$
No solution, absolute value is nonnegative.

29. $|2a| - 5 < -1$
$|2a| < 4$
$-4 < 2a < 4$
$-2 < a < 2$

31. $|3x - 2| - 3 = 1$
$|3x - 2| = 4$
$3x - 2 = 4$ or $3x - 2 = -4$
$3x = 6$ $\quad$ $3x = -2$
$x = 2$ or $x = -\dfrac{2}{3}$

33. $|3x - 2| + 3 = 1$
$|3x - 2| = -2$
No solution, absolute value is nonnegative.

35. $|2(x - 1) + 7| = 5$
$|2x - 2 + 7| = 5$
$|2x + 5| = 5$
$2x + 5 = 5$ or $2x + 5 = -5$
$2x = 0$ $\quad$ $2x = -10$
$x = 0$ or $x = -5$

37. $|3 - a| \le 2$
$-2 \le 3 - a \le 2$
$-5 \le -a \le -1$
$5 \ge a \ge 1$
or $1 \le a \le 5$

39. $|5 - 2a| > 1$
$5 - 2a > 1$ or $5 - 2a < -1$
$-2a > -4$ $\quad$ $-2a < -6$
$a < 2$ or $a > 3$

41. $|4(x - 1) - 5x| < 4$
$|4x - 4 - 5x| < 4$
$|-x - 4| < 4$
$-4 < -x - 4 < 4$
$0 < -x < 8$
$0 > x > -8$
or $-8 < x < 0$

43. $|x - 1| = 5$
$x - 1 = 5$ or $x - 1 = -5$
$x = 6$ or $x = -4$

45. $|3 - x| \le 2$
$-2 \le 3 - x \le 2$
$-5 \le -x \le -1$
$5 \ge x \ge 1$
or $1 \le x \le 5$

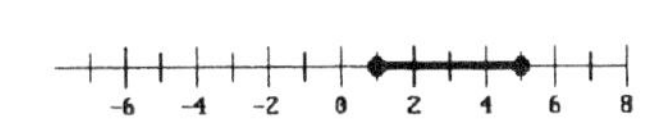

47. $|2x + 7| > 1$
$2x + 7 < -1$ or $2x + 7 > 1$
$2x < -8$ $\quad$ $2x > -6$
$x < -4$ or $x > -3$

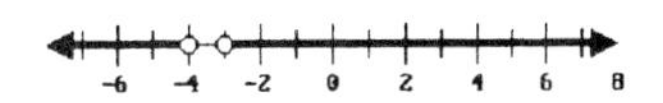

49. $|4x - 5| < 3$
$-3 < 4x - 5 < 3$
$2 < 4x < 8$
$\frac{1}{2} < x < 2$

51. $|3 - 4x| < 1$
$-1 < 3 - 4x < 1$
$-4 < -4x < -2$
$1 > x > \frac{1}{2}$
or $\frac{1}{2} < x < 1$

53. $|5t - 1| = |4t + 3|$
$5t - 1 = 4t + 3$ or $5t - 1 = -(4t + 3)$
$t - 1 = 3$ $\quad$ $5t - 1 = -4t - 3$
$t = 4$ $\quad$ $9t - 1 = -3$
$9t = -2$
$t = 4$ or $t = -\frac{2}{9}$

55. $|4r - 3| = |2r + 9|$
$4r - 3 = 2r + 9$ or $4r - 3 = -(2r + 9)$
$2r - 3 = 9$ $\quad$ $4r - 3 = -2r - 9$
$2r = 12$ $\quad$ $6r - 3 = -9$
$r = 6$ $\quad$ $6r = -6$
$r = 6$ or $r = -1$

57. $|a - 5| = |2 - a|$
$a - 5 = 2 - a$ or $a - 5 = -(2 - a)$
$2a - 5 = 2$ $\quad$ $a - 5 = -2 + a$
$2a = 7$ $\quad$ $-5 = -2$
$a = \frac{7}{2}$ $\quad$ No solution
The only solution is $a = \frac{7}{2}$.

59. $|3x - 4| = |4x - 3|$
$3x - 4 = 4x - 3$ or $3x - 4 = -(4x - 3)$
$-x - 4 = -3$ $\quad$ $3x - 4 = -4x + 3$
$-x = 1$ $\quad$ $7x - 4 = 3$
$x = -1$ $\quad$ $7x = 7$
$x = -1$ or $x = 1$

61. $|x + 1| = |x - 1|$
$x + 1 = x - 1$ or $x + 1 = -(x - 1)$
$1 = -1$ $\quad$ $x + 1 = -x + 1$
No solution $\quad$ $2x + 1 = 1$
$2x = 0$
$x = 0$
The only solution is $x = 0$.

63. reading: x
For 40° F:

$|x - 40| \le 0.75$
$-0.75 \le x - 40 \le 0.75$
$39.25 \le x \le 40.75$
39.25° F to 40.75° F
For 50° F:

$|x - 50| \le 0.75$
$-0.75 \le x - 50 \le 0.75$
$49.25 \le x \le 50.75$
49.25° F to 50.75° F

For 80° F:
$|x - 80| \le 0.75$
$-0.75 \le x - 80 \le 0.75$
$79.25 \le x \le 80.75$
79.25° F to 80.75° F

65. percentage of votes: x
$|x - 64| \le 6$
$-6 \le x - 64 \le 6$
$58 \le x \le 70$
58% to 70%

CHAPTER 2 REVIEW EXERCISES

1. Let s = number of single beds,
$24 - s$ = number of larger beds

money collected $= 15s + 20(24 - s)$
$= 15s + 480 - 20s$
$= 480 - 5s$

s	$480 - 5s$
11	425
12	420
13	415
14	410

They delivered 13 single beds and $24 - 13 = 11$ larger beds.

3. Let w = width
$4w - 2$ = length
$P = 2(w) + 2(4w - 2)$
$P = 2w + 8w - 4$
$P = 10w - 4$

w	$10w - 4$
8	76
9	86
10	96
11	106

The width is between 10 and 11 in. and the length is between 38 and 42 in.

5. $$\begin{aligned} 3x - 4(x - 2) &= 2x - 5 \\ 3x - 4x + 8 &= 2x - 5 \\ -x + 8 &= 2x - 5 \\ -3x + 8 &= -5 \\ -3x &= -13 \\ x &= \frac{13}{3} \end{aligned}$$

7. $$\begin{aligned} \frac{x}{2} - \frac{3}{2} &= \frac{x}{5} + 1 \\ 10\left(\frac{x}{2} - \frac{3}{2}\right) &= 10\left(\frac{x}{5} + 1\right) \\ \frac{10}{1} \cdot \frac{x}{2} - \frac{10}{1} \cdot \frac{3}{2} &= \frac{10}{1} \cdot \frac{x}{5} + 10(1) \\ 5x - 15 &= 2x + 10 \\ 3x - 15 &= 10 \\ 3x &= 25 \\ x &= \frac{25}{3} \end{aligned}$$

9. $$\begin{aligned} 3x - 5x &= 7x - 4x \\ -2x &= 3x \\ 0 &= 5x \\ 0 &= x \end{aligned}$$

11. $$\begin{aligned} 6(8 - a) &= 3(a - 4) + 2(a - 1) \\ 48 - 6a &= 3a - 12 + 2a - 2 \\ 48 - 6a &= 5a - 14 \\ 48 &= 11a - 14 \\ 62 &= 11a \\ \frac{62}{11} &= a \end{aligned}$$

13. $$\begin{aligned} 7[x - 3(x - 3)] &= 4x - 2 \\ 7(x - 3x + 9) &= 4x - 2 \\ 7(-2x + 9) &= 4x - 2 \\ -14x + 63 &= 4x - 2 \\ 63 &= 18x - 2 \\ 65 &= 18x \\ \frac{65}{18} &= x \end{aligned}$$

15. $$\begin{aligned} 5\{x - [2 - (x - 3)]\} &= x - 2 \\ 5[x - (2 - x + 3)] &= x - 2 \\ 5[x - (5 - x)] &= x - 2 \\ 5(x - 5 + x) &= x - 2 \\ 5(2x - 5) &= x - 2 \\ 10x - 25 &= x - 2 \\ 9x - 25 &= -2 \\ 9x &= 23 \\ x &= \frac{23}{9} \end{aligned}$$

17. $|x| = 3$
$x = 3$ or $x = -3$

19. No solution, absolute value is nonnegative.

21. $|2x| = 8$
$2x = 8$ or $2x = -8$
$x = 4$ or $x = -4$

23. $|a + 1| = 4$
$a + 1 = 4$ or $a + 1 = -4$
$a = 3$ or $a = -5$

25. $$\begin{aligned} |4z + 5| &= 0 \\ 4z + 5 &= 0 \\ 4z &= -5 \\ z &= -\frac{5}{4} \end{aligned}$$

27. $|2y - 5| - 8 = -3$
$|2y - 5| = 5$
$2y - 5 = 5$ or $2y - 5 = -5$
$2y = 10$ $\quad 2y = 0$
$y = 5$ or $y = 0$

29. $|x - 5| = |x - 1|$
$x - 5 = x - 1$ or $x - 5 = -(x - 1)$
$-5 = -1$ $\quad x - 5 = -x + 1$
No solution $\quad 2x - 5 = 1$
$2x = 6$
$x = 3$
The only solution is $x = 3$.

31. $|2t - 4| = |t - 2|$

$2t - 4 = t - 2$ or $2t - 4 = -(t - 2)$

$t - 4 = -2$ $\quad$ $2t - 4 = -t + 2$

$t = 2$ $\quad$ $3t - 4 = 2$

$3t = 6$

$t = 2$

33. $3x - 4 \le 5$

$3x \le 9$

$x \le 3$

35. $5x - 4 \le 2x$

$-4 \le -3x$

$\frac{4}{3} \ge x$

or $x \le \frac{4}{3}$

37. $5z + 4 > 2z - 1$

$3z + 4 > -1$

$3z > -5$

$z > -\frac{5}{3}$

39. $5(s - 4) < 3(s - 4)$

$5s - 20 < 3s - 12$

$2s - 20 < -12$

$2s < 8$

$s < 4$

41. $3(r - 2) - 5(r - 1) \ge -2r - 1$

$3r - 6 - 5r + 5 \ge -2r - 1$

$-2r - 1 \ge -2r - 1$

Identity, true for all real numbers

43. $-3 \le 2x - 1 \le 5$

$-2 \le 2x \le 6$

$-1 \le x \le 3$

45. $-5 \le 6 - 2x \le 7$

$-11 \le -2x \le 1$

$\frac{11}{2} \ge x \ge -\frac{1}{2}$

or $-\frac{1}{2} \le x \le \frac{11}{2}$

47. $-3 \le 3(x - 4) \le 6$

$-3 \le 3x - 12 \le 6$

$9 \le 3x \le 18$

$3 \le x \le 6$

49. $3[a - 3(a + 2)] > 0$

$3(a - 3a - 6) > 0$

$3(-2a - 6) > 0$

$-6a - 18 > 0$

$-6a > 18$

$a < -3$

51. $5 - [2 - 3(2 - x)] < -3x + 1$

$5 - (2 - 6 + 3x) < -3x + 1$

$5 - (-4 + 3x) < -3x + 1$

$5 + 4 - 3x < -3x + 1$

$9 - 3x < -3x + 1$

$9 < 1$

Contradiction, no solution

53. $3 - \{2 - 5[q - (2 - q]\} \le 3[q - (q - 5)]$

$3 - [2 - 5(q - 2 + q)] \le 3(q - q + 5)$

$3 - [2 - 5(2q - 2)] \le 3(5)$

$3 - (2 - 10q + 10) \le 15$

$3 - (12 - 10q) \le 15$

$3 - 12 + 10q \le 15$

$-9 + 10q \le 15$

$10q \le 24$

$q \le \frac{12}{5}$

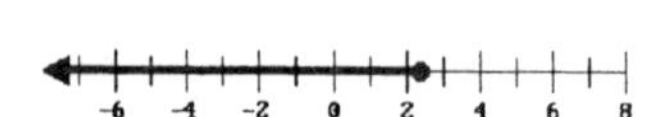

55. $|x| < 3$

$-3 < x < 3$

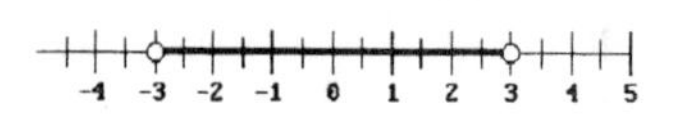

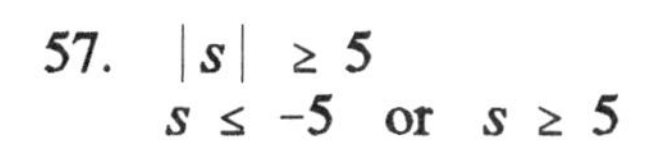

57. $|s| \geq 5$
$s \leq -5$ or $s \geq 5$

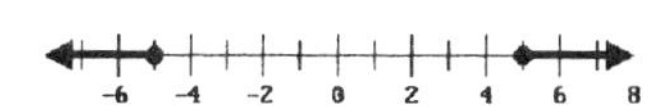

59. $|t| < 0$
Contradiction, absolute value is nonnegative.
No solution

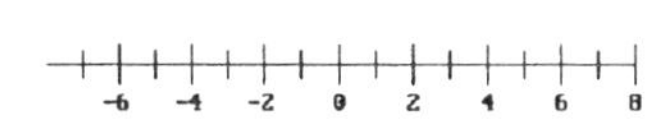

61. $|t - 1| < 2$
$-2 < t - 1 < 2$
$-1 < t < 3$

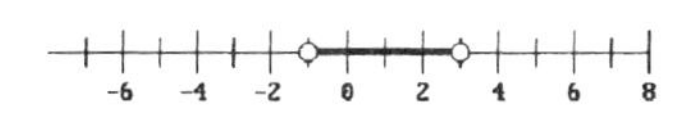

63. $|a - 6| \geq 3$
$a - 6 \geq 3$ or $a - 6 \leq -3$
$a \leq 3$ or $a \geq 9$

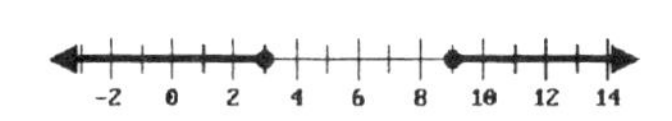

65. $|r + 9| \leq 4$
$-4 \leq r + 9 \leq 4$
$-13 \leq r \leq -5$

67. $|2x - 1| \geq 2$
$2x - 1 \leq -2$ or $2x - 1 \geq 2$
$2x \leq -1$ $\quad$ $2x \geq 3$
$x \leq -\dfrac{1}{2}$ or $x \geq \dfrac{3}{2}$

69. $|3x - 2| < 4$
$-4 < 3x - 2 < 4$
$-2 < 3x < 6$
$-\dfrac{2}{3} < x < 2$

71. $|2x - 3| \leq 5$
$-5 \leq 2x - 3 \leq 5$
$-2 \leq 2x \leq 8$
$-1 \leq x \leq 4$

73. $|3x + 5| + 2 > 7$
$|3x + 5| > 5$
$3x + 5 > 5$ or $3x + 5 < -5$
$3x > 0$ $\quad$ $3x < -10$
$x > 0$ or $x < -\dfrac{10}{3}$

75. $|3 - 2x| + 8 \leq 4$
$|3 - 2x| \leq -4$
No solution, absolute value is nonnegative.

77. the number: x

$3x = 4x - 4$
$-x = -4$
$x = 4$

The number is 4.

79. Let x = U. S. dollars
y = Yen

$y = 114.05x$

$y = 114.05(800)$

$y = 91{,}240$

It is worth 91,240 Yen.

81. gross sales: x

$0.09x = 650$
$x = 7222.22$

His gross sales were \$7222.22.

83. final exam score: x

$$\begin{aligned} 0.40x + 0.60(74) &= 80 \\ 0.40x + 44.4 &= 80 \\ 0.40x &= 35.6 \\ x &= 89 \end{aligned}$$

You need an 89 on the final exam.

85. number of 8-lb packages: x
number of 5-lb packages: $30 - x$

$$\begin{aligned} 8(x) + 5(30 - x) &= 186 \\ 8x + 150 - 5x &= 186 \\ 3x + 150 &= 186 \\ 3x &= 36 \\ x &= 12 \\ 30 - x &= 30 - 12 = 18 \end{aligned}$$

There were 12 8-lb packages and 18 5-lb packages.

87. number of single beds: x
number of larger beds: $23 - x$

$$\begin{aligned} 295 &= 10(x) + 15(23 - x) \\ 295 &= 10x + 345 - 15x \\ 295 &= 345 - 5x \\ -50 &= -5x \\ 10 &= x \end{aligned}$$

$23 - x = 23 - 10 = 13$

They delivered 10 single beds and 13 larger beds.

89. hours editing: $30 - x$
hours tutoring: x

$$\begin{aligned} 12(30 - x) + 20x &\geq 456 \\ 360 - 12x + 20x &\geq 456 \\ 360 + 8x &\geq 456 \\ 8x &\geq 96 \\ 8 &\geq 12 \end{aligned}$$

He must tutor at least 12 hours.

CHAPTER 2 PRACTICE TEST

1. amount in savings account: x
amount in certificate: $5000 - x$

$$\begin{aligned} I &= 0.035(x) + 0.06(5000 - x) \\ I &= 0.035x + 300 - 0.06x \\ I &= -0.025x + 300 \end{aligned}$$

$$\begin{aligned} 220 &= -0.025x + 300 \\ -80 &= -0.025x \\ 3200 &= x \end{aligned}$$

There is \$3200 in the savings account.

3. $$\begin{aligned} 3x - 4(x - 5) &= 7(x - 2) - 8(x - 1) \\ 3x - 4x + 20 &= 7x - 14 - 8x + 8 \\ -x + 20 &= -x - 6 \\ 20 &= -6 \end{aligned}$$

Contradiction, no solution

5. $$\begin{aligned} 3x + 5 &\leq 5x - 3 \\ -2x + 5 &\leq -3 \\ -2x &\leq -8 \\ x &\geq 4 \end{aligned}$$

7. $-3 < 5 - 2x < 4$
$-8 < -2x < -1$
$4 > x > \frac{1}{2}$ or $\frac{1}{2} < x < 4$

9. $|3x - 4| < 2$
$-2 < 3x - 4 < 2$
$2 < 3x < 6$
$\frac{2}{3} < x < 2$

-1 0 1 2 3 4

11. number of Euros: x
number of U. S. dollars: y

$$\begin{aligned} y &= 0.9105x \\ y &= 0.9105(500) \\ y &= 455.25 \end{aligned}$$

It is worth \$455.25(U.S.)

13. number of hours for jogger: x
number of hours for walker: $x + 2$

$$\begin{aligned} 8x &= 3(x + 2) \\ 8x &= 3x + 6 \\ 5x &= 6 \\ x &= \frac{6}{5} \end{aligned}$$

It would take $\frac{6}{5}$ hr.

CHAPTER 3

Exercises 3.1

1.
$$\begin{aligned} 3x - 5y &= 17 \\ 3(4) - 5(1) &= 17 \\ 12 - 5 &= 17 \\ 7 &= 17 \quad \text{False} \end{aligned}$$

3.
$$\begin{aligned} 4y - 3x &= 7 \\ 4(-1) - 3(1) &= 7 \\ -4 - 3 &= 7 \\ -7 &= 7 \quad \text{False} \end{aligned}$$

5.
$$\begin{aligned} 2x + 3y &= 2 \\ 2\left(\frac{3}{2}\right) + 3\left(-\frac{1}{3}\right) &= 2 \\ 3 - 1 &= 2 \\ 2 &= 2 \quad \text{True} \end{aligned}$$

7.
$$\begin{aligned} \frac{2}{3}x - \frac{1}{4}y &= 1 \\ \frac{2}{3}(6) - \frac{1}{4}(12) &= 2 \\ 4 - 3 &= 1 \\ 1 &= 1 \quad \text{True} \end{aligned}$$

9.
$$\begin{aligned} y &= 5 \\ 5 &= 5 \quad \text{True} \end{aligned}$$

11.
$$x + y = 8$$

$$\begin{aligned} -1 + y &= 8 \\ y &= 9 \end{aligned}$$
$(-1, 9)$

$$\begin{aligned} 0 + y &= 8 \\ y &= 8 \end{aligned}$$
$(0, 8)$

$$\begin{aligned} 1 + y &= 8 \\ y &= 7 \end{aligned}$$
$(1, 7)$

$$\begin{aligned} x + (-2) &= 8 \\ x &= 10 \end{aligned}$$
$(10, -2)$

$$\begin{aligned} x + 0 &= 8 \\ x &= 8 \end{aligned}$$
$(8, 0)$

$$\begin{aligned} x + 4 &= 8 \\ x &= 4 \end{aligned}$$
$(4, 4)$

13.
$$5x + 4y = 20$$

$$\begin{aligned} 5(-2) + 4y &= 20 \\ -10 + 4y &= 20 \\ 4y &= 30 \\ y &= \frac{15}{2} \end{aligned}$$
$\left(-2, \frac{15}{2}\right)$

$$\begin{aligned} 5(0) + 4y &= 20 \\ 4y &= 20 \\ y &= 5 \end{aligned}$$
$(0, 5)$

$$\begin{aligned} 5(4) + 4y &= 20 \\ 20 + 4y &= 20 \\ 4y &= 0 \\ y &= 0 \end{aligned}$$
$(4, 0)$

$$\begin{aligned} 5x + 4(-5) &= 20 \\ 5x - 20 &= 20 \\ 5x &= 40 \\ x &= 8 \end{aligned}$$
$(8, -5)$

$$\begin{aligned} 5x + 4(0) &= 20 \\ 5x &= 20 \\ x &= 4 \end{aligned}$$
$(4, 0)$

$$\begin{aligned} 5x + 4(4) &= 20 \\ 5x + 16 &= 20 \\ 5x &= 4 \\ x &= \frac{4}{5} \end{aligned}$$
$\left(\frac{4}{5}, 4\right)$

15. $\frac{x}{3} + \frac{y}{4} = 1$

$12\left(\frac{x}{3} + \frac{y}{4}\right) = 12(1)$

$4x + 3y = 12$

$$\begin{aligned} 4(-3) + 3y &= 12 \\ -12 + 3y &= 12 \\ 3y &= 24 \\ y &= 8 \end{aligned}$$
$(-3, 8)$

$$\begin{aligned} 4(0) + 3y &= 12 \\ 3y &= 12 \\ y &= 4 \end{aligned}$$
$(0, 4)$

$$\begin{aligned} 4(3) + 3y &= 12 \\ 12 + 3y &= 12 \\ 3y &= 0 \\ y &= 0 \end{aligned}$$
$(3, 0)$

$$\begin{aligned} 4x + 3(-4) &= 12 \\ 4x - 12 &= 12 \\ 4x &= 24 \\ x &= 6 \end{aligned}$$
$(6, -4)$

$$\begin{aligned} 4x + 3(0) &= 12 \\ 4x &= 12 \\ x &= 3 \end{aligned}$$
$(3, 0)$

$$\begin{aligned} 4x + 3(4) &= 12 \\ 4x + 12 &= 12 \\ 4x &= 0 \\ x &= 0 \end{aligned}$$
$(0, 4)$

17. $x + y = 6$

x-intercept: Let $y = 0$
$$\begin{aligned} x + 0 &= 6 \\ x &= 6 \end{aligned}$$

y-intercept: Let $x = 0$
$$\begin{aligned} 0 + y &= 6 \\ y &= 6 \end{aligned}$$

19. $x - y = 6$

x-intercept: Let $y = 0$
$$\begin{aligned} x - 0 &= 6 \\ x &= 6 \end{aligned}$$

y-intercept: Let $x = 0$
$$\begin{aligned} 0 - y &= 6 \\ -y &= 6 \\ y &= -6 \end{aligned}$$

21. $y - x = 6$

x-intercept: Let $y = 0$
$$\begin{aligned} 0 - x &= 6 \\ -x &= 6 \\ x &= -6 \end{aligned}$$

y-intercept: Let $x = 0$
$$\begin{aligned} y - 0 &= 6 \\ y &= 6 \end{aligned}$$

23. $2x + 4y = 12$

x-intercept: Let $y = 0$
$$\begin{aligned} 2x + 4(0) &= 12 \\ 2x &= 12 \\ x &= 6 \end{aligned}$$

y-intercept: Let $x = 0$
$$\begin{aligned} 2(0) + 4y &= 12 \\ 4y &= 12 \\ y &= 3 \end{aligned}$$

25. $y = -\frac{4}{3}x + 4$

x-intercept: Let $y = 0$

$0 = -\frac{4}{3}x + 4$

$\frac{4}{3}x = 4$

$x = 3$

y-intercept: Let $x = 0$

$y = -\frac{4}{3}(0) + 4$

$y = 4$

27. $y = \frac{3}{5}x - 3$

x-intercept: Let $y = 0$

$0 = \frac{3}{5}x - 3$

$3 = \frac{3}{5}x$

$5 = x$

y-intercept: Let $x = 0$

$y = \frac{3}{5}(0) - 3$

$y = -3$

29. $2y - 3x = 7$

x-intercept: Let $y = 0$

$$2(0) - 3x = 7$$
$$-3x = 7$$
$$x = -\frac{7}{3}$$

y-intercept: Let $x = 0$

$$2y - 3(0) = 7$$
$$2y = 7$$
$$y = \frac{7}{2}$$

31. $4x + 3y = 0$

Let $x = 0$: $4(0) + 3y = 0$
$3y = 0$
$y = 0$
$(0, 0)$

Let $x = 3$: $4(3) + 3y = 0$
$12 + 3y = 0$
$3y = -12$
$y = -4$
$(3, -4)$

Let $x = -3$: $4(-3) + 3y = 0$
$-12 + 3y = 0$
$3y = 12$
$y = 4$
$(-3, 4)$

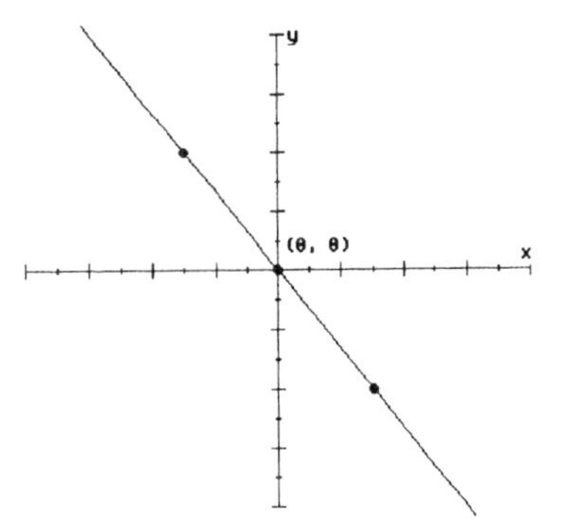

33. $y = x$

Let $x = 0$: $y = 0$
$(0, 0)$

Let $x = 1$: $y = 1$
$(1, 1)$

Let $x = 2$: $y = 2$
$(2, 2)$

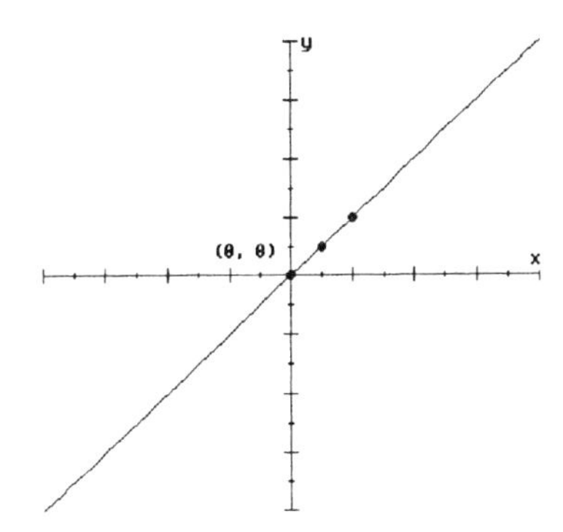

35. $\frac{x}{2} - \frac{y}{3} = 1$

$6\left(\frac{x}{2} - \frac{y}{3}\right) = 6(1)$

$3x - 2y = 6$

Let $x = 0$: $3(0) - 2y = 6$
$-2y = 6$
$y = -3$
$(0, -3)$

Let $y = 0$: $3x - 2(0) = 6$
$3x = 6$
$x = 2$
$(2, 0)$

Let $x = 4$: $3(4) - 2y = 6$
$12 - 2y = 6$
$-2y = -6$
$y = 3$
$(4, 3)$

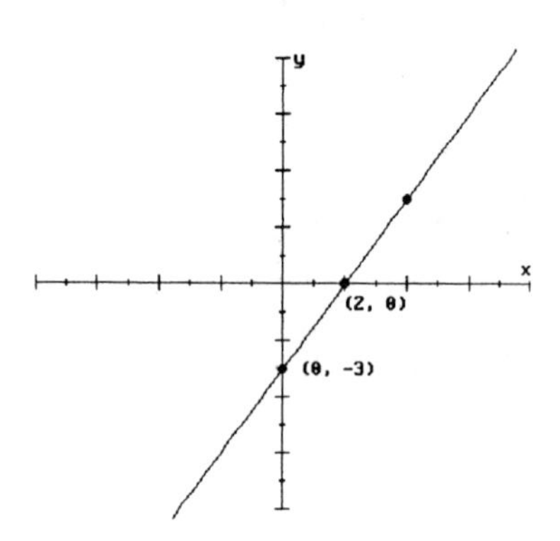

37. $y = 3x - 1$

Let $x = 0$: $y = 3(0) - 1$
$y = -1$
$(0, -1)$

Let $y = 0$: $0 = 3x - 1$
$1 = 3x$

$\frac{1}{3} = x$

$\left(\frac{1}{3}, 0\right)$

Let $x = 1$: $y = 3(1) - 1$
$y = 2$
$(1, 2)$

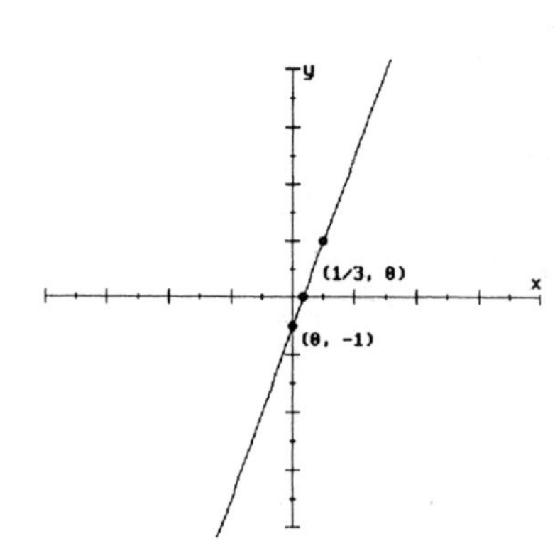

39. $y = -\frac{2}{3}x + 4$

Let $x = 0$: $y = -\frac{2}{3}(0) + 4$
$y = 4$
$(0, 4)$

Let $y = 0$: $0 = -\frac{2}{3}x + 4$

$\frac{2}{3}x = 4$

$x = 6$
$(6, 0)$

Let $x = 3$: $y = -\frac{2}{3}(3) + 4$

$y = -2 + 4$
$y = 2$
$(3, 2)$

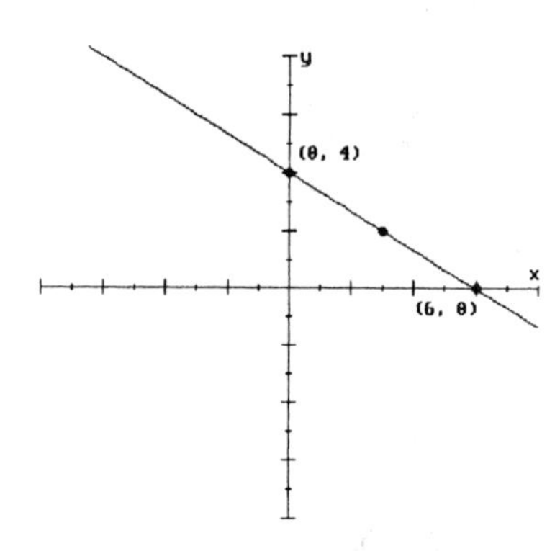

41. $5x - 4y = 20$

Let $x = 0$: $5(0) - 4y = 20$
$-4y = 20$
$y = -5$
$(0, -5)$

Let $y = 0$: $5x - 4(0) = 20$
$5x = 20$
$x = 4$
$(4, 0)$

Let $x = 2$: $5(2) - 4y = 20$
$10 - 4y = 20$
$-4y = 10$
$y = -\frac{5}{2}$
$\left(2, -\frac{5}{2}\right)$

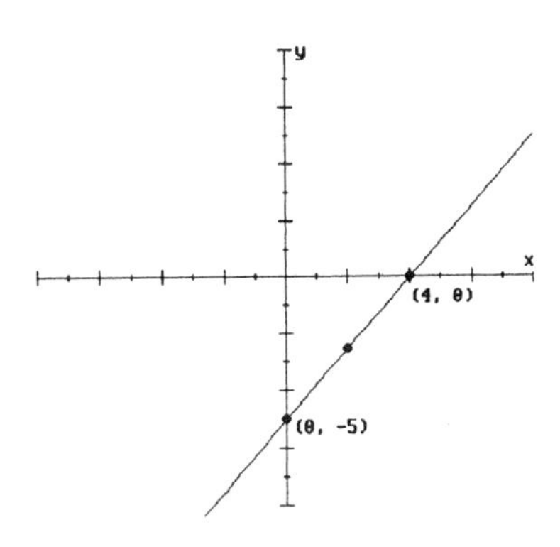

43. $5x - 7y = 30$

Let $x = 0$: $5(0) - 7y = 30$

$$-7y = 30$$

$$y = -\frac{30}{7}$$

$$\left(0, -\frac{30}{7}\right)$$

Let $y = 0$: $5x - 7(0) = 30$

$$5x = 30$$

$$x = 6$$

$$(6, 0)$$

Let $x = 1$: $5(1) - 7y = 30$

$$5 - 7y = 30$$

$$-7y = 25$$

$$y = -\frac{25}{7}$$

$$\left(1, -\frac{25}{7}\right)$$

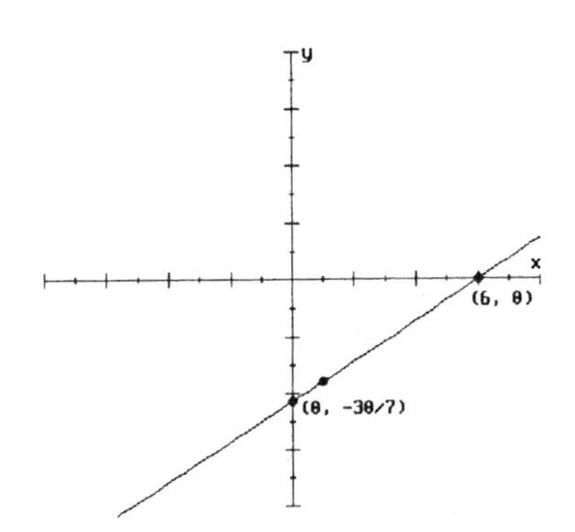

45. $5x + 7y = 30$

Let $x = 0$: $5(0) + 7y = 30$

$$7y = 30$$

$$y = \frac{30}{7}$$

$$\left(0, \frac{30}{7}\right)$$

Let $y = 0$: $5x + 7(0) = 30$

$$5x = 30$$

$$x = 6$$

$$(6, 0)$$

Let $x = 1$: $5(1) + 7y = 30$

$$5 + 7y = 30$$

$$7y = 25$$

$$y = \frac{25}{7}$$

$$\left(1, \frac{25}{7}\right)$$

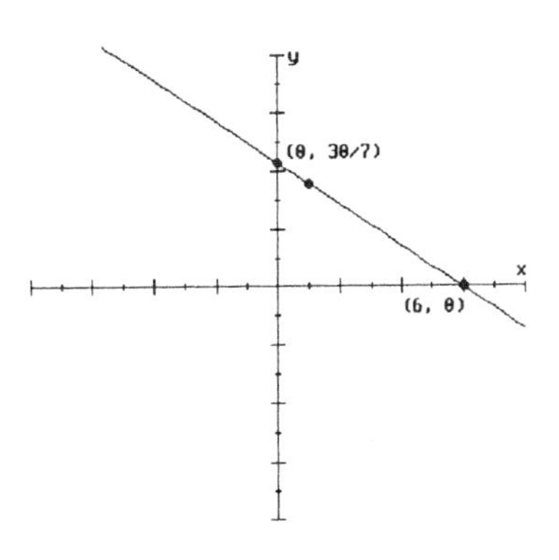

47. $x = 5$

This is a vertical line through (5, 0).

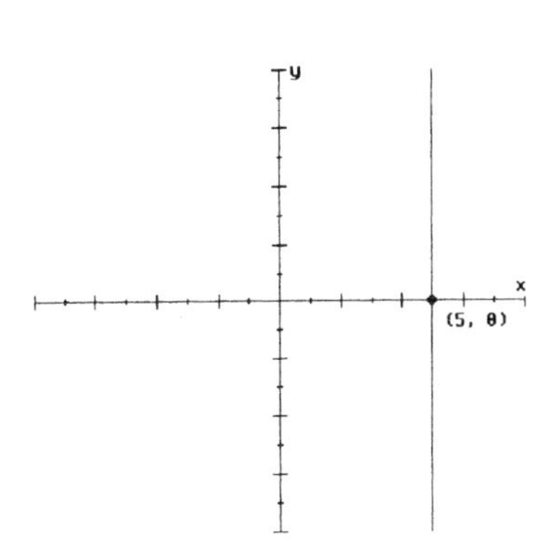

49. $-\frac{3}{4}x + y = 2$

Let $x = 0$: $-\frac{3}{4}(0) + y = 2$

$y = 2$

$(0, 2)$

Let $y = 0$: $-\frac{3}{4}x + 0 = 2$

$-\frac{3}{4}x = 2$

$x = -\frac{8}{3}$

$\left(-\frac{8}{3}, 0\right)$

Let $x = -4$: $-\frac{3}{4}(-4) + y = 2$

$3 + y = 2$

$y = -1$

$(-4, -1)$

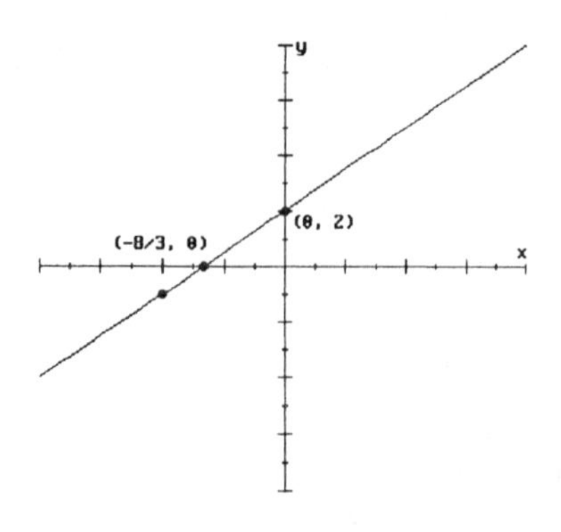

51. $\frac{3}{4}x - y = 2$

Let $x = 0$: $\frac{3}{4}(0) - y = 2$

$-y = 2$

$y = -2$

$(0, -2)$

Let $y = 0$: $\frac{3}{4}x - 0 = 2$

$\frac{3}{4}x = 2$

$x = \frac{8}{3}$

$\left(\frac{8}{3}, 0\right)$

Let $x = 4$; $\frac{3}{4}(4) - y = 2$

$3 - y = 2$

$-y = -1$

$y = 1$

$(4, 1)$

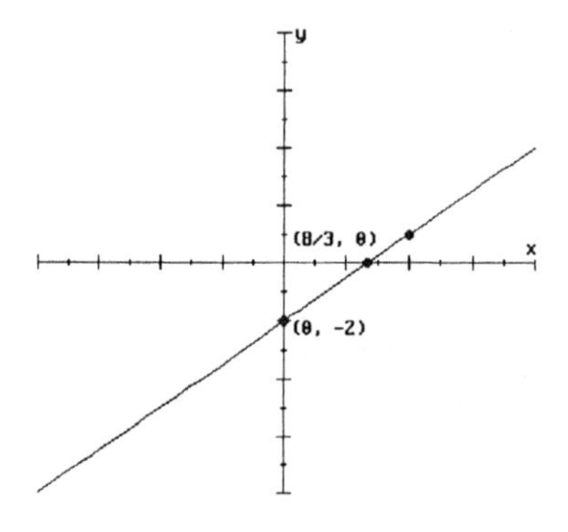

53. $y + 5 = 0$

$y = -5$

This is a horizontal line through $(0, -5)$.

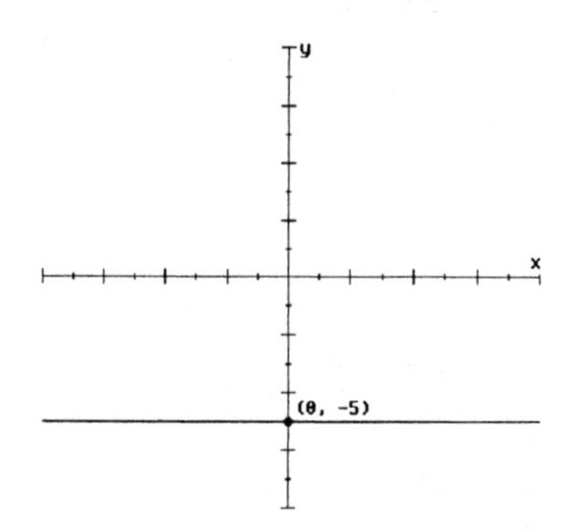

55. $5x - 4y = 0$

Let $x = 0$: $5(0) - 4y = 0$
$$-4y = 0$$
$$y = 0$$
$$(0, 0)$$

Let $x = 4$: $5(4) - 4y = 0$
$$20 - 4y = 0$$
$$-4y = -20$$
$$y = 5$$
$$(4, 5)$$

Let $x = -4$: $5(-4) - 4y = 0$
$$-20 - 4y = 0$$
$$-20 = 4y$$
$$-5 = y$$
$$(-4, -5)$$

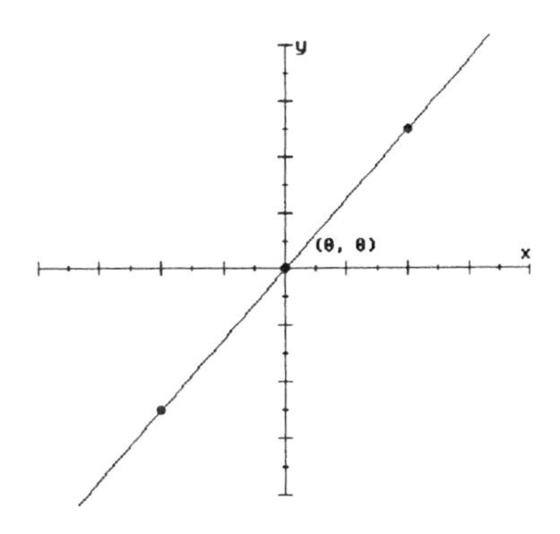

57. $5x - 4 = 0$
$$5x = 4$$
$$x = \frac{4}{5}$$

This is a vertical line through $\left(\frac{4}{5}, 0\right)$.

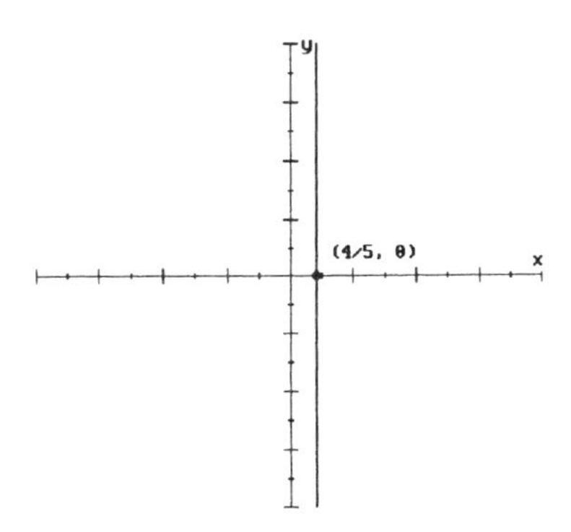

59. $P = 16.75t - 32702$

t	P
1980	463
1985	546.75
1990	630.5
1997	747.75

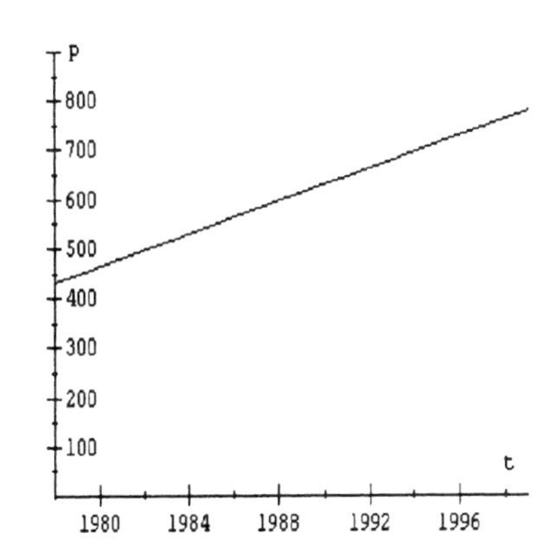

61. (a) Salary = Fixed salary + Commission

$$S = 220 + 0.05v,\ 0 \le v \le 10{,}000$$

(b)

v	S
0	220
100	225
1000	270
10000	720

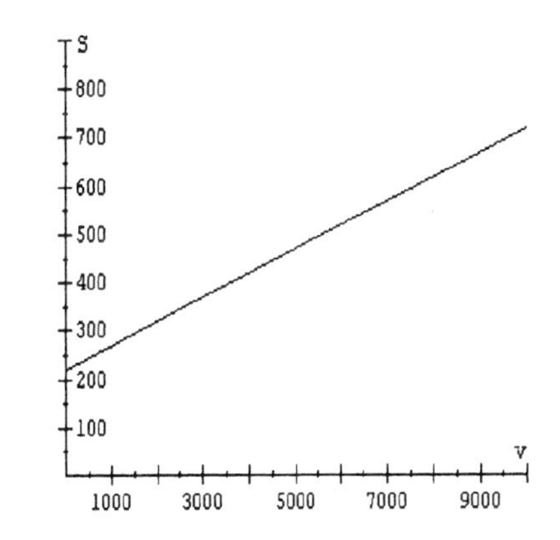

(c)

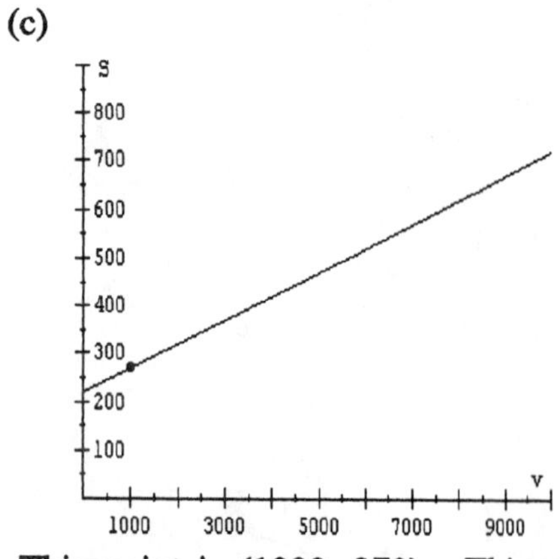

This point is (1000, 270). This means that when the value of items sold is \$1000, her salary will be \$270.

63. (a) distance = distance from home at 11:00 A.M. + rate · time since 11:00 A.M.

$$d = 260 + 52h$$

(b) $0 \le h \le 6$

h	d
0	260
6	572

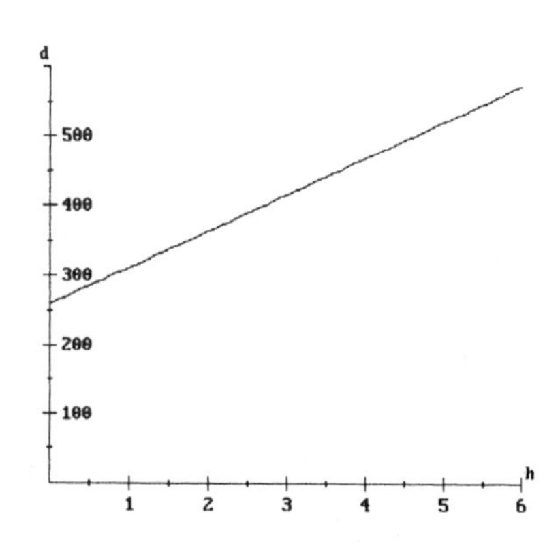

(c)

This point is approximately (2, 360). This means that 2 hours after 11:00 A.M. he will be approximately 360 miles from home.

65. (a) Rental charge

= daily charge + charge per mile · number of miles

$$C = 29 + 0.14n$$

(b)

n	c
0	29
100	43

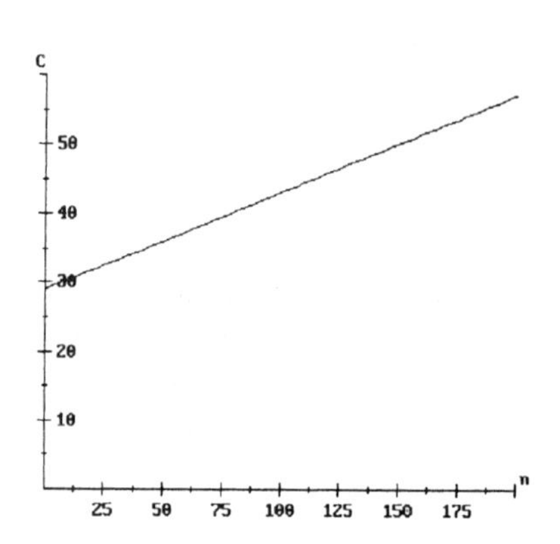

(c)

This point is approximately (100, 45). This means the rental charge for 100 miles is \$43.

Exercises 3.2

1. (a) Find -4 on the x-axis, project vertically until you reach the graph, then move horizontally to the y-axis, and read the y-value, $y = -3$.

   Find 4 on the x-axis, project vertically until you reach the graph, then move horizontally to the y-axis, and read the y-value, $y = 1$.

   (b) Find 2 on the y-axis, project horizontally until you reach the graph, then move vertically to the x-axis, and read the x-value, $x = 6$.

3. (a) Find -3 on the x-axis, project vertically until you reach the graph, then move horizontally to the y-axis, and read the y-value, $y = -4$.

   Find 2 on the x-axis. Notice that the graph crosses the x-axis at 2, hence $y = 0$. Also project vertically and notice that there are two other points on the graph with x coordinate 2. They have y-values of 3 and -4.

   (b) Find 2 on the y-axis, project horizontally until you reach the graph, then move vertically to the x-axis, and read the x-value, $x = 1$.

   Find -6 on the y-axis. Notice that the graph crosses the y-axis at -6, hence $x = 0$. Project horizontally and you find another point on the graph with y coordinate -6. Its x-value is -2.

5. Find -2 on the x-axis, project vertically until you reach the graph, then move horizontally to the y-axis, and read the y-value, $y = -1$.

7. Find -3 on the y-axis, project horizontally until you reach the graph. Notice that you locate 2 points on the graph. Move vertically to the x-axis and read the x-values, $x = -1$ and $x = 3$.

9. (a) $(3, -3)$
   (b) $(-4, 4)$
   (c) $(-2, -3)$ and $(3, -3)$
   (d) $(-4, 4)$ and $(5, 4)$

11. (a) $(2, 3)$ and $(2, -3)$

$$\begin{aligned} x^2 + y^2 &= 13 \\ 2^2 + 3^2 &= 13 \quad ? \\ 4 + 9 &= 13 \quad ? \\ 13 &= 13 \quad \checkmark \end{aligned}$$

$$\begin{aligned} x^2 + y^2 &= 13 \\ 2^2 + (-3)^2 &= 13 \quad ? \\ 4 + 9 &= 13 \quad ? \\ 13 &= 13 \quad \checkmark \end{aligned}$$

   (b) $(-2, -3)$ and $(2, -3)$

$$\begin{aligned} x^2 + y^2 &= 13 \\ (-2)^2 + (-3)^2 &= 13 \quad ? \\ 4 + 9 &= 13 \quad ? \\ 13 &= 13 \quad \checkmark \end{aligned}$$

$$\begin{aligned} x^2 + y^2 &= 13 \\ 2^2 + (-3)^2 &= 13 \quad ? \\ 4 + 9 &= 13 \quad ? \\ 13 &= 13 \quad \checkmark \end{aligned}$$

13. (a) The graph crosses the y-axis at $(0, 4)$

$$\begin{aligned} x + 2y &= 8 \\ 0 + 2(4) &= 8 \quad ? \\ 0 + 8 &= 8 \quad ? \\ 8 &= 8 \quad \checkmark \end{aligned}$$

   (b) The graph crosses the x-axis at $(8, 0)$

$$\begin{aligned} x + 2y &= 8 \\ 8 + 2(0) &= 8 \quad ? \\ 8 + 0 &= 8 \quad ? \\ 8 &= 8 \quad \checkmark \end{aligned}$$

15. (a) The graph crosses the y-axis at $(0, 2)$

$$\begin{aligned} y &= \sqrt{x + 4} \\ 2 &= \sqrt{0 + 4} \quad ? \\ 2 &= \sqrt{4} \quad ? \\ 2 &= 2 \quad \checkmark \end{aligned}$$

(b) The graph meets the x-axis at $(-4, 0)$.

$y = \sqrt{x + 4}$
$0 = \sqrt{-4 + 4}$?
$0 = \sqrt{0}$?
$0 = 0$ ✓

(c) To find the smallest x-coordinate, look for the leftmost point on the graph. This point is $(-4, 0)$ and its x-coordinate is -4.

To find the largest x-coordinate, look for the rightmost point on the graph. The arrowhead on the right end of the graph indicates that the graph goes on indefinitely in that direction. Therefore, there is no largest x-coordinate.

(d) To find the smallest y-coordinate, look for the lowest point on the graph. This point is $(-4, 0)$ and its y-coordinate is 0.

To find the largest y-coordinate, look for the highest point on the graph. However, the arrowhead indicating the graph is going indefinitely upward, thus there is no largest y-coordinate.

17. (a) To find the smallest x-coordinate, look for the leftmost point on the graph. This point is $(-3, 0)$ and its x-coordinate is -3.

To find the largest x-coordinate, look for the rightmost point on the graph. The arrowhead on the right end of the graph indicates that the graph goes on indefinitely in that direction. Therefore, there is no largest x-coordinate.

(b) To find the smallest y-coordinate, look for the lowest point on the graph. However, the arrowhead indicates the graph is going indefinitely downward, thus there is no smallest y-coordinate.

To find the largest y-coordinate, look for the highest point on the graph. This point is $(3, 3)$ and its y-coordinate is 3.

19. (a) $P = 0.5143t - 1003.1$

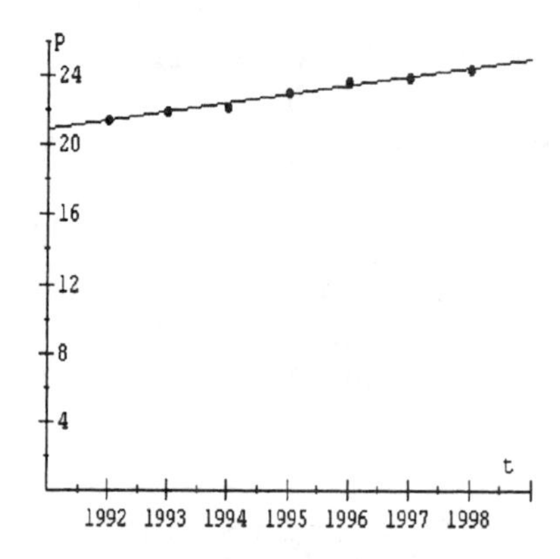

(b) The proposed equation agrees with the observed data. The points are very close to being on the graph of the equation.

21. (a) g = gross sales
$C = 0.10g$

(b)

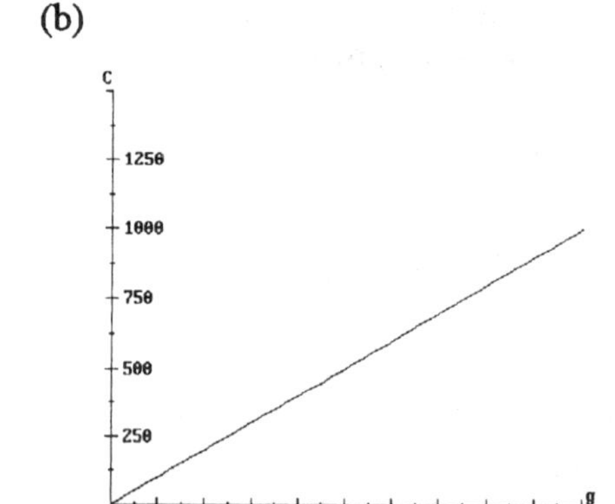

(c) Find $g = 4000$ on the g-axis, project vertically until you reach the graph, then move horizontally to the C-axis, and read the C value, $C = 400$. Her commission is \$400.

(d) Find $C = 850$ on the C-axis, project horizontally until you reach the graph, then move vertically to the g-axis and read the g value, $g = 8500$. Her gross sales were \$8500.

23. (a) C = monthly fee + 0.25 (number of checks)
$C = 10 + 0.25n$

(b)

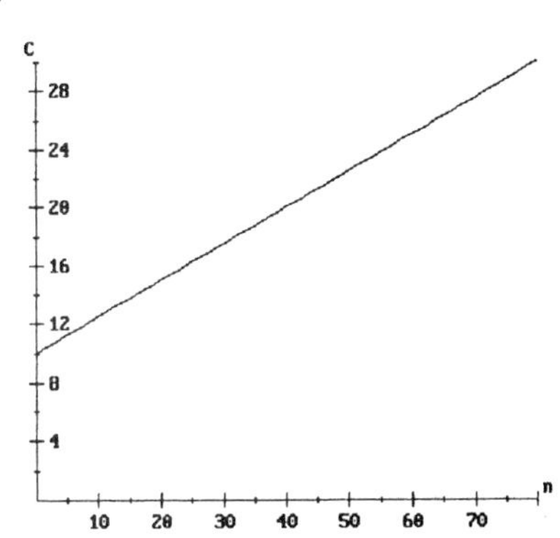

(c) Find 50 on the n-axis, go vertically to the graph and find the corresponding point. Then move horizontally to the C-axis and read the value of C, $C = 22.5$. The charge is \$22.50.

(d) Find 15 on the C-axis, go horizontally to the graph and find the corresponding point. Then move vertically to the n-axis and read the value of n, $n = 20$. 20 checks were processed.

Exercises 3.3

1. $\{(3, 9), (3, 7), (8, 2), (7, 2)\}$

3. $\{(3, a), (3, b), (8, b), (-1, b), (-1, c)\}$

5. domain: $\{3, 4, 5\}$
range: $\{2, 3\}$

7. domain: $\{3, -2, 4\}$
range: $\{-2, -1, 3\}$

9. $\{x \mid x \neq 0\}$

11. $\{x \mid x$ is a real number$\}$

13. $2x + 3 \neq 0$
$2x \neq -3$
$x \neq -\dfrac{3}{2}$
$\left\{x \mid x \neq -\dfrac{3}{2}\right\}$

15. $\{x \mid x$ is a real number$\}$

17. $x - 4 \geq 0$
$x \geq 4$
$\{x \mid x \geq 4\}$

19. $5 - 4x \geq 0$
$5 \geq 4x$
$\dfrac{5}{4} \geq x$
$\left\{x \mid x \leq \dfrac{5}{4}\right\}$

21. $4x \geq 0$
$x \geq 0$
$\{x \mid x \geq 0\}$

23. $x - 3 > 0$
$x > 3$
$\{x \mid x > 3\}$

25. Function, since each element in the domain is assigned only one element in the range.

27. Not a function, since the domain element 6 is assigned two range elements, 3 and 1.

29. Function, since each element in the domain is assigned only one element in the range.

31. Not a function, since the domain element 3 is assigned two range elements, 1 and 2.

33. Function, since each element in the domain is assigned only one element in the range.

35. Function, since each element in the domain is assigned only one element in the range.

37. Not a function, since the domain element 9 is assigned two range elements, -1 and 3.

39. Function, since there corresponds exactly one y-value to each x-value.

41. Not a function. For example, let $x = 0$.

$0^2 + y^2 = 81$
$y^2 = 81$
$y = \pm 9$

There are two range elements corresponding to a domain element.

43. Not a function. For example, let $x = 0$.

$0 = y^2 - 4$
$4 = y^2$
$\pm 2 = y$
There are two range elements corresponding to a domain element.

45. Function, it passes the Vertical Line Test.

47. Not a function, it doesn't pass the Vertical Line Test.

49. Function, it passes the Vertical Line Test.

51. Not a function, it doesn't pass the Vertical Line Test.

53. $y = 0$

55. If $y = -5$, then $x = 7$. The ordered pair is $(7, -5)$.

57. 3 values; -6, -2, and 4

59. 1 value

61. The domain is $\{x \mid -7 \leq x \leq 7\}$.

63. 3

65. $(7, -3)$

67. 3 values; -3, 2, 6

69. 1 value

71. The domain is $\{x \mid -5 \leq x \leq 7\}$

73. Domain: $\{x \mid -4 \leq x \leq 5\}$
Range: $\{y \mid -4 \leq y \leq 6\}$

75. Domain: $\{x \mid 0 \leq x \leq 4\}$
Range: $\{y \mid -3 \leq y \leq 3\}$

77. Domain: $\{x \mid -5 \leq x \leq -2, 1 \leq x \leq 5\}$
Range: $\{y \mid -3 \leq y \leq 1, 3 \leq y \leq 4\}$

79. $A = s^2$

81. Area of shaded region = Total Area - Unshaded Area

$$A = (a)(2a) - (5)(3)$$
$$A = 2a^2 - 15$$

$a > 3$ and $2a > 5$

$a > 3$ and $a > \frac{5}{2}$

a can be any value greater than 3.

83. $C =$ fee per day $\cdot$ number of days $+$ charge per mile $\cdot$ number miles

$C = 29 \cdot 4 + 0.12 \cdot m$
$C = 116 + 0.12m$

85. $C = 10 \cdot d + 12 \cdot (d + 5)$
$C = 10d + 12d + 60$
$C = 22d + 60$

87. $C = 0.02n^2 + 100n + 10000$

$n = 100$: $C = 0.02(100)^2 + 100(100) + 10000$
$= 20{,}200$

\$20,200

$n = 200$: $C = 0.02(200)^2 + 100(200) + 10000$
$= 30{,}800$

\$30,800

$n = 500$: $C = 0.02(500)^2 + 100(500) + 10000$
$= 65{,}000$

\$65,000

$n = 1000$: $C = 0.02(1000)^2 + 100(1000) + 10000$
$= 130{,}000$

\$130,000

89. (a) $C =$ flat fee $+$ fee per minute $\cdot$ number of minutes
$C = 29.95 + 0.33 \cdot m$

(b) $m = 75$: $C = 29.95 + 0.33(75)$
$= 54.7$

\$54.70

$m = 180$: $C = 29.95 + 0.33(180)$
$= 89.35$

\$89.35

$m = 300$: $C = 29.95 + 0.33(300)$
$= 128.95$

\$128.95

91. (a) Yes, it passes the Vertical Line Test.

(b) Domain: The rent varies between \$200 and \$400.
Range: The profit varies between \$35,000 and \$50,000.

(c) A rent of \$300 per unit generates the maximum profit which is \$50,000.

Exercises 3.4

1. $f(0) = 2(0) - 3$
$= 0 - 3$
$= -3$

3. $g(2) = 3(2)^2 - 2 + 1$
$= 3(4) - 2 + 1$
$= 11$

5. $g(-2) = 3(-2)^2 - (-2) + 1$
$= 3(4) + 2 + 1$
$= 15$

7. $h(3) = \sqrt{3 + 5}$
$= \sqrt{8}$
$= 2\sqrt{2}$

9. $h(-3) = \sqrt{-3 + 5}$
$= \sqrt{2}$

11. $h(a) = \sqrt{a + 5}$

13. $f(5) = 5^2 + 2$
$= 25 + 2$
$= 27$

$f(2) = 2^2 + 2$
$= 4 + 2$
$= 6$

$f(5) + f(2) = 27 + 6$
$= 33$

15. $f(6 - 4) = f(2)$
$= 2^2 + 2$
$= 4 + 2$
$= 6$

17. $g(x + 2) = 2(x + 2) - 3$
$= 2x + 4 - 3$
$= 2x + 1$

19. $f(2x) = (2x)^2 + 2$
$= 4x^2 + 2$

21. $g(3x + 2) = 2(3x + 2) - 3$
$= 6x + 4 - 3$
$= 6x + 1$

23. $g(x + 2) = 2(x + 2) - 3$
$= 2x + 4 - 3$
$= 2x + 1$

$g(x) = 2x - 3$

$g(x + 2) - g(x) = (2x + 1) - (2x - 3)$
$= 2x + 1 - 2x + 3$
$= 4$

25. $f(-2) = (-2)^2 + 2(-2) - 3$
$= 4 - 4 - 3$
$= -3$

$f(3) = 3^2 + 2(3) - 3$
$= 9 + 6 - 3$
$= 12$

$f(-2) + f(3) = -3 + 12$
$= 9$

27. $f(x) - 4 = (x^2 + 2x - 3) - 4$
$= x^2 + 2x - 7$

29. $f(x) = x^2 + 2x - 3$

$f(4) = 4^2 + 2(4) - 3$
$= 16 + 8 - 3$
$= 21$

$f(x) - f(4) = (x^2 + 2x - 3) - 21$
$= x^2 + 2x - 24$

31. $f(3x) = (3x)^2 + 2(3x) - 3$
$= 9x^2 + 6x - 3$

33. $3f(x) = 3(x^2 + 2x - 3)$
$= 3x^2 + 6x - 9$

35. $g(-3) = \dfrac{-3}{-3 + 5}$

$= -\dfrac{3}{2}$

37. $g\left(\dfrac{1}{4}\right) = \dfrac{\frac{1}{4}}{\frac{1}{4} + 5}$

$= \dfrac{\frac{1}{4}}{\frac{1}{4} + 5} \cdot \dfrac{4}{4}$

$= \dfrac{1}{1 + 20}$

$= \dfrac{1}{21}$

39. $g(x + 5) = \dfrac{x + 5}{x + 5 + 5}$

$= \dfrac{x + 5}{x + 10}$

41. (a) $g(-5) = 1$; $(-5, 1)$

(b) $g(-2) = -1$; $(-2, -1)$

(c) $g(0) = -2$; $(0, -2)$

(d) $g(1) = -\dfrac{1}{2}$; $\left(1, -\dfrac{1}{2}\right)$

(e) $g(3) = 2$; $(3, 2)$

43. (a) $f(-1) = 3$

(b) $f(0) = 5$

(c) $f(4) = 0$

(d) $x = -5, -3, 5$

(e) 2 values

45. $C(m)$ = daily charge · number of days + charge per mile · number miles

$C(m) = 29 \cdot 3 + 0.14 \cdot m$

$C(m) = 87 + 0.14m$

47. length: L
width: $3L - 12$

A = length · width

$A = L(3L - 12)$

$A = 3L^2 - 12L$

49. number of \$30 tickets sold: x
number of \$24 tickets sold: $600 - x$

C = amount from \$30 tickets + amount from \$24 tickets

$C = 30x + 24(600 - x)$

51. number of minutes for machine A: m
number of minutes for machine B: $m + 35$

N = copies made by machine A + copies made by machine B

$N = 20m + 22(m + 35)$

53. electrician's hours: h
assistant's hours: $h - 2$

A = amount earned by electrician + amount earned by assistant

$A = 45h + 25(h - 2)$

55. number of hours tutoring: t
number of hours as a clerk: $15 - t$

S = amount earned as tutor + amount earned as clerk

$S = 10t + 6.35(15 - t)$

Exercises 3.5

1. (a) 7.5 thousand = 7500 calls

   (b) 4 a.m. (lowest point on graph)

   (c) 28 thousand = 28,000 calls

   (d) $0 \le t \le 4$; $16 \le t \le 22$;
   This corresponds to: between midnight and 4 a.m. and between 4 p.m. and 10 p.m.

   (e) $10 \le t \le 16$; This corresponds to between 10 a.m. and 4 p.m.

   (f) $10 \le t \le 16$; This corresponds to between 10 a.m. and 4 p.m.

3. (a) consumed: approximately 80 quadrillion Btu
   produced: approximately 66 quadrillion Btu
   imported: approximately 14 quadrillion Btu

   (b) Consumption decreased from 1980 to 1982, then increased from 1982 to 1998.

   Production decreased slightly from 1980 to 1982 and then continued to decrease until 1983. It then increased from 1983 to 1984. It remained rather steady with slight decreases and increases until 1992. From 1992 to 1993 it decreased and then from 1993 until 1998 it increased.

   Imports decreased from 1980 to 1982, then had slight decreases and increases between 1982 and 1985. It steadily increased from 1985 to 1998.

5. (a) It denies this belief since as the altitude increases (go right on horizontal axis), the temperature both increases and decreases.

   (b) $0 \le a \le 10$; $50 \le a \le 80$

   This corresponds to between 0 and 10 kilometers and between 50 and 80 kilometers.

   (c) Starting at 45 km, the temperature increases until it reaches approximately 10° C at an altitude of 50 km, then the temperature decreased.

7. (a) At $t = 1$, $p = 30$
Approximately 30%

(b) Find 25 on vertical axis, go horizontally until you intersect the curve, move vertically down to the t-axis, $t \approx 2$.
Approximately 2 days

(c) As time passes material is forgotten. Material is forgotten rapidly during the first 3 hours of the given time period (more than half the material is forgotten during this time), then the rate of forgetting slows down. Only about 20% of the material is remembered by the 6th day.

9. (a) Find 10 on the vertical scale, then go horizontally until you intersect the massed practice curve, move vertically down to the horizontal axis, and note the number of trials.

Approximately 14 trials

(b) Find 10 on the vertical scale, then go horizontally until you intersect the spaced practice curve, move vertically down to the horizontal axis, and note the number of trials.

Approximately 3 trials

(c) Performance with spaced practice is superior to performance with massed practice. Learning with spaced practice occurs more rapidly than with massed practice.

11. Starting at home. Kyle starts his trip by traveling 90 miles away during the first 2 hours of the day. His average rate of speed during that 2-hour period was 45 mph. For the next 2 hours Kyle did not travel. Then Kyle travelled closer to his home between hours 4 and 5. He was travelling an average of 30 mph during this time. Kyle stopped again for about 1 hour, then drove home in 1½ hours at an average rate of 40 mph.

13. The graph shows that children assigned to each group had approximately the same number of aggressive behaviors before treatment. By the end of the treatment period, however, treatments A and C seemed to be the most effective in reducing these aggressive behaviors, where treatment C reduced their behaviors the most quickly.

After the treatment period, however, the children in treatment C reverted back to their agressive behavior. Hence treatment C ends up being the least effective at the end of the time period given following treatment.

The children in treatment B slowly reduced their aggressive behaviors even after the treatment ended. Although it seemed the least effective at the end of the treatment, it actually is shown to be the most effective of the three at the end of the time period given following treatment.

CHAPTER 3 REVIEW EXERCISES

1. $2x + y = 6$

$x = 0$: $2(0) + y = 6$
$y = 6$
$(0, 6)$

$y = 0$: $2x + 0 = 6$
$2x = 6$
$x = 3$
$(3, 0)$

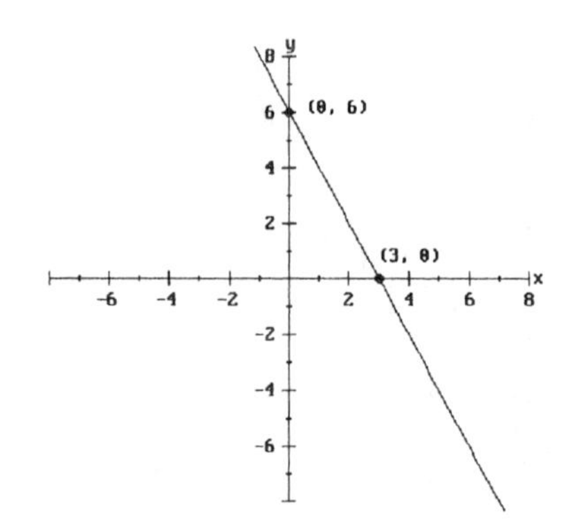

3. $2x - 6y = -6$

$$x = 0:\quad 2(0) - 6y = -6$$
$$-6y = -6$$
$$y = 1$$
$$(0,\ 1)$$

$$y = 0:\quad 2x - 6(0) = -6$$
$$2x = -6$$
$$x = -3$$
$$(-3,\ 0)$$

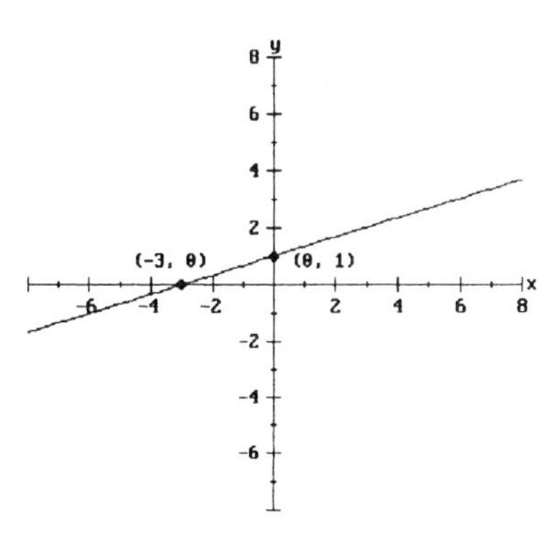

5. $5x - 3y = 10$

$$x = 0:\quad 5(0) - 3y = 10$$
$$-3y = 10$$
$$y = -\frac{10}{3}$$
$$\left(0,\ -\frac{10}{3}\right)$$

$$y = 0:\quad 5x - 3(0) = 10$$
$$5x = 10$$
$$x = 2$$
$$(2,\ 0)$$

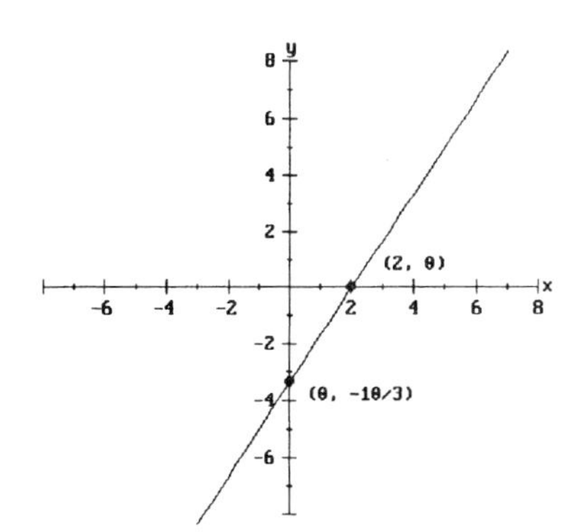

7. $2x + 5y = 7$

$$x = 0:\quad 2(0) + 5y = 7$$
$$5y = 7$$
$$y = \frac{7}{5}$$
$$\left(0,\ \frac{7}{5}\right)$$

$$y = 0:\quad 2x + 5(0) = 7$$
$$2x = 7$$
$$x = \frac{7}{2}$$
$$\left(\frac{7}{2},\ 0\right)$$

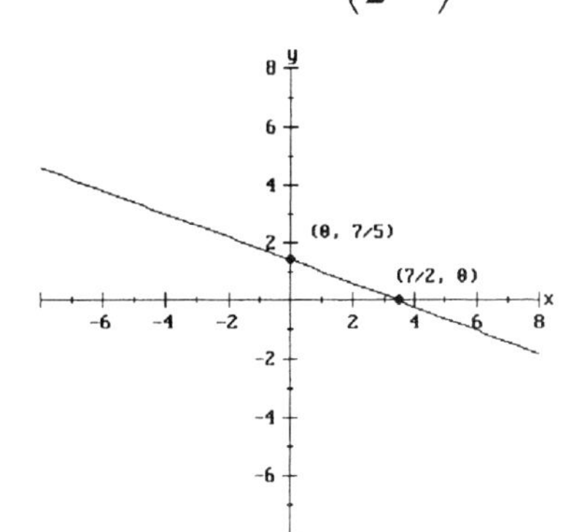

9. $3x - 8y = 11$

$$x = 0:\quad 3(0) - 8y = 11$$
$$-8y = 11$$
$$y = -\frac{11}{8}$$
$$\left(0,\ -\frac{11}{8}\right)$$

$$y = 0:\quad 3x - 8(0) = 11$$
$$3x = 11$$
$$x = \frac{11}{3}$$
$$\left(\frac{11}{3},\ 0\right)$$

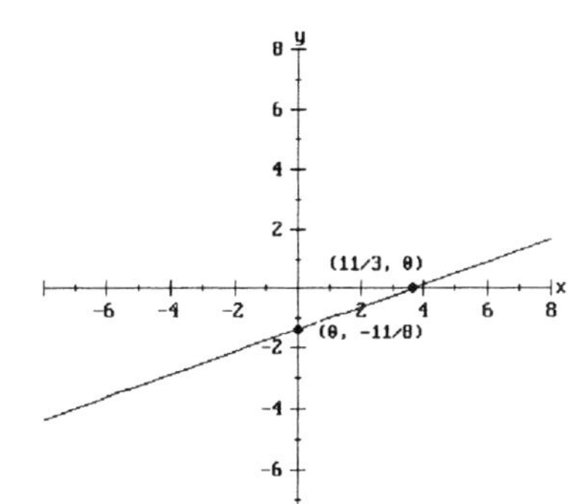

11. $5x + 7y = 21$

$x = 0$: $5(0) + 7y = 21$
$7y = 21$
$y = 3$
$(0, 3)$

$y = 0$: $5x + 7(0) = 21$
$5x = 21$
$x = \frac{21}{5}$

$\left(\frac{21}{5}, 0\right)$

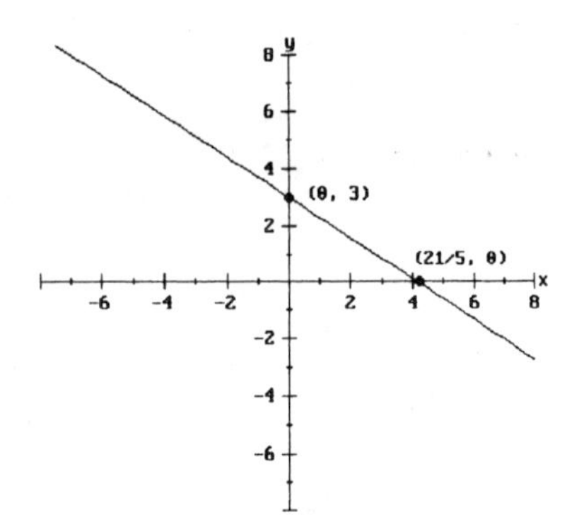

13. $y = x$

$x = 0$: $y = 0$
$(0, 0)$

$x = 1$: $y = 1$
$(1, 1)$

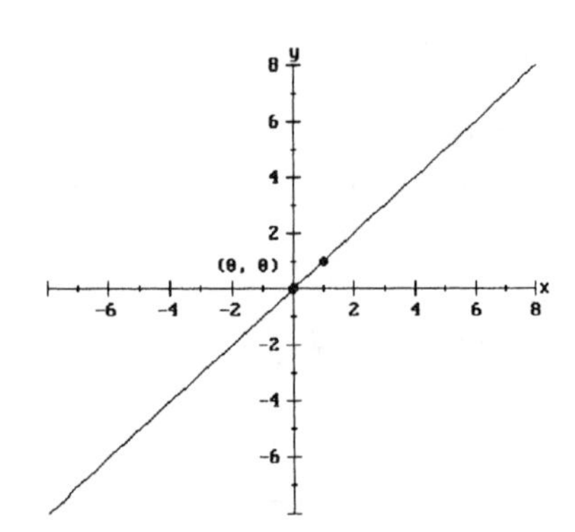

15. $y = -2x$

$x = 0$: $y = 0$
$(0, 0)$

$x = 1$: $y = -2(1)$
$y = -2$
$(1, -2)$

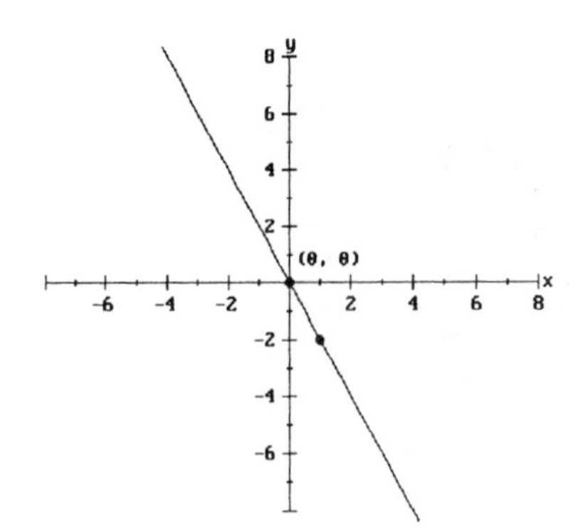

17. $y = \frac{2}{3}x + 2$

$x = 0$: $y = \frac{2}{3}(0) + 2$
$y = 2$
$(0, 2)$

$y = 0$: $0 = \frac{2}{3}x + 2$

$-2 = \frac{2}{3}x$

$-3 = x$
$(-3, 0)$

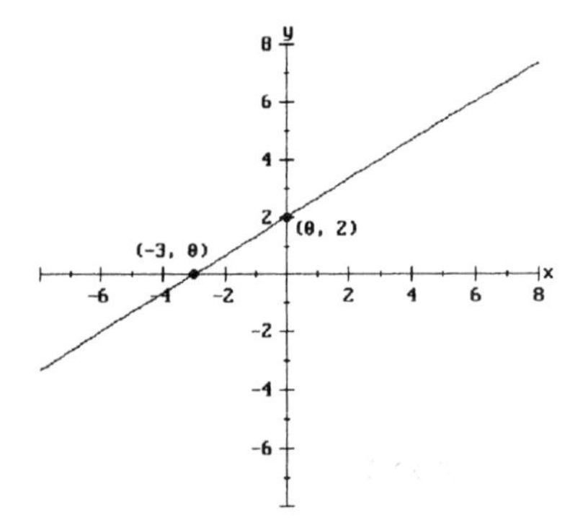

19. $\frac{x}{3} + \frac{y}{2} = 12$

$6\left(\frac{x}{3} + \frac{y}{2}\right) = 6(12)$

$2x + 3y = 72$

$x = 0$: $2(0) + 3y = 72$
$3y = 72$
$y = 24$
$(0, 24)$

$y = 0$: $2x + 3(0) = 72$
$2x = 72$
$x = 36$
$(36, 0)$

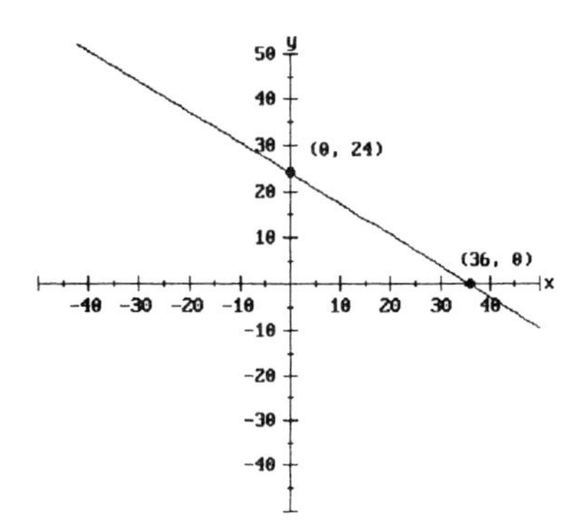

21. $x - 2y = 8$

$$\begin{aligned} x = 0: \quad 0 - 2y &= 8 \\ -2y &= 8 \\ y &= -4 \\ &(0, -4) \end{aligned}$$

$$\begin{aligned} y = 0: \quad x - 2(0) &= 8 \\ x &= 8 \\ &(8, 0) \end{aligned}$$

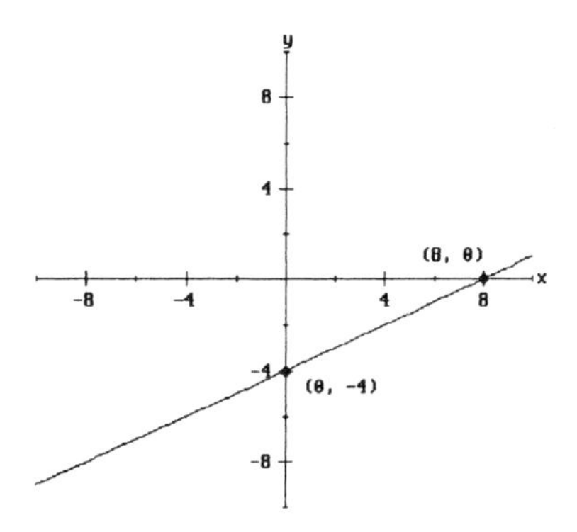

23. $x - 2 = 0$
$\quad x = 2$

This is a vertical line through (2, 0).

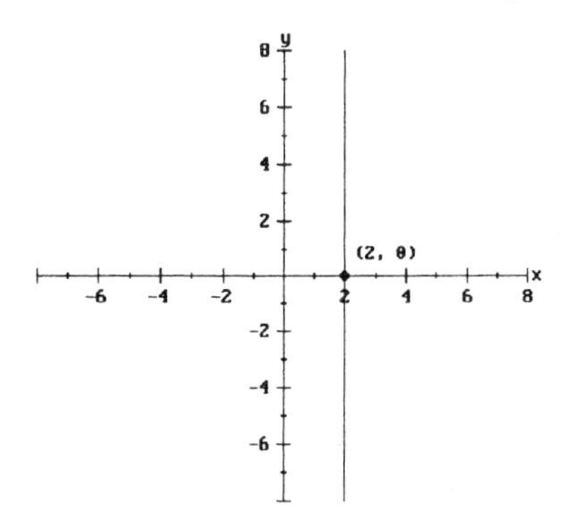

25. $2y = 5$

$$y = \frac{5}{2}$$

This is a horizontal line through $\left(0, \dfrac{5}{2}\right)$.

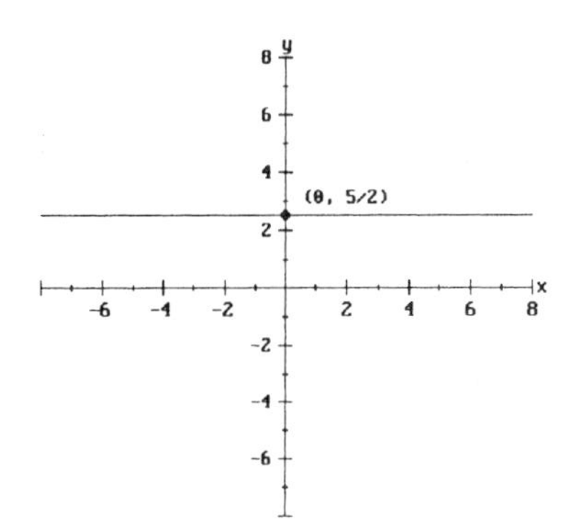

27. (a) Find -2 on the x-axis, project vertically until you reach the graph, then move horizontally to the y-axis, and read the y-value, $y = 3$.

Find 2 on the x-axis, project vertically until you reach the graph, then move horizontally to the y-axis, and read the y-value, $y = -1$.

(b) Find 3 on the y-axis, project horizontally until you reach the graph. Notice that there are two points on the graph with y-coordinate 3. The corresponding x-values are $x = -2$ and $x = 1$.

29. Locate -2 on the x-axis, project vertically until you reach the graph, then move horizontally to the y-axis, and read the y-value, $y = 6$.

31. The graph crosses the x-axis at (1, 0) and (3, 0).

33. (2, 4) and (2, -4)

$$\begin{aligned} x^2 + y^2 &= 20 \\ 2^2 + 4^2 &= 20 \quad ? \\ 4 + 16 &= 20 \quad ? \\ 20 &= 20 \quad \checkmark \end{aligned}$$

$$\begin{aligned} x^2 + y^2 &= 20 \\ 2^2 + (-4)^2 &= 20 \quad ? \\ 4 + 16 &= 20 \quad ? \\ 20 &= 20 \quad \checkmark \end{aligned}$$

35. (a) $I = 150 + 0.05g$

(b)

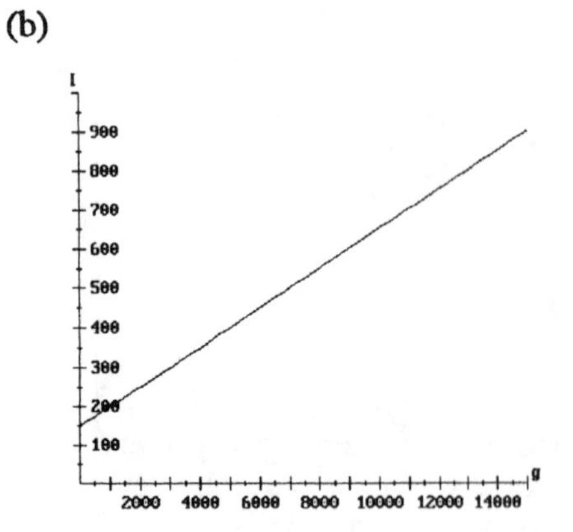

(c) Locate 5000 on the g-axis, project vertically until you reach the graph, move horizontally to the I-axis and read the I-value, $I = 400$. Her income is $400.

(d) Locate 750 on the I-axis, project horizontally until you reach the graph, move vertically to the g-axis and read the g-value, $g = 12000$. Her gross sales were $12,000.

37. $\{x \mid x \text{ is a real number}\}$

39. $4 - x \geq 0$
$4 \geq x$

$\{x \mid x \leq 4\}$

41. $x + 2 \neq 0$
$x \neq -2$

$\{x \mid x \neq -2\}$

43. Function, since each element in the domain is assigned only one element in the range.

45. Not a function, since the domain element 4 is assigned two range elements, 2 and 7.

47. Function, since there corresponds exactly one y-value to each x-value.

49. Function, since there corresponds exactly one y-value to each x-value.

51. Function, it passes the Vertical Line Test.

53. Not a function, it doesn't pass the Vertical Line Test.

55. Domain: $\{-2, 0, 3, 5\}$
Range: $\{3, 5, 7, 4\}$
Function

57. Domain: $\{4, 6, 1, 8\}$
Range: $\{9\}$
Function

59. Domain: $\{x \mid -5 \leq x \leq 6\}$
Range: $\{y \mid -2 \leq y \leq 5\}$
Function

61. Domain: $\{x \mid -3 \leq x \leq 3\}$
Range: $\{y \mid -4 \leq y \leq 4\}$
Not a function, doesn't pass the Vertical Line Test.

63. $f(x) = 3x + 5$
$f(-1) = 3(-1) + 5 = 2$
$f(0) = 3(0) + 5 = 5$
$f(1) = 3(1) + 5 = 8$
$f(2) = 3(2) + 5 = 11$

65. $f(x) = 2x^2 - 3x + 2$

$f(-1) = 2(-1)^2 - 3(-1) + 2$
$= 2(1) - 3(-1) + 2$
$= 7$

$f(0) = 2(0)^2 - 3(0) + 2$
$= 2(0) - 3(0) + 2$
$= 2$

$f(1) = 2(1)^2 - 3(1) + 2$
$= 2(1) - 3(1) + 2$
$= 1$

$f(2) = 2(2)^2 - 3(2) + 2$
$= 2(4) - 3(2) + 2$
$= 4$

67. $h(x) = \sqrt{x - 5}$

$h(6) = \sqrt{6 - 5}$
$= \sqrt{1}$
$= 1$

$h(5) = \sqrt{5 - 5}$
$= \sqrt{0}$
$= 0$

4 is not in the domain of $h(x)$.

69. $h(x) = \frac{x - 1}{x + 3}$

$h(1) = \frac{1 - 1}{1 + 3}$

$= \frac{0}{4}$

$= 0$

$h(3) = \frac{3 - 1}{3 + 3}$

$= \frac{2}{6}$

$= \frac{1}{3}$

-3 is not in the domain of $h(x)$.

71. $f(x) = 2x^2 + 4x - 1$
$f(a) = 2a^2 + 4a - 1$
$f(z) = 2z^2 + 4z - 1$

73. $f(x + 2) = 5(x + 2) + 2$
$= 5x + 10 + 2$
$= 5x + 12$

75. $f(x) + 2 = (5x + 2) + 2$
$= 5x + 4$

77. $f(x) = 5x + 2$
$f(2) = 5(2) + 2 = 12$

$f(x) + f(2) = (5x + 2) + 12$
$= 5x + 14$

79. $g(x + 2) = 6 - (x + 2)$
$= 6 - x - 2$
$= 4 - x$

81. $g(2x) = 6 - 2x$

83. $2g(x) = 2(6 - x)$
$= 12 - 2x$

85. (a) $f(-6) = 0$

(b) $f(0) = 2$

(c) $f(-4) = -2$

(d) $f(6) = -3$

(e) $f(5)$ is smaller

(f) $(-6, 0)$; $(-2, 0)$; $(5, 0)$

87. Rasheed starts his trip by traveling 90 miles away from his home during the first 2 hours of the day. His average rate of speed for the first hour was 40 mph; during the second hour his average rate was 50 mph. During the next hour Rasheed did not travel. Then Rasheed drove closer to his home between hours 3 and 4. He was traveling an average of 60 mph during this time. Rasheed drove another 20 miles away from his home, traveling at 20 mph between the 4th and 5th hour. Finally, during the last hour, Rasheed drove home at a rate of 50 mph.

89.

Total charges	=	delivery fee	+	charge per day	·	number days
C	=	30	+	42	·	n

$C = 30 + 42n$

CHAPTER 3 PRACTICE TEST

1. (a) $3x - 5y = 30$

$x = 0$: $3(0) - 5y = 30$
$-5y = 30$
$y = -6$
$(0, -6)$

$y = 0$: $3x - 5(0) = 30$
$3x = 30$
$x = 10$
$(10, 0)$

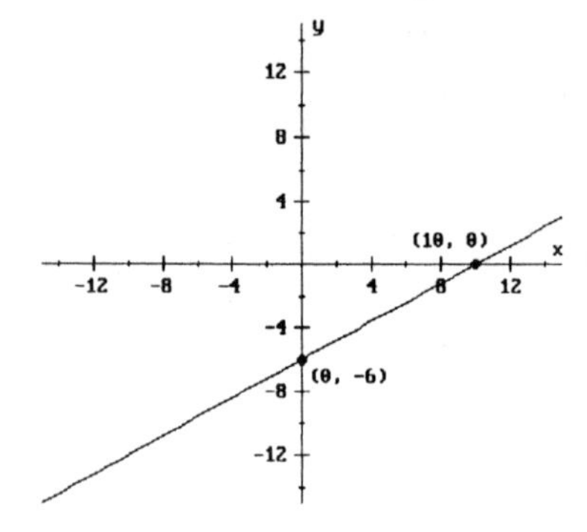

(b) $x - 7 = 0$
$x = 7$

This is a vertical line that passes through $(7, 0)$.

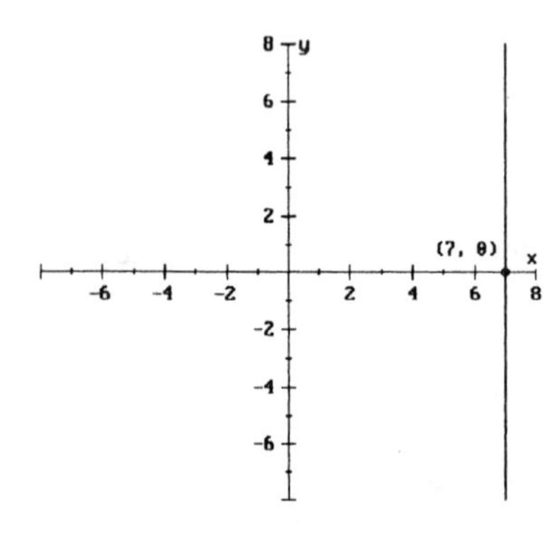

3. (a) domain: $\{2, 3, 4\}$
range: $\{-3, 5, 6\}$

(b) domain: $\{x \mid -4 \le x \le 5\}$
range: $\{y \mid -3 \le y \le 6\}$

5. (a) Not a function, since the domain element 2 is assigned two range elements, 5 and 4.

(b) Function, since each element in the domain is assigned only one element in the range.

(c) Not a function. For example, let $x = 27$.

$27 = 3y^2$
$9 = y^2$
$\pm 3 = y$

There are two range elements corresponding to a domain element.

(d) Function, it passes the Vertical Line Test.

7. (a) $f(-6) = 2$

(b) $f(3) = -2$

(c) $f(0) = 0$

(d) $f(7) = 2$

(e) $f(-4) = 5$; $f(3) = -2$; $f(5) = 0$
$f(-4)$ is the largest.

(f) $f(x) = 0$ when $x = 0$ and $x = 5$.

CHAPTERS 1 - 3 CUMULATIVE REVIEW

1. $A = \{11, 13, 17, 19, 23, 29, 31, 37\}$

3. $A = \{11, 13, 17, 19, 23, 29, 31, 37\}$
   $C = \{10, 15, 20, 25, 30, 35\}$
   $A \cap C = \emptyset$

5. True

7. $\{x \mid -2 \leq x < 8\}$

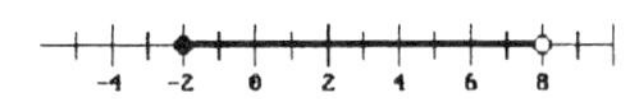

9. Closure property for addition

11. $-2 + 3 - 4 - 8 + 5 = 1 - 4 - 8 + 5$
    $= -3 - 8 + 5$
    $= -11 + 5$
    $= -6$

13. $(-9 + 3)(-4 - 2) = (-6)(-6)$
    $= 36$

15. (a) $|y - x| - (y - x)$
    $= |2 - (-4)| - [2 - (-4)]$
    $= |6| - (6)$
    $= 6 - 6$
    $= 0$

    (b) $\dfrac{3x^2y - 2x}{3xz} = \dfrac{3(-4)^2(2) - 2(-4)}{3(-4)(0)}$
    $= \dfrac{3(16)(2) - 2(-4)}{0}$
    undefined

17. $(-2xy)^2(-3x^2y) = (4x^2y^2)(-3x^2y)$
    $= -12x^4y^3$

19. number: x
    $3 + x \cdot 8$
    or $3 + 8x$

21. width: x
    length: $4x - 3$

    $A = w \cdot l$
    $A = x(4x - 3)$
    $A = 4x^2 - 3x$

    $P = 2w + 2l$
    $P = 2(x) + 2(4x - 3)$
    $P = 2x + 8x - 6$
    $P = 10x - 6$

23. $4x - 3 = 5x + 8$
    $-3 = x + 8$
    $-11 = x$

25. $2x + 3 - (x - 2) = 5 - (x - 4)$
    $2x + 3 - x + 2 = 5 - x + 4$
    $x + 5 = -x + 9$
    $2x + 5 = 9$
    $2x = 4$
    $x = 2$

27. $3(x - 4) + 2(3 - x) = 5[x - (2 - 3x)]$
    $3x - 12 + 6 - 2x = 5(x - 2 + 3x)$
    $x - 6 = 5(4x - 2)$
    $x - 6 = 20x - 10$
    $-6 = 19x - 10$
    $4 = 19x$
    $\dfrac{4}{19} = x$

29. $|2x + 1| = 9$
    $2x + 1 = 9$ or $2x + 1 = -9$
    $2x = 8$ $\quad$ $2x = -10$
    $x = 4$ or $x = -5$

31. $3x - 5 \leq x - 8$
    $2x - 5 \leq -8$
    $2x \leq -3$
    $x \leq -\dfrac{3}{2}$

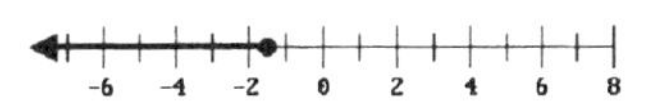

33. $6(x - 2) + (2x - 3) < 2(4x + 1)$
    $6x - 12 + 2x - 3 < 8x + 2$
    $8x - 15 < 8x + 2$
    $-15 < 2$
    Identity, true for all real numbers

35. $|x - 5| < 4$
$-4 < x - 5 < 4$
$1 < x < 9$

-2 0 2 4 6 8 10 12

37. $|5 - 4x| > 7$
$5 - 4x > 7$ or $5 - 4x < -7$
$-4x > 2$ $\quad$ $-4x < -12$
$x < -\frac{1}{2}$ or $x > 3$

-6 -4 -2 0 2 4 6 8

39. Let x = final exam grade

Course grade $= 0.40x + 0.60(63)$
Course grade $= 0.40x + 37.8$

x	$0.40x + 37.8$ = course grade
70	$0.40(70) + 37.8 = 65.8$
80	$0.40(80) + 37.8 = 69.8$
90	$0.40(90) + 37.8 = 73.8$

He would need about an 80 on the final exam to get a course grade of 70.

41. the number: x

$4x = 2x - 3$
$2x = -3$
$x = -\frac{3}{2}$

The number is $-\frac{3}{2}$.

43. number of 300-plate packages: x
number of 100-plate packages: $28 - x$

$2(x) + 1(28 - x) = 33$
$2x + 28 - x = 33$
$x + 28 = 33$
$x = 5$
$28 - x = 28 - 5 = 23$

5 300-plate packages yields
$5(300) = 1500$ plates

23 100-plate packages yields
$23(100) = 2300$ plates

1500 plates + 2300 plates = 3800 plates
He bought 3800 plates.

45. 30 minutes = $30(60) = 1800$ seconds
In 1800 seconds the GL-70 can print
$(1800)(80) = 144{,}000$ characters

$\frac{144000 \text{ characters}}{120 \text{ characters/sec}} = 1200 \text{ sec} = \frac{1200}{60} = 20$
minutes

It would take the VF-44 20 minutes to print the document.

47. width: x
length: $3x + 2$

$P = 2w + 2l$
$P = 2(x) + 2(3x + 2)$
$P = 2x + 6x + 4$
$P = 8x + 4$
$P - 4 = 8x$

$\frac{P - 4}{8} = x$

$5 \le x \le 12$

$5 \le \frac{P - 4}{8} \le 12$

$40 \le P - 4 \le 96$
$44 \le P \le 100$

The perimeter varies from 44 feet to 100 feet.

49. Find 3 on the y-axis, project horizontally until you reach the graph, move vertically to the x-axis and read the x-value, $x = -5$.

$x + y^2 = 4$
$-5 + 3^2 = 4$?
$-5 + 9 = 4$?
$4 = 4$ ✓

51. domain: $\{2, 4, 7\}$
range: $\{3, -1, 5\}$
function

53. domain: $\{4, 3\}$
range: $\{2, 9, 7\}$
Not a function, since the domain element 4 is used twice.

55. $\{x \mid x \text{ is a real number}\}$

57. $5x - 4 \neq 0$
$5x \neq 4$
$x \neq \frac{4}{5}$
$\left\{x \mid x \neq \frac{4}{5}\right\}$

59. function, passes the Vertical Line Test

61. function, passes the Vertical Line Test

63. $f(-3) = -3 + \frac{1}{-3}$
$= \frac{-10}{3}$

65. $h(x^2) = 3x^2 + 1$

67. $f(a + 3) = a + 3 + \frac{1}{a + 3}$

69. $f(3a) = 3a + \frac{1}{3a}$

71. $f(1) = 1 + \frac{1}{1} = 2$
$g(-1) = \frac{1}{(-1)^2} = \frac{1}{1} = 1$
$f(1) + g(-1) = 2 + 1 = 3$

73. $f(7) = 7 + \frac{1}{7} = \frac{50}{7}$
$h\left(-\frac{1}{3}\right) = 3\left(-\frac{1}{3}\right) + 1 = 0$
$= f(7) + h\left(-\frac{1}{3}\right)$
$= \frac{50}{7} + 0$
$= \frac{50}{7}$

75. $5x - 4y = 10$

$x = 0$: $5(0) - 4y = 10$
$-4y = 10$
$y = -\frac{5}{2}$
$\left(0, -\frac{5}{2}\right)$

$y = 0$; $5x - 4(0) = 10$
$5x = 10$
$x = 2$
$(2, 0)$

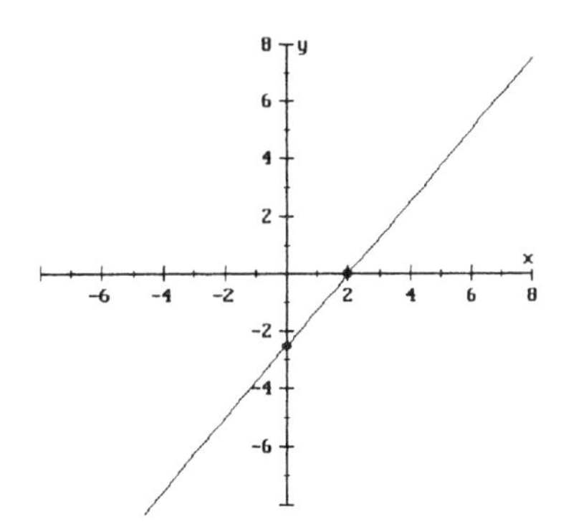

77. $5y = 3x - 10$

$x = 0$: $5y = 3(0) - 10$
$5y = -10$
$y = -2$
$(0, -2)$

$y = 0$: $5(0) = 3x - 10$
$0 = 3x - 10$
$10 = 3x$
$\frac{10}{3} = x$
$\left(\frac{10}{3}, 0\right)$

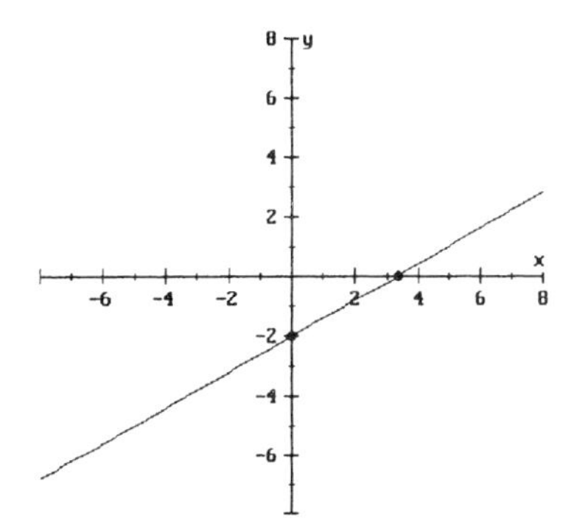

79. $3x - 2 = 8$

$$3x = 10$$

$$x = \frac{10}{3}$$

This is a vertical line passing through $\left(\frac{10}{3}, 0\right)$.

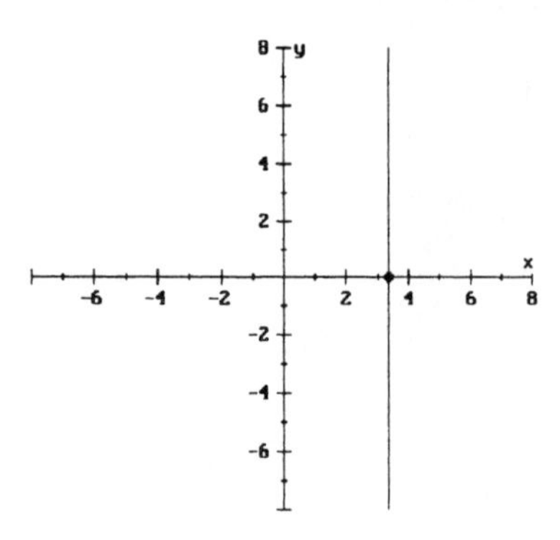

81. (a) Approximately 40,000
    (b) Approximately 4.7 hours
    (c) Between 3.5 hours and 6 hours (graph steepest)
    (d) After approximately 7 hours

83. Charges = flat fee + hourly fee · number of hours

$$C = 350 + 75h$$

85. domain: $\{x \mid x \text{ is a real number}\}$
    range: $\{y \mid y \geq -3\}$

87. Let x = number of hours grading papers then $20 - x$ = number of hours tutoring
    Let y = amount of money he makes

$$y = 15x + 25(20 - x)$$
$$y = 15x + 500 - 25x$$
$$y = 500 - 10x$$

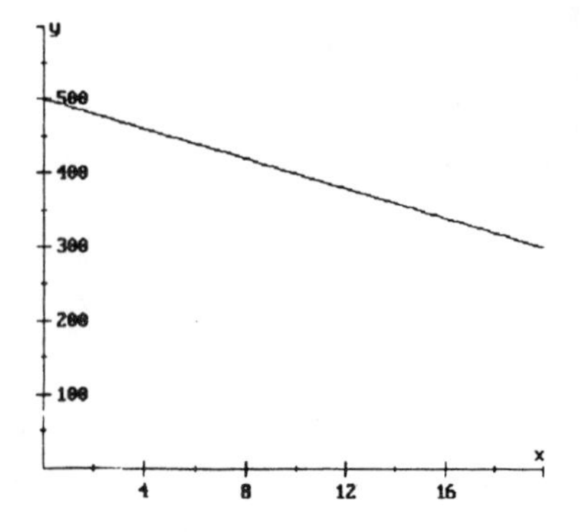

Find 5 on the x-axis, project vertically until you reach the graph, move horizontally to the y-axis and read the y-value, $y = 450$. He makes $450.

Find 380 on the y-axis, project horizontally until you reach the graph, move vertically to the x-axis and read the x-value, $x = 12$. He should grade papers for 12 hours.

CHAPTERS 1 - 3 CUMULATIVE PRACTICE TEST

1. $A = \{4, 8, 12, 16, 20\}$
   $B = \{2, 4, 6, 8, 10, 12, 14, 16, 18, 20, 22\}$

   (a) $A \cap B = \{4, 8, 12, 16, 20\} = A$

   (b) $A \cup B = \{2, 4, 6, 8, 10, 12, 14, 16, 18, 20, 22\} = B$

3. (a)
$$\begin{aligned}(-2)(-2) - (-2)^2(2) &= (-2)(-2) - (4)(2)\\ &= 4 - 8\\ &= -4\end{aligned}$$

   (b)
$$\begin{aligned}5 - \{6 + [2 - 3(4 - 9)]\} &= 5 - \{6 + [2 - 3(-5)]\}\\ &= 5 - [6 + (2 + 15)]\\ &= 5 - (6 + 17)\\ &= 5 - (23)\\ &= -18\end{aligned}$$

5. (a)
$$\begin{aligned}3x - 2 &= 5x + 4\\ -2x - 2 &= 4\\ -2x &= 6\\ x &= -3\end{aligned}$$

   (b)
$$\begin{aligned}3(x - 5) - 2(x - 5) &= 3 - (5 - x)\\ 3x - 15 - 2x + 10 &= 3 - 5 + x\\ x - 5 &= -2 + x\\ -5 &= -2\end{aligned}$$

   Contradiction, no solution

   (c) $|x - 3| = 8$

   $x - 3 = 8$ or $x - 3 = -8$

   $x = 11$ or $x = -5$

7. number of 10-lb packages: x
   number of 30-lb packages: $170 - x$

$$\begin{aligned}10(x) + 30(170 - x) &= 3140\\ 10x + 5100 - 30x &= 3140\\ 5100 - 20x &= 3140\\ -20x &= -1960\\ x &= 98\\ 170 - x &= 170 - 98 = 72\end{aligned}$$

   There are 98 10-lb packages and 72 30-lb packages.

9. **Evan starts his trip by traveling 50 miles away from his home during the first hour of the day; his average rate of speed for the first hour was 50 mph. During the next 2 hours Evan did not travel. Then Evan drove closer to his home between hours 3 and 4. He was traveling an average of 40 mph during this time, and ended up only 10 miles from his home. Evan drove another 60 miles away from his home traveling at 60 mph between the 4th and 5th hour, and stopped traveling for an hour. Finally, during the last hour and a half, Evan drove home at a rate of 47 mph.**

11. Locate the points on the x-axis. They are $(-4, 0)$ and $(4, 0)$.

$$9x^2 + 16y^2 = 144$$
$$9(-4)^2 + 16(0)^2 = 144 \quad ?$$
$$9(16) + 16(0) = 144 \quad ?$$
$$144 = 144 \quad \checkmark$$

$$9x^2 + 16y^2 = 144$$
$$9(4)^2 + 16(0)^2 = 144 \quad ?$$
$$9(16) + 16(0) = 144 \quad ?$$
$$144 = 144 \quad \checkmark$$

13. (a) $f(-5) = \dfrac{-5 - 2}{-5 + 1}$

$= \dfrac{-7}{-4}$

$= \dfrac{7}{4}$

(b) $g(x^2) = 4(x^2) - 1$
$= 4x^2 - 1$

(c) $f(x + 2) = \dfrac{x + 2 - 2}{x + 2 + 1}$

$= \dfrac{x}{x + 3}$

(d) $f(x) + 2 = \dfrac{x - 2}{x + 1} + 2$

(e) $g(5x) = 4(5x) - 1$
$= 20x - 1$

(f) $5g(x) = 5(4x - 1)$
$= 20x - 5$

(g) $f(-1) = \dfrac{-1 - 2}{-1 + 1}$

$= \dfrac{-3}{0}$

undefined

$f(-1) + g(0)$ is undefined.

(h) $g(-1) = 4(-1) - 1$
$= -5$

$g(1) = 4(1) - 1$
$= 3$

$g(-1) - g(1) - 1$
$= -5 - 3 - 1$
$= -9$

15. Let x = gross sales
and y = weekly salary
$y = 200 + 0.10x$

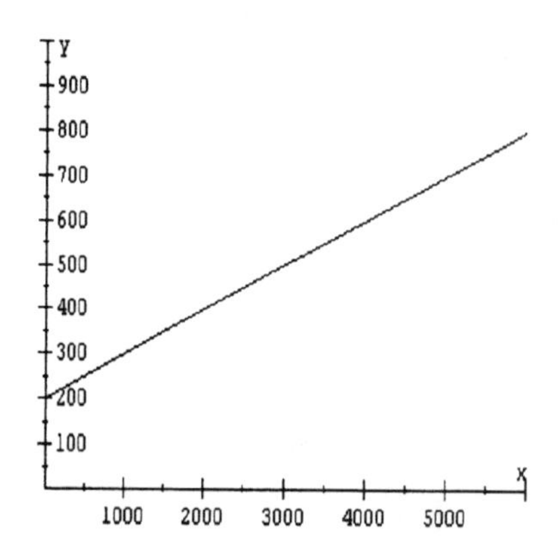

Find 600 on the y-axis, project horizontally until you reach the graph, move vertically to the x-axis and read the x-value, $x \approx 4000$. His gross sales were approximately \$4000.

CHAPTER 4

Exercises 4.1

1. (1, -2) and (-3, 1)

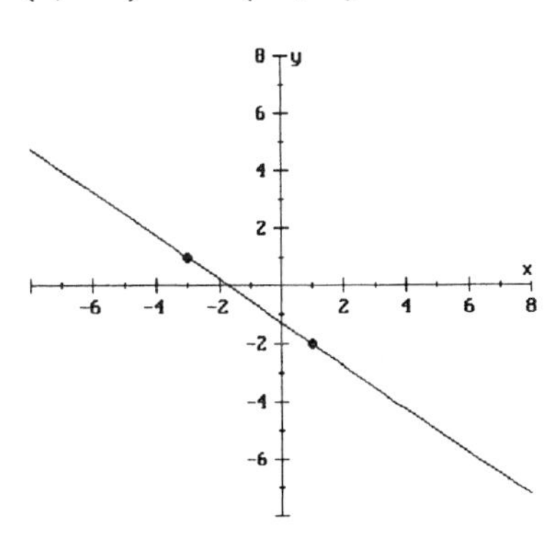

$$m = \frac{y_2 - y_1}{x_2 - x_1}$$

$$= \frac{-2 - 1}{1 - (-3)}$$

$$= \frac{-3}{4} = -\frac{3}{4}$$

3. (0, 2) and (2, 0)

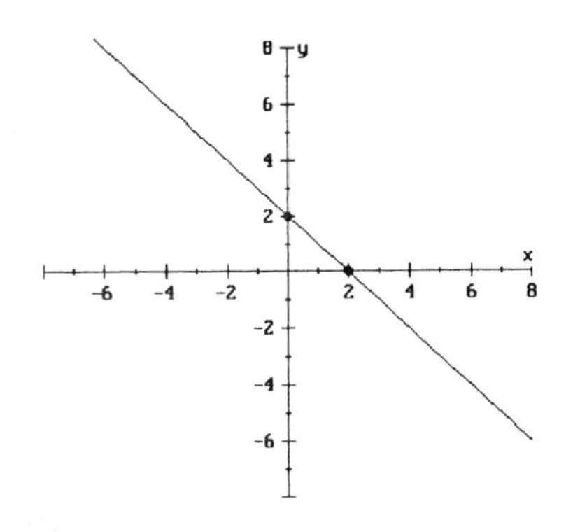

$$m = \frac{y_2 - y_1}{x_2 - x_1}$$

$$= \frac{2 - 0}{0 - 2}$$

$$= \frac{2}{-2}$$

$$= -1$$

5. (-3, -4) and (-2, -5)

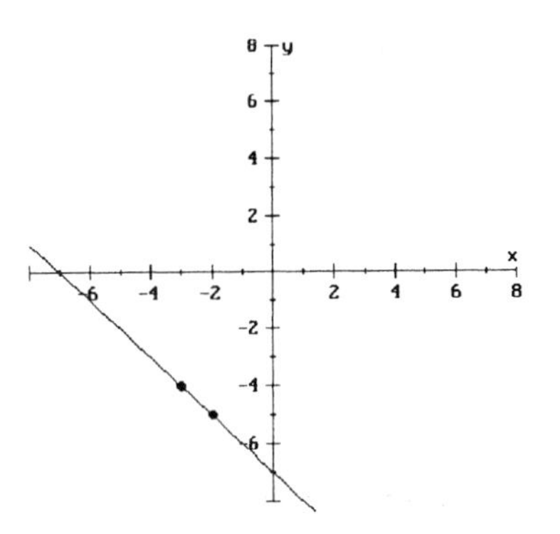

$$m = y_2 - y_1$$

$$= \frac{-4 - (-5)}{-3 - (-2)}$$

$$= \frac{1}{-1}$$

$$= -1$$

7. $m = \dfrac{y_2 - y_1}{x_2 - x_1}$

$$= \frac{4 - 4}{-3 - 2}$$

$$= \frac{0}{-5}$$

$$= 0$$

9. $m = \dfrac{y_2 - y_1}{x_2 - x_1}$

$$= \frac{-3 - 2}{4 - 4}$$

$$= \frac{-5}{0}$$

Undefined

11. $m = \dfrac{y_2 - y_1}{x_2 - x_1}$

$= \dfrac{a - b}{a - b}$

$= 1$

13. $m = \dfrac{y_2 - y_1}{x_2 - x_1}$

$= \dfrac{a^2 - b^2}{a - b}$

$= \dfrac{(a - b)(a + b)}{a - b}$

$= a + b$

15. $m = \dfrac{y_2 - y_1}{x_2 - x_1}$

$= \dfrac{-0.2 - 0.14}{0.7 - 0.06}$

$= \dfrac{-0.34}{0.64}$

$= -\dfrac{17}{32}$

17. $m = \dfrac{y_2 - y_1}{x_2 - x_1}$

$= \dfrac{7 - 0.06}{-0.2 - 0.14}$

$= \dfrac{6.94}{-0.34}$

$= -\dfrac{347}{17}$

19. To get from (0, 0) to (2, 3) (which appear to be points on the line), you go up 3 and to the right 2.

$m = \dfrac{3}{2}$

21. To get from (0, −7) to (2, 0) (which appear to be points on the line), you go up 7 and to the right 2.

$m = \dfrac{7}{2}$

23. (5, 0) and (0, −3)

$m = \dfrac{0 - (-3)}{5 - 0}$

$= \dfrac{3}{5}$

25. $m = 2 = \dfrac{2}{1}$

From (1, 3), go up 2 and to the right 1. This gives us the point (2, 5) on the line.

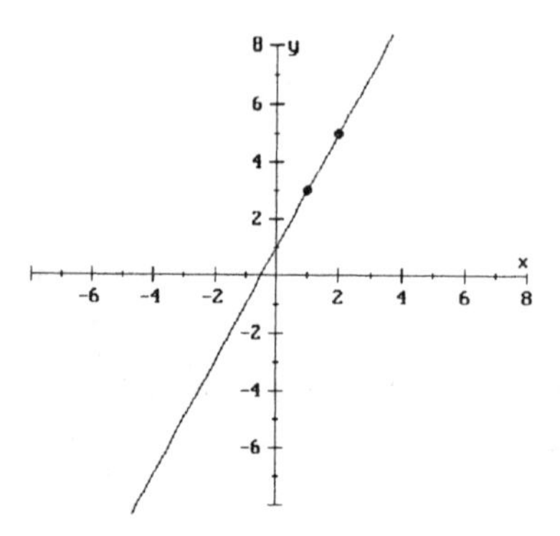

27. $m = -2 = \frac{-2}{1}$

From (1, 3), go down 2 and to the right 1. This gives us the point (2, 1) on the line.

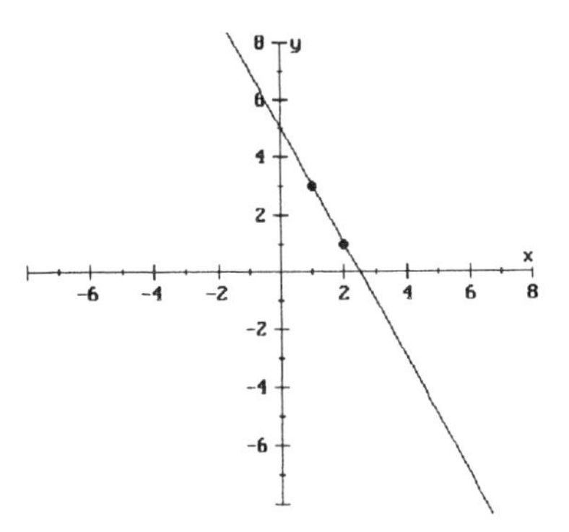

29. $m = -\frac{1}{4} = \frac{-1}{4}$

From (0, 3), go down 1 and to the right 4. This gives us the point (4, 2) on the line.

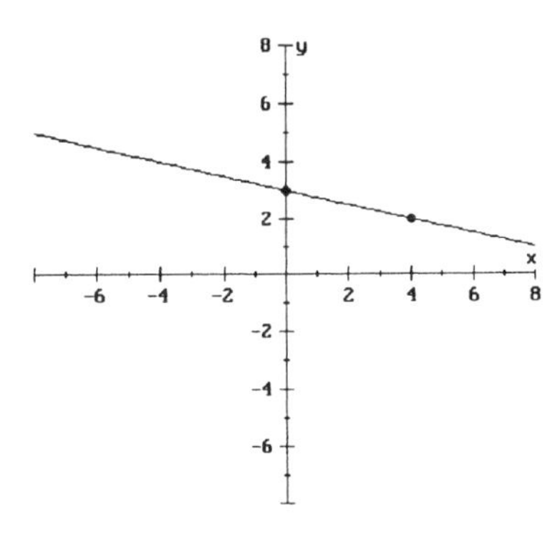

31. $m = \frac{2}{5}$

From (4, 0), go up 2 and to the right 5. This gives us the point (9, 2) on the line.

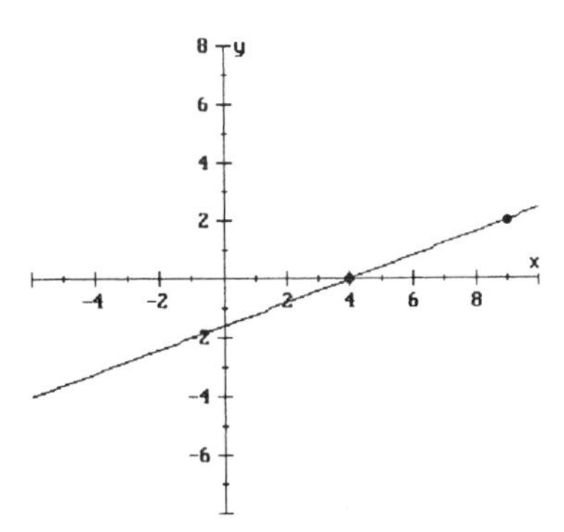

33. $m = 0$: horizontal line through (2, 5)

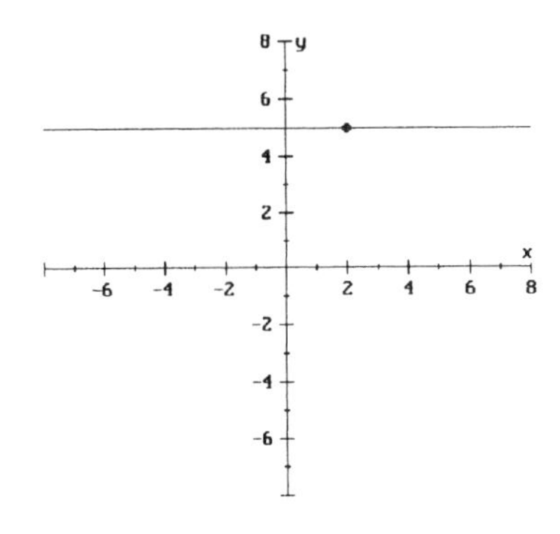

35. undefined slope: vertical line through (2, 5)

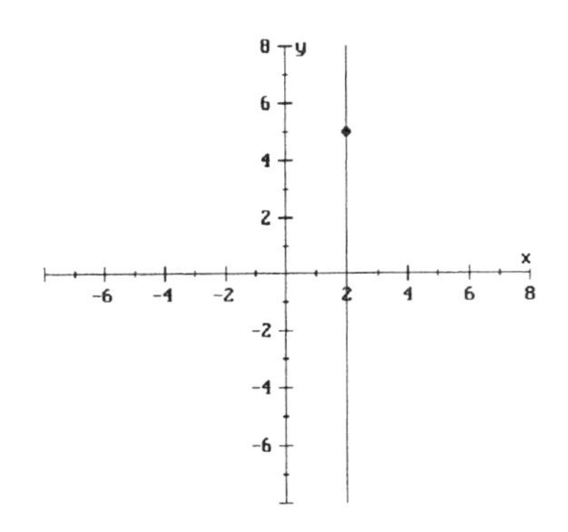

37. $y = 2x - 7$

(0, -7) and (1, -5)

$$m = \frac{-7 - (-5)}{0 - 1}$$

$$= 2$$

39. $y = -0.4x + 5$

(0, 5) and (1, 4.6)

$$m = \frac{5 - 4.6}{0 - 1}$$

$$= \frac{0.4}{-1}$$

$$= -0.4$$

41. $2y - 3x = 8$
$$2y = 3x + 8$$
$$y = \frac{3}{2}x + 4$$

(0, 4) and (2, 7)

$$m = \frac{7 - 4}{2 - 0}$$
$$= \frac{3}{2}$$

43. $m_{P_1P_2} = \frac{4 - 2}{3 - 1} = \frac{2}{2} = 1$

$$m_{P_3P_4} = \frac{-4 - (-2)}{-3 - (-1)} = \frac{-2}{-2} = 1$$

slopes are equal: Parallel

45. $m_{P_1P_2} = \frac{2 - 4}{-1 - 0} = \frac{-2}{-1} = 2$

$$m_{P_3P_4} = \frac{7 - 5}{1 - (-3)} = \frac{2}{4} = \frac{1}{2}$$

Neither

47. $m_{P_1P_2} = \frac{5 - 5}{-2 - 3} = \frac{0}{-5} = 0$

$$m_{P_3P_4} = \frac{-2 - 4}{1 - 1} = \frac{-6}{0} \text{ undefined}$$

Perpendicular, since the first line is horizontal and the second line is vertical.

49. $m = \frac{y_2 - y_1}{x_2 - x_1}$

$$-5 = \frac{h - (-2)}{1 - 4}$$

$$-5 = \frac{h + 2}{-3}$$

$$-3(-5) = (-3)\left(\frac{h + 2}{-3}\right)$$

$$15 = h + 2$$
$$13 = h$$

51. $A(0, 0)$; $B(2, 1)$; $C(-2, 5)$; $D(0, 6)$

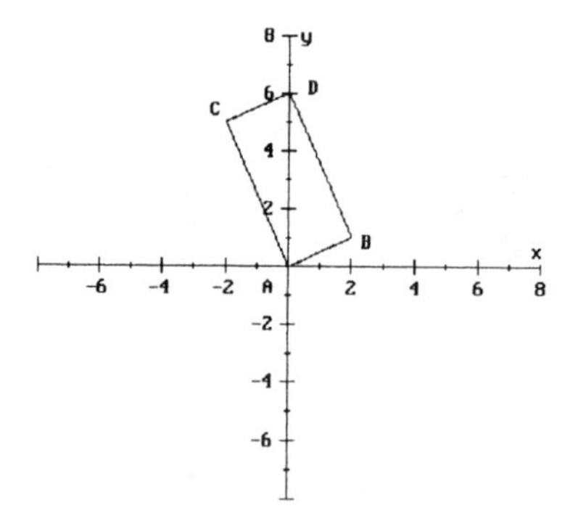

$$M_{AB} = \frac{1 - 0}{2 - 0} = \frac{1}{2}$$

$$M_{AC} = \frac{5 - 0}{-2 - 0} = -\frac{5}{2}$$

$$M_{CD} = \frac{6 - 5}{0 - (-2)} = \frac{1}{2}$$

$$M_{BD} = \frac{6 - 1}{0 - 2} = -\frac{5}{2}$$

Since opposite sides have equal slopes, they are parallel. Hence this is a parallelogram.

53. $A(-3, 2)$; $B(-1, 6)$; $C(3, 4)$

$$M_{AB} = \frac{6 - 2}{-1 - (-3)} = \frac{4}{2} = 2$$

$$M_{BC} = \frac{6 - 4}{-1 - 3} = \frac{2}{-4} - \frac{1}{2}$$

$$M_{AC} = \frac{4 - 2}{3 - (-3)} = \frac{2}{6} = \frac{1}{3}$$

AB is perpendicular to *BC* since their slopes are negative reciprocals. Thus it is a right triangle.

55. 

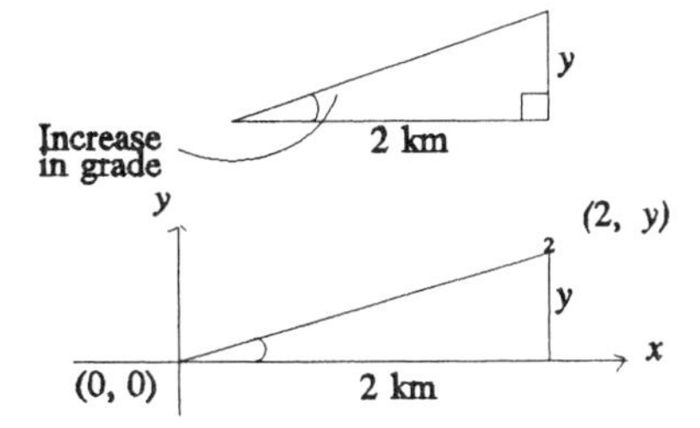

upgrade: $8\% = 0.08 = m$

$$0.08 = \frac{y - 0}{2 - 0}$$

$$0.08 = \frac{y}{2}$$

$$0.16 = y$$

The change in elevation is 0.16 km.

57.

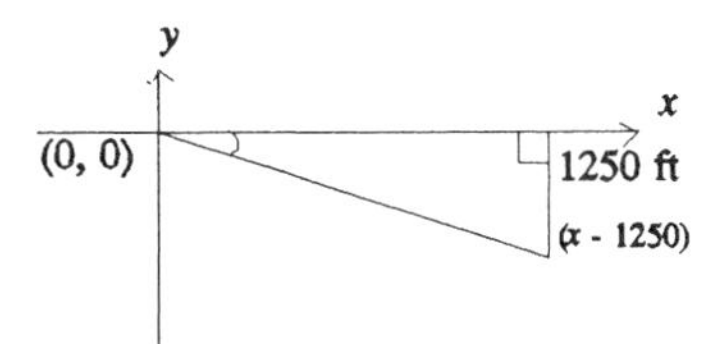

downgrade: $12\% = 0.12$

$$m = -0.12$$

$$\frac{-1250 - 0}{x - 0} = -0.12$$

$$\frac{-1250}{x} = -0.12$$

$$-1250 = -0.12x$$
$$10416.67 = x$$

The horizontal distance is 10,416.67 ft.

59. (2, 15) and (8, 80)

$$m = \frac{80 - 15}{8 - 2}$$

$$= \frac{65}{6}$$

$$= 10.83$$

The sprinter averages approximately 10.83 meters per second during the stretch from 2 seconds to 8 seconds.

61. (1, 56) and (5, 42)

$$m = \frac{56 - 42}{1 - 5}$$

$$= \frac{14}{-4}$$

$$= -3.5$$

The temperature falls an average of approximately 3.5° during the 4 hour period.

63. The slope tells you that the daily profit P increases as the wholesale price p increases. It appears that the points (100, 40) and (135, 100) are on the line.

$$m = \frac{y_2 - y_1}{x_2 - x_1}$$

$$= \frac{100 - 40}{135 - 100}$$

$$= \frac{60}{35}$$

$$= \frac{12}{7}$$

Hence the profit increases 12 thousand dollars when the wholesale price increases \$7.

71. $2xy(3x - 5y) - 6y(2x^2 + xy)$
$= 6x^2y - 10xy^2 - 12x^2y - 6xy^2$
$= -6x^2y - 16xy^2$

73.

$$\frac{t}{3} - 2 = 5$$

$$3\left(\frac{t}{3} - 2\right) = 3(5)$$

$$\frac{3}{1} \cdot \frac{t}{3} - 3(2) = 3(5)$$

$$t - 6 = 15$$
$$t = 21$$

Exercises 4.2

1. $m = 5$; $(1, -3)$

$$y - y_1 = m(x - x_1)$$
$$y - (-3) = 5(x - 1)$$
$$y + 3 = 5x - 5$$
$$y = 5x - 8$$

3. $m = -3$; $(-5, 2)$

$$y - y_1 = m(x - x_1)$$
$$y - 2 = -3[x - (-5)]$$
$$y - 2 = -3(x + 5)$$
$$y - 2 = -3x - 15$$
$$y = -3x - 13$$

5. $m = \frac{2}{3}$; $(6, 1)$

$$y - y_1 = m(x - x_1)$$
$$y - 1 = \frac{2}{3}(x - 6)$$
$$y - 1 = \frac{2}{3}x - 4$$
$$y = \frac{2}{3}x - 3$$

7. $m = -\frac{1}{2}$; $(4, 0)$

$$y - y_1 = m(x - x_1)$$
$$y - 0 = -\frac{1}{2}(x - 4)$$
$$y = -\frac{1}{2}x + 2$$

9. $m = \frac{3}{4}$; $(0, 5)$

$$y = mx + b$$
$$y = \frac{3}{4}x + 5$$

11. $m = 0$; horizontal line through $(-3, -4)$

$$y = -4$$

13. m undefined: vertical line through $(-4, 7)$

$$x = -4$$

15. $(2, 3)$ and $(5, 7)$

$$m = \frac{7 - 3}{5 - 2} = \frac{4}{3}$$
$$y - y_1 = m(x - x_1)$$
$$y - 3 = \frac{4}{3}(x - 2)$$
$$y - 3 = \frac{4}{3}x - \frac{8}{3}$$
$$y = \frac{4}{3}x + \frac{1}{3}$$

17. $(-2, -1)$ and $(-3, -5)$

$$m = \frac{-5 - (-1)}{-3 - (-2)} = \frac{-4}{-1} = 4$$
$$y - y_1 = m(x - x_1)$$
$$y - (-1) = 4[x - (-2)]$$
$$y + 1 = 4(x + 2)$$
$$y + 1 = 4x + 8$$
$$y = 4x + 7$$

19. $(2, 3)$ and $(0, 5)$

$$m = \frac{5 - 3}{0 - 2} = \frac{2}{-2} = -1$$
$$y = mx + b$$
$$y = -x + 5$$

21. $m = 4$; $(0, 6)$

$$y = mx + b$$
$$y = 4x + 6$$

23. $m = -2$; $(-3, 0)$

$$y - y_1 = m(x - x_1)$$
$$y - 0 = -2[x - (-3)]$$
$$y = -2(x + 3)$$
$$y = -2x - 6$$

25. $(2, 3)$ and $(6, 3)$

$$m = \frac{3 - 3}{6 - 2} = \frac{0}{4} = 0$$

Horizontal line: $y = 3$

27. vertical line through $(-2, -4)$

$$x = -2$$

29. $(-3, 0)$ and $(0, 2)$

$$m = \frac{2 - 0}{0 - (-3)} = \frac{2}{3}$$

$$y = mx + b$$

$$y = \frac{2}{3}x + 2$$

31. $y = 3x + 7$
$m = 3$

slope of parallel line: 3

$$y - y_1 = m(x - x_1)$$
$$y - 2 = 3(x - 2)$$
$$y - 2 = 3x - 6$$
$$y = 3x - 4$$

33. $$y = -\frac{2}{3}x - 1$$

$$m = -\frac{2}{3}$$

$$m_\perp = \frac{3}{2}$$

$$y - 2 = \frac{3}{2}[x - (-3)]$$

$$y - 2 = \frac{3}{2}(x + 3)$$

$$y - 2 = \frac{3}{2}x + \frac{9}{2}$$

$$y = \frac{3}{2}x + \frac{13}{2}$$

35. $y = x$
$m = 1$
$m_\perp = -1$

$$y = mx + b$$
$$y = -1x + 0$$
$$y = -x$$

37. $2y - 3x = 12$
$2y = 3x + 12$

$$y = \frac{3}{2}x + 6$$

$$m = \frac{3}{2}$$

parallel line: same slope

$$y - y_1 = m(x - x_1)$$

$$y - (-2) = \frac{3}{2}[x - (-1)]$$

$$y + 2 = \frac{3}{2}(x + 1)$$

$$y + 2 = \frac{3}{2}x + \frac{3}{2}$$

$$y = \frac{3}{2}x - \frac{1}{2}$$

39. $4x - 3y = 9$
$-3y = -4x + 9$

$$y = \frac{4}{3}x - 3$$

$$m = \frac{4}{3}$$

$$m_\perp = -\frac{3}{4}$$

$$y - y_1 = m(x - x_1)$$

$$y - (-3) = -\frac{3}{4}(x - 1)$$

$$y + 3 = -\frac{3}{4}x + \frac{3}{4}$$

$$y = -\frac{3}{4}x - \frac{9}{4}$$

41. $8x - 5y = 20$

$$-5y = -8x + 20$$

$$y = \frac{8}{5}x - 4$$

$$b = -4$$

$$m = \frac{8}{5}$$

$$m_{\perp} = -\frac{5}{8}$$

$$y = mx + b$$

$$y = -\frac{5}{8}x - 4$$

43. slope of line through (3, -6) and (-1, 2)

$$m = \frac{-6 - 2}{3 - (-1)} = \frac{-8}{4} = -2$$

parallel line: same slope

$y = mx + b$

$y = -2x + 4$

45. horizontal: $y = 3$

47. vertical: $x = -1$

49. See Section 4.1 #19. The slope of this line is $\frac{3}{2}$. From the graph we see the y-intercept is (0, 0).

$y = mx + b$

$y = \frac{3}{2}x + 0$

$y = \frac{3}{2}x$

51. See Section 4.1 #21. The slope of this line is $\frac{7}{2}$. From the graph we see the y-intercept is (0, −7).

$$y = mx + b$$

$$y = \frac{7}{2}x - 7$$

53. (a) The line appears to pass through the points (1995, 1000) and (1996, 1050).

$$m = \frac{1050 - 1000}{1996 - 1995} = 50$$

$$H - 1000 = 50(t - 1995)$$

$$H - 1000 = 50t - 99750$$

$$H = 50t - 98750$$

(b) $t = 1999$

$$H = 50(1999) - 98750 = 1200$$

The health expenditures in 1999 are predicted to be approximately $1200 billion.

55. $3x - 2y = 5$

$$-2y = -3x + 5$$

$$y = \frac{3}{2}x - \frac{5}{2}$$

$$m_1 = \frac{3}{2}$$

$$3x - 2y = 6$$

$$-2y = -3x + 6$$

$$y = \frac{3}{2}x - 3$$

$$m_2 = \frac{3}{2}$$

parallel, $m_1 = m_2$

57.
$$2x = 3y - 4$$
$$2x + 4 = 3y$$
$$\frac{2}{3}x + \frac{4}{3} = y$$
$$m_1 = \frac{2}{3}$$

$$2x + 3y = 4$$
$$3y = -2x + 4$$
$$y = -\frac{2}{3}x + \frac{4}{3}$$
$$m_2 = -\frac{2}{3}$$

neither

59.
$$5x + y = 2$$
$$y = -5x + 2$$
$$m_1 = -5$$
$$5y = x + 3$$
$$y = \frac{1}{5}x + \frac{3}{5}$$
$$m_2 = \frac{1}{5}$$

perpendicular, $m_1 = -\frac{1}{m_2}$

61.
$$3x - 7y = 1$$
$$-7y = -3x + 1$$
$$y = \frac{3}{7}x - \frac{1}{7}$$
$$m_1 = \frac{3}{7}$$

$$6x = 14y + 5$$
$$6x - 5 = 14y$$
$$\frac{3}{7}x - \frac{5}{14} = y$$
$$m_2 = \frac{3}{7}$$

parallel, $m_1 = m_2$

63. Cost = flat fee + (cost per mile)(number of miles)
$$C = 29 + 0.12n$$
$$m = 0.12$$

65. Cost = flat fee + (cost per call)(number of calls)
$$B = 23 + 0.825n$$
$$m = 0.825$$

67. (x, P): (18, 200) and (100, 2660)

$$m = \frac{2660 - 200}{100 - 18} = \frac{2460}{82} = 30$$

$$P - 200 = 30(x - 18)$$
$$P - 200 = 30x - 540$$
$$P = 30x - 340$$

Find P when $x = 200$:
$$P = 30(200) - 340$$
$$= 5660$$

The profit is \$5660.

69. (A, B): (35, 75) and (15, 35)

$$m = \frac{75 - 35}{35 - 15} = \frac{40}{20} = 2$$

$$B - 75 = 2(A - 35)$$
$$B - 75 = 2A - 70$$
$$B = 2A + 5$$

Find B when $A = 40$:
$$B = 2(40) + 5$$
$$= 85$$

The score on test B would be 85.

71. (V, E): (35000, 70) and (20000, 85)

$$m = \frac{85 - 70}{20000 - 35000} = \frac{15}{-15000} = -0.001$$

$$E - 70 = -0.001(V - 35000)$$
$$E - 70 = -0.001V + 35$$
$$E = -0.001V + 105$$

Find E when $V = 15000$:

$$E = -0.001(15000) + 105$$
$$= 90$$

The expected score is 90.

75. $\dfrac{-4(-2)(-6)}{-4 - 2(-6)} = \dfrac{8(-6)}{-4 + 12}$

$= \dfrac{-48}{8}$

$= -6$

77. $4x - 5 \neq 0$

$4x \neq 5$

$x \neq \dfrac{5}{4}$

$\left\{x \mid x \neq \dfrac{5}{4}\right\}$

Exercises 4.3

1. $\begin{cases} 2x + y = 12 \\ 3x - y = 13 \end{cases}$

$$\begin{aligned} 2x + y &= 12 \\ \underline{3x - y} &\underline{= 13} \\ 5x \quad &= 25 \\ x &= 5 \end{aligned}$$

$$\begin{aligned} 2(5) + y &= 12 \\ 10 + y &= 12 \\ y &= 2 \end{aligned}$$

$x = 5, y = 2;$ independent

3. $\begin{cases} -x + 5y = 11 \\ x - 2y = -2 \end{cases}$

$$\begin{aligned} -x + 5y &= 11 \\ \underline{x - 2y} &\underline{= -2} \\ 3y &= 9 \\ y &= 3 \end{aligned}$$

$$\begin{aligned} x - 2(3) &= -2 \\ x - 6 &= -2 \\ x &= 4 \end{aligned}$$

$x = 4, y = 3;$ independent

5. $\begin{cases} 3x - y = 0 \\ 2x + 3y = 11 \end{cases}$

1st equation: $y = 3x$
Substitute into 2nd equation:

$$\begin{aligned} 2x + 3(3x) &= 11 \\ 2x + 9x &= 11 \\ 11x &= 11 \\ x &= 1 \\ y &= 3(1) = 3 \end{aligned}$$

$x = 1, y = 3;$ independent

7. $\begin{cases} x + 7y = 20 \\ 5x + 2y = 34 \end{cases}$

1st equation: $x = 20 - 7y$
Substitute into 2nd equation:

$$\begin{aligned} 5(20 - 7y) + 2y &= 34 \\ 100 - 35y + 2y &= 34 \\ 100 - 33y &= 34 \\ -33y &= -66 \\ y &= 2 \\ x &= 20 - 7(2) = 6 \end{aligned}$$

$x = 6, y = 2;$ independent

9. $\begin{cases} 4x + 5y = 0 \\ 2x + 3y = -2 \end{cases}$

Multiply 2nd equation by -2:

$$\begin{aligned} -2(2x + 3y) &= -2(-2) \\ -4x - 6y &= 4 \end{aligned}$$

Add to 1st equation:

$$\begin{aligned} 4x + 5y &= 0 \\ \underline{-4x - 6y} &\underline{= 4} \\ -y &= 4 \\ y &= -4 \end{aligned}$$

$$\begin{aligned} 2x + 3(-4) &= -2 \\ 2x - 12 &= -2 \\ 2x &= 10 \\ x &= 5 \end{aligned}$$

$x = 5, y = -4;$ independent

11. $\begin{cases} 2x + 3y = 7 \\ 4x + 6y = 14 \end{cases}$

Multiply 1st equation by -2:

$$\begin{aligned} -2(2x + 3y) &= -2(7) \\ -4x - 6y &= -14 \end{aligned}$$

Add to 2nd equation:

$$\begin{aligned} -4x - 6y &= -14 \\ 4x + 6y &= 14 \\ \hline 0 &= 0 \end{aligned}$$

dependent

13. $\begin{cases} 5x - 6y = 3 \\ 10x - 12y = 5 \end{cases}$

Multiply 1st equation by −2:

$$\begin{aligned} -2(5x - 6y) &= -2(3) \\ -10x + 12y &= -6 \end{aligned}$$

Add to 2nd equation:

$$\begin{aligned} -10x + 12y &= -6 \\ 10x - 12y &= 5 \\ \hline 0 &= -1 \end{aligned}$$

inconsistent

15. $\begin{cases} 2x - 3y = 10 \\ 3x - 2y = 15 \end{cases}$

Multiply 1st equation by 3:

$$6x - 9y = 30$$

Multiply 2nd equation by −2:

$$-6x + 4y = -30$$

Adding:

$$\begin{aligned} 6x - 9y &= 30 \\ -6x + 4y &= -30 \\ \hline -5y &= 0 \\ y &= 0 \end{aligned}$$

$$\begin{aligned} 2x - 3(0) &= 10 \\ 2x &= 10 \\ x &= 5 \end{aligned}$$

$x = 5.\ y = 0$; independent

17. $\begin{cases} y = 2x + 3 \\ 2x + y = -1 \end{cases}$

Substitute 1st equation into 2nd equation:

$$\begin{aligned} 2x + (2x + 3) &= -1 \\ 4x + 3 &= -1 \\ 4x &= -4 \\ x &= -1 \\ y &= 2(-1) + 3 = 1 \end{aligned}$$

$x = -1,\ y = 1$; independent

19. $\begin{cases} 6a - 3b = 1 \\ 8a + 5b = 7 \end{cases}$

Multiply 1st equation by 5:
$30a - 15b = 5$

Multiply 2nd equation by 3:
$24a + 15b = 21$

Adding:

$$\begin{aligned} 30a - 15b &= 5 \\ 24a + 15b &= 21 \\ \hline 54a &= 26 \\ a &= \frac{13}{27} \end{aligned}$$

$$\begin{aligned} 6\left(\frac{13}{27}\right) - 3b &= 1 \\ \frac{26}{9} - 3b &= 1 \\ -3b &= -\frac{17}{9} \\ b &= \frac{17}{27} \end{aligned}$$

$a = \frac{13}{27},\ b = \frac{17}{27}$; independent

21. $\begin{cases} s = 3t - 5 \\ t = 3s - 5 \end{cases}$

Substitute 1st equation into 2nd equation:

$$\begin{aligned} t &= 3(3t - 5) - 5 \\ t &= 9t - 15 - 5 \\ t &= 9t - 20 \\ -8t &= -20 \\ t &= \frac{5}{2} \\ s &= 3\left(\frac{5}{2}\right) - 5 = \frac{5}{2} \end{aligned}$$

$s = \frac{5}{2},\ t = \frac{5}{2}$; independent

23. $\begin{cases} 3m - 2n = 8 \\ 3n = m - 8 \end{cases}$

2nd equation: $m = 3n + 8$

Substitute into 1st equation:

$$\begin{aligned} 3(3n + 8) - 2n &= 8 \\ 9n + 24 - 2n &= 8 \\ 7n + 24 &= 8 \\ 7n &= -16 \\ n &= -\frac{16}{7} \\ m = 3\left(-\frac{16}{7}\right) + 8 &= \frac{8}{7} \end{aligned}$$

$m = \frac{8}{7}$, $n = -\frac{16}{7}$; independent

25. $\begin{cases} 3p - 4q = 5 \\ 3q - 4p = -9 \end{cases}$

$\begin{cases} 3p - 4q = 5 \\ -4p + 3q = -9 \end{cases}$

Multiply 1st equation by 4:
$12p - 16q = 20$

Multiply 2nd equation by 3:
$-12p + 9q = -27$

Adding:

$$\begin{aligned} 12p - 16q &= 20 \\ \underline{-12p + 9q} &\underline{= -27} \\ -7q &= -7 \\ q &= 1 \end{aligned}$$

$$\begin{aligned} 3p - 4(1) &= 5 \\ 3p - 4 &= 5 \\ 3p &= 9 \\ p &= 3 \end{aligned}$$

$p = 3$, $q = 1$; independent

27. $\begin{cases} \frac{u}{3} - v = 1 \\ u - \frac{v}{2} = 5 \end{cases}$

Multiply 1st equation by 3 and 2nd equation by 2:

$\begin{cases} u - 3v = 3 \\ 2u - v = 10 \end{cases}$

Multiply 1st equation by -2:
$-2u + 6v = -6$

Adding:

$$\begin{aligned} -2u + 6v &= -6 \\ \underline{2u - v} &\underline{= 10} \\ 5v &= 4 \\ v &= \frac{4}{5} \end{aligned}$$

$$\begin{aligned} 2u - \frac{4}{5} &= 10 \\ 5\left(2u - \frac{4}{5}\right) &= 5(10) \\ 10u - 4 &= 50 \\ 10u &= 54 \\ u &= \frac{27}{5} \end{aligned}$$

$u = \frac{27}{5}$, $v = \frac{4}{5}$; independent

29. $\begin{cases} \frac{w}{4} + \frac{z}{6} = 4 \\ \frac{w}{2} - \frac{z}{3} = 4 \end{cases}$

Multiply 1st equation by 12 and 2nd equation by 6:

$\begin{cases} 3w + 2z = 48 \\ 3w - 2z = 24 \end{cases}$

Adding:

$$\begin{aligned} 3w + 2z &= 48 \\ \underline{3w - 2z} &\underline{= 24} \\ 6w &= 72 \\ w &= 12 \\ 3(12) + 2z &= 48 \\ 36 + 2z &= 48 \\ 2z &= 12 \\ z &= 6 \end{aligned}$$

$w = 12$, $z = 6$; independent

31. $\begin{cases} \dfrac{x}{6} + \dfrac{y}{8} = \dfrac{3}{4} \\ \dfrac{x}{4} + \dfrac{y}{3} = \dfrac{17}{12} \end{cases}$

Multiply 1st equation by 24 and 2nd equation by 12:

$$\begin{cases} 4x + 3y = 18 \\ 3x + 4y = 17 \end{cases}$$

Multiply 1st equation by 3:
$12x + 9y = 54$

Multiply 2nd equation by −4:
$-12x - 16y = -68$

Adding:

$$\begin{array}{r} 12x + 9y = 54 \\ \underline{-12x - 16y = -68} \\ -7y = -14 \\ y = 2 \end{array}$$

$$\begin{aligned} 4x + 3(2) &= 18 \\ 4x + 6 &= 18 \\ 4x &= 12 \\ x &= 3 \end{aligned}$$

$x = 3,\ y = 2$; independent

33. $\begin{cases} \dfrac{x+3}{2} + \dfrac{y-4}{3} = \dfrac{19}{6} \\ \dfrac{x-2}{3} + \dfrac{y-2}{2} = 2 \end{cases}$

Multiply 1st equation by 6:

$$\begin{aligned} 3(x + 3) + 2(y - 4) &= 19 \\ 3x + 9 + 2y - 8 &= 19 \\ 3x + 2y + 1 &= 19 \\ 3x + 2y &= 18 \end{aligned}$$

Multiply 2nd equation by 6:

$$\begin{aligned} 2(x - 2) + 3(y - 2) &= 12 \\ 2x - 4 + 3y - 6 &= 12 \\ 2x + 3y - 10 &= 12 \\ 2x + 3y &= 22 \end{aligned}$$

$$\begin{cases} 3x + 2y = 18 \\ 2x + 3y = 22 \end{cases}$$

Multiply 1st equation by −2:
$-6x - 4y = -36$

Multiply 2nd equation by 3:
$6x + 9y = 66$

Adding:

$$\begin{array}{r} -6x - 4y = -36 \\ \underline{6x + 9y = 66} \\ 5y = 30 \\ y = 6 \end{array}$$

$$\begin{aligned} 3x + 2(6) &= 18 \\ 3x + 12 &= 18 \\ 3x &= 6 \\ x &= 2 \end{aligned}$$

$x = 2,\ y = 6$; independent

35. $\begin{cases} 0.1x + 0.01y = 0.37 \\ 0.02x + 0.05y = 0.41 \end{cases}$

Multiply both equations by 100:

$$\begin{cases} 10x + y = 37 \\ 2x + 5y = 41 \end{cases}$$

Multiply 2nd equation by −5:
$-10x - 25y = -205$

Adding:

$$\begin{array}{r} 10x + y = 37 \\ \underline{10x - 25y = -205} \\ -24y = -168 \\ y = 7 \end{array}$$

$$\begin{aligned} 2x + 5(7) &= 41 \\ 2x + 35 &= 41 \\ 2x &= 6 \\ x &= 3 \end{aligned}$$

$x = 3,\ y = 7$; independent

37. $\begin{cases} \dfrac{x}{2} + 0.05y = 0.35 \\ 0.3x + \dfrac{y}{4} = 0.65 \end{cases}$

Multiply both equations by 100:

$$\begin{cases} 50x + 5y = 35 \\ 30x + 25y = 65 \end{cases}$$

Multiply 2nd equation by $-\frac{1}{5}$:

$-6x - 5y = -13$

Adding:

$$\begin{aligned} 50x + 5y &= 35 \\ -6x - 5y &= -13 \\ \hline 44x &= 22 \\ x &= \frac{1}{2} \end{aligned}$$

$$50\left(\frac{1}{2}\right) + 5y = 35$$

$$\begin{aligned} 25 + 5y &= 35 \\ 5y &= 10 \\ y &= 2 \end{aligned}$$

$x = \frac{1}{2}$, $y = 2$; independent

39. amount at 5%: x
amount at 8%: y

$$\begin{cases} x + y = 14000 \\ 0.05x + 0.08y = 835 \end{cases}$$

Multiply 2nd equation by 100:

$$\begin{cases} x + y = 14000 \\ 5x + 8y = 83500 \end{cases}$$

Multiply 1st equation by -5 and add to 2nd equation:

$$\begin{aligned} -5x - 5y &= -70000 \\ 5x + 8y &= 83500 \\ \hline 3y &= 13500 \\ y &= 4500 \end{aligned}$$

$$\begin{aligned} x + 4500 &= 14000 \\ x &= 9500 \end{aligned}$$

She should invest \$9500 at 5% and \$4500 at 8%.

41. amount at 10%: x
amount at 8%: y

$$\begin{cases} x + y = 10000 \\ 0.10x = 0.08y \end{cases}$$

Multiply 2nd equation by 100:

$$\begin{cases} x + y = 10000 \\ 10x = 8y \end{cases}$$

1st equation:
$x = 10000 - y$

Substitute into 2nd equation:

$$\begin{aligned} 10(10000 - y) &= 8y \\ 100000 - 10y &= 8y \\ 100000 &= 18y \\ 5555.56 &= y \end{aligned}$$

$$\begin{aligned} x + 5555.56 &= 10000 \\ x &= 4444.44 \end{aligned}$$

Invest \$4444.44 at 10% and \$5555.56 at 8%.

43. width: x
length: y

$$\begin{cases} 2x + 2y = 36 \\ y = 2 + x \end{cases}$$

Substitute 2nd equation into 1st:

$$\begin{aligned} 2x + 2(2 + x) &= 36 \\ 2x + 4 + 2x &= 36 \\ 4x + 4 &= 36 \\ 4x &= 32 \\ x &= 8 \\ y &= 2 + 8 = 10 \end{aligned}$$

The dimensions are 8 cm by 10 cm.

45. cost per roll of 35-mm film: x
cost per roll of movie film: y

$$\begin{cases} 5x + 3y = 35.60 \\ 3x + 5y = 43.60 \end{cases}$$

Multiply 1st equation by -3 and 2nd equation by 5, then add:

$$\begin{aligned} -15x - 9y &= -106.80 \\ 15x + 25y &= 218 \\ \hline 16y &= 111.2 \\ y &= 6.95 \end{aligned}$$

$$\begin{aligned} 5x + 3(6.95) &= 35.60 \\ 5x + 20.85 &= 35.60 \\ 5x &= 14.75 \\ x &= 2.95 \end{aligned}$$

35-mm film costs \$2.95 per roll and movie film costs \$6.95 per roll.

47. cost of jelly donut: x
cost of cream-filled donut: y

$$\begin{cases} 5x + 7y = 3.16 \\ 8x + 4y = 3.04 \end{cases}$$

Multiply 1st equation by -4 and 2nd equation by 7, then add:

$$\begin{aligned} -20x - 28y &= -12.64 \\ 56x + 28y &= 21.28 \\ \hline 36x &= 8.64 \\ x &= 0.24 \end{aligned}$$

$$\begin{aligned} 5(0.24) + 7y &= 3.16 \\ 1.2 + 7y &= 3.16 \\ 7y &= 1.96 \\ y &= 0.28 \end{aligned}$$

A jelly donut costs \$0.24 and a cream-filled donut costs \$0.28.

49. number of expensive models: x
number of less expensive models: y

$$\begin{cases} 6x + 5y = 730 \\ 3x + 2y = 340 \end{cases}$$

Multiply 2nd equation by -2 and add to 1st:

$$\begin{aligned} 6x + 5y &= 730 \\ -6x - 4y &= 680 \\ \hline y &= 50 \end{aligned}$$

$$\begin{aligned} 3x + 2(50) &= 340 \\ 3x + 100 &= 340 \\ 3x &= 240 \\ x &= 80 \end{aligned}$$

They can produce 80 expensive models and 50 less expensive models.

51. annual medical expenses: x
annual cost: y

husband's plan:

$$\underset{\text{per month}}{\text{cost} = 140} \cdot \underset{\text{months}}{12} + \underset{\text{medical expenses}}{10\% \text{ of}}$$

$$\begin{aligned} y &= (140)(12) + 0.10x \\ y &= 1680 + 0.10x \end{aligned}$$

wife's plan:

$$\text{cost} = \underset{\text{per month}}{\$100} \cdot \underset{\text{months}}{12} + \underset{\text{medical expenses}}{25\% \text{ of}}$$

$$\begin{aligned} y &= 1680 + 0.10x \\ y &= 1200 + 0.25x \end{aligned}$$

Substituting 1st equation into 2nd:

$$\begin{aligned} 1680 + 0.10x &= 1200 + 0.25x \\ 480 + 0.10x &= 0.25x \\ 480 &= 0.15x \\ 3200 &= x \end{aligned}$$

The two plans are equivalent when the annual medical expenses are \$3200.

53. number of miles: x
cost: y

Cheapo
$y = 29 + 0.12x$

Cut-Rate
$y = 22 + 0.15x$

$$\begin{cases} y = 29 + 0.12x \\ y = 22 + 0.15x \end{cases}$$

Substitute 1st equation into 2nd:

$$\begin{aligned} 29 + 0.12x &= 22 + 0.15x \\ 7 + 0.12x &= 0.15x \\ 7 &= 0.03x \\ 233\frac{1}{3} &= x \end{aligned}$$

$233\frac{1}{3}$ miles must be driven.

55.
$$\begin{aligned} C &= R \\ 1.6x + 7200 &= 2.1x \\ 7200 &= 0.5x \\ 14400 &= x \end{aligned}$$

They must sell 14,400 units.

57. number of calls: x
cost: x

1st plan:
$y = 18 + 0.11x$

2nd plan:

$$y = 24 + 50(0) + 0.09(x - 50)$$
$$y = 24 + 0.09x - 4.5$$
$$y = 0.09x + 19.5$$

1st plan = 2nd plan

$$18 + 0.11x = 0.09x + 19.5$$
$$0.11x = 0.09x + 1.5$$
$$0.02x = 1.5$$
$$x = 75$$

75 calls per month make the plans equivalent.

59. number of \$5-bills: x
number of \$10-bills: y

$$\begin{cases} x + y = 43 \\ 5x + 10y = 340 \end{cases}$$

Multiply 1st equation by -5, then add to 2nd:

$$-5x - 5y = -215$$
$$5x + 10y = 340$$
$$5y = 125$$
$$y = 25$$

$$x + 25 = 43$$
$$x = 18$$

There are 18 \$5-bills and 25 \$10-bills.

61. charge per mile: x
flat fee: y

$$\begin{cases} 85x + y = 44.30 \\ 125x + y = 51.50 \end{cases}$$

$$y = 44.30 - 85x$$

Substituting:

$$125x + 44.30 - 85x = 51.50$$
$$40x + 44.30 = 51.50$$
$$40x = 7.2$$
$$x = 0.18$$

$$85(0.18) + y = 44.30$$
$$15.30 + y = 44.30$$
$$y = 29$$

The charge per mile is \$0.18 and the flat fee is \$29.

63. speed of plane: x
speed of wind: y

rate with tailwind: $x + y$
rate with headwind: $x - y$

$$\begin{cases} 6(x + y) = 2310 \\ 6(x - y) = 1530 \end{cases}$$

Divide both equations by 6, then add:

$$x + y = 385$$
$$x - y = 255$$
$$2x = 640$$
$$x = 320$$

$$320 + y = 385$$
$$y = 65$$

The plane's speed is 320 km/hr and the winds speed is 65 km/hr.

65. $$\begin{cases} \dfrac{y - 3}{x - 2} = -1 \\ \dfrac{y - (-2)}{x - 1} = 2 \end{cases}$$

$$\begin{cases} y - 3 = -x + 2 \\ y + 2 = 2x - 2 \end{cases}$$

$$\begin{cases} y = -x + 5 \\ y = 2x - 4 \end{cases}$$

Substituting:

$$-x + 5 = 2x - 4$$
$$5 = 3x - 4$$
$$9 = 3x$$
$$3 = x$$

$$y = -3 + 5 = 2$$

(3, 2)

67. $\begin{cases} P = 201(t - 1990) + 5974 \\ P = 395(t - 1990) + 5190 \end{cases}$

Using substitution,

$$\begin{aligned} 201(t - 1990) + 5974 &= 395(t - 1990) + 5190 \\ 201t - 399990 + 5974 &= 395t - 786050 + 5190 \\ 201t - 394016 &= 395t - 780860 \\ 201t + 386844 &= 395t \\ 386844 &= 194t \\ 1994.04 &= t \\ t &\approx 1994 \end{aligned}$$

The number of dermatologists was equal to the number of gastroenterologists is 1994.

71. $$\begin{aligned} 4(x - 3) - 6(x - 8) &= 10 - (x - 18) \\ 4x - 12 - 6x + 48 &= 10 - x + 18 \\ -2x + 36 &= 28 - x \\ 36 &= 28 + x \\ 8 &= x \end{aligned}$$

73. $$\begin{aligned} f(x) &= 3x^2 - 2x + 1 \\ f(-4) &= 3(-4)^2 - 2(-4) + 1 \\ &= 3(16) - 2(-4) + 1 \\ &= 57 \\ f(a) &= 3a^2 - 2a + 1 \end{aligned}$$

Exercises 4.4

1. $y \le x + 3$
   $y = x + 3$

x	y
0	3
-3	0

Test point: (0, 0)

$0 \le 0 + 3$
$0 \le 3$
True

Shade the half-plane containing (0, 0).

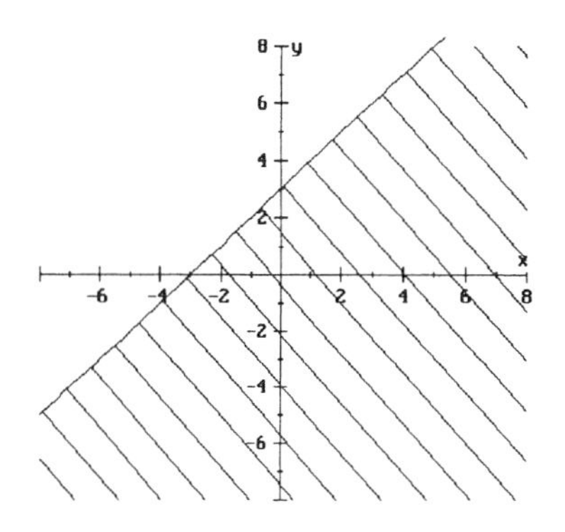

3. $x > y - 2$ (dashed line)

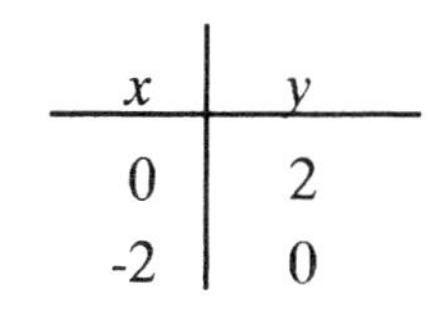

x	y
0	2
-2	0

Test point: (0, 0)

$0 > 0 - 2$
$0 > -2$
True

Shade the half-plane containing (0, 0).

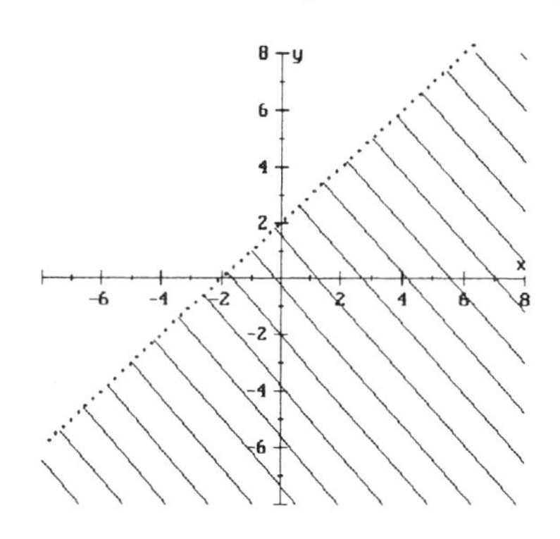

5. $x + y < 3$ (dashed line)
   $x + y = 3$

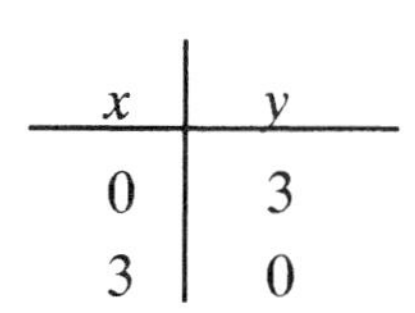

x	y
0	3
3	0

Test point: (0, 0)

$0 + 0 < 3$
$0 < 3$
True

Shade the half-plane containing (0, 0).

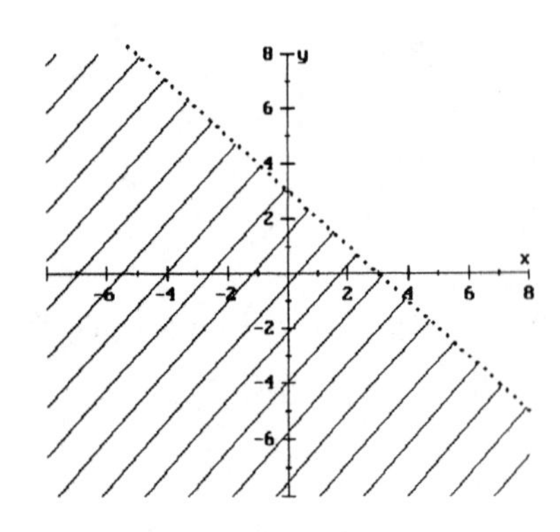

7. $x + y \geq 3$ (solid line)
$x + y = 3$

x	y
0	3
3	0

Test point: (0, 0)

$0 + 0 \geq 3$
$0 \geq 3$
False

Shade the half-plane not containing (0, 0).

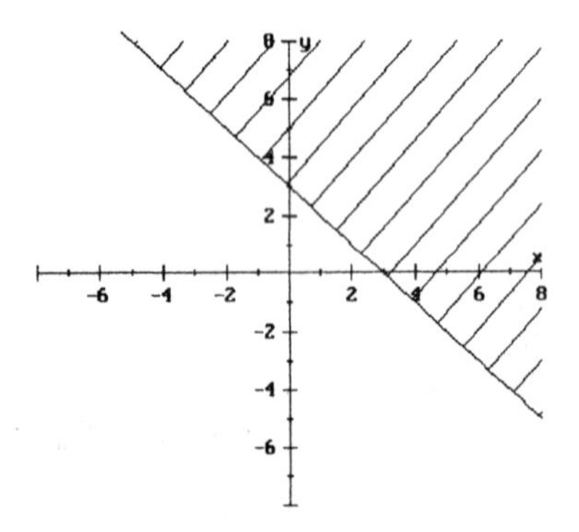

9. $2x + y \leq 6$ (solid line)
$2x + y = 6$

x	y
0	6
3	0

Test point: (0, 0)

$2(0) + 0 \leq 6$
$0 \leq 6$
True

Shade the half-plane containing (0, 0).

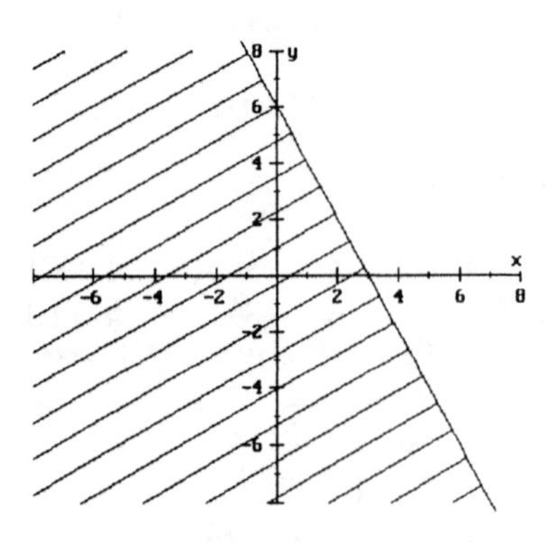

11. $x + 2y > 6$ (dashed line)
$x + 2y = 6$

x	y
0	3
6	0

Test point: (0, 0)

$0 + 2(0) > 6$
$0 > 6$
False

Shade the half-plane not containing (0, 0).

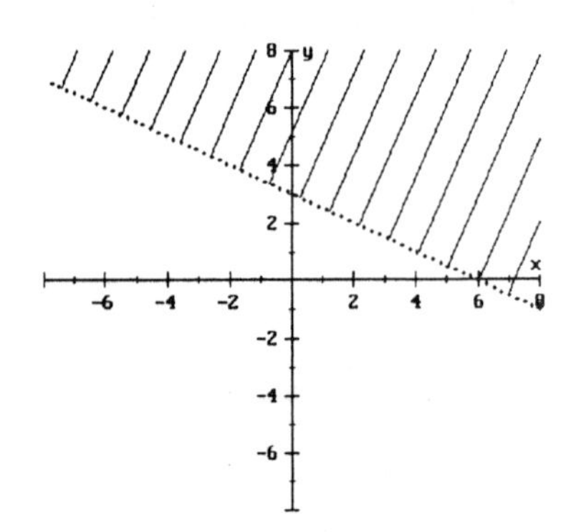

13. $3x + 2y \geq 12$ (solid line)
$3x + 2y = 12$

x	y
0	6
4	0

Test point: (0, 0)

$3(0) + 2(0) \geq 12$
$0 \geq 12$
False

Shade the half-plane not containing (0, 0).

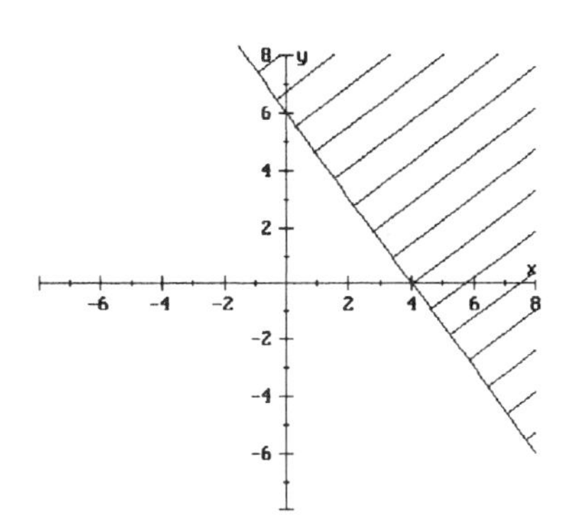

15. $2x + 5y < 10$ (dashed line)
    $2x + 5y = 10$

x	y
0	2
5	0

Test point: (0, 0)

$2(0) + 5(0) < 10$
$0 < 10$
True

Shade the half-plane containing (0, 0).

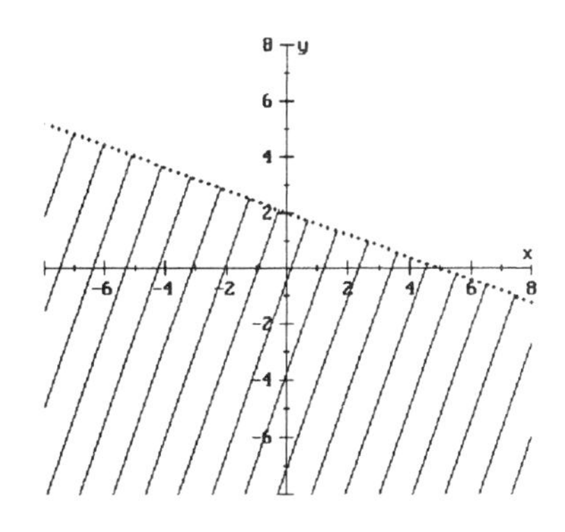

17. $2x + 5y > 10$ (dashed line)
    $2x + 5y = 10$

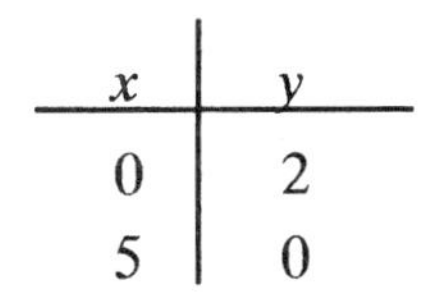

x	y
0	2
5	0

Test point: (0, 0)

$2(0) + 5(0) > 10$
$0 > 10$
False

Shade the half-plane <u>not</u> containing (0, 0).

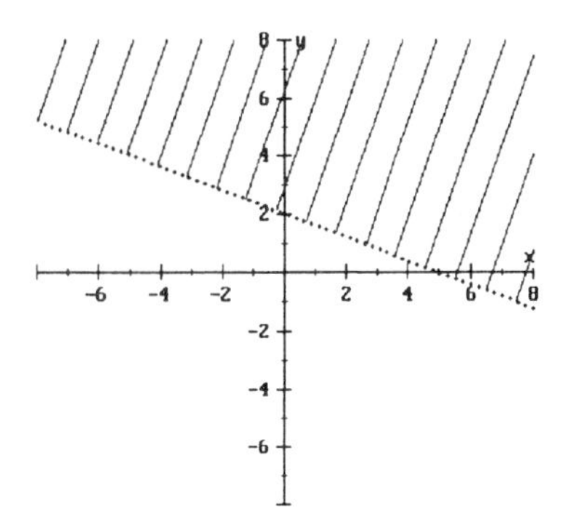

19. $2x + 5y \geq 10$ (solid line)
    $2x + 5y = 10$

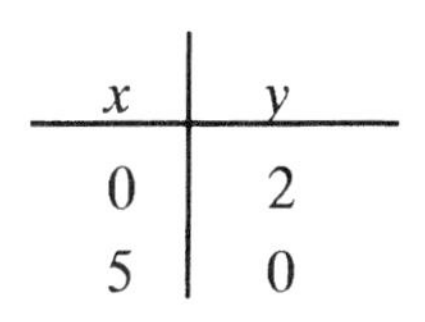

x	y
0	2
5	0

Test point: (0, 0)

$2(0) + 5(0) \geq 10$
$0 \geq 10$
False

Shade the half-plane <u>not</u> containing (0, 0).

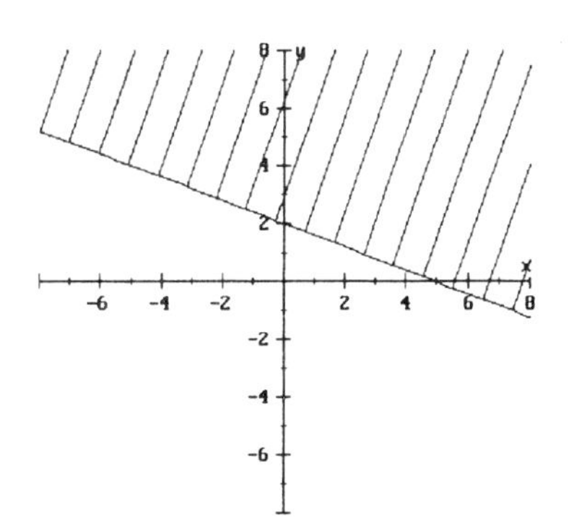

21. $2x - 5y \geq 10$ (solid line)
$2x - 5y = 10$

x	y
0	-2
5	0

Test point: (0, 0)

$2(0) - 5(0) \geq 10$
$0 \geq 10$
False

Shade the half-plane not containing (0, 0).

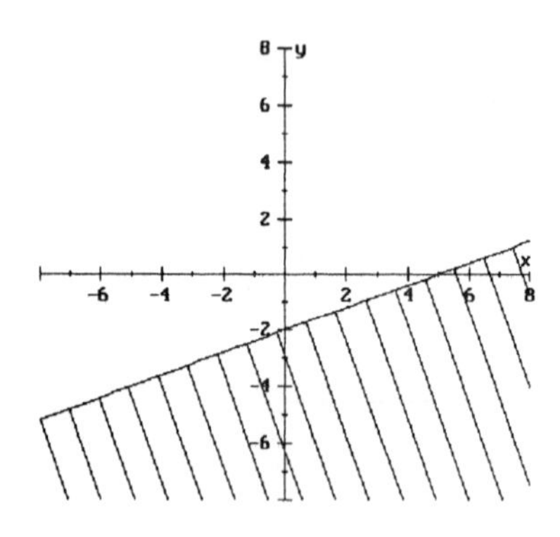

23. $y \leq x$ (solid line)
$y = x$

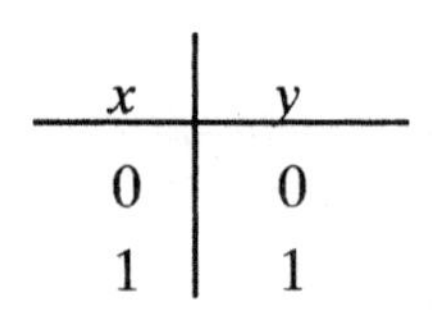

x	y
0	0
1	1

Test point: (0, 1)

$1 \leq 0$
False

Shade the half-plane not containing (0, 1).

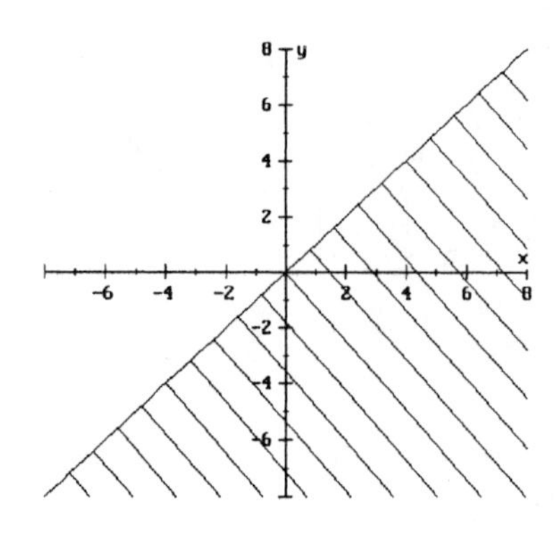

25. $2x - y < 4$ (dashed line)
$2x - y = 4$

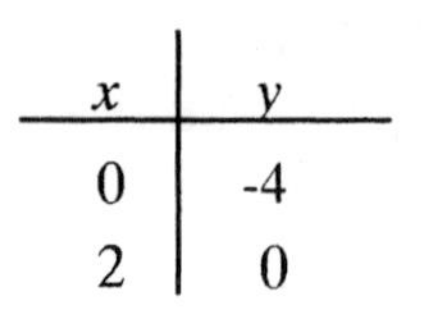

x	y
0	-4
2	0

Test point: (0, 0)

$2(0) - 0 < 4$
$0 < 4$
True

Shade the half-plane containing the point (0, 0).

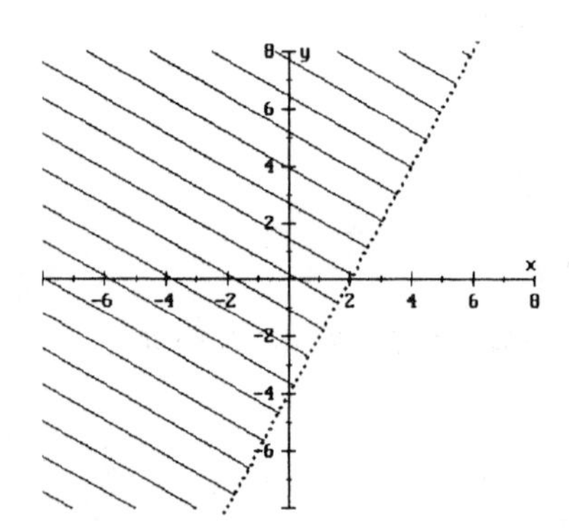

27. $4x - y \geq 8$ (solid line)
$4x - y = 8$

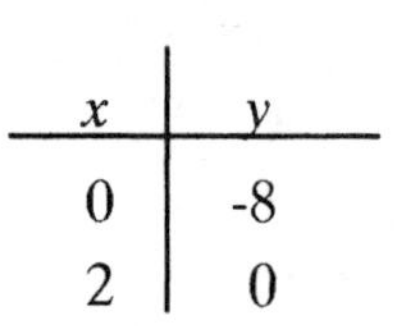

x	y
0	-8
2	0

Test point: (0, 0)

$4(0) - 0 \geq 8$
$0 \geq 8$
False

Shade the half-plane not containing the point (0, 0).

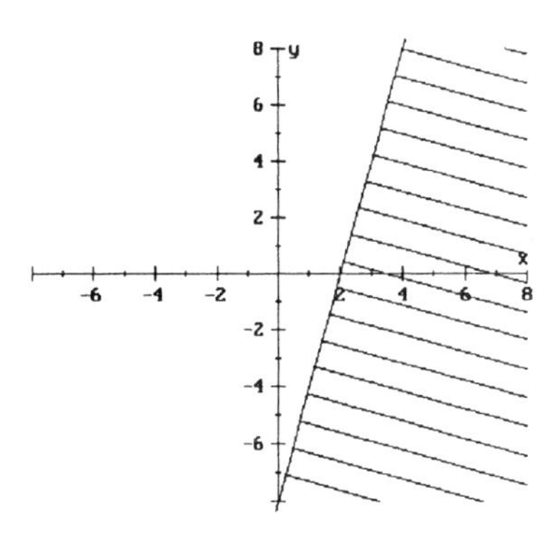

29. $3x - 4y > 12$ (dashed line)

$3x - 4y = 12$

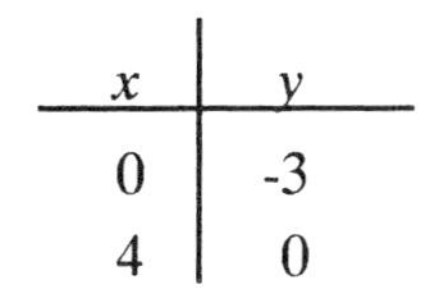

x	y
0	-3
4	0

Test point: (0, 0)

$3(0) - 4(0) > 12$
$0 > 12$
False

Shade the half-plane not containing the point (0, 0).

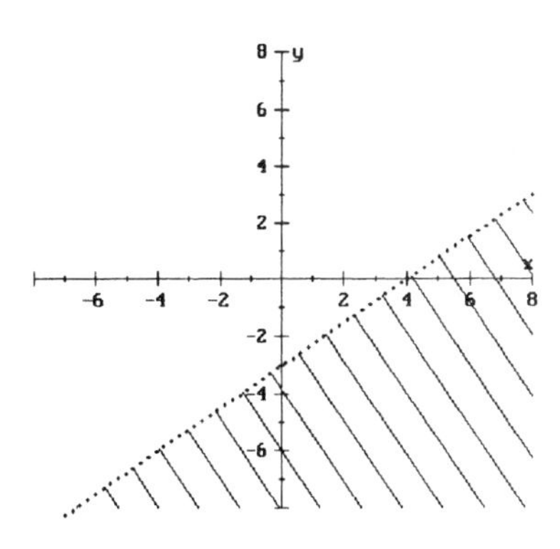

31. $7x - 3y < 15$ (dashed line)

$7x - 3y = 15$

x	y
0	-5
15/7	0

Test point: (0, 0)

$7(0) - 3(0) < 15$
$0 < 15$
True

Shade the half-plane containing (0, 0).

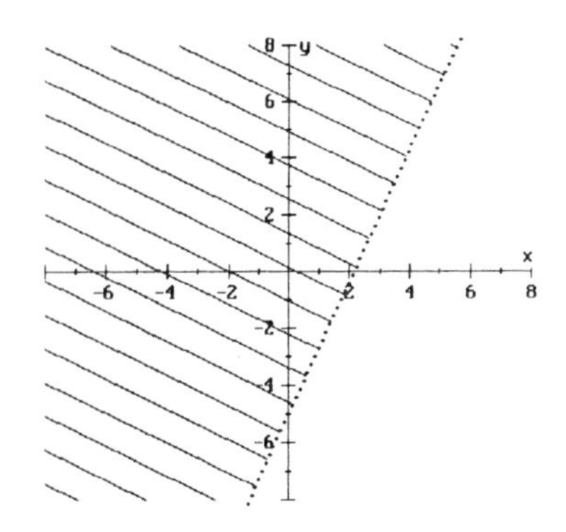

33. $7x + 3y > 15$ (dashed line)

$7x + 3y = 15$

x	y
0	5
15/7	0

Test point: (0, 0)

$7(0) + 3(0) > 15$
$0 > 15$
False

Shade the half-plane not containing the point (0, 0).

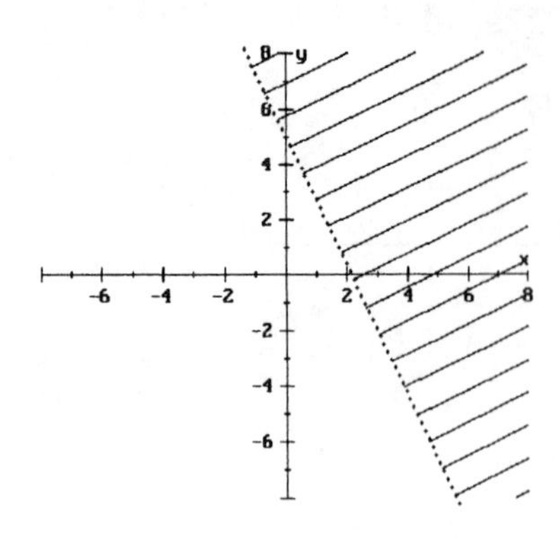

35. $\dfrac{x}{2} + \dfrac{y}{3} < 4$ (dashed line)

$$\frac{x}{2} + \frac{y}{3} = 4$$

$$6\left(\frac{x}{2} + \frac{y}{3}\right) = 6(4)$$

$$3x + 2y = 24$$

x	y
0	12
8	0

Test point: (0, 0)

$$\frac{0}{2} + \frac{0}{3} < 4$$

$$0 < 4$$

True

Shade the half-plane containing (0, 0).

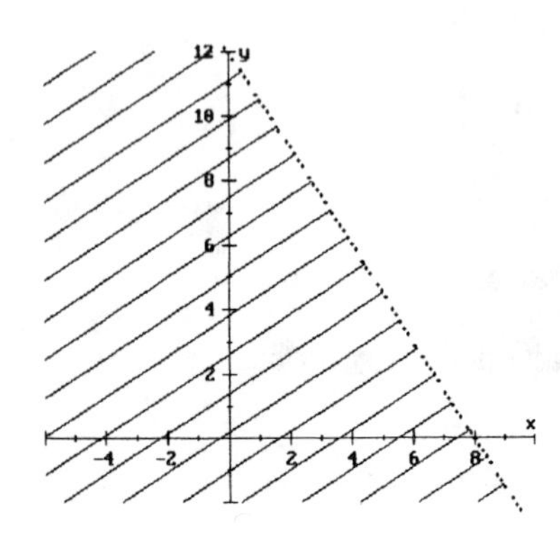

37. $\dfrac{x}{3} - \dfrac{y}{4} \geq 3$ (solid line)

$$\frac{x}{3} - \frac{y}{4} = 3$$

$$12\left(\frac{x}{3} - \frac{y}{4}\right) = 12(3)$$

$$4x - 3y = 36$$

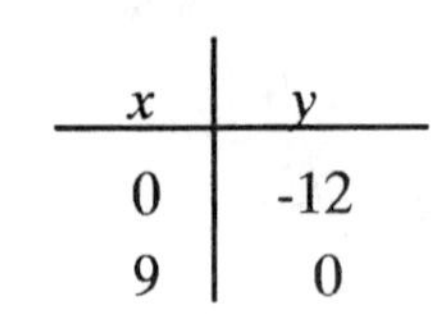

x	y
0	-12
9	0

Test point: (0, 0)

$$\frac{0}{3} - \frac{0}{4} \geq 3$$

$$0 \geq 3$$

False

Shade the half-plane <u>not</u> containing (0, 0).

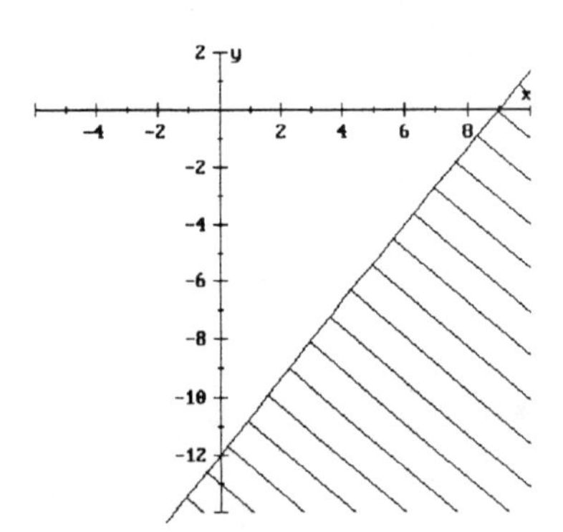

39. $y < 3$ (dashed line)

$y = 3$

Horizontal line passing through (0, 3).

Test point: (0, 0)

$0 < 3$

True

Shade the half-plane containing (0, 0).

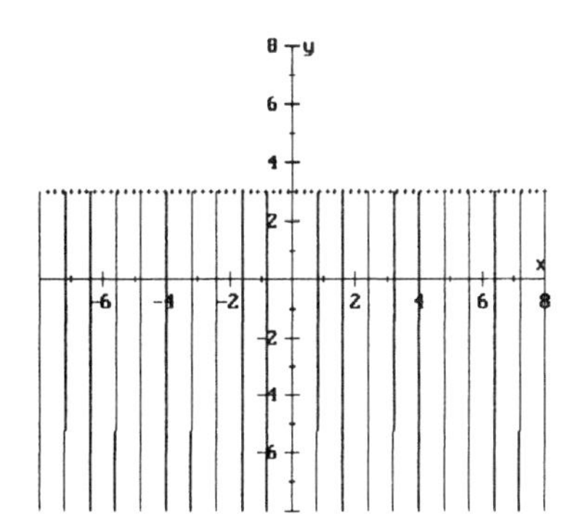

41. $x \geq -2$ (solid line)

$x = -2$

Vertical line passing through the point $(-2, 0)$.

Test point: $(0, 0)$

$0 \geq -2$
True

Shade the half-plane containing the point $(0, 0)$.

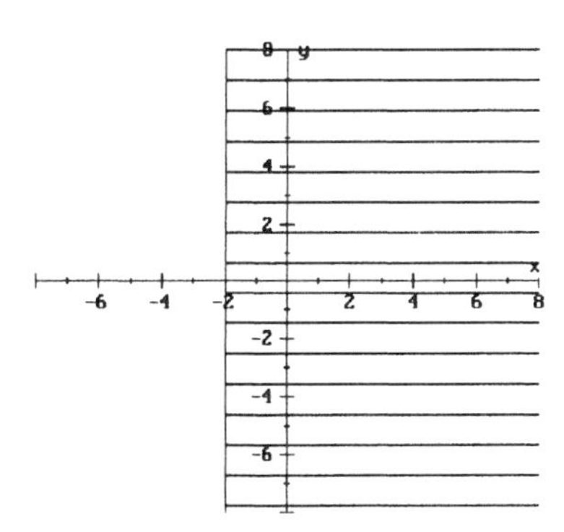

43. $x < -2$ (dashed line)

$x = -2$

Vertical line passing through the point $(-2, 0)$.

Test point: $(0, 0)$

$0 < -2$
False

Shade the half-plane <u>not</u> containing the point $(0, 0)$.

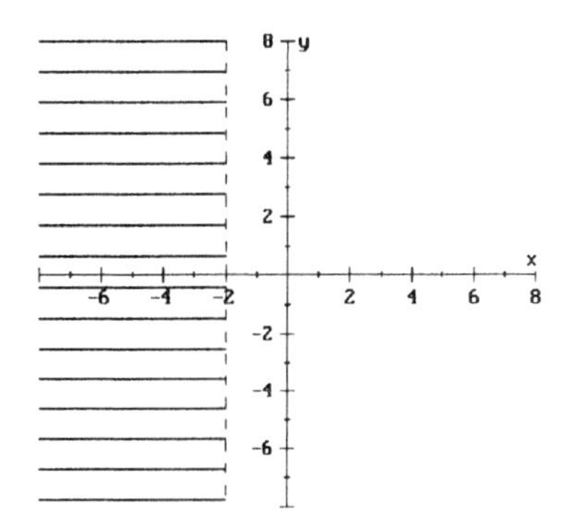

45. $y < 0$ (dashed line)

$y = 0$

Horizontal line passing through the point $(0, 0)$.

Test point: $(1, 1)$

$1 < 0$
False

Shade the half-plane <u>not</u> containing the point $(1, 1)$.

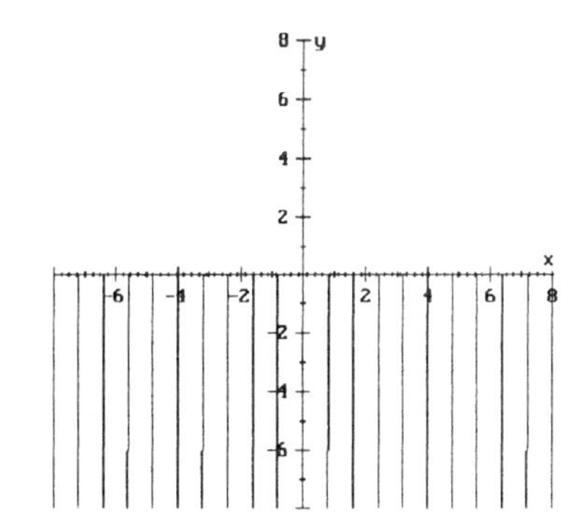

47. $x \leq 0$ (solid line)

$x = 0$

Vertical line containing the point $(0, 0)$.

Test point: $(1, 1)$

$1 < 0$
False

Shade the half-plane <u>not</u> containing the point $(1, 1)$.

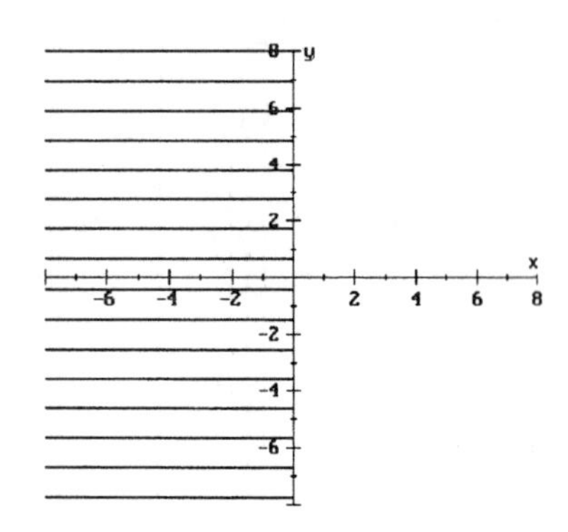

49. $x < \frac{1}{2}$ (dashed line)

$x = \frac{1}{2}$

Vertical line containing the point $\left(\frac{1}{2}, 0\right)$.

Test point: (0, 0)

$0 < \frac{1}{2}$
True

Shade the half-plane containing the point (0, 0).

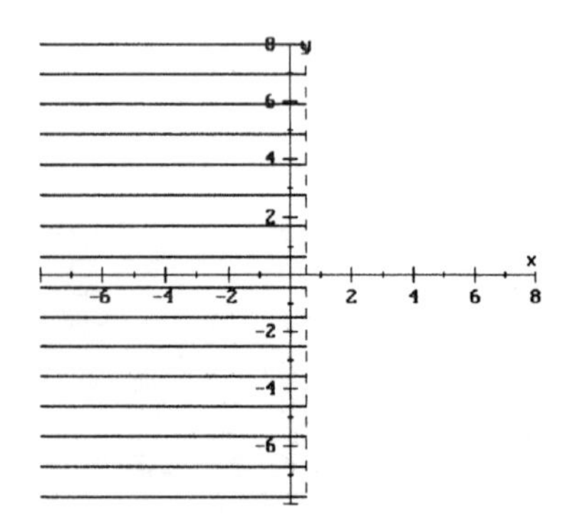

51. $\frac{y}{2} > 0$ (dashed line)

$\frac{y}{2} = 0$

$2\left(\frac{y}{2}\right) = 2(0)$

$y = 0$

Horizontal line containing the point (0, 0).

Test point: (1, 1)

$\frac{1}{2} > 0$
True

Shade the half-plane containing the point (1, 1).

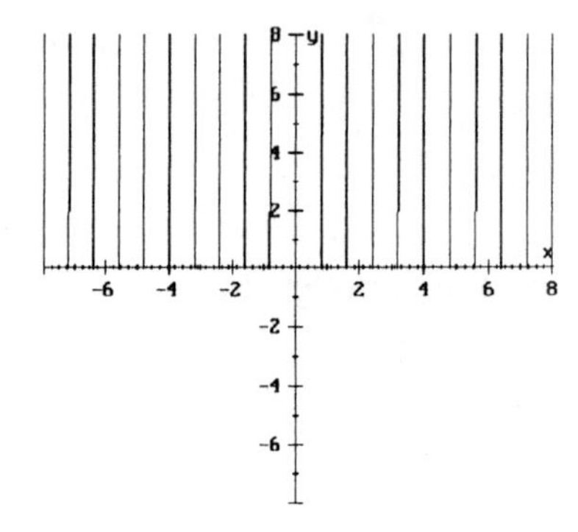

53. $P \leq 80$ (solid line)
$2x + 2y \leq 80$

$2x + 2y = 80$

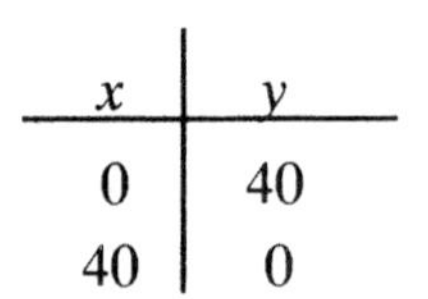

x	y
0	40
40	0

Test point: (0, 0)

$0 + 0 \leq 40$
$0 \leq 40$
True

Shade the half-plane containing the point (0, 0).

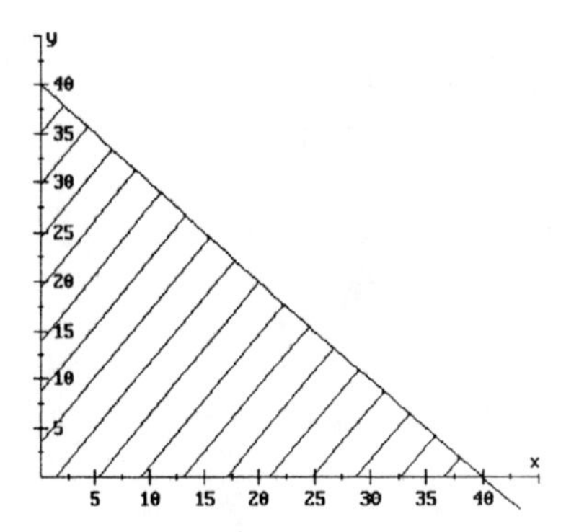

55. $10s + 18p \le 120$ (solid line)

$10s + 18p = 120$

s	p
0	20/3
12	0

Test point: (0, 0)

$$10(0) + 18(0) \le 120$$
$$0 \le 120$$
True

Shade the half-plane containing the point (0, 0).

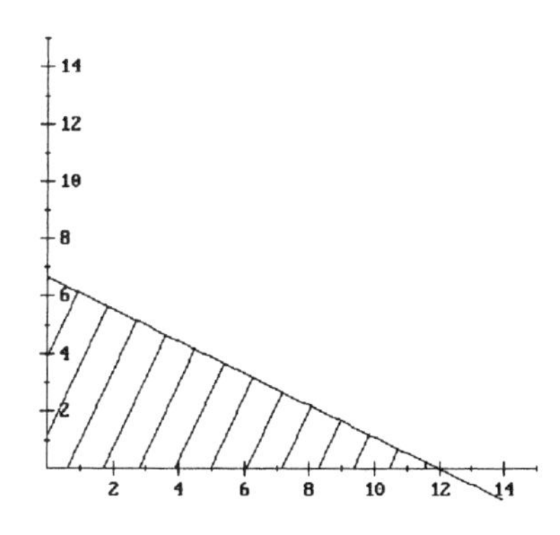

Three possible feasible points: (1, 1), (2, 2), (1, 2)

57. $12r + 8a \le 2500$ (solid line)

$12r + 8a = 2500$

r	a
0	312.5
208.3	0

Test point: (0, 0)

$$12(0) + 8(0) \le 2500$$
$$0 \le 2500$$
True

Shade the half-plane containing the point (0, 0).

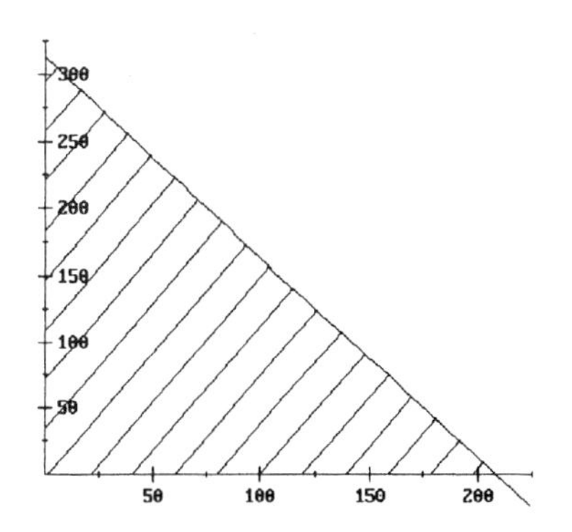

Three possible feasible points: (10, 10), (20, 10), (10, 20)

59. number of regular clerks: r
number of special clerks: s

$15r + 10s \ge 90$ (solid line)

$15r + 10s = 90$

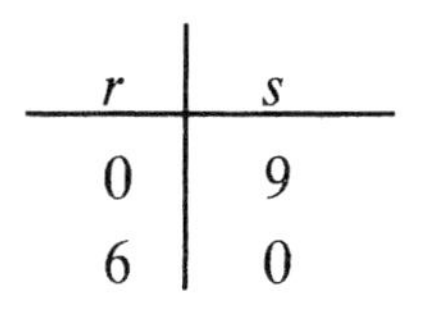

r	s
0	9
6	0

Test point: (0, 0)

$$15(0) + 10(0) \ge 90$$
$$0 \ge 90$$
False

Shade the half-plane not containing the point (0, 0).

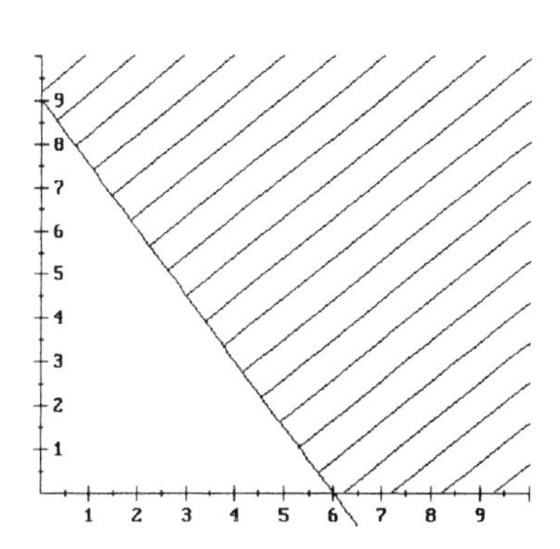

Three possible feasible points: (10, 1), (1, 10), (20, 20)

61. $|-6 - 3| - |6| - |3|$

$= |-9| - |6| - |3|$

$= 9 - 6 - 3$
$= 0$

63. Distributive property

65. Bobby's hours: x
Linda's hours: $x + 1$

$6x = 5(x + 1)$
$6x = 5x + 5$
$x = 5$

It will take 5 hours.

CHAPTER 4 REVIEW EXERCISES

1. $m = \dfrac{y_2 - y_1}{x_2 - x_1}$

$= \dfrac{0 - (-2)}{-1 - 3}$

$= \dfrac{2}{-4}$

$= -\dfrac{1}{2}$

3. $y = 3x - 5$
$m = 3$

5. $4y - 3x = 1$
$4y = 3x + 1$

$y = \dfrac{3}{4}x + \dfrac{1}{4}$

$m = \dfrac{3}{4}$

7. $m = \dfrac{y_2 - y_1}{x_2 - x_1}$

$= \dfrac{5 - 4}{3 - 1}$

$= \dfrac{1}{2}$

parallel lines have equal slopes

9. $m = \dfrac{y_2 - y_1}{x_2 - x_1}$

$= \dfrac{9 - 7}{4 - 4}$

$= \dfrac{2}{0}$
undefined

This line is vertical, hence a line perpendicular must be horizontaland its slope is 0.

11. $m = \dfrac{y_2 - y_1}{x_2 - x_1}$

$= \dfrac{6 - 6}{-7 - 2}$

$= \dfrac{0}{-9}$

$= 0$

13. $y = 3x - 7$
$m = 3$

parallel lines have equal slopes.

15. $3y - 5x + 6 = 0$
$3y = 5x - 6$

$y = \dfrac{5}{3}x - 2$

$m = \dfrac{5}{3}$
Perpendicular lines have slopes that are negative reciprocals.
$m_{\perp} = -\dfrac{3}{5}$

17. $x = 3$ is a vertical line, hence a line parallel will also be vertical and its slope is undefined.

19. $m = \frac{y_2 - y_1}{x_2 - x_1}$

$4 = \frac{a - 2}{4 - 1}$

$4 = \frac{a - 2}{3}$

$12 = a - 2$
$14 = a$

21. $(-3, a)$ and $(0, 3)$

$m_1 = \frac{a - 3}{-3 - 0}$

$m_1 = \frac{a - 3}{-3}$

$(7, a)$ and $(0, 0)$

$m_2 = \frac{a - 0}{7 - 0}$

$m_2 = \frac{a}{7}$

parallel lines have equal slopes.

$m_1 = m_2$

$21\left(\frac{a - 3}{-3}\right) = 21\left(\frac{a}{7}\right)$

$-7(a - 3) = 3a$
$-7a + 21 = 3a$
$21 = 10a$

$\frac{21}{10} = a$

23. $m = \frac{-4 - 3}{1 - (-2)}$

$= -\frac{7}{3}$

$y - y_1 = m(x - x_1)$

$y - 3 = -\frac{7}{3}[x - (-2)]$

$y - 3 = -\frac{7}{3}(x + 2)$

$y - 3 = -\frac{7}{3}x - \frac{14}{3}$

$y = -\frac{7}{3}x - \frac{5}{3}$

25. $m = \frac{5 - (-5)}{3 - (-3)}$

$= \frac{10}{6}$

$= \frac{5}{3}$

$y - y_1 = m(x - x_1)$

$y - 5 = \frac{5}{3}(x - 3)$

$y - 5 = \frac{5}{3}x - 5$

$y = \frac{5}{3}x$

27. $y - y_1 = m(x - x_1)$

$y - 5 = \frac{2}{5}(x - 2)$

$y - 5 = \frac{2}{5}x - \frac{4}{5}$

$y = \frac{2}{5}x + \frac{21}{5}$

29. $y - y_1 = m(x - x_1)$
$y - 7 = 5(x - 4)$
$y - 7 = 5x - 20$
$y = 5x - 13$

31. $y = mx + b$
$y = 5x + 3$

33. $y = 3$

35. $y = \frac{3}{2}x - 1$

$m = \frac{3}{2}$

parallel lines have equal slopes

$m = \frac{3}{2}$; (0, 0)

$y = mx + b$

$y = \frac{3}{2}x + 0$

$y = \frac{3}{2}x$

37. $2y - 5x = 1$

$2y = 5x + 1$

$y = \frac{5}{2}x + \frac{1}{2}$

$m = \frac{5}{2}$

$m_{\perp} = -\frac{2}{5}$

$m_{\perp} = -\frac{2}{5}$; (0, 6)

$y = mx + b$

$y = -\frac{2}{5}x + 6$

39. $3x = -5y$

$-\frac{3}{5}x = y$

$m = -\frac{3}{5}$

$m_{\perp} = \frac{5}{3}$

$m_{\perp} = \frac{5}{3}$; (0, 0)

$y = mx + b$

$y = \frac{5}{3}x + 0$

$y = \frac{5}{3}x$

41. (3, 0) and (0, -5)

$m = \frac{-5 - 0}{0 - 3} = \frac{5}{3}$

$y = mx + b$

$y = \frac{5}{3}x - 5$

43. $3x - 2y = 5$

$-2y = -3x + 5$

$y = \frac{3}{2}x - \frac{5}{2}$

$m = \frac{3}{2}$

$5y = x + 3$

$y = \frac{1}{5}x + \frac{3}{5}$

$b = \frac{3}{5}$

$m = \frac{3}{2}$; $\left(0, \frac{3}{5}\right)$

$y = mx + b$

$y = \frac{3}{2}x + \frac{3}{5}$

45. (x, P); (250, 12000) and (300, 20000)

$m = \frac{20000 - 12000}{300 - 250}$

$= \frac{8000}{50}$

$= 160$

$$P - 12000 = 160(x - 250)$$
$$P - 12000 = 160x - 40000$$
$$P = 160x - 28000$$

If $x = 400$:
$$P = 160(400) - 28000 = 36000$$

He would make $36,000.

47. (a) The line appears to pass through the points (1985, 550) and (1988, 600).

$$m = \frac{600 - 550}{1988 - 1985} = \frac{50}{3}$$

$$P - 550 = \frac{50}{3}(t - 1985)$$

$$P - 550 = \frac{50}{3}t - \frac{99250}{3}$$

$$P = \frac{50}{3}t - \frac{97600}{3}$$

(b) $t = 1999$
$$P = \frac{50}{3}(1999) - \frac{97600}{3} = 783.3$$

In 1999 there will be approximately 783.3 thousand physicians.

49. $\begin{cases} x - y = 4 \\ 2x - 3y = 7 \end{cases}$

$\begin{cases} -2(x - y) = -2(4) \\ 2x - 3y = 7 \end{cases}$

$$\begin{aligned} -2x + 2y &= -8 \\ 2x - 3y &= 7 \\ \hline -y &= -1 \\ y &= 1 \end{aligned}$$

$$x - 1 = 4$$
$$x = 5$$

$x = 5, \ y = 1$

51. $\begin{cases} \dfrac{x}{6} - \dfrac{y}{4} = \dfrac{4}{3} \\ \dfrac{x}{5} - \dfrac{y}{2} = \dfrac{8}{5} \end{cases}$

$\begin{cases} 12\left(\dfrac{x}{6} - \dfrac{y}{4}\right) = 12\left(\dfrac{4}{3}\right) \\ -10\left(\dfrac{x}{5} - \dfrac{y}{2}\right) = -10\left(\dfrac{8}{5}\right) \end{cases}$

$$\begin{aligned} 2x - 3y &= 16 \\ -2x + 5y &= -16 \\ \hline 2y &= 0 \\ y &= 0 \end{aligned}$$

$$2x - 3(0) = 16$$
$$2x = 16$$
$$x = 8$$

$x = 8, \ y = 0$

53. $\begin{cases} 3x - \dfrac{y}{4} = 2 \\ 6x - \dfrac{y}{2} = 4 \end{cases}$

$\begin{cases} 4\left(3x - \dfrac{y}{4}\right) = 4(2) \\ 2\left(6x - \dfrac{y}{2}\right) = 2(4) \end{cases}$

$\begin{cases} 12x - y = 8 \\ 12x - y = 8 \end{cases}$

Dependent

55. $\begin{cases} x = 2y - 3 \\ y = 3x + 2 \end{cases}$

Substituting:
$$y = 3(2y - 3) + 2$$
$$y = 6y - 9 + 2$$
$$y = 6y - 7$$
$$-5y = -7$$

$$y = \frac{7}{5}$$

$$x = 2\left(\frac{7}{5}\right) - 3 = -\frac{1}{5}$$

$x = -\dfrac{1}{5}, \ y = \dfrac{7}{5}$

57. amount in CD: x
amount in bond: y

$$\begin{cases} x + y = 8500 \\ 0.0475x + 0.0665y = 512.05 \end{cases}$$

1st equation: $x = 8500 - y$
Substituting:

$$\begin{aligned} 0.0475(8500 - y) + 0.0665y &= 512.05 \\ 403.75 - 0.0475y + 0.0665y &= 512.05 \\ 403.75 + 0.019y &= 512.05 \\ 0.019y &= 108.3 \\ y &= 5700 \end{aligned}$$

$$\begin{aligned} x + 5700 &= 8500 \\ x &= 2800 \end{aligned}$$

\$2800 is invested in the CD and \$5700 is invested in the bond.

59. $y - 2x < 4$ (dashed line)

$y - 2x = 4$

x	y
0	4
-2	0

Test point: (0, 0)

$0 - 2(0) < 4$
$0 < 4$
True

Shade half-plane containing (0, 0).

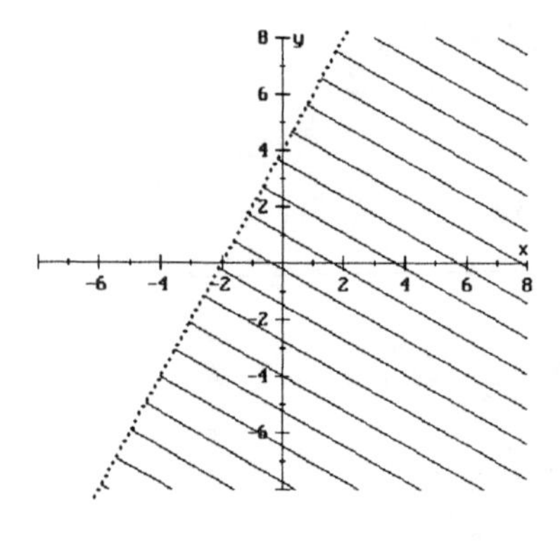

61. $2y - 3x > 6$ (dashed line)

$2y - 3x = 6$

x	y
0	3
-2	0

Test point: (0, 0)

$2(0) - 3(0) > 6$
$0 > 6$
False

Shade half-plane <u>not</u> containing (0, 0).

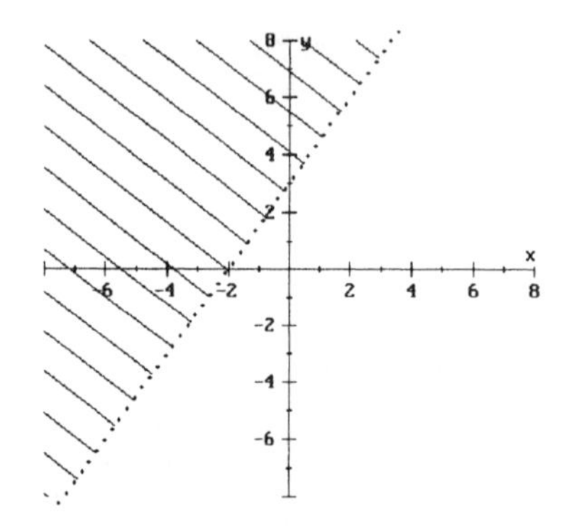

63. $5y - 8x \leq 20$ (solid line)

$5y - 8x = 20$

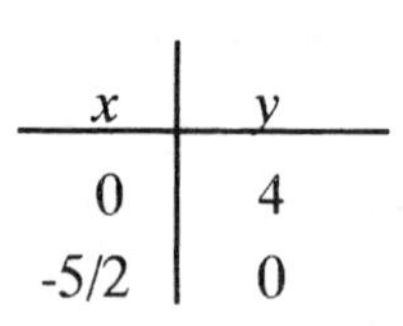

x	y
0	4
-5/2	0

Test point: (0, 0)

$5(0) - 8(0) \leq 20$
$0 \leq 20$
True

Shade half-plane containing (0, 0).

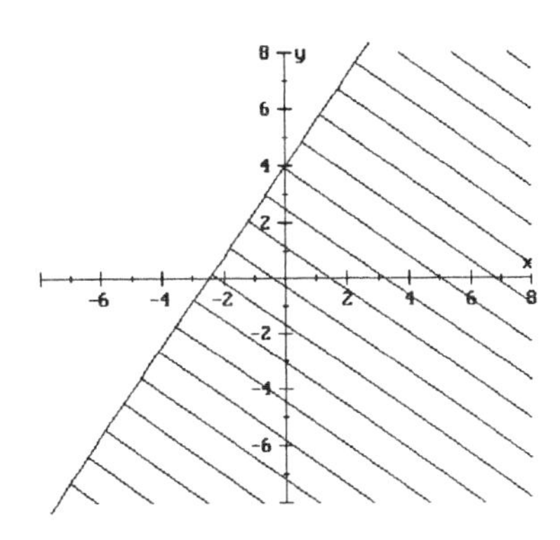

65. $\dfrac{x}{2} + \dfrac{y}{3} \geq 6$ (solid line)

$6\left(\dfrac{x}{2} + \dfrac{y}{3}\right) \geq 6(6)$

$3x + 2y \geq 36$

$3x + 2y = 36$

x	y
0	18
12	0

Test point: (0, 0)

$3(0) + 2(0) \geq 36$

$0 \geq 36$

False

Shade half-plane not containing (0, 0).

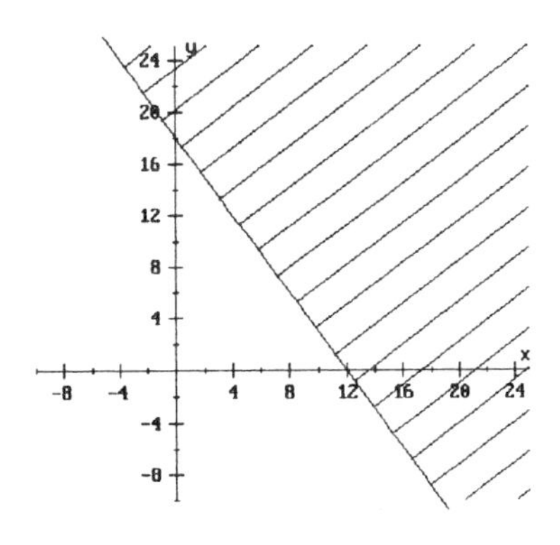

67. $y < 5$ (dashed line)

$y = 5$

Horizontal line passing through (0, 5).

Test point: (0, 0)

$0 < 5$

True

Shade half-plane containing (0, 0).

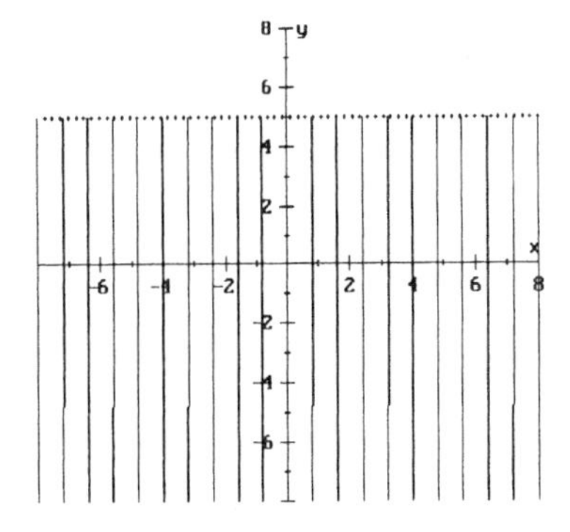

69. $2x + 2y > 100$ (dashed line)

$2x + 2y = 100$

x	y
0	50
50	0

Test point: (0, 0)

$2(0) + 2(0) > 100$

$0 > 100$

False

Shade half-plane not containing (0, 0).

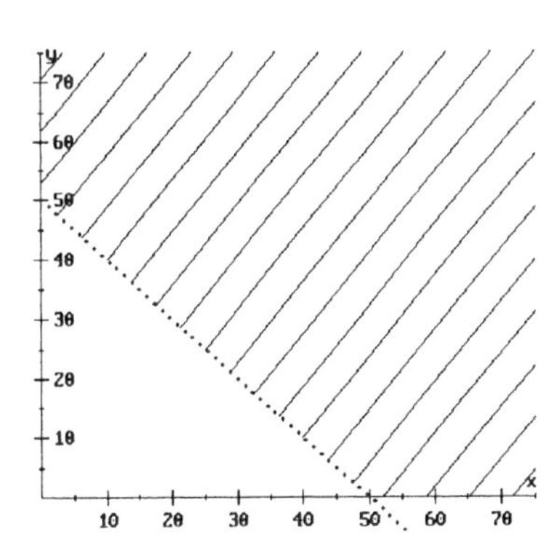

Three feasible points: (60, 60), (60, 70), (70, 60)

CHAPTER 4 PRACTICE TEST

1. (a) $4x + 3y - 24 = 0$
$4x + 3y = 24$

x	y
0	8
6	0

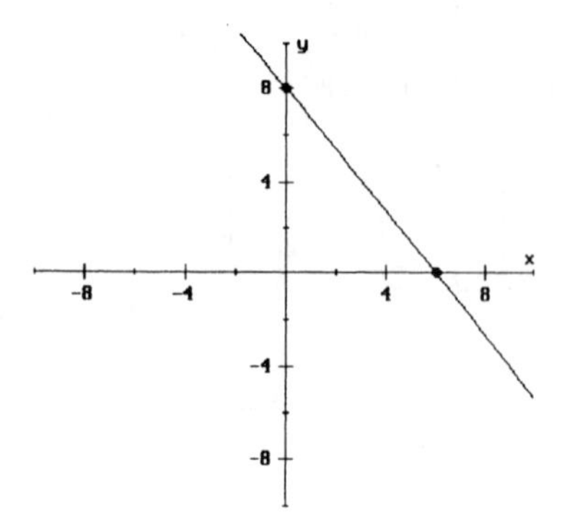

(b) $2x = 8$
$x = 4$

Vertical line passing through (4, 0).

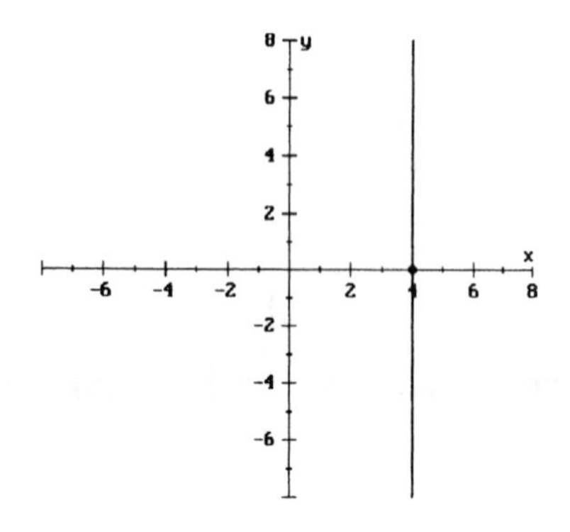

(c) $y = \frac{1}{2}x - 6$

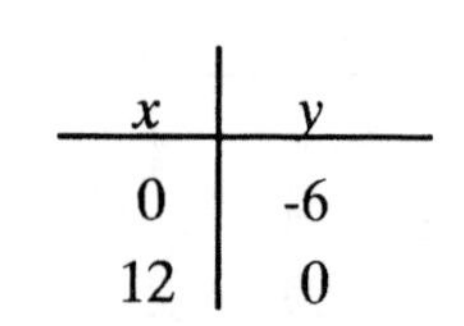

x	y
0	-6
12	0

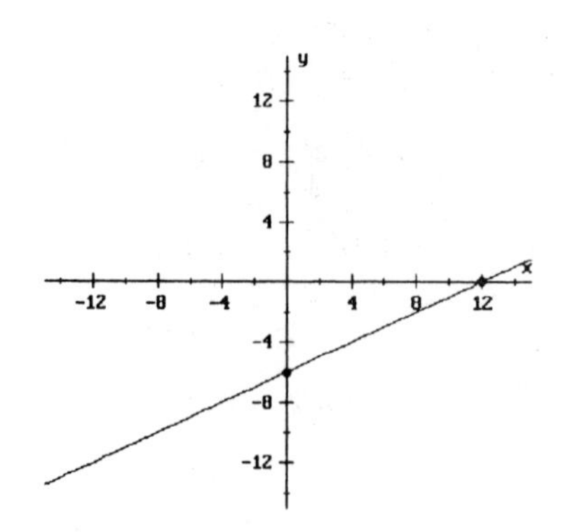

3. $$m = \frac{y_2 - y_1}{x_2 - x_1}$$
$$2 = \frac{5 - 2}{2 - a}$$
$$2 = \frac{3}{2 - a}$$
$$2(2 - a) = \left(\frac{3}{2 - a}\right)(2 - a)$$
$$4 - 2a = 3$$
$$-2a = -1$$
$$a = \frac{1}{2}$$

5. (A, B): (60, 90); (80, 150)

$$m = \frac{150 - 90}{80 - 60} = \frac{60}{20} = 3$$

$$B - 90 = 3(A - 60)$$
$$B - 90 = 3A - 180$$
$$B = 3A - 90$$

If $A = 85$:
$$B = 3(85) - 90 = 165$$

They should score 165 on test B.

7. amount at 8½%: x
amount at 9%: y

$$\begin{cases} x + y = 3500 \\ 0.085x + 0.09y = 309 \end{cases}$$

1st equation: $x = 3500 - y$
Substituting into 2nd equation:

$$\begin{aligned} 0.085(3500 - y) + 0.09y &= 309 \\ 297.5 - 0.085y + 0.09y &= 309 \\ 297.5 + 0.005y &= 309 \\ 0.005y &= 11.5 \\ y &= 2300 \end{aligned}$$

$$\begin{aligned} x + 2300 &= 3500 \\ x &= 1200 \end{aligned}$$

He invested \$1200 at 8½% and \$2300 at 9%.

9. $315c + 425s \leq 7200$ (solid line)

$315c + 425s = 7200$

c	s
0	16.9
22.9	0

Test point: (0, 0)

$$\begin{aligned} 315(0) + 425(0) &\leq 7200 \\ 0 &\leq 7200 \end{aligned}$$
True

Shade half-plane containing (0, 0).

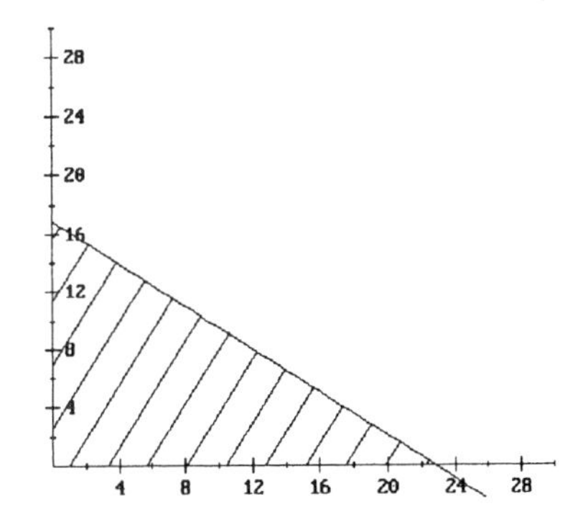

Three feasible points: (5, 5), (2, 5), (5, 2)

CHAPTER 5

Exercises 5.1

1. polynomial

(a) binomial
(b) degree: 2
(c) 1 variable
(d) 5; 4

3. polynomial

(a) monomial
(b) degree: 0
(c) 0 variables
(d) 59

5. not a polynomial

7. polynomial

(a) monomial
(b) degree: $1 + 3 + 7 = 11$
(c) 3 variables
(d) −4

9. not a polynomial

11. polynomial

(a) 4 terms
(b) degree: 3
(c) 1 variable
(d) 4; −1; −2; 1

13. $P = P(t) = 0.035t^2 + 0.38t + 32.6$

1970: $t = 1$
$P(1) = 0.035(1)^2 + 0.38(1) + 32.6$
$= 33.015$

1975: $t = 6$
$P(6) = 0.035(6)^2 + 0.38(6) + 32.6$
$= 36.14$

1980: $t = 11$
$P(11) = 0.035(11)^2 + 0.38(11) + 32.6$
$= 41.015$

1985: $t = 16$
$P(16) = 0.035(16)^2 + 0.38(16) + 32.6$
$= 47.64$

1990: $t = 21$
$P(21) = 0.035(21)^2 + 0.38(21) + 32.6$
$= 56.015$

1995: $t = 26$
$P(26) = 0.035(26)^2 + 0.38(26) + 32.6$
$= 66.14$

The proposed equation agrees fairly well with the observed data.

15. (a) Revenue = cost per pound · number of pounds

$R = d(4300 - 10d^2)$
$R = 4300d - 10d^3$

(b)

d	$R = 4300d - 10d^3$
8	$4300(8) - 10(8)^3 = 29{,}280$
10	$4300(10) - 10(10)^3 = 33{,}000$
12	$4300(12) - 10(12)^3 = 34{,}320$
14	$4300(14) - 10(14)^3 = 32{,}760$

17. (a) Revenue = number of items · price per item

$R = n(2 + 0.45n - 0.001n^2)$
$R = 2n + 0.45n^2 - 0.001n^3$

(b)

n	$2 + 0.001n^2 = \text{price}$	$n \cdot \text{price} = R$
100	$2 + 0.45(100) - 0.001(100)^2 = 37$	$100 \cdot 37 = 3700$
200	$2 + 0.45(200) - 0.001(200)^2 = 52$	$200 \cdot 52 = 10{,}400$
300	$2 + 0.45(300) - 0.001(300)^2 = 47$	$300 \cdot 47 = 14{,}100$

19. width: w
length: $3w$
height: $w - 6$

$S = 2 \cdot \text{width} \cdot \text{length} + 2 \cdot \text{length} \cdot \text{height}$
$\quad + 2 \cdot \text{width} \cdot \text{height}$
$S = 2w(3w) + 2(3w)(w - 6) + 2w(w - 6)$
$S = 6w^2 + 6w^2 - 36w + 2w^2 - 12w$
$S = 14w^2 - 48w$

When $w = 20$:

$S = 14(20)^2 - 48(20)$
$= 4640$ sq. in.

When $w = 26$:

$S = 14(26)^2 - 48(26)$
$= 8216$ sq. in.

When $w = 42$:

$S = 14(42)^2 - 48(42)$
$= 22{,}680$ sq. in.

21. $A = 0.28m + 0.39m^2 - 0.02m^3$

(a)

M	A
0	0
5	8.7
10	21.8
15	24.5
20	1.6
25	0
30	0

(b) It is at a maximum in approximately 13 minutes.

23. $-3 - (-4) + (-5) - (-8)$
$= -3 + 4 + (-5) + 8$
$= 1 + (-5) + 8$
$= -4 + 8$
$= 4$

25. 1st side: $2x$
2nd side: x
3rd side: 24

$P = \text{1st side} + \text{2nd side} + \text{3rd side}$
$75 = 2x + x + 24$
$75 = 3x + 24$
$51 = 3x$
$17 = x$
$2x = 2(17) = 34$

The 1st side is 34 in. and the 2nd side is 17 in.

Exercises 5.2

1. $(3x^2 - 2x + 5) + (2x^2 - 7x + 4)$
$= 3x^2 - 2x + 5 + 2x^2 - 7x + 4$
$= 5x^2 - 9x + 9$

3. $(15x^2 - 3xy - 4y^2) - (16x^2 - 3x + 2)$
$= 15x^2 - 3xy - 4y^2 - 16x^2 + 3x - 2$
$= -x^2 - 3xy - 4y^2 + 3x - 2$

5. $(3a^2 - 2ab + 4b^2) + [(3a^2 - b^2) - (3ab)]$
$= (3a^2 - 2ab + 4b^2) + (3a^2 - b^2 - 3ab)$
$= 3a^2 - 2ab + 4b^2 + 3a^2 - b^2 - 3ab$
$= 6a^2 - 5ab + 3b^2$

7. $(3a^2 - 2ab + 4b^2) - [(3a^2 - b^2) - (3ab)]$
$= (3a^2 - 2ab + 4b^2) - (3a^2 - b^2 - 3ab)$
$= 3a^2 - 2ab + 4b^2 - 3a^2 + b^2 + 3ab$
$= ab + 5b^2$

9. $[(3a^2 - b^2) - (3ab)] - (3a^2 - 2ab + 4b^2)$
$= (3a^2 - b^2 - 3ab) - (3a^2 - 2ab + 4b^2)$
$= 3a^2 - b^2 - 3ab - 3a^2 + 2ab - 4b^2$
$= -ab - 5b^2$

11. $(3xy + 2y^2) + (-7y^2 - 3y + 4)$
$= 3xy + 2y^2 - 7y^2 - 3y + 4$
$= 3xy - 5y^2 - 3y + 4$

13. $(-8x^2 - 5xy + 9y^2) - (3x^2 - 2xy + 7y^2)$
$= -8x^2 - 5xy + 9y^2 - 3x^2 + 2xy - 7y^2$
$= -11x^2 - 3xy + 2y^2$

15. $(3a^2 - 2b^2) - [(2a^2 - 3ab + 4b^2) + (6a^2 + 2ab - 2b^2)]$
$= (3a^2 - 2b^2) - (2a^2 - 3ab + 4b^2 + 6a^2 + 2ab - 2b^2)$
$(3a^2 - 2b^2) - (8a^2 - ab + 2b^2)$
$= 3a^2 - 2b^2 - 8a^2 + ab - 2b^2$
$= -5a^2 + ab - 4b^2$

17. $x^2y(3x^2 - 2xy + 2y^2)$
$= x^2y(3x^2) + x^2y(-2xy) + x^2y(2y^2)$
$= 3x^4y - 2x^3y^2 + 2x^2y^3$

19. $(3x^2 - 2xy + 2y^2)(x^2y)$
$= 3x^2(x^2y) - 2xy(x^2y) + 2y^2(x^2y)$
$= 3x^4y - 2x^3y^2 + 2x^2y^3$

21. $(x + 3)(x + 4) = x^2 + 4x + 3x + 12$
$= x^2 + 7x + 12$

23. $(3x - 1)(2x + 1) = 6x^2 + 3x - 2x - 1$
$= 6x^2 + x - 1$

25. $(3x + 1)(2x - 1) = 6x^2 - 3x + 2x - 1$
$= 6x^2 - x - 1$

27. $(3a - b)(a + b)$
$= 3a^2 + 3ab - ab - b^2$
$= 3a^2 + 2ab - b^2$

29. $(2r - s)^2 = (2r - s)(2r - s)$
$= 4r^2 - 2rs - 2rs + s^2$
$= 4r^s - 4rs + s^2$

31. $(2r + s)(2r - s)$
$= 4r^2 - 2rs + 2rs - s^2$
$= 4r^2 - s^2$

33. $(3x - 2y)(3x + 2y)$
$= 9x^2 + 6xy - 6xy - 4y^2$
$= 9x^2 - 4y^2$

35. $(3x - 2y)^2 = (3x - 2y)(3x - 2y)$
$= 9x^2 - 6xy - 6xy + 4y^2$
$= 9x^2 - 12xy + 4y^2$

37. $(x - y)(a - b) = xa - xb - ya + yb$

39. $(2r + 3s)(2a + 3b)$
$= 4ra + 6rb + 6sa + 9sb$

41. $(2y^2 - 3)(y^2 + 1) = 2y^4 + 2y^2 - 3y^2 - 3$
$= 2y^4 - y^2 - 3$

43. $(5x^2 - 3x + 4)(x + 3)$
$= (5x^2 - 3x + 4)(x) + (5x^2 - 3x + 4)(3)$
$= 5x^3 - 3x^2 + 4x + 15x^2 - 9x + 12$
$= 5x^3 + 12x^2 - 5x + 12$

45. $(9a^2 + 3a - 5)(3a + 1)$
$= (9a^2 + 3a - 5)(3a) + (9a^2 + 3a - 5)(1)$
$= 27a^3 + 9a^2 - 15a + 9a^2 + 3a - 5$
$= 27a^3 + 18a^2 - 12a - 5$

47. $(a^2 - ab + b^2)(a + b)$
$= (a^2 - ab + b^2)(a) + (a^2 - ab + b^2)(b)$
$= a^3 - a^2b + ab^2 + a^2b - ab^2 + b^3$
$= a^3 + b^3$

49. $(x - y - z)^2$
$= (x - y - z)(x - y - z)$
$= x(x - y - z) - y(x - y - z) - z(x - y - z)$
$= x^2 - xy - xz - xy + y^2 + yz - xz + yz + z^2$
$= x^2 - 2xy - 2xz + 2yz + y^2 + z^2$

51. $(x^2 - 2x + 4)(x + 2)$
$= (x^2 - 2x + 4)(x) + (x^2 - 2x + 4)(2)$
$= x^3 - 2x^2 + 4x + 2x^2 - 4x + 8$
$= x^3 + 8$

53. $(a + b + c + d)(a + b)$
$= (a + b + c + d)(a) + (a + b + c + d)(b)$
$= a^2 + ab + ac + ad + ab + b^2 + bc + bd$
$= a^2 + 2ab + ac + ad + b^2 + bc + bd$

55. width: w
length: $3w + 5$

Area = width · length
$A = w(3w + 5)$
$A = 3w^2 + 5w$

Perimeter = 2 · width + 2 · length
$P = 2w + 2(3w + 5)$
$P = 2w + 6w + 10$
$P = 8w + 10$

57.

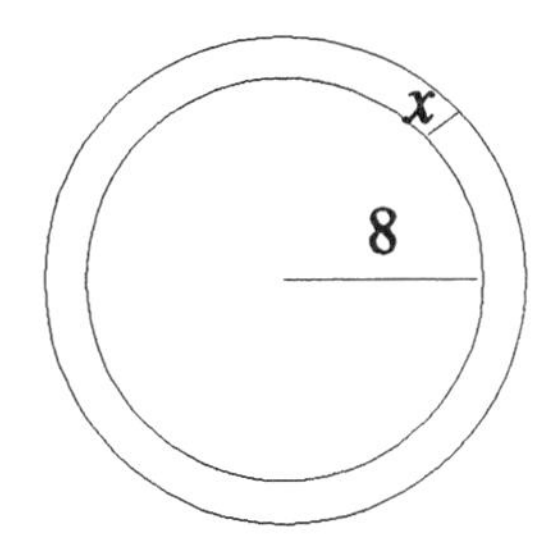

radius of inner circle: 8
area of inner circle: $\pi(8)^2 = 64\pi$
radius of outer circle: $x + 8$
area of outer circle: $\pi(x + 8)^2$

Area of walkway	=	area of outer circle	−	area of inner circle

$A = \pi(x + 8)^2 - 64\pi$
$A = \pi(x^2 + 16x + 64) - 64\pi$
$A = \pi x^2 + 16\pi x + 64\pi - 64\pi$
$A = \pi x^2 + 16\pi x$ sq. ft

59.

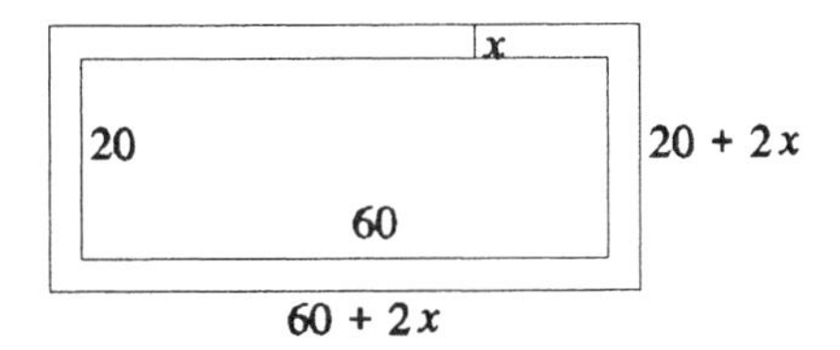

Area of walkway	=	area of outer rectangle	−	area of inner rectangle

$A = (60 + 2x)(20 + 2x) - (60)(20)$
$A = 1200 + 160x + 4x^2 - 1200$
$A = 4x^2 + 160x$ sq. ft

61. $f(x) - g(x) = (2x - 3) - (3x^2 - 5x + 2)$
$= 2x - 3 - 3x^2 + 5x - 2$
$= -3x^2 + 7x - 5$

63. $h(x) - [f(x) - g(x)]$
$= (x^3 - x) - [(2x - 3) - (3x^2 - 5x + 2)]$
$= (x^3 - x) - (2x - 3 - 3x^2 + 5x - 2)$
$= (x^3 - x) - (-3x^2 + 7x - 5)$
$= x^3 - x + 3x^2 - 7x + 5$
$= x^3 + 3x^2 - 8x + 5$

65. $f(x) \cdot g(x)$
$= (2x - 3)(3x^2 - 5x + 2)$
$= 2x(3x^2 - 5x + 2) - 3(3x^2 - 5x + 2)$
$= 6x^3 - 10x^2 + 4x - 9x^2 + 15x - 6$
$= 6x^3 - 19x^2 + 19x - 6$

67. $h(x)[f(x) + g(x)]$
$= (x^3 - x)[(2x - 3) + (3x^2 - 5x + 2)]$
$= (x^3 - x)(2x - 3 + 3x^2 - 5x + 2)$
$= (x^3 - x)(3x^2 - 3x - 1)$
$= x^3(3x^2 - 3x - 1) - x(3x^2 - 3x - 1)$
$= 3x^5 - 3x^4 - x^3 - 3x^3 + 3x^2 + x$
$= 3x^5 - 3x^4 - 4x^3 + 3x^2 + x$

69. $g(x) = 3x^2 - 5x + 2$
$g(2) = 3(2)^2 - 5(2) + 2$
$= 3(4) - 5(2) + 2$
$= 4$

$g(x) + g(2) = 3x^2 - 5x + 2 + 4$
$= 3x^2 - 5x + 6$

71. $g(x + 2) = 3(x + 2)^2 - 5(x + 2) + 2$
$= 3(x^2 + 4x + 4) - 5(x + 2) + 2$
$= 3x^2 + 12x + 12 - 5x - 10 + 2$
$= 3x^2 + 7x + 4$

73. $h(x) - 1 = (x^3 - x) - 1$
$= x^3 - x - 1$

77. $\dfrac{-4 - [4 - 5(2 - 8)]}{4 - 5 \cdot 2 - 8} = \dfrac{-4 - [4 - 5(-6)]}{4 - 10 - 8}$
$= \dfrac{-4 - (4 + 30)}{-14}$
$= \dfrac{-4 - 34}{-14}$
$= \dfrac{-38}{-14}$
$= \dfrac{19}{7}$

79. $|3 - 2x| \le 5$

$-5 \le 3 - 2x \le 5$
$-8 \le -2x \le 2$
$4 \ge x \ge -1$ or $-1 \le x \le 4$

Exercises 5.3

1. $(x + 4)(x + 5) = x^2 + (4 + 5)x + 20$
$= x^2 + 9x + 20$

3. $(x + 3)(x - 7) = x^2 + (3 - 7)x - 21$
$= x^2 - 4x - 21$

5. $(x - 8)(x - 11) = x^2 + (-8 - 11)x + 88$
$= x^2 - 19x + 88$

7. $(x + 5)(x - 6) = x^2 + (-6 + 5)x - 30$
$= x^2 - x - 30$

9. $x + 5(x - 6) = x + 5x - 30$
$= 6x - 30$

11. $(x - 7) - (x - 4) = x - 7 - x + 4$
$= -3$

13. $(2x + 1)(x - 4) = 2x^2 + (-8 + 1)x - 4$
$= 2x^2 - 7x - 4$

15. $(5a - 4)(5a + 4) = (5a)^2 - (4)^2$
$= 25a^2 - 16$

17. $(3z + 5)^2 = (3z)^2 + 2(3z)(5) + 5^2$
$= 9z^2 + 30z + 25$

19. $(3z - 5)^2 = (3z)^2 - 2(3z)(5) + 5^2$
$= 9z^2 - 30z + 25$

21. $(3z - 5)(3z + 5) = (3z)^2 - 5^2$
$= 9z^2 - 25$

23. $(5r^2 + 3s)(3r + 5s) = 15r^3 + 25r^2s + 9rs + 15s^2$

25. $(3s - 2y)^2 = (3s)^2 - 2(3s)(2y) + (2y)^2$
$= 9s^2 - 12sy + 4y^2$

27. $(3y + 10z)^2 = (3y)^2 + 2(3y)(10z) + (10z)^2$
$= 9y^2 + 60yz + 100z^2$

29. $(3y + 10z)(3y - 10z) = (3y)^2 - (10z)^2$
$= 9y^2 - 100z^2$

31. $(3a + 2b)(3a + 4b)$
$= 9a^2 + (12 + 6)ab + 8b^2$
$= 9a^2 + 18ab + 8b^2$

33. $(8a - 1)^2 = (8a)^2 - 2(8a)(1) + 1^2$
$= 64a^2 - 16a + 1$

35. $(5r - 2s)(5r - 2s)$
$= (5r)^2 - 2(5r)(2s) + (2s)^2$
$= 25r^2 - 20rs + 4s^2$

37. $(7x - 8)(8x - 7)$
$= 56x^2 + (-49 - 64)x + 56$
$= 56x^2 - 113x + 56$

39. $(5t - 3s^2)(5t + 3s) = 25t^2 + 15st - 15s^2t - 9s^3$

41. $(3y^3 - 4x)^2$
$= (3y^3)^2 - 2(3y^3)(4x) + (4x)^2$
$= 9y^6 - 24xy^3 + 16x^2$

43. $(3y^3 - 4x)(3y^3 + 4x) = (3y^3)^2 - (4x)^2$
$= 9y^6 - 16x^2$

45. $(2rst - 7xyz)^2$
$= (2rst)^2 - 2(2rst)(7xyz) + (7xyz)^2$
$= 4r^2s^2t^2 - 28rstxyz + 49x^2y^2z^2$

47. $(x - 4) - (3x + 1)^2$
$= x - 4 - [(3x)^2 + 2(3x)(1) + 1^2]$
$= x - 4 - (9x^2 + 6x + 1)$
$= x - 4 - 9x^2 - 6x - 1$
$= -9x^2 - 5x - 5$

49. $(a - b)^2 - (b + a)^2$
$= a^2 - 2ab + b^2 - (b^2 + 2ab + a^2)$
$= a^2 - 2ab + b^2 - b^2 - 2ab - a^2$
$= -4ab$

51. $(3a + 2)(-5a)(2a - 1)$
$= (-15a^2 - 10a)(2a - 1)$
$= -30a^3 + 15a^2 - 20a^2 + 10a$
$= -30a^3 - 5a^2 + 10a$

53. $(3a + 2) - 5a(2a - 1)$
$= 3a + 2 - 10a^2 + 5a$
$= -10a^2 + 8a + 2$

55. $3r^3 - 3r(r - s)(r + s)$
$= 3r^3 - 3r(r^2 - s^2)$
$= 3r^3 - 3r^3 + 3rs^2$
$= 3rs^2$

57. $(3y + 1)(y + 2) - (2y - 3)^2$
$= 3y^2 + (6 + 1)y + 2 - [(2y)^2 - 2(2y)(3) + 3^2]$
$= 3y^2 + 7y + 2 - (4y^2 - 12y + 9)$
$= 3y^2 + 7y + 2 - 4y^2 + 12y - 9$
$= -y^2 + 19y - 7$

59. $(5a - 3b)^3$
$= (5a - 3b)(5a - 3b)(5a - 3b)$
$= [(5a)^2 - 2(5a)(3b) + (3b)^2](5a - 3b)$
$= (25a^2 - 30ab + 9b^2)(5a - 3b)$
$= 125a^3 - 150a^2b + 45ab^2 - 75a^2b + 90ab^2$
$- 27b^3$
$= 125a^3 - 225a^2b + 135ab^2 - 27b^3$

61. $3x(x + 1) - 2x(x + 1)^2$
$= 3x^2 + 3x - 2x(x^2 + 2x + 1)$
$= 3x^2 + 3x - 2x^3 - 4x^2 - 2x$
$= -2x^3 - x^2 + x$

63. $[(a + b) + 1][(a + b) - 1]$
$= (a + b)^2 - 1^2$
$= a^2 + 2ab + b^2 - 1$

65. $[(a - 2b) + 5z]^2$
$= (a - 2b)^2 + 2(a - 2b)(5z) + (5z)^2$
$= a^2 - 2a(2b) + (2b)^2 + 10z(a - 2b) + 25z^2$
$= a^2 - 4ab + 4b^2 + 10az - 20bz + 25z^2$

67. $[a - 2b + 5z][a - 2b - 5z]$
$= [(a - 2b) + 5z][(a - 2b) - 5z]$
$= (a - 2b)^2 - (5z)^2$
$= a^2 - 2a(2b) + (2b)^2 - 25z^2$
$= a^2 - 4ab + 4b^2 - 25z^2$

69. $[(a + b) + (2x + 1)][(a + b) - (2x + 1)]$
$= (a + b)^2 - (2x + 1)^2$
$= a^2 + 2ab + b^2 - [(2x)^2 + 2(2x)(1) + 1^2]$
$= a^2 + 2ab + b^2 - (4x^2 + 4x + 1)$
$= a^2 + 2ab + b^2 - 4x^2 - 4x - 1$

71. $(a^n - 3)(a^n + 3) = (a^n)^2 - 3^2$
$= a^{2n} - 9$

73. $f(x) = 3x^2 - 2x + 1$

$f(x + 2) = 3(x + 2)^2 - 2(x + 2) + 1$
$= 3(x^2 + 4x + 4) - 2x - 4 + 1$
$= 3x^2 + 12x + 12 - 2x - 3$
$= 3x^2 + 10x + 9$

75. $f(x) = 3x^2 - 2x + 1$
$f(2) = 3(2)^2 - 2(2) + 1 = 9$
$f(x) + f(2) = 3x^2 - 2x + 1 + 9$
$= 3x^2 - 2x + 10$

77. $f(x) = 3x^2 - 2x + 1$
$f(2x) = 3(2x)^2 - 2(2x) + 1$
$= 3(4x^2) - 4x + 1$
$= 12x^2 - 4x + 1$

79. $2f(x) = 2(3x^2 - 2x + 1)$
$= 6x^2 - 4x + 2$

81. $3f(x) - 2g(x)$
$= 3(3x^2 - 2x + 1) - 2(x^2 - 4)$
$= 9x^2 - 6x + 3 - 2x^2 + 8$
$= 7x^2 - 6x + 11$

83.

x
1 - 2x
1 - 2x

$V = wlh$
$V = (1 - 2x)(1 - 2x)x$
$V = (1 - 4x + 4x^2)x$
$V = x - 4x^2 + 4x^3$

85. Shaded Area = Area of Outer rectangle − Area of Inner rectangle

$A = x(2x) - 5(6)$
$A = 2x^2 - 30$

87. Shaded Area = Area of Outer rectangle − Area of Inner rectangle

$A = (x + 14 + x + 3)(4 + 12 + 2) - 14(12)$
$A = (2x + 17)(18) - 168$
$A = 36x + 306 - 168$
$A = 36x + 138$

93. $|3x - 2| > 5$
$3x - 2 > 5$ or $3x - 2 < -5$
$3x > 7$ $\quad 3x < -3$
$x > \frac{7}{3}$ or $x < -1$

95. width: x
length: $5x + 4$

$P = 2w + 2l$
$41.6 = 2x + 2(5x + 4)$
$41.6 = 2x + 10x + 8$
$41.6 = 12x + 8$
$33.6 = 12x$
$2.8 = x$
$5x + 4 = 5(2.8) + 4 = 18$

The dimensions are 2.8 in. by 18 in.

Exercises 5.4

1. $4x^2 + 2x = 2x(2x + 1)$

3. $x^2 + x = x(x + 1)$

5. $3xy^2 - 6x^2y^3 = 3xy^2(1 - 2xy)$

7. $6x^4 + 9x^3 - 21x^2 = 3x^2(2x^2 + 3x - 7)$

9. $35x^4y^4z - 15x^3y^5z + 10x^2y^3z^2$
$= 5x^2y^3z(7x^2y - 3xy^2 + 2z)$

11. $24r^3s^4 - 18r^3s^5 - 6r^2s^3$
$= 6r^2s^3(4rs - 3rs^2 - 1)$

13. $35a^2b^3 - 21a^3b^2 + 7ab$
$= 7ab(5ab^2 - 3a^2b + 1)$

15. $3x(x + 2) + 5(x + 2)$
$= (x + 2)(3x + 5)$

17. $3x(x + 2) - 5(x + 2)$
$= (x + 2)(3x - 5)$

19. $2x(x + 3y) + 5y(x + 3y)$
$= (x + 3y)(2x + 5y)$

21. $3x(x + 4y) - 5y(x + 4y)$
$= (x + 4y)(3x - 5y)$

23. $(2r + 1)(a - 2) + 5(a - 2)$
$= (a - 2)[(2r + 1) + 5]$
$= (a - 2)(2r + 6)$
$= (a - 2)2(r + 3)$
$= 2(a - 2)(r + 3)$

25. $2x(a - 3)^2 + 2(a - 3)^2$
$= 2(a - 3)^2(x + 1)$

27. $16a^2(b - 4)^2 - 4a(b - 4)$
$= 4a(b - 4)[4a(b - 4) - 1]$
$= 4a(b - 4)(4ab - 16a - 1)$

29. $2x^2 - 8x + 3x - 12$
$= 2x(x - 4) + 3(x - 4)$
$= (x - 4)(2x + 3)$

31. $3x^2 - 12xy + 5xy - 20y^2$
$= 3x(x - 4y) + 5y(x - 4y)$
$= (x - 4y)(3x + 5y)$

33. $7ax - 7bx + 3ay - 3by$
$= 7x(a - b) + 3y(a - b)$
$= (a - b)(7x + 3y)$

35. $7ax + 7bx - 3ay - 3by$
$= 7x(a + b) - 3y(a + b)$
$= (a + b)(7x - 3y)$

37. $7ax - 7bx - 3ay + 3by$
$= 7x(a - b) - 3y(a - b)$
$= (a - b)(7x - 3y)$

39. $7ax - 7bx - 3ay - 3by$
$= 7x(a - b) - 3y(a + b)$
not factorable

41. $2r^2 + 2rs - sr - s^2$
$= 2r(r + s) - s(r + s)$
$= (r + s)(2r - s)$

43. $2r^2 + 2rs - sr + s^2$
$= 2r(r + s) - s(r - s)$
not factorable

45. $5a^2 - 5ab - 2ab + 2b^2$
$= 5a(a - b) - 2b(a - b)$
$= (a - b)(5a - 2b)$

47. $3a^2 - 6a - a + 2$
$= 3a(a - 2) - 1(a - 2)$
$= (a - 2)(3a - 1)$

49. $3a^2 - 6a + a - 2$
$= 3a(a - 2) + 1(a - 2)$
$= (a - 2)(3a + 1)$

51. $3a^2 + 6a - a - 2$
$= 3a(a + 2) - 1(a + 2)$
$= (a + 2)(3a - 1)$

53. $a^3 + 2a^2 + 4a + 8$
$= a^2(a + 2) + 4(a + 2)$
$= (a + 2)(a^2 + 4)$

59. $|9 - 3a| \le 6$
$-6 \le 9 - 3a \le 6$
$-15 \le -3a \le -3$
$5 \ge a \ge 1$ or $1 \le a \le 5$

61. $m = \frac{-2}{3}$; From (4, −1), go down 2 and right 3 to the point (7, −3) on the line.

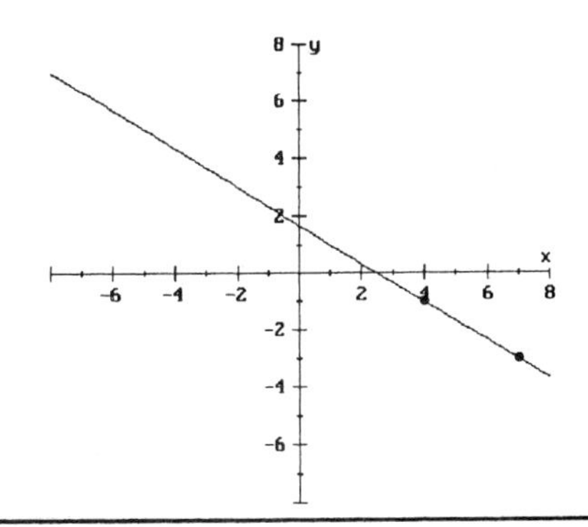

Exercises 5.5

1. Factors of −45 whose sum is 4: 9, −5
$x^2 + 4x - 45 = (x + 9)(x - 5)$

3. Factors of −10 whose sum is −3: −5, 2
$y^2 - 3y - 10 = (y - 5)(y + 2)$

5. Factors of −15 whose sum is −2: −5, 3
$x^2 - 2xy - 15y^2 = (x - 5y)(x + 3y)$

7. $r^2 - 81 = r^2 - 9^2$
$= (r - 9)(r + 9)$

9. Factors of −6 whose sum is −1: −3, 2
$x^2 - xy - 6y^2 = (x - 3y)(x + 2y)$

11. Factors of 6 whose sum is −1: none
not factorable

13. Factors of 12 whose sum is −7: −3, −4
$r^2 - 7rs + 12s^2 = (r - 3s)(r - 4s)$

15. Factors of −12 whose sum is −1: −4, 3
$r^2 - rs - 12s^2 = (r - 4s)(r + 3s)$

17. $9x^2 - 49y^2 = (3x)^2 - (7y)^2$
$= (3x - 7y)(3x + 7y)$

19. Factors of $15 \cdot 4 = 60$ whose sum is 17: 5, 12

$15x^2 + 17x + 4 = 15x^2 + 5x + 12x + 4$
$= 5x(3x + 1) + 4(3x + 1)$
$= (3x + 1)(5x + 4)$

21. Factors of $15(-6) = -90$ whose sum is −1: −10, 9

$15x^2 - xy - 6y^2$
$= 15x^2 - 10xy + 9xy - 6y^2$
$= 5x(3x - 2y) + 3y(3x - 2y)$
$= (3x - 2y)(5x + 3y)$

23. Factors of $15(-6) = -90$ whose sum is 1: 10, −9

$15x^2 + xy - 6y^2$
$= 15x^2 + 10xy - 9xy - 6y^2$
$= 5x(3x + 2y) - 3y(3x + 2y)$
$= (3x + 2y)(5x - 3y)$

25. $10xy + 3x^2 - 25y^2 = 3x^2 + 10xy - 25y^2$

Factors of $3(-25) = -75$ whose sum is 10: 15, −5

$3x^2 + 10xy - 25y^2$
$= 3x^2 + 15xy - 5xy - 25y^2$
$= 3x(x + 5y) - 5y(x + 5y)$
$= (x + 5y)(3x - 5y)$

27. Factors of $10(-10) = -100$ whose sum is 21: 25, −4

$10a^2 + 21ab - 10b^2$
$= 10a^2 + 25ab - 4ab - 10b^2$
$= 5a(2a + 5b) - 2b(2a + 5b)$
$= (2a + 5b)(5a - 2b)$

29. $25 - 5y - 2y^2 = -(2y^2 + 5y - 25)$

Factors of $2(-25) = -50$ whose sum is 5: 10, −5

$-(2y^2 + 5y - 25)$
$= -(2y^2 + 10y - 5y - 25)$
$= -[2y(y + 5) - 5(y + 5)]$
$= -(y + 5)(2y - 5)$

31. $25 - 5y - y^2 = -(y^2 + 5y - 25)$

Factors of −25 whose sum is 5: none

Not factorable

33. $2x^3 + 4x^2 - 16x = 2x(x^2 + 2x - 8)$
$= 2x(x + 4)(x - 2)$

35. $x^3 + 3x^2 - 28x = x(x^2 + 3x - 28)$
$= x(x + 7)(x - 4)$

37. $2x(x - 3) + (x - 1)(x + 2)$
$= 2x^2 - 6x + x^2 + x - 2$
$= 3x^2 - 5x - 2$
$= 3x^2 - 6x + x - 2$
$= 3x(x - 2) + 1(x - 2)$
$= (x - 2)(3x + 1)$

39. $18ab - 15abx - 18abx^2$
$= 3ab(6 - 5x - 6x^2)$
$= -3ab(6x^2 + 5x - 6)$
$= -3ab(6x^2 + 9x - 4x - 6)$
$= -3ab[3x(2x + 3) - 2(2x + 3)]$
$= -3ab(2x + 3)(3x - 2)$

41. $90y^4 - 114y^3 + 36y^2$
$= 6y^2(15y^2 - 19y + 6)$
$= 6y^2(15y^2 - 10y - 9y + 6)$
$= 6y^2[5y(3y - 2) - 3(3y - 2)]$
$= 6y^2(3y - 2)(5y - 3)$

43. $6r^4 - r^2 - 2$
$= 6r^4 - 4r^2 + 3r^2 - 2$
$= 2r^2(3r^2 - 2) + 1(3r^2 - 2)$
$= (3r^2 - 2)(2r^2 + 1)$

45. $3x^2 - 27 = 3(x^2 - 9)$
$= 3(x^2 - 3^2)$
$= 3(x - 3)(x + 3)$

47. $20xy^5 + 2x^2y^3 - 8x^3y$
$= 2xy(10y^4 + xy^2 - 4x^2)$

49. $108x^3y + 72x^2y^3 - 15xy^5$
$= 3xy(36x^2 + 24xy^2 - 5y^4)$
$= 3xy(36x^2 + 30xy^2 - 6xy^2 - 5y^4)$
$= 3xy[6x(6x + 5y^2) - y^2(6x + 5y^2)]$
$= 3xy(6x + 5y^2)(6x - y^2)$

51. $12a^7b + a^4b - 6ab$
$= ab(12a^6 + a^3 - 6)$
$= ab(12a^6 + 9a^3 - 8a^3 - 6)$
$= ab[3a^3(4a^3 + 3) - 2(4a^3 + 3)]$
$= ab(4a^3 + 3)(3a^3 - 2)$

53. $15a^8 + 19a^4 - 10$
$= 15a^8 + 25a^4 - 6a^4 - 10$
$= 5a^4(3a^4 + 5) - 2(3a^4 + 5)$
$= (3a^4 + 5)(5a^4 - 2)$

55. $16x^2 - 9y^2 = (4x)^2 - (3y)^2$
$= (4x - 3y)(4x + 3y)$

57. $x^2 + 4xy + 4y^2 = (x)^2 + 2(x)(2y) + (2y)^2$
$= (x + 2y)^2$

59. $x^2 - 4xy + 4y^2 = (x)^2 - 2(x)(2y) + (2y)^2$
$= (x - 2y)^2$

61. Factors of −4 whose sum is −4: none
not factorable

63. $6x^2y^2 - 5xy + 1$
$= 6x^2y^2 - 3xy - 2xy + 1$
$= 3xy(2xy - 1) - 1(2xy - 1)$
$= (2xy - 1)(3xy - 1)$

65. $81r^2s^2 - 16 = (9rs)^2 - 4^2$
$= (9rs - 4)(9rs + 4)$

67. $9x^2y^2 - 4z^2 = (3xy)^2 - (2z)^2$
$= (3xy - 2z)(3xy + 2z)$

69. $25a^2 - 4a^2b^2 = a^2(25 - 4b^2)$
$= a^2[5^2 - (2b)^2]$
$= a^2(5 - 2b)(5 + 2b)$

71. $1 - 14a + 49a^2 = 1^2 - 2(1)(7a) + (7a)^2$
$= (1 - 7a)^2$

73. $8a^3 - b^3 = (2a)^3 - b^3$
$= (2a - b)[(2a)^2 + (2a)(b) + b^2]$
$= (2a - b)(4a^2 + 2ab + b^2)$

75. $x^3 + 125y^3 = x^3 + (5y)^3$
$= (x + 5y)[x^2 - x(5y) + (5y)^2]$
$= (x + 5y)(x^2 - 5xy + 25y^2)$

77. not factorable

79. $4y^2 - 20y + 10 = 2(2y^2 - 10y + 5)$

81. $12a^3c + 36a^2c^2 + 27ac^3$
$= 3ac(4a^2 + 12ac + 9c^2)$
$= 3ac[(2a)^2 + 2(2a)(3c) + (3c)^2]$
$= 3ac(2a + 3c)^2$

83. $4x^4 - 81y^4 = (2x^2)^2 - (9y^2)^2$
$= (2x^2 - 9y^2)(2x^2 + 9y^2)$

85. $25a^6 + 10a^3b + b^2$
$= (5a^3)^2 + 2(5a^3)(b) + b^2$
$= (5a^3 + b)^2$

87. $12y^6 + 27y^2 = 3y^2(4y^4 + 9)$

89. $(a + b)^2 - 4 = (a + b)^2 - 2^2$
$= (a + b - 2)(a + b + 2)$

91. $a^2 - 2ab + b^2 - 16$
$= (a - b)^2 - 16$
$= (a - b)^2 - 4^2$
$= (a - b - 4)(a - b + 4)$

93. $x^2 + 6x + 9 - r^2$
$= [x^2 + 2(x)(3) + 3^2] - r^2$
$= (x + 3)^2 - r^2$
$= (x + 3 - r)(x + 3 + r)$

95. $a^3 + a^2 - 4a - 4$
$= a^2(a + 1) - 4(a + 1)$
$= (a + 1)(a^2 - 4)$
$= (a + 1)(a^2 - 2^2)$
$= (a + 1)(a - 2)(a + 2)$

97. $x^4 + x^3 - x - 1$
$= x^3(x + 1) - 1(x + 1)$
$= (x + 1)(x^3 - 1)$
$= (x + 1)(x^3 - 1^3)$
$= (x + 1)(x - 1)(x^2 + x + 1)$

99. $P = n^3 - 3n^2 + 2n$
$P = n(n^2 - 3n + 2)$
$P = n(n - 1)(n - 2)$

111. $x - 6y = 6$
$-6y = -x + 6$
$y = \frac{1}{6}x - 1$
$m = \frac{1}{6}$
$m_\perp = -6$
$y - y_1 = m(x - x_1)$
$y - 1 = -6(x - 5)$
$y - 1 = -6x + 30$
$y = -6x + 31$

113. (T, n): (74, 20) and (84, 40)
$m = \frac{40 - 20}{84 - 74}$
$= 2$
$n - 20 = 2(T - 74)$
$n - 20 = 2T - 148$
$n = 2T - 128$

When $T = 90$:
$$n = 2(90) - 128 = 52$$
He should expect to sell 52 pairs.

Exercises 5.6

1. $(x + 2)(x - 3) = 0$
$x + 2 = 0$ or $x - 3 = 0$
$x = -2$ or $x = 3$

3. $0 = (2y - 1)(y - 4)$
$2y - 1 = 0$ or $y - 4 = 0$
$2y = 1$ or $y = 4$
$y = \frac{1}{2}$ or $y = 4$

5. $x^2 = 25$
$x^2 - 25 = 0$
$(x - 5)(x + 5) = 0$
$x - 5 = 0$ or $x + 5 = 0$
$x = 5$ or $x = -5$

7. $x(x - 4) = 0$
$x = 0$ or $x - 4 = 0$
$x = 0$ or $x = 4$

9. $12 = x(x - 4)$
$12 = x^2 - 4x$
$0 = x^2 - 4x - 12$
$0 = (x - 6)(x + 2)$
$x - 6 = 0$ or $x + 2 = 0$
$x = 6$ or $x = -2$

11. $5y(y - 7) = 0$
$5y = 0$ or $y - 7 = 0$
$y = 0$ or $y = 7$

13. $x^2 - 16 = 0$
$(x - 4)(x + 4) = 0$
$x - 4 = 0$ or $x + 4 = 0$
$x = 4$ or $x = -4$

15. $0 = 9c^2 - 16$
$0 = (3c - 4)(3c + 4)$
$3c - 4 = 0$ or $3c + 4 = 0$
$3c = 4$ $\quad$ $3c = -4$
$c = \frac{4}{3}$ or $c = -\frac{4}{3}$

17. $8a^2 - 18 = 0$
$2(4a^2 - 9) = 0$
$4a^2 - 9 = 0$
$(2a - 3)(2a + 3) = 0$
$2a - 3 = 0$ or $2a + 3 = 0$
$2a = 3$ $\quad$ $2a = -3$
$a = \frac{3}{2}$ or $a = -\frac{3}{2}$

19. $0 = x^2 - x - 6$
$0 = (x - 3)(x + 2)$
$x - 3 = 0$ or $x + 2 = 0$
$x = 3$ or $x = -2$

21. $2y^2 - 3y + 1 = 0$
$(2y - 1)(y - 1) = 0$
$2y - 1 = 0$ or $y - 1 = 0$
$2y = 1$ $\quad$ $y = 1$
$y = \frac{1}{2}$ or $y = 1$

23. $0 = 8x^2 + 4x - 112$
$0 = 4(2x^2 + x - 28)$
$0 = 2x^2 + x - 28$
$0 = (2x - 7)(x + 4)$
$2x - 7 = 0$ or $x + 4 = 0$
$2x = 7$ $\quad$ $x = -4$
$x = \frac{7}{2}$ or $x = -4$

25. $x^3 + x^2 - 6x = 0$
$x(x^2 + x - 6) = 0$
$x(x + 3)(x - 2) = 0$
$x = 0$ or $x + 3 = 0$ or $x - 2 = 0$
$x = 0$ or $x = -3$ or $x = 2$

27. $t^4 + 2t^3 + t^2 = 0$
$t^2(t^2 + 2t + 1) = 0$
$t^2(t + 1)(t + 1) = 0$
$t^2 = 0$ or $t + 1 = 0$
$t = 0$ or $t = -1$

29. $20m^3 = 5m$
$20m^3 - 5m = 0$
$5m(4m^2 - 1) = 0$
$5m(2m - 1)(2m + 1) = 0$
$5m = 0$ or $2m - 1 = 0$ or $2m + 1 = 0$
$m = 0$ $\quad$ $2m = 1$ $\quad$ $2m = -1$
$m = 0$ or $m = \frac{1}{2}$ or $m = -\frac{1}{2}$

31. $x(x - 3) = x^2 - 10$
$$x^2 - 3x = x^2 - 10$$
$$-3x = -10$$
$$x = \frac{10}{3}$$

33. $(t - 4)(t + 1) = (t - 3)(t - 2)$
$$t^2 - 3t - 4 = t^2 - 5t + 6$$
$$-3t - 4 = -5t + 6$$
$$2t - 4 = 6$$
$$2t = 10$$
$$t = 5$$

35. $(2t - 4)(t + 1) = (t - 3)(t - 2)$
$$2t^2 - 2t - 4 = t^2 - 5t + 6$$
$$t^2 + 3t - 10 = 0$$
$$(t + 5)(t - 2) = 0$$
$$t + 5 = 0 \quad \text{or} \quad t - 2 = 0$$
$$t = -5 \quad \text{or} \quad t = 2$$

37. $(x + 6)^2 = 16$
$$x^2 + 12x + 36 = 16$$
$$x^2 + 12x + 20 = 0$$
$$(x + 10)(x + 2) = 0$$
$$x + 10 = 0 \quad \text{or} \quad x + 2 = 0$$
$$x = -10 \quad \text{or} \quad x = -2$$

39. $(3x - 4)^2 = 20 - 24x$
$$9x^2 - 24x + 16 = 20 - 24x$$
$$9x^2 - 4 = 0$$
$$(3x - 2)(3x + 2) = 0$$
$$3x - 2 = 0 \quad \text{or} \quad 3x + 2 = 0$$
$$3x = 2 \qquad 3x = -2$$
$$x = \frac{2}{3} \quad \text{or} \quad x = -\frac{2}{3}$$

41. $2(x - 3)(x + 2) = 0$
$$(x - 3)(x + 2) = 0$$
$$x - 3 = 0 \quad \text{or} \quad x + 2 = 0$$
$$x = 3 \quad \text{or} \quad x = -2$$

43. $2x - 3(x + 2) = 4$
$$2x - 3x - 6 = 4$$
$$-x - 6 = 4$$
$$-x = 10$$
$$x = -10$$

45. (a) $f(x) = 0$
$$x^2 - 2x - 15 = 0$$
$$(x - 5)(x + 3) = 0$$
$$x - 5 = 0 \quad \text{or} \quad x + 3 = 0$$
$$x = 5 \quad \text{or} \quad x = -3$$

(b) $f(x) = 9$
$$x^2 - 2x - 15 = 9$$
$$x^2 - 2x - 24 = 0$$
$$(x - 6)(x + 4) = 0$$
$$x - 6 = 0 \quad \text{or} \quad x + 4 = 0$$
$$x = 6 \quad \text{or} \quad x = -4$$

47. the number: x
$$x^2 - 5 = 1 + 5x$$
$$x^2 - 5x - 6 = 0$$
$$(x - 6)(x + 1) = 0$$
$$x - 6 = 0 \quad \text{or} \quad x + 1 = 0$$
$$x = 6 \quad \text{or} \quad x = -1$$
Since the number is positive, it is 6.

49. the number: x
$$x^2 + 8^2 = 68$$
$$x^2 + 64 = 68$$
$$x^2 - 4 = 0$$
$$(x - 2)(x + 2) = 0$$
$$x - 2 = 0 \quad \text{or} \quad x + 2 = 0$$
$$x = 2 \quad \text{or} \quad x = -2$$
Since the number is positive, it is 2.

51. the number: x
$$(x + 6)^2 = 169$$
$$x^2 + 12x + 36 = 169$$
$$x^2 + 12x - 133 = 0$$
$$(x + 19)(x - 7) = 0$$
$$x + 19 = 0 \quad \text{or} \quad x - 7 = 0$$
$$x = -19 \quad \text{or} \quad x = 7$$
The numbers are -19 and 7.

53. the number: x
$$x + \frac{1}{x} = \frac{13}{6}$$
$$6x\left(x + \frac{1}{x}\right) = 6x\left(\frac{13}{6}\right)$$
$$6x^2 + 6 = 13x$$
$$6x^2 - 13x + 6 = 0$$
$$(3x - 2)(2x - 3) = 0$$

$$3x - 2 = 0 \quad \text{or} \quad 2x - 3 = 0$$
$$3x = 2 \qquad\qquad 2x = 3$$
$$x = \frac{2}{3} \quad \text{or} \quad x = \frac{3}{2}$$

The numbers are $\frac{2}{3}$ and $\frac{3}{2}$.

55. the number: x

$$x + 2\left(\frac{1}{x}\right) = 3$$

$$x + \frac{2}{x} = 3$$

$$x\left(x + \frac{2}{x}\right) = x(3)$$

$$x^2 + 2 = 3x$$
$$x^2 - 3x + 2 = 0$$
$$(x - 2)(x - 1) = 0$$
$$x - 2 = 0 \quad \text{or} \quad x - 1 = 0$$
$$x = 2 \quad \text{or} \quad x = 1$$

The numbers are 2 and 1.

57. $P = 10000(-d^2 + 12d - 35)$

(a) When $d = 5$:
$$P = 10000\left[-5^2 + 12(5) - 35\right]$$
$$= 0$$

The profit is \$0.

(b)
$$10000 = 10000(-d^2 + 12d - 35)$$
$$1 = -d^2 + 12d - 35$$
$$d^2 - 12d + 36 = 0$$
$$(d - 6)(d - 6) = 0$$
$$d - 6 = 0$$
$$d = 6$$

The price must be \$6.

59. $s = s(t) = -16t^2 + 64$

(a) When $t = 1$:
$$s = s(1) = -16(1)^2 + 64$$
$$= 48$$

He is 48 ft above the pool.

(b)
$$0 = -16t^2 + 64$$
$$16t^2 = 64$$

$$t^2 = 4$$
$$t = 2$$

He will hit the water in 2 seconds.

(c) When $t = 0$:
$$s = s(0) = -16(0)^2 + 64$$
$$= 64$$

The board is 64 feet high.

61. width: x
length: $x + 2$

$$A = wl$$
$$80 = x(x + 2)$$
$$80 = x^2 + 2x$$
$$0 = x^2 + 2x - 80$$

$$0 = (x + 10)(x - 8)$$
$$x + 10 = 0 \quad \text{or} \quad x - 8 = 0$$
$$x = -10 \quad \text{or} \quad x = 8$$

Since x is a dimension it must be positive.

$$x = 8$$
$$x + 2 = 8 + 2 = 10$$

The dimensions are 8 feet by 10 feet.

63.

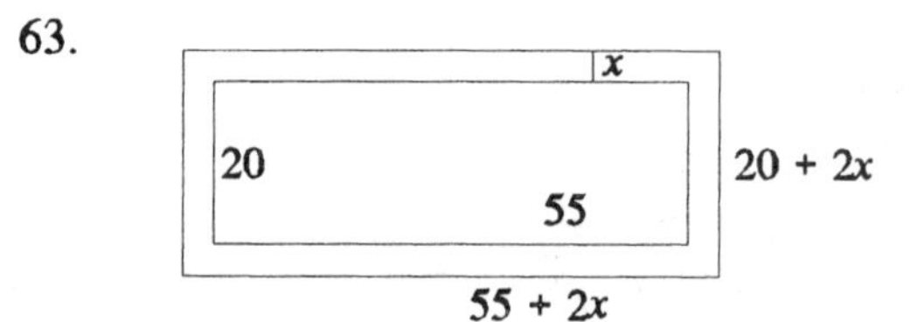

Area of walkway = area of outer rectangle − area of inner rectangle

$$400 = (20 + 2x)(55 + 2x) - (20)(55)$$
$$400 = 1100 + 150x + 4x^2 - 1100$$
$$400 = 4x^2 + 150x$$
$$0 = 4x^2 + 150x - 400$$
$$0 = 2x^2 + 75x - 200$$
$$0 = (2x - 5)(x + 40)$$
$$2x - 5 = 0 \quad \text{or} \quad x + 40 = 0$$
$$2x = 5 \qquad\qquad x = -40$$
$$x = \frac{5}{2}$$

Since $x > 0$, the width of the walkway is 5/2 feet.

67. 1st car's hours: x
2nd car's hours: $x + 1$

$52x = 40(x + 1)$
$52x = 40x + 40$
$12x = 40$
$x = \frac{10}{3}$

It will take 10/3 hours.

69. $3x - 2y = 8$
$-2y = -3x + 8$
$y = \frac{3}{2}x - 4$

$m = \frac{3}{2}$

Exercises 5.7

1.
$$\begin{array}{r} x + 4 \\ x - 5 \overline{)\, x^2 - x - 20} \\ \underline{-(x^2 - 5x)} \\ 4x - 20 \\ \underline{-(4x - 20)} \\ 0 \end{array}$$

3.
$$\begin{array}{r} a + 2 \\ 3a + 4 \overline{)\, 3a^2 + 10a + 8} \\ \underline{-(3a^2 + 4a)} \\ 6a + 8 \\ \underline{-(6a + 8)} \\ 0 \end{array}$$

5.
$$\begin{array}{r} 7z - 2 \\ 3z + 1 \overline{)\, 21z^2 + z - 16} \\ \underline{-(21z^2 + 7z)} \\ -6z - 16 \\ \underline{-(6z - 2)} \\ -14 \end{array}$$

$7z - 2$, R -14

7.
$$\begin{array}{r} a^2 + 2a + 3 \\ a - 1 \overline{)\, a^3 + a^2 + a - 8} \\ \underline{-(a^3 - a^2)} \\ 2a^2 + a \\ \underline{-(2a^2 - 2a)} \\ 3a - 8 \\ \underline{-(3a - 3)} \\ -5 \end{array}$$

$a^2 + 2a + 3$, R -5

9.
$$\begin{array}{r} 3x^2 + x - 1 \\ x - 4 \overline{)\, 3x^3 - 11x^2 - 5x + 12} \\ \underline{-(3x^3 - 12x^2)} \\ x^2 - 5x \\ \underline{-(x^2 - 4x)} \\ -x + 12 \\ \underline{-(-x + 4)} \\ 8 \end{array}$$

$3x^2 + x - 1$, R 8

11.
$$\begin{array}{r} 2z - 3 \\ 2z + 3 \overline{)\, 4z^2 + 0z - 15} \\ \underline{-(4z^2 + 6z)} \\ -6z - 15 \\ \underline{-(-6z - 9)} \\ -6 \end{array}$$

$2z - 3$, R -6

13.
$$\begin{array}{r} x^3 + 3x^2 + 2x + 1 \\ x - 2 \overline{)\, x^4 + x^3 - 4x^2 - 3x - 2} \\ \underline{-(x^4 - 2x^3)} \\ 3x^3 - 4x^2 \\ \underline{-(3x^3 - 6x^2)} \\ 2x^2 - 3x \\ \underline{-(2x^2 - 4x)} \\ x - 2 \\ \underline{-(x - 2)} \\ 0 \end{array}$$

15. $y - 1 \overline{)3y^4 - 5y^3 + 2y^2 + 5y - 10}$ quotient: $3y^3 - 2y^2 + 5$

$-(3y^4 - 3y^3)$

$-2y^3 + 2y^2$

$-(-2y^3 + 2y^2)$

$5y - 10$

$-(5y - 5)$

-5

$3y^3 - 2y^2 + 5,\ \text{R } -5$

17. $a + 2 \overline{)2a^4 + 4a^3 - 5a + 6}$ quotient: $2a^3 - 5$

$-(2a^4 + 4a^3)$

$-5a + 6$

$-(-5a - 10)$

16

$2a^3 - 5,\ \text{R } 16$

19. $y + 1 \overline{)y^3 + 0y^2 + 0y - 1}$ quotient: $y^2 - y + 1$

$-(y^3 + y^2)$

$-y^2 + 0y$

$-(-y^2 - y)$

$y - 1$

$-(y + 1)$

-2

$y^2 - y + 1,\ \text{R } -2$

21. $2a - 1 \overline{)8a^3 + 0a^2 + 0a + 1}$ quotient: $4a^2 + 2a + 1$

$-(8a^3 - 4a^2)$

$4a^2 + 0a$

$-(4a^2 - 2a)$

$2a + 1$

$-(2a - 1)$

2

$4a^2 + 2a + 1,\ \text{R } 2$

23. $y^2 + 1 \overline{)2y^5 - 3y^4 + 0y^3 - y^2 + y + 4}$ quotient: $2y^3 - 3y^2 - 2y + 2$

$-(2y^5 + 2y^3)$

$-3y^4 - 2y^3 - y^2$

$-(-3y^4 - 3y^2)$

$-2y^3 + 2y^2 + y$

$-(-2y^3 - 2y)$

$2y^2 + 3y + 4$

$-(2y^2 + 2)$

$3y + 2$

$2y^3 - 3y^2 - 2y + 2,\ \text{R } 3y + 2$

25. $z^2 + 2z - 1 \overline{)3z^5 + 0z^4 - 4z^3 + 3z^2 + 12z - 10}$ quotient: $3z^3 - 6z^2 + 11z - 25$

$-(3z^5 + 6z^4 - 3z^3)$

$-6z^4 - z^3 + 3z^3$

$-(-6z^4 - 12z^3 + 6z^2)$

$11z^3 - 3z^2 + 12z$

$-(11z^3 + 22z^2 - 11z)$

$-25z^2 + 23z - 10$

$-(-25z^2 - 50z + 25)$

$73z - 35$

$3z^3 - 6z^2 + 11z - 25,\ \text{R } 73z - 35$

27. $x^2 - 2 \overline{)7x^6 - 14x^4 - 3x^2 + x + 1}$ quotient: $7x^4 - 3$

$-(7x^6 - 14x^4)$

$3x^2 + x + 1$

$-(-3x^2 + 6)$

$x - 5$

$7x^4 - 3,\ \text{R } x - 5$

29. $f(x) = 5x^2 - 3x + 2$

$$\begin{aligned} f(4) &= 5(4)^2 - 3(4) + 2 \\ &= 5(16) - 3(4) + 2 \\ &= 70 \end{aligned}$$

$$\begin{aligned} f(x^2) &= 5(x^2)^2 - 3(x^2) + 2 \\ &= 5x^4 - 3x^2 + 2 \end{aligned}$$

CHAPTER 5 REVIEW EXERCISES

1. $(2x^3 - 4) - [(2x^2 - 3x + 4) + (5x^2 - 3)]$
$= (2x^3 - 4) - (2x^2 - 3x + 4 + 5x^2 - 3)$
$= (2x^3 - 4) - (7x^2 - 3x + 1)$
$= 2x^3 - 4 - 7x^2 + 3x - 1$
$= 2x^3 - 7x^2 + 3x - 5$

3. $3x^2(2xy - 3y + 1)$
$= 6x^3y - 9x^2y + 3x^2$

5. $3ab(2a - 3b) - 2a(3ab - 4b^2)$
$= 6a^2b - 9ab^2 - 6a^2b + 8ab^2$
$= -ab^2$

7. $3x - 2(x - 3) - [x - 2(5 - x)]$
$= 3x - 2x + 6 - (x - 10 + 2x)$
$= x + 6 - (3x - 10)$
$= x + 6 - 3x + 10$
$= -2x + 16$

9. $(x - 3)(x + 2)$
$= x^2 + (2 - 3)x - 6$
$= x^2 - x - 6$

11. $(3y - 2)(2y - 1)$
$= 6y^2 + (-3 - 4)y + 2$
$= 6y^2 - 7y + 2$

13. $(3x - 4y)^2$
$= (3x)^2 - 2(3x)(4y) + (4y)^2$
$= 9x^2 - 24xy + 16y^2$

15. $(4x^2 - 5y)^2$
$= (4x^2)^2 - 2(4x^2)(5y) + (5y)^2$
$= 16x^4 - 40x^2y + 25y^2$

17. $(7x^2 - 5y^3)(7x^2 + 5y^3)$
$= (7x^2)^2 - (5y^3)^2$
$= 49x^4 - 25y^6$

19. $(5y^2 - 3y + 7)(2y - 3)$
$= 10y^3 - 6y^2 + 14y - 15y^2 + 9y - 21$
$= 10y^3 - 21y^2 + 23y - 21$

21. $(2x + 3y + 4)(2x + 3y - 5)$
$= [(2x + 3y) + 4][(2x + 3y) - 5]$
$= (2x + 3y)^2 + (-5 + 4)(2x + 3y) - 20$
$= (2x)^2 + 2(2x)(3y) + (3y)^2 - (2x + 3y) - 20$
$= 4x^2 + 12xy + 9y^2 - 2x - 3y - 20$

23. $[(x - y) + 5]^2$
$= (x - y)^2 + 2(x - y)(5) + 5^2$
$= x^2 - 2xy + y^2 + 10x - 10y + 25$

25. $(a + b - 4)^2$
$= (a + b - 4)(a + b - 4)$
$= a^2 + ab - 4a + ab + b^2 - 4b$
$- 4a - 4b + 16$
$= a^2 + 2ab + b^2 - 8a - 8b + 16$

27. $6x^2y - 12xy^2 + 9xy = 3xy(2x - 4y + 3)$

29. $3x(a + b) - 2(a + b) = (a + b)(3x - 2)$

31. $2(a - b)^2 + 3(a - b)$
$= (a - b)[2(a - b) + 3]$
$= (a - b)(2a - 2b + 3)$

33. $5ax - 5a + 3bx - 3b$
$= 5a(x - 1) + 3b(x - 1)$
$= (x - 1)(5a + 3b)$

35. $5y^2 - 5y + 3y - 3 = 5y(y - 1) + 3(y - 1)$
$= (y - 1)(5y + 3)$

37. $a^2 + a(a - 10) = a^2 + a^2 - 10a$
$= 2a^2 - 10a$
$= 2a(a - 5)$

39. $2t(t + 2) - (t - 2)(t + 4)$
$= 2t^2 + 4t - (t^2 + 2t - 8)$
$= 2t^2 + 4t - t^2 - 2t + 8$
$= t^2 + 2t + 8$
not factorable

41. Factors of -35 whose sum is -2: $-7, 5$

$x^2 - 2x - 35 = (x - 7)(x + 5)$

43. Factors of -14 whose sum is 5: $-2, 7$

$a^2 + 5ab - 14b^2 = (a - 2b)(a + 7b)$

45. Factors of $35 \cdot 2 = 70$ whose sum is 17: 7, 10

$35a^2 + 17ab + 2b^2$
$= 35a^2 + 7ab + 10ab + 2b^2$
$= 7a(5a + b) + 2b(5a + b)$
$= (5a + b)(7a + 2b)$

47. $a^2 - 6ab^2 + 9b^4$
$= (a)^2 - 2(a)(3b^2) + (3b^2)^2$
$= (a - 3b^2)^2$

49. $3a^3 - 21a^2 + 30a = 3a(a^2 - 7a + 10)$
$= 3a(a - 5)(a - 2)$

51. $2x^3 - 50xy^2 = 2x(x^2 - 25y^2)$
$= 2x[(x)^2 - (5y)^2]$
$= 2x(x - 5y)(x + 5y)$

53. $6x^2 + 5x - 6$
$= 6x^2 + 9x - 4x - 6$
$= 3x(2x + 3) - 2(2x + 3)$
$= (2x + 3)(3x - 2)$

55. $8x^3 + 125y^3$
$= (2x)^3 + (5y)^3$
$= (2x + 5y)[(2x)^2 - (2x)(5y) + (5y)^2]$
$= (2x + 5y)(4x^2 - 10xy + 25y^2)$

57. $6a^2 - 17ab - 3b^2$
$= 6a^2 - 18ab + ab - 3b^2$
$= 6a(a - 3b) + b(a - 3b)$
$= (a - 3b)(6a + b)$

59. $21a^4 + 41a^2b^2 + 10b^4$
$= 21a^4 + 6a^2b^2 + 35a^2b^2 + 10b^4$
$= 3a^2(7a^2 + 2b^2) + 5b^2(7a^2 + 2b^2)$
$= (7a^2 + 2b^2)(3a^2 + 5b^2)$

61. $25x^4 - 40x^2y^2 + 16y^4$
$= (5x^2)^2 - 2(5x^2)(4y^2) + (4y^2)^2$
$= (5x^2 - 4y^2)^2$

63. $20x^3y - 60x^2y^2 + 45xy^3$
$= 5xy(4x^2 - 12xy + 9y^2)$
$= 5xy[(2x)^2 - 2(2x)(3y) + (3y)^2]$
$= 5xy(2x - 3y)^2$

65. $6a^4b - 8a^3b^2 - 8a^2b^3$
$= 2a^2b(3a^2 - 4ab - 4b^2)$
$= 2a^2b(3a^2 - 6ab + 2ab - 4b^2)$
$= 2a^2b[3a(a - 2b) + 2b(a - 2b)]$
$= 2a^2b(a - 2b)(3a + 2b)$

67. $6x^5 - 10x^3 - 4x$
$= 2x(3x^4 - 5x^2 - 2)$
$= 2x(3x^4 - 6x^2 + x^2 - 2)$
$= 2x[3x^2(x^2 - 2) + 1(x^2 - 2)]$
$= 2x(x^2 - 2)(3x^2 + 1)$

69. $(a - b)^2 - 4 = (a - b)^2 - 2^2$
$= (a - b - 2)(a - b + 2)$

71. $9y^2 + 30y + 25 - 9x^2$
$= [(3y)^2 + 2(3y)(5) + 5^2] - (3x)^2$
$= (3y + 5)^2 - (3x)^2$
$= (3y + 5 - 3x)(3y + 5 + 3x)$

73.
$$\begin{array}{r} 3x^2 + 2x + 11 \\ x - 2\overline{)3x^3 - 4x^2 + 7x - 5} \\ \underline{-(3x^3 - 6x^2)} \\ 2x^2 + 7x \\ \underline{-(2x^2 - 4x)} \\ 11x - 5 \\ \underline{-(11x - 22)} \\ 17 \end{array}$$

$3x^2 + 2x + 11$, R 17

75. Long division of $8a^3 + 0a^2 + 0a - 27$ by $2a - 3$:

$$\begin{array}{r} 4a^2 + 6a + 9 \\ 2a - 3 \overline{)8a^3 + 0a^2 + 0a - 27} \\ \underline{-(8a^3 - 12a^2)} \\ 12a^2 + 0a \\ \underline{-(12a^2 - 18a)} \\ 18a - 27 \\ \underline{-(18a - 27)} \\ 0 \end{array}$$

77. $x(x + 5)(3x - 4) = 0$

$x = 0$ or $x + 5 = 0$ or $3x - 4 = 0$

$x = 0$ $\quad x = -5$ $\quad 3x = 4$

$x = 0$ or $x = -5$ or $x = \frac{4}{3}$

79.
$$x^4 = 25x^2$$
$$x^4 - 25x^2 = 0$$
$$x^2(x^2 - 25) = 0$$
$$x^2(x - 5)(x + 5) = 0$$
$x^2 = 0$ or $x - 5 = 0$ or $x + 5 = 0$

$x = 0$ or $x = 5$ or $x = -5$

81. $(x + 6)(x - 3) = (4x - 2)(x + 4)$
$$x^2 + 3x - 18 = 4x^2 + 14x - 8$$
$$0 = 3x^2 + 11x + 10$$
$$0 = (3x + 5)(x + 2)$$
$3x + 5 = 0$ or $x + 2 = 0$

$3x = -5$ $\quad x = -2$

$x = -\frac{5}{3}$ or $x = -2$

83. $f(x)g(x) = (2x^2 - 5x + 3)(3x - 1)$
$$= 6x^3 - 15x^2 + 9x - 2x^2 + 5x - 3$$
$$= 6x^3 - 17x^2 + 14x - 3$$

85. $f(x) = 2x^2 - 5x + 3$

$$f(x + 1) = 2(x + 1)^2 - 5(x + 1) + 3$$
$$= 2(x^2 + 2x + 1) - 5x - 5 + 3$$
$$= 2x^2 + 4x + 2 - 5x - 2$$
$$= 2x^2 - x$$

87. $S(n) = \frac{1}{6}(2n^3 + 3n^2 + n)$
$$= \frac{1}{6}n(2n^2 + 3n + 1)$$
$$= \frac{1}{6}n(2n + 1)(n + 1)$$

89. (a) R = price of item $\cdot$ number of items

$$R = c(20000 - 1000c^2)$$
$$R = 20000c - 1000c^3$$

(b)

c	$20000c - 1000c^3 = R$
1	$20000(1) - 1000(1)^3 = 19{,}000$
2	$20000(2) - 1000(2)^3 = 32{,}000$
4	$20000(4) - 1000(4)^3 = 16{,}000$

CHAPTER 5 PRACTICE TEST

1. $(9a^2 + 6ab + 4b^2)(3a - 2b)$
$$= 27a^3 + 18a^2b + 12ab^2 - 18a^2b - 12ab^2 - 8b^3$$
$$= 27a^3 - 8b^3$$

3. $(x - y - 2)(x - y + 2)$
$$= [(x - y) - 2][(x - y) + 2]$$
$$= (x - y)^2 - 2^2$$
$$= x^2 - 2xy + y^2 - 4$$

5. $10x^3y^2 - 6x^2y^3 + 2xy$
$$= 2xy(5x^2y - 3xy^2 + 1)$$

7. $x^2 + 11xy - 60y^2 = (x + 15y)(x - 4y)$

9. Factors of $2 \cdot 7 = 14$ whose sum is -3: none

not factorable

11. $r^3s - 10r^2s^2 + 25rs^3$
$$= rs(r^2 - 10rs + 25s^2)$$
$$= rs[(r)^2 - 2(r)(5s) + (5s)^2]$$
$$= rs(r - 5s)^2$$

13. $(2x + y)^2 - 64$
$$= (2x + y)^2 - 8^2$$
$$= [(2x + y) - 8][(2x + y) + 8]$$
$$= (2x + y - 8)(2x + y + 8)$$

15. Long division of $x^3 + 0x^2 + 7x - 8$ by $x - 3$:

$$\begin{array}{r} x^2 + 3x + 16 \\ x - 3 \overline{)\, x^3 + 0x^2 + 7x - 8} \\ \underline{-(x^3 - 3x^2)} \\ 3x^2 + 7x \\ \underline{-(3x^2 - 9x)} \\ 16x - 8 \\ \underline{-(16x - 48)} \\ 40 \end{array}$$

$x^2 + 3x + 16$, R 40

17. $s = s(t) = -16t^2 + 32t + 48$

(a) When $t = 2$,

$$s = s(2) = -16(2)^2 + 32(2) + 48$$
$$s(2) = 48 \text{ feet}$$

(b) When $s = 0$,

$$0 = -16t^2 + 32t + 48$$
$$0 = -16(t^2 - 2t - 3)$$
$$0 = -16(t - 3)(t + 1)$$

$t - 3 = 0$ or $t + 1 = 0$

$t = 3$ or $t = -1$

Since $t > 0$, $t = 3$ seconds.

CHAPTER 6

Exercises 6.1

1. $x - 2 \neq 0$
   $x \neq 2$

3. x and y can be any number

5. $5x \neq 0$
   $x \neq 0$

7. $3x - y = 0$
   $3x \neq y$

9. $x - 8 = 0$
   $x = 8$
   Domain: $\{x \mid x \neq 8\}$

11. $5x - 8 = 0$
    $5x = 8$
    $x = \dfrac{8}{5}$

    Domain: $\left\{x \mid x \neq \dfrac{8}{5}\right\}$

13. $x^2 - 25 = 0$
    $(x + 5)(x - 5) = 0$
    $x + 5 = 0$ or $x - 5 = 0$
    $x = -5$ $\quad x = 5$
    Domain: $\{x \mid x \neq -5, x \neq 5\}$

15. $x^2 - x - 6 = 0$
    $(x - 3)(x + 2) = 0$
    $x - 3 = 0$ or $x + 2 = 0$
    $x = 3$ $\quad x = -2$
    Domain: $\{x \mid x \neq 3, x \neq -2\}$

17. (a) $A = lw$
    $20 = lw$

    $\dfrac{20}{w} = l$

    (b) $L(w) = \dfrac{20}{w}$

w	$L(w)$
0.5	$20/0.5 = 40$
1	$20/1 = 20$
2	$20/2 = 10$
4	$20/4 = 5$
5	$20/5 = 4$
10	$20/10 = 2$
15	$20/15 = 4/3$
20	$20/20 = 1$

As the width increases the length decreases.

19. $A(x) = \dfrac{0.4x + 8000}{x}$

    (a) $A(5) = \dfrac{0.4(5) + 8000}{5} = \$1600.40$

    $A(10) = \dfrac{0.4(10) + 8000}{10} = \$800.40$

    $A(50) = \dfrac{0.4(50) + 8000}{50} = \$160.40$

    $A(100) = \dfrac{0.4(100) + 8000}{100} = \$80.40$

    $A(200) = \dfrac{0.4(200) + 8000}{200} = \$40.40$

    $A(300) = \dfrac{0.4(300) + 8000}{300} = \$27.07$

    (b)

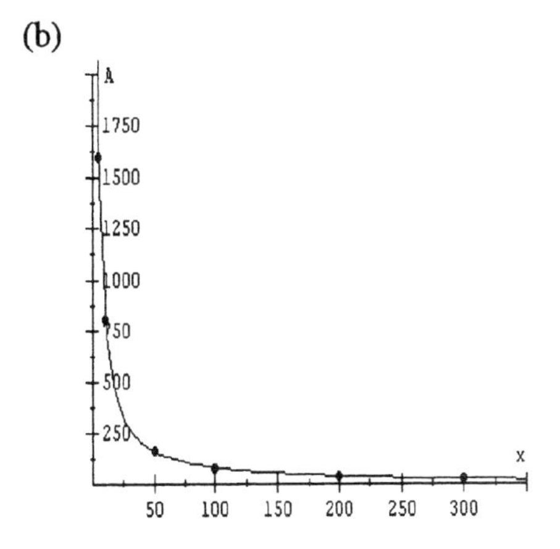

As the number of items increases, the average cost per item decreases and appears to level off at approximately \$0.40.

21. $n(a) = \dfrac{40a}{0.4a + 8} \qquad 0 \le a \le 200$

(a) $n(1) = \dfrac{40(1)}{0.4(1) + 8} = 4.8$

Approximately 5 deer can be supported on 1 acre.

$n(10) = \dfrac{40(10)}{0.4(10) + 8} = 33.3$

Approximately 33 deer can be supported on 10 acres.

$n(50) = \dfrac{40(50)}{0.4(50) + 8} = 71.4$

Approximately 71 deer can be supported on 50 acres.

$n(100) = \dfrac{40(100)}{0.4(100) + 8} = 83.3$

Approximately 83 deer can be supported on 100 acres.

(b)

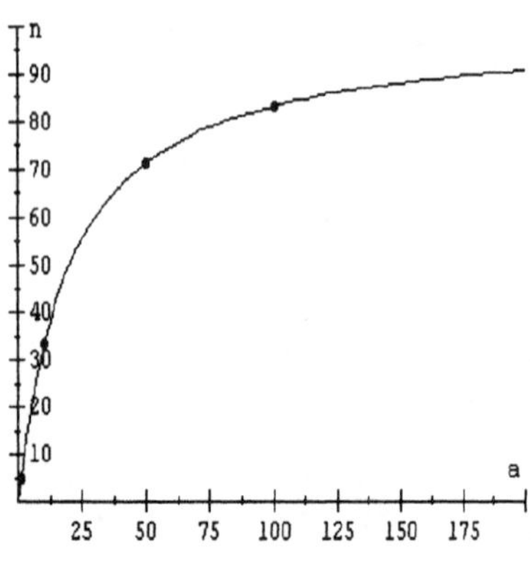

As the number of acres increases, the number of deer that can be supported increases.

23. $C(g) = 2g + \dfrac{5000}{g}$

(a) $g > 0$

(b) $C(1) = 2(1) + \dfrac{5000}{1} = 5002$

It costs $5002 to store 1 gal of a certain chemical.

$C(5) = 2(5) + \dfrac{5000}{5} = 1010$

It costs $1010 to store 5 gal of a certain chemical.

$C(10) = 2(10) + \dfrac{5000}{1} = 520$

It costs $520 to store 10 gal of a certain chemical.

$C(20) = 2(20) + \dfrac{5000}{20} = 290$

It costs $290 to store 20 gal of a certain chemical.

$C(40) = 2(40) + \dfrac{5000}{40} = 205$

It costs $205 to store 40 gal of a certain chemical.

$C(100) = 2(100) + \dfrac{5000}{100} = 250$

It costs $250 to store 100 gal of a certain chemical.

$C(200) = 2(200) + \dfrac{5000}{200} = 425$

It costs $425 to store 200 gal of a certain chemical.

$C(300) = 2(300) + \dfrac{5000}{300} = 617$

It costs $617 to store 300 gal of a certain chemical.

(c)

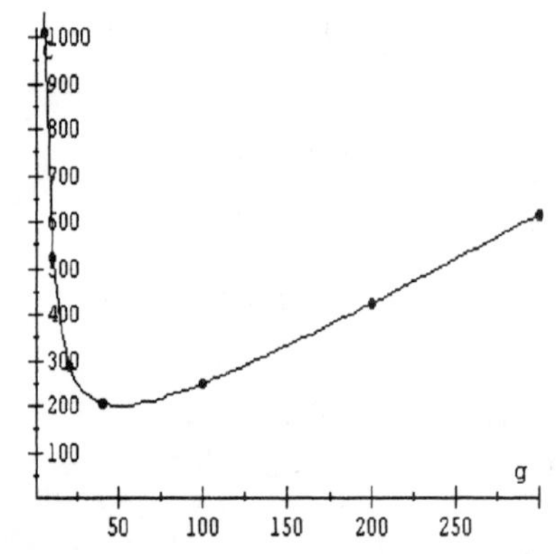

At first as the number of gallons increases the cost decreases. However after 40 gallons the cost begins to increase again.

Exercises 6.2

1. $\dfrac{3x^2y}{15xy} = \dfrac{3x \cdot x \cdot y}{3 \cdot 5 \cdot x \cdot y}$

$= \dfrac{x}{5}$

3. $\dfrac{16x^2y^3a}{18xy^4a^3} = \dfrac{2 \cdot 8 \cdot x \cdot x \cdot y^3 \cdot a}{2 \cdot 9 \cdot x \cdot y^3 \cdot y \cdot a \cdot a^2}$

$= \dfrac{8x}{9ya^2}$

5. $\dfrac{(x - 5)(x + 4)}{(x - 3)(x + 4)} = \dfrac{x - 5}{x - 3}$

7. $\dfrac{(2x - 3)(x - 5)}{(2x + 3)(x + 3)}$ cannot be reduced.

9. $\dfrac{(2x - 3)(x - 5)}{(2x - 3)(5 - x)} = \dfrac{(2x - 3)(-1)(5 - x)}{(2x - 3)(5 - x)}$

$= -1$

11. $\dfrac{3x^2 - 3x}{6x^2 + 18x} = \dfrac{3x(x - 1)}{6x(x + 3)}$

$= \dfrac{x - 1}{2(x + 3)}$

13. $\dfrac{x^2 - 9}{x^2 - 6x + 9} = \dfrac{(x - 3)(x + 3)}{(x - 3)(x - 3)}$

$= \dfrac{x + 3}{x - 3}$

15. $\dfrac{w^2 - 8wz + 7z^2}{w^2 + 8wz + 7z^2} = \dfrac{(w - 7z)(w - z)}{(w + 7z)(w + z)}$

cannot be reduced

17. $\dfrac{x^2 + x - 12}{x - 3} = \dfrac{(x + 4)(x - 3)}{x - 3}$

$= x + 4$

19. $\dfrac{x^2 + x - 12}{3 - x} = \dfrac{(x + 4)(x - 3)}{-(x - 3)}$

$= \dfrac{x + 4}{-1}$

$= -(x + 4)$

21. $\dfrac{3a^2 - 13a - 30}{15a^2 + 28a + 5} = \dfrac{(3a + 5)(a - 6)}{(3a + 5)(5a + 1)}$

$= \dfrac{a - 6}{5a + 1}$

23. $\dfrac{2x^3 - 2x^2 - 12x}{3x^2 - 6x} = \dfrac{2x(x - 3)(x + 2)}{3x(x - 2)}$

$= \dfrac{2(x - 3)(x + 2)}{3(x - 2)}$

25. $\dfrac{2x^2}{3x^2 - 6} = \dfrac{2x^2}{3(x^2 - 2)}$

cannot be reduced

27. $\dfrac{3a^2 - 7a - 6}{3 + 5a - 2a^2} = \dfrac{3a^2 - 7a - 6}{-(2a^2 - 5a - 3)}$

$= \dfrac{(3a + 2)(a - 3)}{-(2a + 1)(a - 3)}$

$= -\dfrac{3a + 2}{2a + 1}$

29. $\dfrac{6r^2 - r - 2}{2 + r - 6r^2} = \dfrac{6r^2 - r - 2}{-(6r^2 - r - 2)}$

$= \dfrac{1}{-1}$

$= -1$

31. $\dfrac{a^3 + b^3}{(a + b)^3} = \dfrac{(a + b)(a^2 - ab + b^2)}{(a + b)^3}$

$= \dfrac{a^2 - ab + b^2}{(a + b)^2}$

33. $\dfrac{ax + bx - 2ay - 2by}{4xy - 2x^2} = \dfrac{(a + b)(x - 2y)}{-2x(-2y + x)}$

$= \dfrac{(a + b)(x - 2y)}{-2x(x - 2y)}$

$= -\dfrac{a + b}{2x}$

35. $\dfrac{(a+b)^2 - (x+y)^2}{a+b+x+y} = \dfrac{(a+b-x-y)(a+b+x+y)}{a+b+x+y}$

$= a + b - x - y$

37. $\dfrac{2x}{3y} = \dfrac{2x(3x^2)}{3y(3x^2)}$

$= \dfrac{6x^3}{9x^2y}$

39. $\dfrac{5}{3a^2b} = \dfrac{5(5a^2b)}{3a^2b(5a^2b)}$

$= \dfrac{25a^2b}{15a^4b^2}$

41. $\dfrac{3x}{x - 5} = \dfrac{3x(x + 5)}{(x - 5)(x + 5)}$

$= \dfrac{3x^2 + 15x}{(x - 5)(x + 5)}$

43. $\dfrac{a - b}{x + y} = \dfrac{?}{x^2 + 2xy + y^2}$

$\dfrac{a - b}{x + y} = \dfrac{?}{(x + y)^2}$

$\dfrac{a - b}{x + y} = \dfrac{(a - b)(x + y)}{(x + y)(x + y)}$

45. $\dfrac{y - 2}{2y - 3} = \dfrac{?}{12 - 5y - 2y^2}$

$\dfrac{y - 2}{2y - 3} = \dfrac{?}{-(2y - 3)(y + 4)}$

$\dfrac{y - 2}{2y - 3} = \dfrac{(y - 2)(-1)(y + 4)}{(2y - 3)(-1)(y + 4)}$

$\dfrac{y - 2}{2y - 3} = \dfrac{-y^2 - 2y + 8}{12 - 5y - 2y^2}$

47. $\dfrac{x - y}{a^2 - b^2} = \dfrac{?}{a^2x + a^2y - b^2x - b^2y}$

$\dfrac{x - y}{a^2 - b^2} = \dfrac{?}{(a^2 - b^2)(x + y)}$

$\dfrac{x - y}{a^2 - b^2} = \dfrac{(x - y)(x + y)}{(a^2 - b^2)(x + y)}$

51.
$$\begin{aligned} 10(x + 24) - 6x &= 400 \\ 10x + 240 - 6x &= 400 \\ 4x + 240 &= 400 \\ 4x &= 160 \\ x &= 40 \end{aligned}$$

53.
$$\begin{aligned} f(x) &= 2x^3 - x^2 + 3 \\ f(x^2) &= 2(x^2)^3 - (x^2)^2 + 3 \\ &= 2x^6 - x^4 + 3 \end{aligned}$$

Exercises 6.3

1. $\dfrac{10a^2b}{3xy^3} \cdot \dfrac{9xy^4}{5a^4b^7}$

$= \dfrac{2 \cdot 5 \cdot a^2 \cdot b \cdot 3 \cdot 3 \cdot x \cdot y^3 \cdot y}{3xy^3 \cdot 5a^2 \cdot a^2 \cdot b \cdot b^6}$

$= \dfrac{2 \cdot 3 \cdot y}{a^2b^6}$

$= \dfrac{6y}{a^2b^6}$

3. $$\frac{17a^2b^3}{18yx} \div \frac{34a^2}{9xy} = \frac{17a^2b^3}{18yx} \cdot \frac{9xy}{34a^2}$$

$$= \frac{17a^2b^3 \cdot 9xy}{2 \cdot 9yx \cdot 17 \cdot 2a^2}$$

$$= \frac{b^3}{2 \cdot 2}$$

$$= \frac{b^3}{4}$$

5. $$\frac{16x^3b}{9a} \cdot 24a^2b^3 = \frac{16x^3b}{9a} \cdot \frac{24a^2b^3}{1}$$

$$= \frac{16x^3b \cdot 3 \cdot 8 \cdot a \cdot a \cdot b^3}{3 \cdot 3a}$$

$$= \frac{16x^3b \cdot 8 \cdot ab^3}{3}$$

$$= \frac{128x^3ab^4}{3}$$

7. $$\frac{32r^2s^3}{12a^2b} \cdot \left(\frac{15ab^2}{16r} \cdot \frac{24rs^2}{5ab}\right)$$

$$= \frac{32r^2s^3}{12a^2b} \cdot \left(\frac{3 \cdot 5ab \cdot b \cdot 8 \cdot 3rs^2}{8 \cdot 2r \cdot 5ab}\right)$$

$$= \frac{32r^2s^3}{12a^2b} \cdot \frac{9bs^2}{2}$$

$$= \frac{4 \cdot 2 \cdot 4r^2s^3 \cdot 3 \cdot 3bs^2}{4 \cdot 3a^2b \cdot 2}$$

$$= \frac{12r^2s^5}{a^2}$$

9. $$\frac{32r^2s^3}{12a^2b} \div \left(\frac{15ab^2}{16r} \cdot \frac{24rs^2}{5ab}\right)$$

$$= \frac{32r^2s^3}{12a^2b} \div \left(\frac{5 \cdot 3a \cdot b \cdot b \cdot 8 \cdot 3rs^2}{8 \cdot 2r \cdot 5ab}\right)$$

$$= \frac{32r^2s^3}{12a^2b} \div \frac{9bs^2}{2}$$

$$= \frac{32r^2s^3}{12a^2b} \cdot \frac{2}{9bs^2}$$

$$= \frac{4 \cdot 8r^2 \cdot s^2 \cdot s \cdot 2}{4 \cdot 3a^2\ b \cdot 9bs^2}$$

$$= \frac{16r^2s}{27a^2b^2}$$

11. $$\frac{x^2 - x - 2}{x + 3} \div \frac{3x + 9}{2x + 2}$$

$$= \frac{x^2 - x - 2}{x + 3} \cdot \frac{2x + 2}{3x + 9}$$

$$= \frac{(x - 2)(x + 1)}{x + 3} \cdot \frac{2(x + 1)}{3(x + 3)}$$

$$= \frac{2(x + 1)^2(x - 2)}{3(x + 3)^2}$$

13. $$\frac{x^2 - x - 2}{x + 3} \cdot \frac{3x + 9}{2x + 2}$$

$$= \frac{(x - 2)(x + 1)}{x + 3} \cdot \frac{3(x + 3)}{2(x + 1)}$$

$$= \frac{3(x - 2)}{2}$$

15. $$\frac{5a^3 - 5a^2b}{3a^2 + 3ab} \cdot (a + b)$$

$$= \frac{5a^2(a - b)}{3a(a + b)} \cdot \frac{(a + b)}{1}$$

$$= \frac{5a(a - b)}{3}$$

17. $$\frac{5a^3 - 5a^2b}{3a^2 + 3ab} \div (a + b)$$

$$= \frac{5a^3 - 5a^2b}{3a^2 + 3ab} \cdot \frac{1}{a + b}$$

$$= \frac{5a^2(a - b)}{3a(a + b)} \cdot \frac{1}{a + b}$$

$$= \frac{5a(a - b)}{3(a + b)^2}$$

19. $\dfrac{2x^2 + 3x - 5}{2x + 5} \cdot \dfrac{1}{1 - x}$

$= \dfrac{(2x + 5)(x - 1)}{2x + 5} \cdot \dfrac{1}{-1(x - 1)}$

$= \dfrac{1}{-1}$

$= -1$

21. $\dfrac{2a^2 - 7a + 6}{4a^2 - 9} \cdot \dfrac{4a^2 + 12a + 9}{a^2 - a - 2}$

$= \dfrac{(2a - 3)(a - 2)}{(2a - 3)(2a + 3)} \cdot \dfrac{(2a + 3)(2a + 3)}{(a - 2)(a + 1)}$

$= \dfrac{2a + 3}{a + 1}$

23. $\dfrac{x^2 - y^2}{(x + y)^3} \cdot \dfrac{(x + y)^2}{(x - y)^2}$

$= \dfrac{(x + y)(x - y)}{(x + y)^3} \cdot \dfrac{(x + y)^2}{(x - y)^2}$

$= \dfrac{(x + y)^3(x - y)}{(x + y)^3(x - y)^2}$

$= \dfrac{1}{x - y}$

25. $\dfrac{9x^2 + 3x - 2}{6x^2 - 2x} \div \dfrac{3x + 2}{6x^2}$

$= \dfrac{9x^2 + 3x - 2}{6x^2 - 2x} \cdot \dfrac{6x^2}{3x + 2}$

$= \dfrac{(3x + 2)(3x - 1)}{2x(3x - 1)} \cdot \dfrac{6x^2}{3x + 2}$

$= 3x$

27. $\dfrac{2x^2 + x - 3}{x^2 - 1} \div \dfrac{2x^2 + 5x + 3}{2 - 2x}$

$= \dfrac{2x^2 + x - 3}{x^2 - 1} \cdot \dfrac{2 - 2x}{2x^2 + 5x + 3}$

$= \dfrac{(2x + 3)(x - 1)}{(x + 1)(x - 1)} \cdot \dfrac{-2(x - 1)}{(2x + 3)(x + 1)}$

$= \dfrac{-2(x - 1)}{(x + 1)^2}$ or $\dfrac{-2x + 2}{(x + 1)^2}$

29. $\dfrac{6q^2 - q - 2}{8q^2 + 4q} \cdot \dfrac{8q^2}{6q^2 - 4q}$

$= \dfrac{(3q - 2)(2q + 1)}{4q(2q + 1)} \cdot \dfrac{8q^2}{2q(3q - 2)}$

$= \dfrac{8q^2(3q - 2)(2q + 1)}{8q^2(3q - 2)(2q + 1)}$

$= 1$

31. $\dfrac{m^2 - 10m + 25}{m^2 - 25} \div (5m^2 - 25m)$

$= \dfrac{m^2 - 10m + 25}{m^2 - 25} \cdot \dfrac{1}{5m^2 - 25m}$

$= \dfrac{(m - 5)(m - 5)}{(m - 5)(m + 5)} \cdot \dfrac{1}{5m(m - 5)}$

$= \dfrac{1}{5m(m + 5)}$

33. $\left(\dfrac{2t^2 + 3t + 1}{3t^2 - t - 2}\right) \cdot \left(\dfrac{t - 1}{2t - 1}\right) \cdot \left(\dfrac{3t + 2}{t + 1}\right)$

$= \dfrac{(2t + 1)(t + 1)}{(3t + 2)(t - 1)} \cdot \dfrac{(t - 1)}{(2t - 1)} \cdot \dfrac{(3t + 2)}{(t + 1)}$

$= \dfrac{2t + 1}{2t - 1}$

35. $\dfrac{9a^2 + 9a + 2}{3a^2 - 2a - 1} \cdot \left(\dfrac{a - 1}{3a^2 + 4a + 1} \div \dfrac{3a + 2}{a + 1}\right)$

$= \dfrac{9a^2 + 9a + 2}{3a^2 - 2a - 1} \cdot \left(\dfrac{a - 1}{3a^2 + 4a + 1} \cdot \dfrac{a + 1}{3a + 2}\right)$

$= \dfrac{(3a + 2)(3a + 1)}{(3a + 1)(a - 1)} \cdot \left(\dfrac{a - 1}{(3a + 1)(a + 1)} \cdot \dfrac{a + 1}{3a + 2}\right.$

$= \dfrac{(3a + 2)}{(a - 1)} \cdot \dfrac{(a - 1)}{(3a + 1)(3a + 2)}$

$= \dfrac{1}{3a + 1}$

37. $\dfrac{2r^2 - 5r - 3}{6r - 2} \div \left(\dfrac{r - 3}{2} \div \dfrac{3r - 1}{2r + 1}\right)$

$= \dfrac{2r^2 - 5r - 3}{6r - 2} \div \left(\dfrac{r - 3}{2} \cdot \dfrac{2r + 1}{3r - 1}\right)$

$= \dfrac{(2r + 1)(r - 3)}{2(3r - 1)} \div \left[\dfrac{(r - 3)(2r + 1)}{2(3r - 1)}\right]$

$= \dfrac{(2r + 1)(r - 3)}{2(3r - 1)} \cdot \dfrac{2(3r - 1)}{(r - 3)(2r + 1)}$

$= 1$

39. $\dfrac{x^2 - y^2}{y^3x - x^3y} \div (x^2 + 2xy + y^2)$

$= \dfrac{x^2 - y^2}{y^3x - x^3y} \cdot \dfrac{1}{x^2 + 2xy + y^2}$

$= \dfrac{(x - y)(x + y)}{-xy(x - y)(x + y)} \cdot \dfrac{1}{(x + y)^2}$

$= -\dfrac{1}{xy(x + y)^2}$

41. $f(x) \cdot g(x) = \dfrac{x + 2}{x - 3} \cdot 2x^2 - 2x - 12$

$= \dfrac{x + 2}{x - 3} \cdot \dfrac{2(x - 3)(x + 2)}{1}$

$= 2(x + 2)^2$

43. $f(x) \div g(x) = \dfrac{x + 2}{x - 3} \div (2x^2 - 2x - 12)$

$= \dfrac{x + 2}{x - 3} \cdot \dfrac{1}{2x^2 - 2x - 12}$

$= \dfrac{x + 2}{x - 3} \cdot \dfrac{1}{2(x - 3)(x + 2)}$

$= \dfrac{1}{2(x - 3)^2}$

45. $h(x) \div f(x) \cdot g(x)$

$= \dfrac{1}{3x^2 + 6x} \div \dfrac{x + 2}{x - 3} \cdot (2x^2 - 2x - 12)$

$= \dfrac{1}{3x^2 + 6x} \cdot \dfrac{x - 3}{x + 2} \cdot \dfrac{2x^2 - 2x - 12}{1}$

$= \dfrac{1}{3x(x + 2)} \cdot \dfrac{x - 3}{x + 2} \cdot \dfrac{2(x - 3)(x + 2)}{1}$

$= \dfrac{2(x - 3)^2}{3x(x + 2)}$

47. $\begin{cases} 6x - 2y = 7 \\ 16x + 3y = 2 \end{cases}$

Multiply the 1st equation by 3 and the 2nd equation by 2, then add:

$18x - 6y = 21$
$\underline{32x + 6y = 4}$
$50x = 25$
$x = \dfrac{1}{2}$

$6\left(\dfrac{1}{2}\right) - 2y = 7$
$3 - 2y = 7$
$-2y = 4$
$y = -2$

$x = \dfrac{1}{2}, \quad y = -2$

49. degree: 3

Exercises 6.4

1. $\dfrac{x + 2}{x} - \dfrac{2 - x}{x} = \dfrac{x + 2 - (2 - x)}{x}$

$= \dfrac{x + 2 - 2 + x}{x}$

$= \dfrac{2x}{x}$

$= 2$

3. $\dfrac{10a}{a - b} - \dfrac{10b}{a - b} = \dfrac{10a - 10b}{a - b}$

$= \dfrac{10(a - b)}{a - b}$

$= 10$

5. $\dfrac{3a}{3a^2 + a - 2} - \dfrac{2}{3a^2 + a - 2}$

$= \dfrac{3a - 2}{3a^2 + a - 2}$

$= \dfrac{3a - 2}{(3a - 2)(a + 1)}$

$= \dfrac{1}{a + 1}$

7. $\dfrac{x^2}{x^2 - y^2} + \dfrac{y^2}{x^2 - y^2} - \dfrac{2xy}{x^2 - y^2}$

$= \dfrac{x^2 + y^2 - 2xy}{x^2 - y^2}$

$= \dfrac{x^2 - 2xy + y^2}{x^2 - y^2}$

$= \dfrac{(x - y)(x - y)}{(x - y)(x + y)}$

$= \dfrac{x - y}{x + y}$

9. $\dfrac{5x}{x + 3} + \dfrac{3x}{x + 3} + 4$

$= \dfrac{5x}{x + 3} + \dfrac{3x}{x + 3} + \dfrac{4(x + 3)}{x + 3}$

$= \dfrac{5x + 3x + 4(x + 3)}{x + 3}$

$= \dfrac{5x + 3x + 4x + 12}{x + 3}$

$= \dfrac{12x + 12}{x + 3}$

11. $\dfrac{3x}{2y^2} - \dfrac{4x^2}{9y} = \dfrac{3x(9)}{18y^2} - \dfrac{4x^2(2y)}{18y^2}$

$= \dfrac{3x(9) - 4x^2(2y)}{18y^2}$

$= \dfrac{27x - 8x^2y}{18y^2}$

13. $\dfrac{7y}{6x^2} - \dfrac{8x^2}{9y^2} = \dfrac{7y(3y^2)}{18x^2y^2} - \dfrac{8x^2(2x^2)}{18x^2y^2}$

$= \dfrac{7y(3y^2) - 8x^2(2x^2)}{18x^2y^2}$

$= \dfrac{21y^3 - 16x^4}{18x^2y^2}$

15. $\dfrac{7y}{6x^2} \cdot \dfrac{8x^2}{9y^2} = \dfrac{7y \cdot 2 \cdot 4x^2}{2 \cdot 3x^2 \cdot 9y \cdot y}$

$= \dfrac{28}{27y}$

17. $\dfrac{36a^2}{b^2c} + \dfrac{24}{bc^3} - \dfrac{3}{7bc}$

$= \dfrac{36a^2(7c^2)}{7b^2c^3} + \dfrac{24(7b)}{7b^2c^3} - \dfrac{3(bc^2)}{7b^2c^3}$

$= \dfrac{36a^2(7c^2) + 24(7b) - 3(bc^2)}{7b^2c^3}$

$= \dfrac{252a^2c^2 + 168b - 3bc^2}{7b^2c^3}$

19. $\dfrac{3}{x} + \dfrac{x}{x + 2}$

$= \dfrac{3(x + 2)}{x(x + 2)} + \dfrac{x(x)}{x(x + 2)}$

$= \dfrac{3(x + 2) + x(x)}{x(x + 2)}$

$= \dfrac{3x + 6 + x^2}{x(x + 2)}$

$= \dfrac{x^2 + 3x + 6}{x(x + 2)}$

21. $\dfrac{a}{a - b} - \dfrac{b}{a}$

$= \dfrac{a(a)}{a(a - b)} - \dfrac{b(a - b)}{a(a - b)}$

$= \dfrac{a(a) - b(a - b)}{a(a - b)}$

$= \dfrac{a^2 - ab + b^2}{a(a - b)}$

23. $\dfrac{5}{x+7} - \dfrac{2}{x-3}$

$= \dfrac{5(x-3)}{(x+7)(x-3)} - \dfrac{2(x+7)}{(x+7)(x-3)}$

$= \dfrac{5(x-3) - 2(x+7)}{(x+7)(x-3)}$

$= \dfrac{5x - 15 - 2x - 14}{(x+7)(x-3)}$

$= \dfrac{3x - 29}{(x+7)(x-3)}$

25. $\dfrac{2}{x-7} + \dfrac{3x+1}{x+2}$

$= \dfrac{2(x+2)}{(x-7)(x+2)} + \dfrac{(3x+1)(x-7)}{(x-7)(x+2)}$

$= \dfrac{2(x+2) + (3x+1)(x-7)}{(x-7)(x+2)}$

$= \dfrac{2x + 4 + 3x^2 - 20x - 7}{(x-7)(x+2)}$

$= \dfrac{3x^2 - 18x - 3}{(x-7)(x+2)}$

27. $\dfrac{2r+s}{r-s} - \dfrac{r-2s}{r+s}$

$= \dfrac{(2r+s)(r+s)}{(r-s)(r+s)} - \dfrac{(r-2s)(r-s)}{(r-s)(r+s)}$

$= \dfrac{(2r+s)(r+s) - (r-2s)(r-s)}{(r-s)(r+s)}$

$= \dfrac{2r^2 + 3rs + s^2 - (r^2 - 3rs + 2s^2)}{(r-s)(r+s)}$

$= \dfrac{2r^2 + 3rs + s^2 - r^2 + 3rs - 2s^2}{(r-s)(r+s)}$

$= \dfrac{r^2 + 6rs - s^2}{(r-s)(r+s)}$

29. $\dfrac{4}{x-4} + \dfrac{x}{4-x}$

$= \dfrac{4}{x-4} + \dfrac{x}{(-1)(x-4)}$

$= \dfrac{4}{x-4} + \dfrac{(-1)(x)}{x-4}$

$= \dfrac{4 + (-1)(x)}{x-4}$

$= \dfrac{4-x}{x-4}$

$= \dfrac{-1(x-4)}{x-4}$

$= -1$

31. $\dfrac{7a+3}{2a-1} + \dfrac{5a+4}{1-2a}$

$= \dfrac{7a+3}{2a-1} + \dfrac{5a+4}{(-1)(2a-1)}$

$= \dfrac{7a+3}{2a-1} + \dfrac{(-1)(5a+4)}{2a-1}$

$= \dfrac{7a + 3 + (-1)(5a+4)}{2a-1}$

$= \dfrac{7a + 3 - 5a - 4}{2a-1}$

$= \dfrac{2a-1}{2a-1}$

$= 1$

33. $\dfrac{a}{a-b} - \dfrac{b}{a^2-b^2}$

$= \dfrac{a}{a-b} - \dfrac{b}{(a-b)(a+b)}$

$= \dfrac{a(a+b)}{(a-b)(a+b)} - \dfrac{b}{(a-b)(a+b)}$

$= \dfrac{a(a+b) - b}{(a-b)(a+b)}$

$= \dfrac{a^2 + ab - b}{(a-b)(a+b)}$

35. $\dfrac{a}{a - b} \div \dfrac{b}{a^2 - b^2}$

$= \dfrac{a}{a - b} \cdot \dfrac{a^2 - b^2}{b}$

$= \dfrac{a}{a - b} \cdot \dfrac{(a - b)(a + b)}{b}$

$= \dfrac{a(a + b)}{b}$

37. $\dfrac{2y + 1}{y + 2} + \dfrac{3y}{y + 3} + y$

$= \dfrac{(2y + 1)(y + 3)}{(y + 2)(y + 3)} + \dfrac{3y(y + 2)}{(y + 2)(y + 3)}$

$+ \dfrac{y(y + 2)(y + 3)}{(y + 2)(y + 3)}$

$= \dfrac{(2y + 1)(y + 3) + 3y(y + 2) + y(y + 2)(y + 3)}{(y + 2)(y + 3)}$

$= \dfrac{2y^2 + 7y + 3 + 3y^2 + 6y + y^3 + 5y^2 + 6y}{(y + 2)(y + 3)}$

$= \dfrac{y^3 + 10y^2 + 19y + 3}{(y + 2)(y + 3)}$

39. $\dfrac{5x + 1}{x + 2} + \dfrac{2x + 6}{x^2 + 5x + 6}$

$= \dfrac{5x + 1}{x + 2} + \dfrac{2(x + 3)}{(x + 2)(x + 3)}$

$= \dfrac{5x + 1}{x + 2} + \dfrac{2}{x + 2}$

$= \dfrac{5x + 1 + 2}{x + 2}$

$= \dfrac{5x + 3}{x + 2}$

41. $\dfrac{5x + 1}{x + 2} \div \dfrac{2x + 6}{x^2 + 5x + 6}$

$= \dfrac{5x + 1}{x + 2} \cdot \dfrac{x^2 + 5x + 6}{2x + 6}$

$= \dfrac{5x + 1}{x + 2} \cdot \dfrac{(x + 2)(x + 3)}{2(x + 3)}$

$= \dfrac{5x + 1}{2}$

43. $\dfrac{x + 1}{x^2 - 3x} + \dfrac{x - 2}{x^2 - 6x + 9}$

$= \dfrac{x + 1}{x(x - 3)} + \dfrac{x - 2}{(x - 3)^2}$

$= \dfrac{(x + 1)(x - 3)}{x(x - 3)^2} + \dfrac{(x - 2)(x)}{x(x - 3)^2}$

$= \dfrac{(x + 1)(x - 3) + (x - 2)(x)}{x(x - 3)^2}$

$= \dfrac{x^2 - 2x - 3 + x^2 - 2x}{x(x - 3)^2}$

$= \dfrac{2x^2 - 4x - 3}{x(x - 3)^2}$

45. $\dfrac{a + 3}{a - 3} + \dfrac{a}{a + 4} - \dfrac{3}{a^2 + a - 12}$

$= \dfrac{a + 3}{a - 3} + \dfrac{a}{a + 4} - \dfrac{3}{(a + 4)(a - 3)}$

$= \dfrac{(a + 3)(a + 4)}{(a - 3)(a + 4)} + \dfrac{a(a - 3)}{(a - 3)(a + 4)}$

$- \dfrac{3}{(a - 3)(a + 4)}$

$= \dfrac{(a + 3)(a + 4) + a(a - 3) - 3}{(a - 3)(a + 4)}$

$= \dfrac{a^2 + 7a + 12 + a^2 - 3a - 3}{(a - 3)(a + 4)}$

$= \dfrac{2a^2 + 4a + 9}{(a - 3)(a + 4)}$

47. $\dfrac{y+7}{y+5} - \dfrac{y}{y-3} + \dfrac{16}{y^2+2y-15}$

$= \dfrac{y+7}{y+5} - \dfrac{y}{y-3} + \dfrac{16}{(y+5)(y-3)}$

$= \dfrac{(y+7)(y-3)}{(y+5)(y-3)} - \dfrac{y(y+5)}{(y+5)(y-3)}$

$+ \dfrac{16}{(y+5)(y-3)}$

$= \dfrac{(y+7)(y-3) - y(y+5) + 16}{(y+5)(y-3)}$

$= \dfrac{y^2+4y-21-y^2-5y+16}{(y+5)(y-3)}$

$= \dfrac{-y-5}{(y+5)(y-3)}$

$= \dfrac{-(y+5)}{(y+5)(y-3)}$

$= \dfrac{-1}{y-3}$

49. $\dfrac{3x-4}{x-5} + \dfrac{4x}{10+3x-x^2}$

$= \dfrac{3x-4}{x-5} + \dfrac{4x}{-(x-5)(x+2)}$

$= \dfrac{3x-4}{x-5} + \dfrac{(-1)(4x)}{(x-5)(x+2)}$

$= \dfrac{(3x-4)(x+2)}{(x-5)(x+2)} + \dfrac{-4x}{(x-5)(x+2)}$

$= \dfrac{(3x-4)(x+2)-4x}{(x-5)(x+2)}$

$= \dfrac{3x^2+2x-8-4x}{(x-5)(x+2)}$

$= \dfrac{3x^2-2x-8}{(x-5)(x+2)}$

51. $\dfrac{2a-1}{a^2+a-6} + \dfrac{a+2}{a^2-2a-15} - \dfrac{a+1}{a^2-7a+10}$

$= \dfrac{2a-1}{(a+3)(a-2)} + \dfrac{a+2}{(a-5)(a+3)} - \dfrac{a+1}{(a-5)(a-2)}$

$= \dfrac{(2a-1)(a-5)}{(a+3)(a-2)(a-5)} + \dfrac{(a+2)(a-2)}{(a+3)(a-2)(a-5)}$

$- \dfrac{(a+1)(a+3)}{(a+3)(a-2)(a-5)}$

$= \dfrac{2a^2-11a+5+a^2-4-(a^2+4a+3)}{(a+3)(a-2)(a-5)}$

$= \dfrac{3a^2-11a+1-a^2-4a-3}{(a+3)(a-2)(a-5)}$

$= \dfrac{2a^2-15a-2}{(a+3)(a-2)(a-5)}$

53. $\dfrac{5a}{3a-1} + \dfrac{2a+1}{5a+2} + 5a + 1$

$= \dfrac{5a(5a+2)}{(3a-1)(5a+2)} + \dfrac{(2a+1)(3a-1)}{(3a-1)(5a+2)}$

$+ \dfrac{(5a+1)(3a-1)(5a+2)}{(3a-1)(5a+2)}$

$= \dfrac{5a(5a+2)+(2a+1)(3a-1)+(5a+1)(3a-1)(5a+2)}{(3a-1)(5a+2)}$

$= \dfrac{25a^2+10a+6a^2+a-1+75a^3+20a^2-9a-2}{(3a-1)(5a+2)}$

$= \dfrac{75a^3+51a^2+2a-3}{(3a-1)(5a+2)}$

55. $\dfrac{2r+xs}{r+s} + \dfrac{2s+xr}{r+s}$

$= \dfrac{2r+xs+2s+xr}{r+s}$

$= \dfrac{2r+2s+xs+xr}{r+s}$

$= \dfrac{(r+s)(2+x)}{r+s}$

$= 2 + x$

57. $\dfrac{r^2}{r^3 - s^3} + \dfrac{rs}{r^3 - s^3} + \dfrac{s^2}{r^3 - s^3}$

$= \dfrac{r^2 + rs + s^2}{r^3 - s^3}$

$= \dfrac{r^2 + rs + s^2}{(r - s)(r^2 + rs + s^2)}$

$= \dfrac{1}{r - s}$

59. $\dfrac{2s + t}{s^3 - t^3} + \dfrac{3s}{s^2 + st + t^2}$

$\dfrac{2s + t}{(s - t)(s^2 + st + t^2)} + \dfrac{3s}{s^2 + st + t^2}$

$\dfrac{2s + t}{(s - t)(s^2 + st + t^2)} + \dfrac{3s(s - t)}{(s - t)(s^2 + st + t^2)}$

$= \dfrac{2s + t + 3s(s - t)}{(s - t)(s^2 + st + t^2)}$

$= \dfrac{2s + t + 3s^2 - 3st}{(s - t)(s^2 + st + t^2)}$

$= \dfrac{3s^2 - 3st + 2s + t}{s^3 - t^3}$

61. (a) $\dfrac{x^2 + 4x}{4x} = \dfrac{x(x + 4)}{4x}$

$= \dfrac{x + 4}{4}$

(b) $\dfrac{x^2 + 4x}{4x} = \dfrac{x^2}{4x} + \dfrac{4x}{4x}$

$= \dfrac{x}{4} + 1$

63. (a) $\dfrac{15x^3y^2 - 10x^2y^3}{5x^2y^2} = \dfrac{5x^2y^2(3x - 2y)}{5x^2y^2}$

$= 3x - 2y$

(b) $\dfrac{15x^3y^2 - 10x^2y^3}{5x^2y^2} = \dfrac{15x^3y^2}{5x^2y^2} - \dfrac{10x^2y^3}{5x^2y^2}$

$= 3x - 2y$

65. (a) $\dfrac{6m^2n - 4m^3n^2 - 9mn}{15mn^2}$

$= \dfrac{mn(6m - 4m^2n - 9)}{15mn^2}$

$= \dfrac{6m - 4m^2n - 9}{15n}$

(b) $\dfrac{6m^2n - 4m^3n^2 - 9mn}{15mn}$

$= \dfrac{6m^2n}{15mn^2} - \dfrac{4m^3n^2}{15mn^2} - \dfrac{9mn}{15mn^2}$

$= \dfrac{2m}{5n} - \dfrac{4m^2}{15} - \dfrac{3}{5n}$

67. $f(x) + g(x)$

$= \dfrac{x - 2}{x + 2} + \dfrac{2x}{x^2 - 4}$

$= \dfrac{x - 2}{x + 2} + \dfrac{2x}{(x + 2)(x - 2)}$

$= \dfrac{(x - 2)(x - 2)}{(x + 2)(x - 2)} + \dfrac{2x}{(x + 2)(x - 2)}$

$= \dfrac{(x - 2)(x - 2) + 2x}{(x + 2)(x - 2)}$

$= \dfrac{x^2 - 4x + 4 + 2x}{(x + 2)(x - 2)}$

$= \dfrac{x^2 - 2x + 4}{(x + 2)(x - 2)}$

69. $f(x) \cdot h(x) = \dfrac{x - 2}{x + 2} \cdot \dfrac{x}{x - 2}$

$= \dfrac{x}{x + 2}$

71. $g(x) + h(x)$

$$= \frac{2x}{x^2 - 4} + \frac{x}{x - 2}$$

$$= \frac{2x}{(x - 2)(x + 2)} + \frac{x}{x - 2}$$

$$= \frac{2x}{(x - 2)(x + 2)} + \frac{x(x + 2)}{(x - 2)(x + 2)}$$

$$= \frac{2x + x(x + 2)}{(x - 2)(x + 2)}$$

$$= \frac{2x + x^2 + 2x}{(x - 2)(x + 2)}$$

$$= \frac{x^2 + 4x}{(x - 2)(x + 2)}$$

73. $h(x) \div g(x) = \frac{x}{x - 2} \div \frac{2x}{x^2 - 4}$

$$= \frac{x}{x - 2} \cdot \frac{x^2 - 4}{2x}$$

$$= \frac{x}{x - 2} \cdot \frac{(x - 2)(x + 2)}{2x}$$

$$= \frac{x + 2}{2}$$

75. $f(x) = \frac{3}{x}$

$$f(x + h) = \frac{3}{x + h}$$

$$f(x + h) - f(x) = \frac{3}{x + h} - \frac{3}{x}$$

$$= \frac{3x}{x(x + h)} - \frac{3(x + h)}{x(x + h)}$$

$$= \frac{3x - 3(x + h)}{x(x + h)}$$

$$= \frac{3x - 3x - 3h}{x(x + h)}$$

$$= \frac{-3h}{x(x + h)}$$

77. $f(x) = \dfrac{x}{x - 3}$

$f(x + h) = \dfrac{x + h}{x + h - 3}$

$$f(x + h) - f(x) = \frac{x + h}{x + h - 3} - \frac{x}{x - 3}$$

$$= \frac{(x + h)(x - 3)}{(x + h - 3)(x - 3)} - \frac{x(x + h - 3)}{(x - 3)(x + h - 3)}$$

$$= \frac{(x + h)(x - 3) - x(x + h - 3)}{(x + h - 3)(x - 3)}$$

$$= \frac{x^2 - 3x + xh - 3h - x^2 - xh + 3x}{(x + h - 3)(x - 3)}$$

$$= \frac{-3h}{(x + h - 3)(x - 3)}$$

81. $m = \dfrac{y_2 - y_1}{x_2 - x_1}$

$$= \frac{2 - (-2)}{-4 - 5}$$

$$= -\frac{4}{9}$$

83. $2x - 3y \geq 12$ (solid line)

$2x - 3y = 12$

x	y
0	-4
6	0

Test point: (0, 0)

$2(0) - 3(0) \geq 12$

$0 \geq 12$

False

Shade half-plane not containing (0, 0).

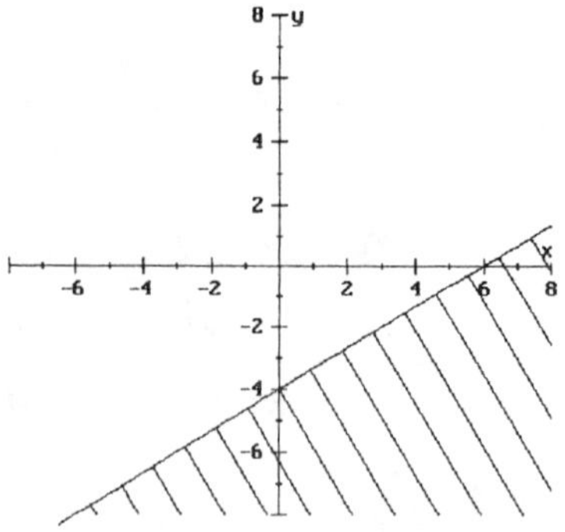

8 y 6 4 2 -6 -4 -2 2 4 6 8 x -2 -4 -6

Exercises 6.5

1. $\dfrac{\frac{3}{xy^2}}{\frac{15}{x^2y}} = \dfrac{3}{xy^2} \div \dfrac{15}{x^2y}$

$$= \frac{3}{xy^2} \cdot \frac{x^2y}{15}$$

$$= \frac{x}{5y}$$

3. $\dfrac{\frac{3}{x - y}}{\frac{x - y}{3}} = \dfrac{3}{x - y} \div \dfrac{x - y}{3}$

$$= \frac{3}{x - y} \cdot \frac{3}{x - y}$$

$$= \frac{9}{(x - y)^2}$$

5. $$\frac{a - \frac{1}{3}}{\frac{9a^2 - 1}{3a}}$$

$$= \frac{a - \frac{1}{3}}{\frac{9a^2 - 1}{3a}} \cdot \frac{3a}{3a}$$

$$= \frac{3a^2 - \frac{1}{3} \cdot \frac{3a}{1}}{\frac{9a^2 - 1}{3a} \cdot \frac{3a}{1}}$$

$$= \frac{3a^2 - a}{9a^2 - 1}$$

$$= \frac{a(3a - 1)}{(3a - 1)(3a + 1)}$$

$$= \frac{a}{3a + 1}$$

7. $$\frac{x + \frac{2}{x^2}}{\frac{1}{x} + 2}$$

$$= \frac{x + \frac{2}{x^2}}{\frac{1}{x} + 2} \cdot \frac{x^2}{x^2}$$

$$= \frac{x(x^2) + \frac{2}{x^2} \cdot \frac{x^2}{1}}{\frac{1}{x} \cdot \frac{x^2}{1} + 2x^2}$$

$$= \frac{x^3 + 2}{x + 2x^2}$$

$$= \frac{x^3 + 2}{x(1 + 2x)}$$

9. $$\left(1 - \frac{4}{x^2}\right) \div \left(\frac{1}{x} - \frac{2}{x^2}\right)$$

$$= \frac{1 - \frac{4}{x^2}}{\frac{1}{x} - \frac{2}{x^2}} \cdot \frac{x^2}{x^2}$$

$$= \frac{x^2 - \frac{4}{x^2} \cdot \frac{x^2}{1}}{\frac{1}{x} \cdot \frac{x^2}{1} - \frac{2}{x^2} \cdot \frac{x^2}{1}}$$

$$= \frac{x^2 - 4}{x - 2}$$

$$= \frac{(x - 2)(x + 2)}{x - 2}$$

$$= x + 2$$

11. $$\frac{1 - \frac{5}{y}}{y + 3 - \frac{40}{y}}$$

$$= \frac{1 - \frac{5}{y}}{y + 3 - \frac{40}{y}} \cdot \frac{y}{y}$$

$$= \frac{y - \frac{5}{y} \cdot \frac{y}{1}}{y^2 + 3y - \frac{40}{y} \cdot \frac{y}{1}}$$

$$= \frac{y - 5}{y^2 + 3y - 40}$$

$$= \frac{y - 5}{(y + 8)(y - 5)}$$

$$= \frac{1}{y + 8}$$

13. $$\frac{1 - \frac{4}{z} + \frac{4}{z^2}}{\frac{1}{z^2} - \frac{2}{z^3}}$$

$$= \frac{1 - \frac{4}{z} + \frac{4}{z^2}}{\frac{1}{z^2} - \frac{2}{z^3}} \cdot \frac{z^3}{z^3}$$

$$= \frac{z^3 - \frac{4}{z} \cdot \frac{z^3}{1} + \frac{4}{z^2} \cdot \frac{z^3}{1}}{\frac{1}{z^2} \cdot \frac{z^3}{1} - \frac{2}{z^3} \cdot \frac{z^3}{1}}$$

$$= \frac{z^3 - 4z^2 + 4z}{z - 2}$$

$$= \frac{z(z - 2)^2}{z - 2}$$

$$= z(z - 2)$$

15. $$\frac{\frac{4}{y^2} - \frac{12}{xy} + \frac{9}{x^2}}{\frac{4}{y^2} - \frac{9}{x^2}}$$

$$= \frac{\frac{4}{y^2} - \frac{12}{xy} + \frac{9}{x^2}}{\frac{4}{y^2} - \frac{9}{x^2}} \cdot \frac{x^2y^2}{x^2y^2}$$

$$= \frac{\frac{4}{y^2} \cdot \frac{x^2y^2}{1} - \frac{12}{xy} \cdot \frac{x^2y^2}{1} + \frac{9}{x^2} \cdot \frac{x^2y^2}{1}}{\frac{4}{y^2} \cdot \frac{x^2y^2}{1} - \frac{9}{x^2} \cdot \frac{x^2y^2}{1}}$$

$$= \frac{4x^2 - 12xy + 9y^2}{4x^2 - 9y^2}$$

$$= \frac{(2x - 3y)(2x - 3y)}{(2x - 3y)(2x + 3y}$$

$$= \frac{2x - 3y}{2x + 3y}$$

17. $$\frac{\frac{2x}{y} + 7 + \frac{5y}{x}}{3x + 2y - \frac{y^2}{x}}$$

$$= \frac{\frac{2x}{y} + 7 + \frac{5y}{x}}{3x + 2y - \frac{y^2}{x}} \cdot \frac{xy}{xy}$$

$$= \frac{\frac{2x}{y} \cdot \frac{xy}{1} + 7xy + \frac{5y}{x} \cdot \frac{xy}{1}}{3x(xy) + 2y(xy) - \frac{y^2}{x} \cdot \frac{xy}{1}}$$

$$= \frac{2x^2 + 7xy + 5y^2}{3x^2y + 2xy^2 - y^3}$$

$$= \frac{(2x + 5y)(x + y)}{y(3x - y)(x + y)}$$

$$= \frac{2x + 5y}{y(3x - y)}$$

19. $$\frac{1 - \frac{3}{x} - \frac{10}{x^2}}{4 + \frac{8}{x}}$$

$$= \frac{1 - \frac{3}{x} - \frac{10}{x^2}}{4 + \frac{8}{x}} \cdot \frac{x^2}{x^2}$$

$$= \frac{x^2 - \frac{3}{x} \cdot \frac{x^2}{1} - \frac{10}{x^2} \cdot \frac{x^2}{1}}{4x^2 + \frac{8}{x} \cdot \frac{x^2}{1}}$$

$$= \frac{x^2 - 3x - 10}{4x^2 + 8x}$$

$$= \frac{(x - 5)(x + 2)}{4x(x + 2)}$$

$$= \frac{x - 5}{4x}$$

21. $$\frac{9 - \frac{25}{t^2}}{6 + \frac{10}{t}}$$

$$= \frac{9 - \frac{25}{t^2}}{6 + \frac{10}{t}} \cdot \frac{t^2}{t^2}$$

$$= \frac{9t^2 - \frac{25}{t^2} \cdot \frac{t^2}{1}}{6t^2 + \frac{10}{t} \cdot \frac{t^2}{1}}$$

$$= \frac{9t^2 - 25}{6t^2 + 10t}$$

$$= \frac{(3t - 5)(3t + 5)}{2t(3t + 5)}$$

$$= \frac{3t - 5}{2t}$$

23. $$\frac{\frac{1}{x - 3} - \frac{1}{x}}{3}$$

$$= \frac{\frac{1}{x - 3} - \frac{1}{x}}{3} \cdot \frac{x(x - 3)}{x(x - 3)}$$

$$= \frac{\frac{1}{x - 3} \cdot \frac{x(x - 3)}{1} - \frac{1}{x} \cdot \frac{x(x - 3)}{1}}{3x(x - 3)}$$

$$= \frac{x - (x - 3)}{3x(x - 3)}$$

$$= \frac{x - x + 3}{3x(x - 3)}$$

$$= \frac{3}{3x(x - 3)}$$

$$= \frac{1}{x(x - 3)}$$

25. $$\frac{\frac{h}{h + 1} - \frac{h}{h - 1}}{2}$$

$$= \frac{\frac{h}{h + 1} - \frac{h}{h - 1}}{2} \cdot \frac{(h + 1)(h - 1)}{(h + 1)(h - 1)}$$

$$= \frac{\frac{h}{h + 1} \cdot \frac{(h + 1)(h - 1)}{1} - \frac{h}{h - 1} \cdot \frac{(h + 1)(h - 1)}{1}}{2(h + 1)(h - 1)}$$

$$= \frac{h(h - 1) - h(h + 1)}{2(h + 1)(h - 1)}$$

$$= \frac{h^2 - h - h^2 - h}{2(h + 1)(h - 1)}$$

$$= \frac{-2h}{2(h + 1)(h - 1)}$$

$$= \frac{-h}{(h + 1)(h - 1)}$$

27. $$\left(3 + \frac{1}{2x - 1}\right) \div \left(5 + \frac{x}{2x - 1}\right)$$

$$= \frac{3 + \frac{1}{2x - 1}}{5 + \frac{x}{2x - 1}}$$

$$= \frac{3 + \frac{1}{2x - 1}}{5 + \frac{x}{2x - 1}} \cdot \frac{2x - 1}{2x - 1}$$

$$= \frac{3(2x - 1) + \frac{1}{2x - 1} \cdot \frac{2x - 1}{1}}{5(2x - 1) + \frac{x}{2x - 1} \cdot \frac{2x - 1}{1}}$$

$$= \frac{6x - 3 + 1}{10x - 5 + x}$$

$$= \frac{6x - 2}{11x - 5}$$

29. $$\frac{\frac{4}{x+3} - \frac{4}{x}}{3}$$

$$= \frac{\frac{4}{x+3} - \frac{4}{x}}{3} \cdot \frac{x(x+3)}{x(x+3)}$$

$$= \frac{\frac{4}{x+3} \cdot \frac{x(x+3)}{1} - \frac{4}{x} \cdot \frac{x(x+3)}{1}}{3x(x+3)}$$

$$= \frac{4x - 4(x+3)}{3x(x+3)}$$

$$= \frac{4x - 4x - 12}{3x(x+3)}$$

$$= \frac{-12}{3x(x+3)}$$

$$= \frac{-4}{x(x+3)}$$

31. $$f(x) = \frac{2x}{x-3}$$

$$f\left(\frac{2}{3}\right) = \frac{2\left(\frac{2}{3}\right)}{\frac{2}{3} - 3}$$

$$= \frac{\frac{4}{3}}{\frac{2}{3} - 3}$$

$$= \frac{\frac{4}{3}}{-\frac{7}{3}}$$

$$= -\frac{4}{7}$$

33. $$g(x) = -\frac{1}{x}$$

$$g\left(\frac{2}{x}\right) = -\frac{1}{\frac{2}{x}} = -\frac{x}{2}$$

35. $$f(x) = \frac{2x}{x-3}$$

$$f(x+3) = \frac{2(x+3)}{x+3-3}$$

$$= \frac{2x+6}{x}$$

$$\frac{f(x+3) - f(x)}{3}$$

$$= \frac{\frac{2x+6}{x} - \frac{2x}{x-3}}{3}$$

$$= \frac{\frac{2x+6}{x} - \frac{2x}{x-3}}{3} \cdot \frac{x(x-3)}{x(x-3)}$$

$$= \frac{\frac{2x+6}{x} \cdot \frac{x(x-3)}{1} - \frac{2x}{x-3} \cdot \frac{x(x-3)}{1}}{3x(x-3)}$$

$$= \frac{(2x+6)(x-3) - 2x(x)}{3x(x-3)}$$

$$= \frac{2x^2 - 18 - 2x^2}{3x(x-3)}$$

$$= \frac{-18}{3x(x-3)}$$

$$= \frac{-6}{x(x-3)}$$

37. $$h(x) = \frac{x+2}{x}$$

$$h(x-2) = \frac{x-2+2}{x-2}$$

$$= \frac{x}{x-2}$$

$$\frac{h(x-2)-h(x)}{2}$$

$$= \frac{\frac{x}{x-2} - \frac{x+2}{x}}{2}$$

$$= \frac{\frac{x}{x-2} - \frac{x+2}{x}}{2} \cdot \frac{x(x-2)}{x(x-2)}$$

$$= \frac{\frac{x}{x-2} \cdot \frac{x(x-2)}{1} - \frac{x+2}{x} \cdot \frac{x(x-2)}{1}}{2x(x-2)}$$

$$= \frac{x^2 - (x+2)(x-2)}{2x(x-2)}$$

$$= \frac{x^2 - (x^2-4)}{2x(x-2)}$$

$$= \frac{x^2 - x^2 + 4}{2x(x-2)}$$

$$= \frac{4}{2x(x-2)}$$

$$= \frac{2}{x(x-2)}$$

39. $\frac{f(x)+g(x)}{h(x)}$

$$= \frac{\frac{2x}{x-3} + \left(-\frac{1}{x}\right)}{\frac{x+2}{x}}$$

$$= \frac{\frac{2x}{x-3} - \frac{1}{x}}{\frac{x+2}{x}} \cdot \frac{x(x-3)}{x(x-3)}$$

$$= \frac{\frac{2x}{x-3} \cdot \frac{x(x-3)}{1} - \frac{1}{x} \cdot \frac{x(x-3)}{1}}{\frac{x+2}{x} \cdot \frac{x(x-3)}{1}}$$

$$= \frac{2x^2 - (x-3)}{(x+2)(x-3)}$$

$$= \frac{2x^2 - x + 3}{x^2 - x - 6}$$

41. $\frac{g(x)+1}{h(x)}$

$$= \frac{-\frac{1}{x} + 1}{\frac{x+2}{x}}$$

$$= \frac{-\frac{1}{x} + 1}{\frac{x+2}{x}} \cdot \frac{x}{x}$$

$$= \frac{-\frac{1}{x} \cdot \frac{x}{1} + x}{\frac{x+2}{x} \cdot \frac{x}{1}}$$

$$= \frac{-1 + x}{x+2}$$

43. $\frac{f(x)-h(x)}{3}$

$$= \frac{\frac{2x}{x-3} - \frac{x+2}{x}}{3}$$

$$= \frac{\frac{2x}{x-3} - \frac{x+2}{x}}{3} \cdot \frac{x(x-3)}{x(x-3)}$$

$$= \frac{\frac{2x}{x-3} \cdot \frac{x(x-3)}{1} - \frac{x+2}{x} \cdot \frac{x(x-3)}{1}}{3x(x-3)}$$

$$= \frac{2x^2 - (x+2)(x-3)}{3x(x-3)}$$

$$= \frac{2x^2 - (x^2 - x - 6)}{3x(x-3)}$$

$$= \frac{2x^2 - x^2 + x + 6}{3x(x-3)}$$

$$= \frac{x^2 + x + 6}{3x(x-3)}$$

45. $(x - 4)(x + 1) = (x - 3)(x - 2)$
$$x^2 - 3x - 4 = x^2 - 5x + 6$$
$$-3x - 4 = -5x + 6$$
$$2x - 4 = 6$$
$$2x = 10$$
$$x = 5$$

47. $3x - 2y = 5$
$$-2y = -3x + 5$$
$$y = \frac{3}{2}x - \frac{5}{2}$$
$$m = \frac{3}{2}$$

Parallel lines have equal slopes.

$$y - y_1 = m(x - x_1)$$

$$y - 5 = \frac{3}{2}[x - (-4)]$$

$$y - 5 = \frac{3}{2}(x + 4)$$
$$y - 5 = \frac{3}{2}x + 6$$
$$y = \frac{3}{2}x + 11$$

Exercises 6.6

1. $$\frac{x}{3} - \frac{x}{2} + \frac{x}{4} = 1$$

$$12\left(\frac{x}{3} - \frac{x}{2} + \frac{x}{4}\right) = 12(1)$$

$$\frac{12}{1} \cdot \frac{x}{3} - \frac{12}{1} \cdot \frac{x}{2} + \frac{12}{1} \cdot \frac{x}{4} = 12$$

$$4x - 6x + 3x = 12$$
$$x = 12$$

3. $$\frac{t}{3} + \frac{t}{5} < \frac{t}{6} - 11$$

$$30\left(\frac{t}{3} + \frac{t}{5}\right) < 30\left(\frac{t}{6} - 11\right)$$

$$\frac{30}{1} \cdot \frac{t}{3} + \frac{30}{1} \cdot \frac{t}{5} < \frac{30}{1} \cdot \frac{t}{6} - 30(11)$$

$$10t + 6t < 5t - 330$$
$$16t < 5t - 330$$
$$11t < -330$$
$$t < -30$$

5. $$\frac{a - 1}{6} + \frac{a + 1}{10} = a - 3$$

$$30\left(\frac{a - 1}{6} + \frac{a + 1}{10}\right) = 30(a - 3)$$

$$\frac{30}{1} \cdot \frac{a - 1}{6} + \frac{30}{1} \cdot \frac{a + 1}{10} = 30(a - 3)$$

$$5(a - 1) + 3(a + 1) = 30(a - 3)$$
$$5a - 5 + 3a + 3 = 30a - 90$$
$$8a - 2 = 30a - 90$$
$$-2 = 22a - 90$$
$$88 = 22a$$
$$4 = a$$

7. $$\frac{y - 5}{2} = \frac{y - 2}{5}$$

$$\frac{10}{1} \cdot \frac{y - 5}{2} = \frac{10}{1} \cdot \frac{y - 2}{5}$$

$$5(y - 5) = 2(y - 2)$$
$$5y - 25 = 2y - 4$$
$$3y - 25 = -4$$
$$3y = 21$$
$$y = 7$$

9. $$\frac{y - 5}{2} \le \frac{y - 2}{5} + 3$$

$$10\left(\frac{y - 5}{2}\right) \le 10\left(\frac{y - 2}{5} + 3\right)$$

$$\frac{10}{1} \cdot \frac{y - 5}{2} \le \frac{10}{1} \cdot \frac{y - 2}{5} + 10(3)$$

$$5(y - 5) \le 2(y - 2) + 30$$
$$5y - 25 \le 2y - 4 + 30$$
$$5y - 25 \le 2y + 26$$
$$3y - 25 \le 26$$
$$3y \le 51$$
$$y \le 17$$

11.
$$\frac{x-3}{4}-\frac{x-4}{3}=2$$
$$12\left(\frac{x-3}{4}-\frac{x-4}{3}\right)=12(2)$$
$$\frac{12}{1}\cdot\frac{x-3}{4}-\frac{12}{1}\cdot\frac{x-4}{3}=24$$
$$\begin{aligned}3(x-3)-4(x-4)&=24\\3x-9-4x+16&=24\\-x+7&=24\\-x&=17\\x&=-17\end{aligned}$$

13.
$$\frac{x-3}{4}-\frac{x-4}{3}\ge 2$$
$$12\left(\frac{x-3}{4}-\frac{x-4}{3}\right)\ge 12(2)$$
$$\frac{12}{1}\cdot\frac{x-3}{4}-\frac{12}{1}\cdot\frac{x-4}{3}\ge 24$$
$$\begin{aligned}3(x-3)-4(x-4)&\ge 24\\3x-9-4x+16&\ge 24\\-x+7&\ge 24\\-x&\ge 17\\x&\le -17\end{aligned}$$

15.
$$\frac{3x+11}{6}-\frac{2x+1}{3}=x+5$$
$$18\left(\frac{3x+11}{6}-\frac{2x+1}{3}\right)=18(x+5)$$
$$\frac{18}{1}\cdot\frac{3x+11}{6}-\frac{18}{1}\cdot\frac{2x+1}{3}=18x+90$$
$$\begin{aligned}3(3x+11)-6(2x+1)&=18x+90\\9x+33-12x-6&=18x+90\\-3x+27&=18x+90\\27&=21x+90\\-63&=21x\\-3&=x\end{aligned}$$

17.
$$\frac{x-3}{5}-\frac{3x+1}{4}<8$$
$$20\left(\frac{x-3}{5}-\frac{3x+1}{4}\right)<20(8)$$
$$\frac{20}{1}\cdot\frac{x-3}{5}-\frac{20}{1}\cdot\frac{3x+1}{4}<160$$
$$\begin{aligned}4(x-3)-5(3x+1)&<160\\4x-12-15x-5&<160\\-11x-17&<160\\-11x&<177\\x&>-\frac{177}{11}\end{aligned}$$

19.
$$\frac{5}{x}-\frac{1}{2}=\frac{3}{x}$$
$$2x\left(\frac{5}{x}-\frac{1}{2}\right)=2x\left(\frac{3}{x}\right)$$
$$\frac{2x}{1}\cdot\frac{5}{x}-\frac{2x}{1}\cdot\frac{1}{2}=\frac{2x}{1}\cdot\frac{3}{x}$$
$$\begin{aligned}10-x&=6\\-x&=-4\\x&=4\end{aligned}$$

21. $\frac{4}{x}-\frac{1}{5}+\frac{7}{2x}$
$$=\frac{4(10)}{10x}-\frac{1(2x)}{10x}+\frac{7(5)}{10x}$$
$$=\frac{4(10)-2x+7(5)}{10x}$$
$$=\frac{40-2x+35}{10x}$$
$$=\frac{-2x+75}{10x}$$

23.
$$\frac{1}{t-3} + \frac{2}{t} = \frac{5}{3t}$$

$$3t(t-3)\left(\frac{1}{t-3} + \frac{2}{t}\right) = 3t(t-3)\left(\frac{5}{3t}\right)$$

$$\frac{3t(t-3)}{1} \cdot \frac{1}{t-3} + \frac{3t(t-3)}{1} \cdot \frac{2}{t} = \frac{3t(t-3)}{1} \cdot \frac{5}{3t}$$

$$\begin{aligned} 3t + 3(t-3)(2) &= 5(t-3) \\ 3t + 6t - 18 &= 5t - 15 \\ 9t - 18 &= 5t - 15 \\ 4t - 18 &= -15 \\ 4t &= 3 \\ t &= \frac{3}{4} \end{aligned}$$

25.
$$\frac{6}{a-3} - \frac{3}{8} = \frac{21}{4a-12}$$

$$\frac{6}{a-3} - \frac{3}{8} = \frac{21}{4(a-3)}$$

$$8(a-3)\left(\frac{6}{a-3} - \frac{3}{8}\right) = 8(a-3)\left[\frac{21}{4(a-3)}\right]$$

$$\frac{8(a-3)}{1} \cdot \frac{6}{a-3} - \frac{8(a-3)}{1} \cdot \frac{3}{8} = \frac{8(a-3)}{1} \cdot \frac{21}{4(a-3)}$$

$$\begin{aligned} 48 - 3(a-3) &= 2(21) \\ 48 - 3a + 9 &= 42 \\ -3a + 57 &= 42 \\ -3a &= -15 \\ a &= 5 \end{aligned}$$

27.
$$\frac{7}{x-5} + 2 = \frac{x+2}{x-5}$$

$$(x-5)\left(\frac{7}{x-5} + 2\right) = (x-5)\left(\frac{x+2}{x-5}\right)$$

$$\frac{x-5}{1} \cdot \frac{7}{x-5} + (x-5)(2) = \frac{x-5}{1} \cdot \frac{x+2}{x-5}$$

$$\begin{aligned} 7 + 2x - 10 &= x + 2 \\ 2x - 3 &= x + 2 \\ x - 3 &= 2 \\ x &= 5 \end{aligned}$$

No solution since $x = 5$ causes a denominator to equal 0 in the original equation.

29.
$$\frac{4}{y^2-2y}-\frac{3}{2y}=\frac{17}{6y}$$
$$\frac{4}{y(y-2)}-\frac{3}{2y}=\frac{17}{6y}$$
$$6y(y-2)\left[\frac{4}{y(y-2)}-\frac{3}{2y}\right]=6y(y-2)\left(\frac{17}{6y}\right)$$
$$\frac{6y(y-2)}{1}\cdot\frac{4}{y(y-2)}-\frac{6y(y-2)}{1}\cdot\frac{3}{2y}=\frac{6y(y-2)}{1}\cdot\frac{17}{6y}$$
$$\begin{aligned}24-9(y-2)&=17(y-2)\\24-9y+18&=17y-34\\-9y+42&=17y-34\\42&=26y-34\\76&=26y\\\frac{38}{13}&=y\end{aligned}$$

31.
$$\frac{2x}{x-5}-\frac{2x+1}{x+2}=\frac{3}{x+2}$$
$$(x-5)(x+2)\left(\frac{2x}{x-5}-\frac{2x+1}{x+2}\right)=(x-5)(x+2)\left(\frac{3}{x+2}\right)$$
$$(x-5)(x+2)\left(\frac{2x}{x-5}\right)-(x-5)(x+2)\left(\frac{2x+1}{x+2}\right)=(x-5)(x+2)\left(\frac{3}{x+2}\right)$$
$$\begin{aligned}(x+2)(2x)-(x-5)(2x+1)&=(x-5)(3)\\2x^2+4x-(2x^2-9x-5)&=3x-15\\2x^2+4x-2x^2+9x+5&=3x-15\\10x+5&=-15\\10x&=-20\\x&=-2\end{aligned}$$

No solution since $x=-2$ causes a denominator to equal 0 in the original equation.

33.
$$\frac{5}{y^2+3y}-\frac{4}{3y}+\frac{1}{2}$$
$$=\frac{5}{y(y+3)}-\frac{4}{3y}+\frac{1}{2}$$
$$=\frac{5(6)}{6y(y+3)}-\frac{4(2)(y+3)}{6y(y+3)}+\frac{3y(y+3)}{6y(y+3)}$$
$$=\frac{5(6)-4(2)(y+3)+3y(y+3)}{6y(y+3)}$$
$$=\frac{30-8y-24+3y^2+9y}{6y(y+3)}$$
$$=\frac{3y^2+y+6}{6y(y+3)}$$

35.

$$\frac{9}{x^2 + 4x} = \frac{6}{x^2 + 2x}$$

$$\frac{9}{x(x + 4)} = \frac{6}{x(x + 2)}$$

$$\frac{x(x+4)(x+2)}{1} \cdot \frac{9}{x(x+4)} = \frac{x(x+4)(x+2)}{1} \cdot \frac{6}{x(x+2)}$$

$$9(x + 2) = 6(x + 4)$$
$$9x + 18 = 6x + 24$$
$$3x + 18 = 24$$
$$3x = 6$$
$$x = 2$$

37.

$$x + \frac{1}{x} = 2$$

$$x\left(x + \frac{1}{x}\right) = x(2)$$

$$x^2 + \frac{x}{1} \cdot \frac{1}{x} = 2x$$

$$x^2 + 1 = 2x$$
$$x^2 - 2x + 1 = 0$$
$$(x - 1)(x - 1) = 0$$
$$x - 1 = 0$$
$$x = 1$$

39.

$$\frac{3x + 2}{x^2 - 4x - 5} + \frac{x - 4}{x + 1} = \frac{x}{x - 5}$$

$$\frac{3x + 2}{(x - 5)(x + 1)} + \frac{x - 4}{x + 1} = \frac{x}{x - 5}$$

$$(x - 5)(x + 1)\left[\frac{3x + 2}{(x - 5)(x + 1)} + \frac{x - 4}{x + 1}\right] = (x - 5)(x + 1)\left(\frac{x}{x - 5}\right)$$

$$\frac{(x - 5)(x + 1)}{1}\left[\frac{3x + 2}{(x - 5)(x + 1)}\right] + \frac{(x - 5)(x + 1)}{1}\left(\frac{x - 4}{x + 1}\right) = \frac{(x - 5)(x + 1)}{1}\left(\frac{x}{x - 5}\right)$$

$$3x + 2 + (x - 5)(x - 4) = (x + 1)x$$
$$3x + 2 + x^2 - 9x + 20 = x^2 + x$$
$$x^2 - 6x + 22 = x^2 + x$$
$$-6x + 22 = x$$
$$22 = 7x$$
$$\frac{22}{7} = x$$

41. $\dfrac{5x+1}{x^2-4} - \dfrac{2x+3}{x+2} + \dfrac{2x+3}{x-2}$

$= \dfrac{5x+1}{(x+2)(x-2)} - \dfrac{2x+3}{x+2} + \dfrac{2x+3}{x-2}$

$= \dfrac{5x+1}{(x+2)(x-2)} - \dfrac{(2x+3)(x-2)}{(x+2)(x-2)} + \dfrac{(2x+3)(x+2)}{(x+2)(x-2)}$

$= \dfrac{5x+1-(2x+3)(x-2)+(2x+3)(x+2)}{(x+2)(x-2)}$

$= \dfrac{5x+1-(2x^2-x-6)+2x^2+7x+6}{(x+2)(x-2)}$

$= \dfrac{5x+1-2x^2+x+6+2x^2+7x+6}{(x+2)(x-2)}$

$= \dfrac{13x+13}{(x+2)(x-2)}$

43.

$$\frac{1}{x^2-x-2} + \frac{2}{x^2-1} = \frac{1}{x^2-3x+2}$$

$$\frac{1}{(x-2)(x+1)} + \frac{2}{(x-1)(x+1)} = \frac{1}{(x-2)(x-1)}$$

$$(x-2)(x+1)(x-1)\left[\frac{1}{(x-2)(x+1)} + \frac{2}{(x-1)(x+1)}\right] = (x-2)(x+1)(x-1)\left[\frac{1}{(x-2)(x-1)}\right]$$

$$\frac{(x-2)(x+1)(x-1)}{1} \cdot \frac{1}{(x-2)(x+1)} + \frac{(x-2)(x+1)(x-1)}{1} \cdot \frac{2}{(x-1)(x+1)} = (x-2)(x+1)(x-1)\left[\frac{1}{(x-2)(x-1)}\right]$$

$$\begin{aligned} x-1+2(x-2) &= x+1 \\ x-1+2x-4 &= x+1 \\ 3x-5 &= x+1 \\ 2x-5 &= 1 \\ 2x &= 6 \\ x &= 3 \end{aligned}$$

45.

$$\frac{1}{x-4} - \frac{5}{x+2} = \frac{6}{x^2-2x-8}$$

$$\frac{1}{x-4} - \frac{5}{x+2} = \frac{6}{(x-4)(x+2)}$$

$$(x-4)(x+2)\left(\frac{1}{x-4} - \frac{5}{x+2}\right) = (x-4)(x+2)\left[\frac{6}{(x-4)(x+2)}\right]$$

$$\frac{(x-4)(x+2)}{1} \cdot \frac{1}{x-4} - \frac{(x-4)(x+2)}{1} \cdot \frac{5}{x+2} = \frac{(x-4)(x+2)}{1} \cdot \frac{6}{(x-4)(x+2)}$$

$$\begin{aligned} x + 2 - 5(x-4) &= 6 \\ x + 2 - 5x + 20 &= 6 \\ -4x + 22 &= 6 \\ -4x &= -16 \\ x &= 4 \end{aligned}$$

No solution since $x = 4$ causes a denominator to equal 0 in the original equation.

47. $\dfrac{n}{3n+2} + \dfrac{6}{9n^2-4} - \dfrac{2}{3n-2}$

$= \dfrac{n}{3n+2} + \dfrac{6}{(3n+2)(3n-2)} - \dfrac{2}{3n-2}$

$= \dfrac{n(3n-2)}{(3n+2)(3n-2)} + \dfrac{6}{(3n+2)(3n-2)} - \dfrac{2(3n+2)}{(3n+2)(3n-2)}$

$= \dfrac{n(3n-2) + 6 - 2(3n+2)}{(3n+2)(3n-2)}$

$= \dfrac{3n^2 - 2n + 6 - 6n - 4}{(3n+2)(3n-2)}$

$= \dfrac{3n^2 - 8n + 2}{(3n+2)(3n-2)}$

49.

$$\frac{1}{3n+4} + \frac{8}{9n^2-16} = \frac{1}{3n-4}$$

$$\frac{1}{3n+4} + \frac{8}{(3n+4)(3n-4)} = \frac{1}{3n-4}$$

$$(3n+4)(3n-4)\left[\frac{1}{3n+4} + \frac{8}{(3n+4)(3n-4)}\right] = (3n+4)(3n-4)\left(\frac{1}{3n-4}\right)$$

$$\frac{(3n+4)(3n-4)}{1} \cdot \frac{1}{3n-4} + \frac{(3n+4)(3n-4)}{1} \cdot \frac{8}{(3n+4)(3n-4)} = \frac{(3n+4)(3n-4)}{1} \cdot \frac{1}{(3n-4)}$$

$$\begin{aligned} 3n - 4 + 8 &= 3n + 4 \\ 3n + 4 &= 3n + 4 \end{aligned}$$

Identity

all reals except $n = \pm\dfrac{4}{3}$

51.

$$\frac{4}{2x-1}+\frac{2}{x+3}=\frac{5}{2x^2+5x-3}$$

$$\frac{4}{2x-1}+\frac{2}{x+3}=\frac{5}{(2x-1)(x+3)}$$

$$(2x-1)(x+3)\left(\frac{4}{2x-1}+\frac{2}{x+3}\right)=(2x-1)(x+3)\left[\frac{5}{(2x-1)(x+3)}\right]$$

$$\frac{(2x-1)(x+3)}{1}\cdot\frac{4}{2x-1}+\frac{(2x-1)(x+3)}{1}\cdot\frac{2}{x+3}=\frac{(2x-1)(x+3)}{1}\cdot\frac{5}{(2x-1)(x+3)}$$

$$\begin{aligned}4(x+3)+2(2x-1)&=5\\4x+12+4x-2&=5\\8x+10&=5\\8x&=-5\\x&=-\frac{5}{8}\end{aligned}$$

53.

$$\frac{6}{x}-\frac{2}{x-1}=1$$

$$x(x-1)\left(\frac{6}{x}-\frac{2}{x-1}\right)=x(x-1)(1)$$

$$\frac{x(x-1)}{1}\cdot\frac{6}{x}-\frac{x(x-1)}{1}\cdot\frac{2}{x-1}=x(x-1)$$

$$\begin{aligned}6(x-1)-2x&=x^2-x\\6x-6-2x&=x^2-x\\4x-6&=x^2-x\\0&=x^2-5x+6\\0&=(x-2)(x-3)\\x-2=0 \quad &\text{or} \quad x-3=0\\x=2 \quad &\text{or} \quad x=3\end{aligned}$$

55.

$$\frac{x}{x-1}=\frac{2x}{x+1}$$

$$\frac{(x-1)(x+1)}{1}\cdot\frac{x}{x-1}=\frac{(x-1)(x+1)}{1}\cdot\frac{2x}{x+1}$$

$$\begin{aligned}x(x+1)&=2x(x-1)\\x^2+x&=2x^2-2x\\0&=x^2-3x\\0&=x(x-3)\\x=0 \quad &\text{or} \quad x-3=0\\x=0 \quad &\text{or} \quad x=3\end{aligned}$$

57.

$$\frac{2x}{x+2}=x-1$$

$$\frac{x+2}{1}\cdot\frac{2x}{x+2}=(x+2)(x-1)$$

$$\begin{aligned}2x&=x^2+x-2\\0&=x^2-x-2\\0&=(x-2)(x+1)\\x-2=0 \quad &\text{or} \quad x+1=0\\x=2 \quad &\text{or} \quad x=-1\end{aligned}$$

59.
$$\frac{5}{x} + \frac{9}{x+2} = 4$$
$$x(x+2)\left(\frac{5}{x} + \frac{9}{x+2}\right) = x(x+2)(4)$$
$$\frac{x(x+2)}{1} \cdot \frac{5}{x} + \frac{x(x+2)}{1} \cdot \frac{9}{x+2} = 4x(x+2)$$
$$5(x+2) + 9x = 4x(x+2)$$
$$5x + 10 + 9x = 4x^2 + 8x$$
$$14x + 10 = 4x^2 + 8x$$
$$0 = 4x^2 - 6x - 10$$
$$0 = 2x^2 - 3x - 5$$
$$0 = (2x-5)(x+1)$$
$$2x - 5 = 0 \quad \text{or} \quad x + 1 = 0$$
$$2x = 5 \qquad x = -1$$
$$x = \frac{5}{2} \quad \text{or} \quad x = -1$$

61.
$$\frac{6}{3a+5} - \frac{2}{a-4} = \frac{10}{3a^2 - 7a - 20}$$
$$\frac{6}{3a+5} - \frac{2}{a-4} = \frac{10}{(3a+5)(a-4)}$$
$$(3a+5)(a-4)\left(\frac{6}{3a+5} - \frac{2}{a-4}\right) = (3a+5)(a-4)\left[\frac{10}{(3a+5)(a-4)}\right]$$
$$\frac{(3a+5)(a-4)}{1} \cdot \frac{6}{3a+5} - \frac{(3a+5)(a-4)}{1} \cdot \frac{2}{a-4} = \frac{(3a+5)(a-4)}{1} \cdot \frac{10}{(3a+5)(a-4)}$$
$$6(a-4) - 2(3a+5) = 10$$
$$6a - 24 - 6a - 10 = 10$$
$$-34 = 10$$

No solution

63.
$$\frac{2x+3}{x^2-x-2} - \frac{x+4}{x^2+3x+2} = \frac{x}{x^2-4}$$
$$\frac{2x+3}{(x-2)(x+1)} - \frac{x+4}{(x+2)(x+1)} = \frac{x}{(x+2)(x-2)}$$
$$(x-2)(x+1)(x+2)\left[\frac{2x+3}{(x-2)(x+1)} - \frac{x+4}{(x+2)(x+1)}\right] = (x-2)(x+1)(x+2)\left[\frac{x}{(x+2)(x-2)}\right]$$
$$\frac{(x-2)(x+1)(x+2)}{1} \cdot \frac{(2x+3)}{(x-2)(x+1)} - \frac{(x-2)(x+1)(x+2)}{1} \cdot \frac{(x+4)}{(x+2)(x+1)} = (x+1)x$$

$$\begin{aligned}
(x + 2)(2x + 3) - (x - 2)(x + 4) &= (x + 1)x \\
2x^2 + 7x + 6 - (x^2 + 2x - 8) &= x^2 + x \\
2x^2 + 7x + 6 - x^2 - 2x + 8 &= x^2 + x \\
x^2 + 5x + 14 &= x^2 + x \\
5x + 14 &= x \\
14 &= -4x \\
-\frac{7}{2} &= x
\end{aligned}$$

65.

$$\frac{3}{x^2 - x - 6} + \frac{2}{2x^2 - 5x - 3} = \frac{5}{2x^2 + 5x + 2}$$

$$\frac{3}{(x-3)(x+2)} + \frac{2}{(2x+1)(x-3)} = \frac{5}{(2x+1)(x+2)}$$

$$(x-3)(x+2)(2x+1)\left[\frac{3}{(x-3)(x+2)} + \frac{2}{(2x+1)(x-3)}\right] = (x-3)(x+2)(2x+1)\left[\frac{5}{(2x+1)(x+2)}\right]$$

$$\frac{(x-3)(x+2)(2x+1)}{1} \cdot \frac{3}{(x-3)(x+2)} + \frac{(x-3)(x+2)(2x+1)}{1} \cdot \frac{2}{(2x+1)(x-3)} = \frac{(x-3)(x+2)(2x+1)}{1} \cdot \frac{5}{(2x+1)(x+2)}$$

$$\begin{aligned}
3(2x+1) + 2(x+2) &= 5(x-3) \\
6x + 3 + 2x + 4 &= 5x - 15 \\
8x + 7 &= 5x - 15 \\
3x + 7 &= -15 \\
3x &= -22 \\
x &= \frac{-22}{3}
\end{aligned}$$

67.

$$\frac{2x+1}{x^2+x-2} + \frac{4x}{x^2-1} = \frac{15x-1}{x^2+3x+2}$$

$$\frac{2x+1}{(x+2)(x-1)} + \frac{4x}{(x+1)(x-1)} = \frac{15x-1}{(x+2)(x+1)}$$

$$(x + 1)(x+2)(x-1)\left[\frac{2x+1}{(x+2)(x-1)} + \frac{4x}{(x+1)(x-1)}\right] = (x+2)(x-1)(x+1)\left[\frac{15x-1}{(x+2)(x+1)}\right]$$

$$\frac{(x+2)(x-1)(x+1)}{1} \cdot \frac{(2x+1)}{(x+2)(x-1)} + \frac{(x+2)(x-1)(x+1)}{1} \cdot \frac{4x}{(x+1)(x-1)} = (x-1)(15x-1)$$

$$\begin{aligned}
(x + 1)(2x + 1) + (x + 2)4x &= (x - 1)(15x - 1) \\
2x^2 + 3x + 1 + 4x^2 + 8x &= 15x^2 - 16x + 1 \\
6x^2 + 11x + 1 &= 15x^2 - 16x + 1 \\
0 &= 9x^2 - 27x \\
0 &= 9x(x - 3)
\end{aligned}$$

$$9x = 0 \quad \text{or} \quad x - 3 = 0$$
$$x = 0 \qquad\qquad x = 3$$

69.

$$\frac{4}{4x^2-9}-\frac{5}{4x^2-8x+3}=\frac{8}{4x^2+4x-3}$$

$$\frac{4}{(2x-3)(2x+3)}-\frac{5}{(2x-1)(2x-3)}=\frac{8}{(2x+3)(2x-1)}$$

$$(2x-3)(2x+3)(2x-1)\left[\frac{4}{(2x-3)(2x+3)}-\frac{5}{(2x-1)(2x-3)}\right]=$$

$$(2x-3)(2x+3)(2x-1)\left[\frac{8}{(2x+3)(2x-1)}\right]$$

$$\frac{(2x-3)(2x+3)(2x-1)}{1}\cdot\frac{4}{(2x-3)(2x+3)}-\frac{(2x-3)(2x+3)(2x-1)}{1}\cdot\frac{5}{(2x-1)(2x-3)}=\frac{(2x-3)(2x+3)(2x-1)}{1}\cdot\frac{8}{(2x+3)(2x-1)}$$

$$\begin{aligned} 4(2x-1)-5(2x+3) &= 8(2x-3) \\ 8x-4-10x-15 &= 16x-24 \\ -2x-19 &= 16x-24 \\ -19 &= 18x-24 \\ 5 &= 18x \\ \frac{5}{18} &= x \end{aligned}$$

71. $f(x)+g(x)$

$$=\frac{3x+1}{(x-2)^2}+\frac{x+5}{x-2}$$

$$=\frac{3x+1}{(x-2)^2}+\frac{(x+5)(x-2)}{(x-2)^2}$$

$$=\frac{3x+1+(x+5)(x-2)}{(x-2)^2}$$

$$=\frac{3x+1+x^2+3x-10}{(x-2)^2}$$

$$=\frac{x^2+6x-9}{(x-2)^2}$$

73. $2f(x)-3g(x)$

$$=2\left[\frac{3x+1}{(x-2)^2}\right]-3\left[\frac{x+5}{x-2}\right]$$

$$=\frac{2(3x+1)}{(x-2)^2}-\frac{3(x+5)(x-2)}{(x-2)^2}$$

$$=\frac{2(3x+1)-3(x+5)(x-2)}{(x-2)^2}$$

$$=\frac{6x+2-3(x^2+3x-10)}{(x-2)^2}$$

$$=\frac{6x+2-3x^2-9x+30}{(x-2)^2}$$

$$=\frac{-3x^2-3x+32}{(x-2)^2}$$

75. $$\frac{f(x)}{g(x)} = \frac{\dfrac{3x+1}{(x-2)^2}}{\dfrac{x+5}{x-2}}$$

$$= \frac{3x+1}{(x-2)^2} \cdot \frac{x-2}{x+5}$$

$$= \frac{3x+1}{(x-2)(x+5)}$$

77. $$3x^2y(2x^3y^2) = 3 \cdot 2x^{2+3}y^{1+2} = 6x^5y^3$$

79. Answers may vary.
One example: $f(x) = x^2$

$$f(x) = x^2$$
$$f(x+2) = (x+2)^2$$
$$f(2) = 2^2 = 4$$

$$f(x) + f(2) = x^2 + 4$$
$$f(x+2) = (x+2)^2 = x^2 + 4x + 4$$
$$f(x) + f(2) \neq f(x+2)$$

Exercises 6.7

1. $$5x + 7y = 4$$
$$5x + 7y - 7y = 4 - 7y$$
$$5x = 4 - 7y$$
$$\frac{5x}{5} = \frac{4-7y}{5}$$
$$x = \frac{4-7y}{5}$$

3. $$2x - 9y = 11$$
$$2x - 9y - 2x = 11 - 2x$$
$$-9y = 11 - 2x$$
$$\frac{-9y}{-9} = \frac{11-2x}{-9}$$
$$y = \frac{11-2x}{-9}$$
$$y = \frac{2x-11}{9}$$

5. $$w + 4z - 1 = 2w - z + 3$$
$$w + 4z - 1 - 4z + 1 - 2w = 2w - z + 3 - 4z + 1 - 2w$$
$$-w = -5z + 4$$
$$\frac{-w}{-1} = \frac{-5z+4}{-1}$$
$$w = 5z - 4$$

7. $$2(6r - 5t) > 5(2r + t)$$
$$12r - 10t > 10r + 5t$$
$$12r - 10t - 10r + 10t > 10r + 5t - 10r + 10t$$
$$2r > 15t$$
$$\frac{2r}{2} > \frac{15t}{2}$$
$$r > \frac{15t}{2}$$

9. $$3m - 4n + 6p = 5n + 2p - 8$$
$$3m - 4n + 6p + 4n - 2p + 8 = 5n + 2p - 8 + 4n - 2p + 8$$
$$3m + 4p + 8 = 9n$$
$$\frac{3m + 4p + 8}{9} = \frac{9n}{9}$$
$$\frac{3m + 4p + 8}{9} = n$$

11. $$\frac{a}{5} - \frac{b}{3} = \frac{a}{2} - \frac{b}{6}$$
$$30\left(\frac{a}{5} - \frac{b}{3}\right) = 30\left(\frac{a}{2} - \frac{b}{6}\right)$$
$$\frac{30}{1} \cdot \frac{a}{5} - \frac{30}{1} \cdot \frac{b}{3} = \frac{30}{1} \cdot \frac{a}{2} - \frac{30}{1} \cdot \frac{b}{6}$$
$$6a - 10b = 15a - 5b$$
$$6a - 10b - 6a + 5b = 15a - 5b - 6a + 5b$$
$$-5b = 9a$$
$$\frac{-5b}{9} = \frac{9a}{9}$$
$$\frac{-5b}{9} = a$$

13.
$$\frac{x+y}{3} - \frac{x}{2} + \frac{y}{6} = 3(x-y)$$
$$6\left(\frac{x+y}{3} - \frac{x}{2} + \frac{y}{6}\right) = 6[3(x-y)]$$
$$\frac{6}{1} \cdot \frac{x+y}{3} - \frac{6}{1} \cdot \frac{x}{2} + \frac{6}{1} \cdot \frac{y}{6} = 18(x-y)$$
$$2(x+y) - 3x + y = 18x - 18y$$
$$2x + 2y - 3x + y = 18x - 18y$$
$$-x + 3y = 18x - 18y$$
$$-x + 3y + x + 18y = 18x - 18y + x + 18y$$
$$21y = 19x$$
$$\frac{21y}{19} = \frac{19x}{19}$$
$$\frac{21y}{19} = x$$

15.
$$ax + b = cx + d$$
$$ax + b - cx - b = cx + d - cx - b$$
$$ax - cx = d - b$$
$$x(a-c) = d - b$$
$$\frac{x(a-c)}{a-c} = \frac{d-b}{a-c}$$
$$x = \frac{d-b}{a-c}$$

17.
$$3x + 2y - 5 = ax + by + 1$$
$$3x + 2y - 5 - ax - 2y + 5 = ax + by + 1 - ax - 2y + 5$$
$$3x - ax = by - 2y + 6$$
$$x(3-a) = by - 2y + 6$$
$$\frac{x(3-a)}{3-a} = \frac{by - 2y + 6}{3-a}$$
$$x = \frac{by - 2y + 6}{3-a}$$

19. $(x+3)(y+7) = a$
$$\frac{(x+3)(y+7)}{y+7} = \frac{a}{y+7}$$
$$x + 3 = \frac{a}{y+7}$$
$$x + 3 - 3 = \frac{a}{y+7} - 3$$
$$x = \frac{a}{y+7} - 3 = \frac{a - 3y - 21}{y+7}$$

21.
$$y = \frac{u-1}{u+1}$$
$$(u+1)(y) = (u+1)\left(\frac{u-1}{u+1}\right)$$
$$uy + y = u - 1$$
$$uy + y - uy + 1 = u - 1 - uy + 1$$
$$y + 1 = u - uy$$
$$y + 1 = u(1-y)$$
$$\frac{y+1}{1-y} = \frac{u(1-y)}{1-y}$$
$$\frac{y+1}{1-y} = u$$

23.
$$x = \frac{2t-3}{3t-2}$$
$$(3t-2)x = (3t-2)\left(\frac{2t-3}{3t-2}\right)$$
$$3tx - 2x = 2t - 3$$
$$3tx - 2x - 2t + 2x = 2t - 3 - 2t + 2x$$
$$3tx - 2t = 2x - 3$$
$$t(3x-2) = 2x - 3$$
$$\frac{t(3x-2)}{3x-2} = \frac{2x-3}{3x-2}$$
$$t = \frac{2x-3}{3x-2}$$

25. $A = \frac{1}{2}bh$
$$2A = bh$$
$$\frac{2A}{h} = \frac{bh}{h}$$
$$\frac{2A}{h} = b$$

27.
$$A = \frac{1}{2}h(b_1 + b_2)$$
$$2A = h(b_1 + b_2)$$
$$\frac{2A}{h} = \frac{h(b_1 + b_2)}{h}$$
$$\frac{2A}{h} = b_1 + b_2$$
$$\frac{2A}{h} - b_2 = b_1 + b_2 - b_2$$
$$\frac{2A}{h} - b_2 = b_1$$

29.
$$A = P(1 + rt)$$
$$A = P + Prt$$
$$A - P = P + Prt - P$$
$$A - P = Prt$$
$$\frac{A - P}{Pt} = \frac{Prt}{Pt}$$
$$\frac{A - P}{Pt} = r$$

31.
$$C = \frac{5}{9}(F - 32)$$
$$\frac{9}{5}(C) = \frac{9}{5}\left[\frac{5}{9}(F - 32)\right]$$
$$\frac{9}{5}C = F - 32$$
$$\frac{9}{5}C + 32 = F - 32 + 32$$
$$\frac{9}{5}C + 32 = F$$

33.
$$\frac{P_1}{V_1} = \frac{P_2}{V_2}$$
$$\frac{V_2}{1} \cdot \frac{P_1}{V_1} = \frac{V_2}{1} \cdot \frac{P_2}{V_2}$$
$$\frac{P_1 V_2}{V_1} = P_2$$

35.
$$S = s_0 + v_0 t + \frac{1}{2}gt^2$$
$$S - s_0 - v_0 t = s_0 + v_0 t + \frac{1}{2}gt^2 - s_0 - v_0 t$$
$$S - s_0 - v_0 t = \frac{1}{2}gt^2$$
$$2(S - s_0 - v_0 t) = gt^2$$
$$\frac{2(S - s_0 - v_0 t)}{t^2} = \frac{gt^2}{t^2}$$
$$\frac{2(S - s_0 - v_0 t)}{t^2} = g$$

37.
$$\frac{x - \mu}{s} < 1.96 \quad , \quad (s > 0)$$
$$x - \mu < 1.96s$$
$$x - \mu + \mu < 1.96s + \mu$$
$$x < 1.96s + \mu$$

39.
$$\frac{1}{f} = \frac{1}{f_1} + \frac{1}{f_2}$$
$$\frac{ff_1f_2}{1} \cdot \frac{1}{f} = \frac{ff_1f_2}{1}\left(\frac{1}{f_1} + \frac{1}{f_2}\right)$$
$$f_1f_2 = \frac{ff_1f_2}{1} \cdot \frac{1}{f_1} + \frac{ff_1f_2}{1} \cdot \frac{1}{f_2}$$
$$f_1f_2 = ff_2 + ff_1$$
$$f_1f_2 - ff_1 = ff_2 + ff_1 - ff_1$$
$$f_1f_2 - ff_1 = ff_2$$
$$f_1(f_2 - f) = ff_2$$
$$\frac{f_1(f_2 - f)}{f_2 - f} = \frac{ff_2}{f_2 - f}$$
$$f_1 = \frac{ff_2}{f_2 - f}$$

41.
$$S = 2\pi r^2 + 2\pi rh$$
$$S - 2\pi r^2 = 2\pi r^2 + 2\pi rh - 2\pi r^2$$
$$S - 2\pi r^2 = 2\pi rh$$
$$\frac{S - 2\pi r^2}{2\pi r} = \frac{2\pi rh}{2\pi r}$$
$$\frac{S - 2\pi r^2}{2\pi r} = h$$

43.
$$\begin{aligned}(x-3)^2 &= 4\\ x^2-6x+9 &= 4\\ x^2-6x+5 &= 0\\ (x-5)(x-1) &= 0\\ x-5=0 \quad &\text{or}\quad x-1=0\\ x=5 \quad &\text{or}\quad x=1\end{aligned}$$

45. number of heavy-duty batteries: x
number of regular batteries: $18 - x$

$$\begin{aligned}60(x) + 50(18-x) &= 940\\ 60x + 900 - 50x &= 940\\ 10x + 900 &= 940\\ 10x &= 40\\ x &= 4\\ 18 - x = 18 - 4 &= 14\end{aligned}$$

They bought 4 heavy-duty batteries and 14 regular batteries.

Exercises 6.8

1. Let x = the number

$$\frac{3}{4}(x) = \frac{2}{5}(x) - 7$$

$$20\left(\frac{3}{4}x\right) = 20\left(\frac{2}{5}x - 7\right)$$

$$\begin{aligned}15x &= 8x - 140\\ 7x &= -140\\ x &= -20\end{aligned}$$

The number is -20.

3. Let x = the number of men

$$\frac{7 \text{ men}}{9 \text{ women}} = \frac{x \text{ men}}{810 \text{ women}}$$

$$\frac{7}{9} = \frac{x}{810}$$

$$810\left(\frac{7}{9}\right) = 810\left(\frac{x}{810}\right)$$

$$630 = x$$

There are 630 men.

5. Let x = one number
then $x - 21$ = the other number

$$\frac{x-21}{x} = \frac{5}{12}$$

$$12x\left(\frac{x-21}{x}\right) = 12x\left(\frac{5}{12}\right)$$

$$\begin{aligned}12(x-21) &= 5x\\ 12x - 252 &= 5x\\ -252 &= -7x\\ 36 &= x\\ x - 21 = 36 - 21 &= 15\end{aligned}$$

The numbers are 15 and 36.

7. Let x = the number of inches in 52 cm

$$\frac{1 \text{ inch}}{2.54 \text{ cm}} = \frac{x \text{ inches}}{52 \text{ cm}}$$

$$\frac{1}{2.54} = \frac{x}{52}$$

$$52\left(\frac{1}{2.54}\right) = 52\left(\frac{x}{52}\right)$$

$$20.47 = x$$

There are 20.47 inches in 52 cm.

9. Let x = the number of dribbles in 28 droogs
and let y = the number of dreeps in 28 droogs

$$\frac{5 \text{ droogs}}{4 \text{ dreeps}} = \frac{28 \text{ droogs}}{y \text{ dreeps}}$$

$$\frac{5}{4} = \frac{28}{y}$$

$$4y\left(\frac{5}{4}\right) = 4y\left(\frac{28}{y}\right)$$

$$\begin{aligned}5y &= 112\\ y &= 22.4\end{aligned}$$

22.4 dreeps is equivalent to 28 droogs.

$$\frac{7 \text{ dreeps}}{25 \text{ dribbles}} = \frac{22.4 \text{ dreeps}}{x \text{ dribbles}}$$

$$\frac{7}{25} = \frac{22.4}{x}$$

$$25x\left(\frac{7}{25}\right) = 25x\left(\frac{22.4}{x}\right)$$

$$7x = 560$$
$$x = 80$$

There are 80 dribbles in 28 droogs.

11. Let x = 2nd side's length,

then $\frac{1}{2}x$ = 1st side's length

and $x + 2$ = 3rd side's length

Perimeter = sum of side lengths

$$22 = x + \frac{1}{2}x + x + 2$$

$$22 = \frac{5}{2}x + 2$$

$$2(22) = 2\left(\frac{5}{2}x + 2\right)$$

$$44 = 5x + 4$$
$$40 = 5x$$
$$8 = x$$

$$\frac{1}{2}x = \frac{1}{2}(8) = 4$$
$$x + 2 = 8 + 2 = 10$$

The sides have lengths 4 cm, 8 cm and 10 cm.

13. Let x = the width

then the length $= \frac{5}{2}x$

$$P = 2 \cdot \text{width} + 2 \cdot \text{length}$$

$$50 = 2x + 2\left(\frac{5}{2}x\right)$$

$$50 = 2x + 5x$$
$$50 = 7x$$

$$\frac{50}{7} = x$$

$$\frac{5}{2}x = \frac{5}{2}\left(\frac{50}{7}\right) = \frac{125}{7}$$

The rectangle is $\frac{50}{7}$ cm by $\frac{125}{7}$ cm.

15.

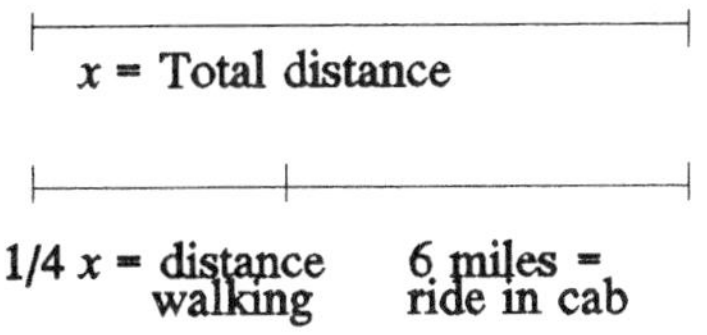

x = Total distance

1/4 x = distance walking

6 miles = ride in cab

$$\frac{1}{4}x + 6 = x$$

$$4\left(\frac{1}{4}x + 6\right) = 4(x)$$

$$x + 24 = 4x$$
$$24 = 3x$$
$$8 = x$$

His home is 8 miles from the ballfield.

17.

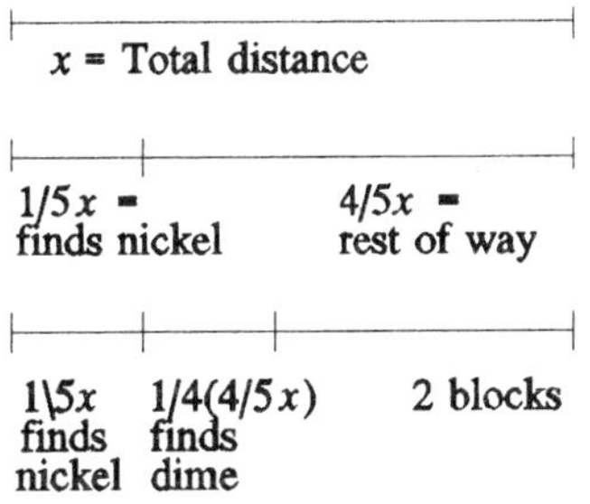

x = Total distance

1/5x = finds nickel

4/5x = rest of way

1\5x finds nickel

1/4(4/5x) finds dime

2 blocks

$$\frac{1}{5}x + \frac{1}{4}\left(\frac{4}{5}x\right) + 2 = x$$

$$\frac{1}{5}x + \frac{1}{5}x + 2 = x$$

$$5\left(\frac{1}{5}x + \frac{1}{5}x + 2\right) = 5(x)$$

$$x + x + 10 = 5x$$
$$2x + 10 = 5x$$
$$10 = 3x$$
$$\frac{10}{3} = x$$

She walked $\frac{10}{3} = 3\frac{1}{3}$ blocks.

19. $$\frac{1}{R} = \frac{1}{R_1} + \frac{1}{R_2} + \frac{1}{R_3}$$

$$\frac{1}{1\frac{1}{4}} = \frac{1}{2} + \frac{1}{5} + \frac{1}{R_3}$$

$$\frac{1}{\frac{5}{4}} = \frac{1}{2} + \frac{1}{5} + \frac{1}{R_3}$$

$$\frac{4}{5} = \frac{1}{2} + \frac{1}{5} + \frac{1}{R_3}$$

$$10R_3\left(\frac{4}{5}\right) = 10R_3\left(\frac{1}{2} + \frac{1}{5} + \frac{1}{R_3}\right)$$

$$8R_3 = 5R_3 + 2R_3 + 10$$
$$8R_3 = 7R_3 + 10$$
$$R_3 = 10$$

The third resistance is 10 ohms.

21. amount in certificate of deposit: $x + 3000$
amount in bond: x

$$0.06(x + 3000) + 0.10(x) = 580$$
$$0.06x + 180 + 0.10x = 580$$
$$0.16x + 180 = 580$$
$$0.16x = 400$$
$$x = 2500$$
$$x + 3000 = 2500 + 3000 = 5500$$

She invested \$5500 in the certificate of deposit and \$2500 in the bond.

23. amount invested at $5\frac{1}{2}\%$: x

amount invested at 7%: $25000 - x$

$$0.055(x) + 0.07(25000 - x) = 1465$$
$$0.055x + 1750 - 0.07x = 1465$$
$$1750 - 0.015x = 1465$$
$$-0.015x = -285$$
$$x = 19000$$
$$25000 - x = 25000 - 19000 = 6000$$

He has \$19,000 in the $5\frac{1}{2}\%$ account and \$6000 in the 7% account.

25. amount in $8\frac{1}{2}\%$ bond: x

amount in 11% bond: $18000 - x$

$$
\begin{aligned}
0.085(x) + 0.11(18000 - x) &= 0.10(18000) \\
0.085x + 1980 - 0.11x &= 1800 \\
-0.025x + 1980 &= 1800 \\
-0.025x &= -180 \\
x &= 7200 \\
18000 - x &= 18000 - 7200 = 10800
\end{aligned}
$$

He should invest \$7200 in the $8\frac{1}{2}\%$ bond and \$10,800 in the 11% bond.

27. Let x = number of hours it takes Carol and Bill to paint the room working together

Portion completed by Carol in x hours	+	Portion completed by Bill in x hours	=	1 complete job
$\frac{x}{3}$	+	$\frac{x}{5}$	=	1

$$
\begin{aligned}
15\left(\frac{x}{3} + \frac{x}{5}\right) &= 15(1) \\
5x + 3x &= 15 \\
8x &= 15 \\
x &= \frac{15}{8}
\end{aligned}
$$

It will take them $\frac{15}{8} = 1\frac{7}{8}$ hours to paint the room.

29. Let x = number of hours it takes the two cleaning services working together to clean the building

Portion completed by QCS in x hours	+	Portion completed by SQCS in x hours	=	1 complete job
$\frac{x}{30}$	+	$\frac{x}{20}$	=	1

$$
\begin{aligned}
60\left(\frac{x}{30} + \frac{x}{20}\right) &= 60(1) \\
2x + 3x &= 60 \\
5x &= 60 \\
x &= 12
\end{aligned}
$$

It will take them 12 hours working together.

31. Let x = number of days for the bricklayer and assistant to complete the wall when working together

Portion completed by bricklayer in x days	+	Portion completed by assistant in x days	=	1 complete job
$\dfrac{x}{2\frac{2}{3}}$	+	$\dfrac{x}{5}$	=	1

$$\frac{x}{\frac{8}{3}} + \frac{x}{5} = 1$$

$$\frac{3x}{8} + \frac{x}{5} = 1$$

$$40\left(\frac{3x}{8} + \frac{x}{5}\right) = 40(1)$$

$$15x + 8x = 40$$
$$23x = 40$$
$$x = \frac{40}{23}$$

It will take them $\frac{40}{23} = 1\frac{17}{23}$ days to complete the wall.

33. Let x = number of hours SQCS works

Portion completed by QCS	+	Portion completed by SQCS	=	1 complete job
$\dfrac{10}{30}$	+	$\dfrac{x}{20}$	=	1

$$\frac{1}{3} + \frac{x}{20} = 1$$

$$60\left(\frac{1}{3} + \frac{x}{20}\right) = 60(1)$$

$$20 + 3x = 60$$
$$3x = 40$$
$$x = \frac{40}{3} = 13\frac{1}{3}$$

To complete the entire job it will take $10 + 13\frac{1}{3} = 23\frac{1}{3}$ hours.

35. Let x = number of minutes for tub to fill

$$\frac{x}{10} - \frac{x}{15} = 1$$

$$30\left(\frac{x}{10} - \frac{x}{15}\right) = 30(1)$$

$$3x - 2x = 30$$
$$x = 30$$

It will take 30 minutes.

37. Let x = number of hours they work together.

Aaron alone	+	Aaron and Kimberly together	=	whole job

$$3\left(\frac{1}{10}\right) + x\left(\frac{1}{10}\right) + x\left(\frac{1}{8}\right) = 1$$

$$40\left(\frac{3}{10} + \frac{x}{10} + \frac{x}{8}\right) = 40(1)$$

$$12 + 4x + 5x = 40$$
$$12 + 9x = 40$$
$$9x = 28$$

$$x = \frac{28}{9}$$

$\frac{28}{9}$ hours = 3 hours 7 minutes

3 PM + 3 hours 7 minutes = 6:07 PM

39. Let x = number of hours it takes Megan to do the job alone

Lori alone	+	Lori and Megan together	=	whole job

$$1\left(\frac{1}{6}\right) + 2\left(\frac{1}{6}\right) + 2\left(\frac{1}{x}\right) = 1$$

$$6x\left(\frac{1}{6} + \frac{1}{3} + \frac{2}{x}\right) = 6x(1)$$

$$x + 2x + 12 = 6x$$
$$3x + 12 = 6x$$
$$12 = 3x$$
$$4 = x$$

It would take Megan 4 hours alone.

41. Let r = rate for Bill
then $r + 10$ = rate for Jill

$$t_{\text{Bill}} = t_{Jill}$$

$$\frac{d_{\text{Bill}}}{r_{\text{Bill}}} = \frac{d_{\text{Jill}}}{r_{\text{Jill}}}$$

$$\frac{10}{r} = \frac{15}{r + 10}$$

$$r(r + 10)\left(\frac{10}{r}\right) = r(r + 10)\left(\frac{15}{r + 10}\right)$$

$$\begin{aligned} 10(r + 10) &= 15r \\ 10r + 100 &= 15r \\ 100 &= 5r \\ 20 &= r \end{aligned}$$

Bill rides 20 kph.

43. hours at slower speed: x
hours at faster speed: $14 - x$

distance at slower speed	+	distance at faster speed	=	total distance
$20x$	+	$50(14 - x)$	=	600

$$\begin{aligned} 20x + 700 - 50x &= 600 \\ -30x + 700 &= 600 \\ -30x &= -100 \\ x &= \frac{10}{3} \end{aligned}$$

She drove at a slower speed for $\frac{10}{3}$ hours.

Her distance at this speed was $20\left(\frac{10}{3}\right) = 66\frac{2}{3}$ miles.

45. Let x = amount of 20% solution

Amount of alcohol in the 20% solution	+	Amount of alcohol in the 50% solution	=	Total amount of alcohol in final solution
$0.20(x)$	+	$0.50(5)$	=	$0.30(x + 5)$

$$\begin{aligned} 0.20x + 2.5 &= 0.30x + 1.5 \\ 10(0.20x + 2.5) &= 10(0.30x + 1.5) \\ 2x + 25 &= 3x + 15 \\ 25 &= x + 15 \\ 10 &= x \end{aligned}$$

10 oz. of the 20% solution should be used.

47. Let x = amount of 30% solution
Let y = amount of 75% solution

$$\begin{cases} x + y = 80 \\ 0.30x + 0.75y = 0.50(80) \end{cases}$$

$x + y = 80$
$x = 80 - y$
Substitute into 2nd equation:

$$\begin{aligned} 0.30(80 - y) + 0.75y &= 40 \\ 24 - 0.30y + 0.75y &= 40 \\ 24 + 0.45y &= 40 \\ 0.45y &= 16 \\ y &= 35\frac{5}{9} \\ x &= 80 - y \\ &= 80 - 35\frac{5}{9} \\ &= 44\frac{4}{9} \end{aligned}$$

$44\frac{4}{9}$ ml of the 30% solution should be mixed with $35\frac{5}{9}$ ml of the 75% solution.

49. Let x = amount of 40% alloy
Let y = amount of 60% alloy

$$\begin{cases} x + y = 80 \\ 0.40x + 0.60y = 0.55(80) \end{cases}$$

$x = 80 - y$
Substitute into 2nd equation:

$$\begin{aligned} 0.40(80 - y) + 0.60y &= 44 \\ 32 - 0.40y + 0.60y &= 44 \\ 32 + 0.20y &= 44 \\ 0.20y &= 12 \\ y &= 60 \\ x &= 80 - y \\ &= 80 - 60 \\ &= 20 \end{aligned}$$

20 tons of the 40% alloy and 60 tons of the 60% alloy should be used.

51. Let x = amount of pure alcohol

Amount of alcohol in 60% solution	+	Amount of alcohol in pure solution	=	Total amount of alcohol in final solution
$0.60(2)$	+	$1(x)$	=	$0.80(2 + x)$

$$\begin{aligned} 1.2 + x &= 1.6 + 0.80x \\ 10(1.2. + x) &= 10(1.6 + 0.80x) \\ 12 + 10x &= 16 + 8x \\ 12 + 2x &= 16 \\ 2x &= 4 \\ x &= 2 \end{aligned}$$

He must add 2 liters of pure alcohol.

53. Let x = amount drained off = water added

$$\begin{aligned} 0.30(3 - x) + 0(x) &= 0.20(3) \\ 0.9 - 0.3x &= 0.6 \\ 10(0.9 - 0.3x) &= 10(0.6) \\ 9 - 3x &= 6 \\ -3x &= -3 \\ x &= 1 \end{aligned}$$

He drained off 1 gallon.

55. Let x = number of advance tickets
Let y = number of door tickets

$$\begin{cases} x + y = 3600 \\ 25x + 30.50y = 97700 \end{cases}$$

$y = 3600 - x$
Substitute into 2nd equation:

$$\begin{aligned} 25x + 30.50(3600 - x) &= 97700 \\ 25x + 109800 - 30.50x &= 97700 \\ -5.5x + 109800 &= 97700 \\ -5.5x &= -12100 \\ x &= 2200 \end{aligned}$$

2200 advance tickets were sold.

57. Let x = number of nickels,
then $x + 5$ = number of dimes
and $2x$ = number of quarters

$$\begin{aligned} 5(x) + 10(x + 5) + 25(2x) &= 2000 \\ 5x + 10x + 50 + 50x &= 2000 \\ 65x + 50 &= 2000 \\ 65x &= 1950 \\ x &= 30 \\ x + 5 &= 30 + 5 = 35 \\ 2x &= 2(30) = 60 \end{aligned}$$

He has 30 nickels, 35 dimes and 60 quarters.

59. Let x = number of orchestra tickets,
then $2x$ = number of general admission tickets
and $900 - (x + 2x) = 900 - 3x$ = number of balcony tickets

$$\begin{aligned} 25(x) + 20.50(900 - 3x) + 16(2x) &= 17325 \\ 25x + 18450 - 61.5x + 32x &= 17325 \\ -4.5x + 18450 &= 17325 \\ -4.5x &= -1125 \\ x &= 250 \\ 2x &= 2(250) = 500 \\ 900 - 3x &= 900 - 3(250) = 150 \end{aligned}$$

They sold 250 orchestra tickets, 500 general admission tickets and 150 balcony tickets.

61. Let x = score on final exam

$$\begin{aligned} 0.20(85) + 0.20(65) + 0.20(72) + 0.40(x) &\geq 80 \\ 17 + 13 + 14.4 + 0.40x &\geq 80 \\ 44.4 + 0.40x &\geq 80 \\ 0.40x &\geq 35.6 \\ x &\geq 89 \end{aligned}$$

He must receive at least a grade of 89.

63. Let x = amount in high-risk bond
then $20000 - x$ = amount in savings

$$\begin{aligned} 0.082(x) + 0.039(20000 - x) &\geq 1000 \\ 0.082x + 780 - 0.039x &\geq 1000 \\ 0.043x + 780 &\geq 1000 \\ 0.043x &\geq 220 \\ x &\geq 5116.28 \end{aligned}$$

She must invest at least \$5116.28 in the high-risk bond.

65. Let x = final exam score

$$0.20(85) + 0.20(92) + 0.20(86) + 0.40(x) \ge 90$$
$$17 + 18.4 + 17.2 + 0.40x \ge 90$$
$$52.6 + 0.40x \ge 90$$
$$0.40x \ge 37.4$$
$$x \ge 93.5$$

He must score at least 93.5.

67. $R(x) = \dfrac{6x}{6 + x}$

(a) $3 = \dfrac{6x}{6 + x}$

$$(6 + x)3 = (6 + x)\left(\frac{6x}{6 + x}\right)$$

$$18 + 3x = 6x$$
$$18 = 3x$$
$$6 = x$$

It should be 6 ohms.

(b) $A = \dfrac{6x}{6 + x}$

$$(6 + x)A = (6 + x)\left(\frac{6x}{6 + x}\right)$$

$$6A + Ax = 6x$$
$$6A = 6x - Ax$$
$$6A = x(6 - A)$$
$$\frac{6A}{6 - A} = x$$

It should be $\dfrac{6A}{6 - A}$ ohms.

69. $A(x) = \dfrac{0.2x + 6000}{x}$

(a) $100 = \dfrac{0.2x + 6000}{x}$

$$x(100) = x\left(\frac{0.2x + 6000}{x}\right)$$

$$100x = 0.2x + 6000$$
$$99.8x = 6000$$
$$x = 60.1$$

They should produce approximately 60 items.

(b) $T = \dfrac{0.2x + 6000}{x}$

$$x(T) = x\left(\frac{0.2x + 6000}{x}\right)$$

$$Tx = 0.2x + 6000$$
$$Tx - 0.2x = 6000$$
$$x(T - 0.2) = 6000$$
$$x = \frac{6000}{T - 0.2}$$

$\dfrac{6000}{t - 0.2}$ items should be produced.

73. $3x^3y - 6x^2y^2 + 3xy^3$
$= 3xy(x^2 - 2xy + y^2)$
$= 3xy(x - y)^2$

75. $|2x + 8| \le 10$
$-10 \le 2x + 8 \le 10$
$-18 \le 2x \le 2$
$-9 \le x \le 1$

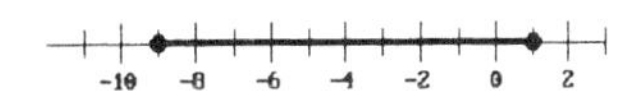

CHAPTER 6 REVIEW EXERCISES

1. $x + 5 \ne 0$
$x \ne -5$

Domain: $\{x \mid x \ne -5\}$

3. $3x^2 - 11x - 4 \ne 0$
$(3x + 1)(x - 4) \ne 0$
$3x + 1 \ne 0$ or $x - 4 \ne 0$
$3x \ne -1$ $\quad x \ne 4$
$z \ne -\dfrac{1}{3}$

Domain: $\left\{x \mid x \ne -\dfrac{1}{3}, x \ne 4\right\}$

5. $\dfrac{4x^2y^3}{16xy^5} = \dfrac{4 \cdot x \cdot x \cdot y^3}{4 \cdot 4 \cdot x \cdot y^3 \cdot y^2}$

$= \dfrac{x}{4y^2}$

7. $\dfrac{x^2 + 2x - 8}{x^2 + 3x - 10} = \dfrac{(x + 4)(x - 2)}{(x + 5)(x - 2)}$

$= \dfrac{x + 4}{x + 5}$

9. $\dfrac{x^4 - 2x^3 + 3x^2}{x^2} = \dfrac{x^2(x^2 - 2x + 3)}{x^2}$

$= x^2 - 2x + 3$

11. $\dfrac{5xa - 7a + 5xb - 7b}{3xa - 2a + 3xb - 2b} = \dfrac{(5x - 7)(a + b)}{(3x - 2)(a + b)}$

$= \dfrac{5x - 7}{3x - 2}$

13. $\dfrac{4x^2y^3z^2}{12xy^4} \cdot \dfrac{24xy^5}{16xy}$

$= \dfrac{4 \cdot 24x^3y^8z^2}{12 \cdot 16x^2y^5}$

$= \dfrac{4 \cdot 3 \cdot 8x \cdot x^2y^5 \cdot y^3z^2}{4 \cdot 3 \cdot 8 \cdot 2x^2y^5}$

$= \dfrac{xy^3z^2}{2}$

15. $\dfrac{5}{3x^2y} + \dfrac{1}{3x^2y} = \dfrac{5 + 1}{3x^2y}$

$= \dfrac{6}{3x^2y}$

$= \dfrac{2}{x^2y}$

17. $\dfrac{3x}{x - 1} + \dfrac{3}{x - 1} = \dfrac{3x + 3}{x - 1}$

19. $\dfrac{2x^2}{x^2 + x - 6} + \dfrac{2x}{x^2 + x - 6} - \dfrac{12}{x^2 + x - 6}$

$= \dfrac{2x^2 + 2x - 12}{x^2 + x - 6}$

$= \dfrac{2(x^2 + x - 6)}{x^2 + x - 6}$

$= 2$

21. $\dfrac{5}{3a^2b} - \dfrac{7}{4ab^4}$

$= \dfrac{5(4b^3)}{12a^2b^4} - \dfrac{7(3a)}{12a^2b^4}$

$= \dfrac{5(4b^3) - 7(3a)}{12a^2b^4}$

$= \dfrac{20b^3 - 21a}{12a^2b^4}$

23. $\dfrac{3x + 1}{2x^2} - \dfrac{3x - 2}{5x}$

$= \dfrac{(3x + 1)(5)}{10x^2} - \dfrac{(3x - 2)(2x)}{10x^2}$

$= \dfrac{(3x + 1)(5) - (3x - 2)(2x)}{10x^2}$

$= \dfrac{15x + 5 - 6x^2 + 4x}{10x^2}$

$= \dfrac{-6x^2 + 19x + 5}{10x^2}$

25. $\dfrac{x - 7}{5 - x} + \dfrac{3x + 3}{x - 5}$

$= \dfrac{x - 7}{(-1)(x - 5)} + \dfrac{3x + 3}{x - 5}$

$= \dfrac{(-1)(x - 7)}{x - 5} + \dfrac{3x + 3}{x - 5}$

$= \dfrac{(-1)(x - 7) + 3x + 3}{x - 5}$

$= \dfrac{-x + 7 + 3x + 3}{x - 5}$

$= \dfrac{2x + 10}{x - 5}$

27. $\dfrac{x^2 + x - 6}{x + 4} \cdot \dfrac{2x^2 + 8x}{x^2 + x - 6}$

$= \dfrac{(x + 3)(x - 2)}{x + 4} \cdot \dfrac{2x(x + 4)}{(x + 3)(x - 2)}$

$= 2x$

29. $\dfrac{a^2 - 2ab + b^2}{a + b} \div \dfrac{(a - b)^3}{a + b}$

$= \dfrac{a^2 - 2ab + b^2}{a + b} \cdot \dfrac{a + b}{(a - b)^3}$

$= \dfrac{(a - b)^2}{a + b} \cdot \dfrac{a + b}{(a - b)^3}$

$= \dfrac{1}{a - b}$

31. $\dfrac{3x}{2x + 3} - \dfrac{5}{x - 4}$

$= \dfrac{3x(x - 4)}{(2x + 3)(x - 4)} - \dfrac{5(2x + 3)}{(2x + 3)(x - 4)}$

$= \dfrac{3x(x - 4) - 5(2x + 3)}{(2x + 3)(x - 4)}$

$= \dfrac{3x^2 - 12x - 10x - 15}{(2x + 3)(x - 4)}$

$= \dfrac{3x^2 - 22x - 15}{(2x + 3)(x - 4)}$

33. $\dfrac{3x - 2}{2x - 7} + \dfrac{5x + 2}{2x - 3}$

$= \dfrac{(3x - 2)(2x - 3)}{(2x - 7)(2x - 3)} + \dfrac{(5x + 2)(2x - 7)}{(2x - 7)(2x - 3)}$

$= \dfrac{(3x - 2)(2x - 3) + (5x + 2)(2x - 7)}{(2x - 7)(2x - 3)}$

$= \dfrac{6x^2 - 13x + 6 + 10x^2 - 31x - 14}{(2x - 7)(2x - 3)}$

$= \dfrac{16x^2 - 44x - 8}{(2x - 7)(2x - 3)}$

35. $\dfrac{5a}{a^2 - 3a} + \dfrac{2}{4a^3 + 4a^2}$

$= \dfrac{5a}{a(a - 3)} + \dfrac{2}{4a^2(a + 1)}$

$= \dfrac{5a(4a)(a + 1)}{4a^2(a - 3)(a + 1)} + \dfrac{2(a - 3)}{4a^2(a - 3)(a + 1)}$

$= \dfrac{5a(4a)(a + 1) + 2(a - 3)}{4a^2(a - 3)(a + 1)}$

$= \dfrac{20a^3 + 20a^2 + 2a - 6}{4a^2(a - 3)(a + 1)}$

$= \dfrac{2(10a^3 + 10a^2 + a - 3)}{4a^2(a - 3)(a + 1)}$

$= \dfrac{10a + 10a^2 + a - 3}{2a^2(a - 3)(a + 1)}$

37. $\dfrac{5}{x^2 - 4x + 4} + \dfrac{3}{x^2 - 4}$

$= \dfrac{5}{(x - 2)(x - 2)} + \dfrac{3}{(x - 2)(x + 2)}$

$= \dfrac{5(x + 2)}{(x - 2)^2(x + 2)} + \dfrac{3(x - 2)}{(x - 2)^2(x + 2)}$

$= \dfrac{5(x + 2) + 3(x - 2)}{(x - 2)^2(x + 2)}$

$= \dfrac{5x + 10 + 3x - 6}{(x - 2)^2(x + 2)}$

$= \dfrac{8x + 4}{(x - 2)^2(x + 2)}$

39. $\dfrac{2x}{7x^2 - 14x - 21} + \dfrac{2x}{14x - 42}$

$= \dfrac{2x}{7(x - 3)(x + 1)} + \dfrac{2x}{14(x - 3)}$

$= \dfrac{2x(2)}{14(x - 3)(x + 1)} + \dfrac{2x(x + 1)}{14(x - 3)(x + 1)}$

$= \dfrac{2x(2) + 2x(x + 1)}{14(x - 3)(x + 1)}$

$= \dfrac{4x + 2x^2 + 2x}{14(x - 3)(x + 1)}$

$= \dfrac{2x^2 + 6x}{14(x - 3)(x + 1)}$

$= \dfrac{2x(x + 3)}{14(x - 3)(x + 1)}$

$= \dfrac{x(x + 3)}{7(x - 3)(x + 1)}$

41. $\dfrac{5x}{x-2} + \dfrac{3x}{x+2} - \dfrac{2x+3}{x^2-4}$

$= \dfrac{5x}{x-2} + \dfrac{3x}{x+2} - \dfrac{2x+3}{(x-2)(x+2)}$

$= \dfrac{5x(x+2)}{(x-2)(x+2)} + \dfrac{3x(x-2)}{(x-2)(x+2)} - \dfrac{2x+3}{(x-2)(x+2)}$

$= \dfrac{5x(x+2) + 3x(x-2) - (2x+3)}{(x-2)(x+2)}$

$= \dfrac{5x^2 + 10x + 3x^2 - 6x - 2x - 3}{(x-2)(x+2)}$

$= \dfrac{8x^2 + 2x - 3}{(x-2)(x+2)}$

43. $\left(\dfrac{2x+y}{5x^2y - xy^2}\right)\left(\dfrac{25x^2 - y^2}{10x^2 + 3xy - y^2}\right)\left(\dfrac{5x^2 - xy}{5x+y}\right)$

$= \dfrac{2x+y}{xy(5x-y)} \cdot \dfrac{(5x-y)(5x+y)}{(5x-y)(2x+y)} \cdot \dfrac{x(5x-y)}{5x+y}$

$= \dfrac{1}{y}$

45. $\dfrac{4x+11}{x^2+x-6} - \dfrac{x+2}{x^2+4x+3}$

$= \dfrac{4x+11}{(x+3)(x-2)} - \dfrac{x+2}{(x+3)(x+1)}$

$= \dfrac{(4x+11)(x+1)}{(x+3)(x-2)(x+1)} - \dfrac{(x+2)(x-2)}{(x+3)(x-2)(x+1)}$

$= \dfrac{(4x+11)(x+1) - (x+2)(x-2)}{(x+3)(x-2)(x+1)}$

$= \dfrac{4x^2 + 15x + 11 - (x^2 - 4)}{(x+3)(x-2)(x+1)}$

$= \dfrac{4x^2 + 15x + 11 - x^2 + 4}{(x+3)(x-2)(x+1)}$

$= \dfrac{3x^2 + 15x + 15}{(x+3)(x-2)(x+1)}$

47. $\dfrac{5x}{x^2-x-2} + \dfrac{4x+3}{x^3+x^2} - \dfrac{x-6}{x^3-2x^2}$

$= \dfrac{5x}{(x-2)(x+1)} + \dfrac{4x+3}{x^2(x+1)} - \dfrac{x-6}{x^2(x-2)}$

$= \dfrac{5x(x^2)}{x^2(x-2)(x+1)} + \dfrac{(4x+3)(x-2)}{x^2(x-2)(x+1)} - \dfrac{(x-6)(x+1)}{x^2(x-2)(x+1)}$

$= \dfrac{5x(x^2) + (4x+3)(x-2) - (x-6)(x+1)}{x^2(x-2)(x+1)}$

$= \dfrac{5x^3 + 4x^2 - 5x - 6 - x^2 + 5x + 6}{x^2(x-2)(x+1)}$

$= \dfrac{5x^3 + 4x^2 - 5x - 6 - x^2 + 5x + 6}{x^2(x-2)(x+1)}$

$= \dfrac{5x^3 + 3x^2}{x^2(x-2)(x+1)}$

$= \dfrac{x^2(5x+3)}{x^2(x-2)(x+1)}$

$= \dfrac{5x+3}{(x-2)(x+1)}$

49. $\dfrac{4x^2 + 12x + 9}{8x^3 + 27} \cdot \dfrac{12x^3 - 18x^2 + 27x}{4x^2 - 9}$

$= \dfrac{(2x+3)(2x+3)}{(2x+3)(4x^2 - 6x + 9)} \cdot \dfrac{3x(4x^2 - 6x + 9)}{(2x-3)(2x+3)}$

$= \dfrac{3x}{2x-3}$

51. $4x \div \left(\dfrac{8x^2 - 8xy}{2ax + bx - 2ay - by} \div \dfrac{2ax + 2bx + 3ay + 3by}{2a^2 + 3ab + b^2}\right)$

$= 4x \div \left(\dfrac{8x^2 - 8xy}{2ax + bx - 2ay - by} \cdot \dfrac{2a^2 + 3ab + b^2}{2ax + 2bx + 3ay + 3by}\right)$

$= 4x \div \left[\dfrac{8x(x-y)}{(2a+b)(x-y)} \cdot \dfrac{(2a+b)(a+b)}{(a+b)(2x+3y)}\right]$

$= 4x \div \left(\dfrac{8x}{2x+3y}\right)$

$= \dfrac{4x}{1} \cdot \dfrac{2x+3y}{8x}$

$= \dfrac{2x+3y}{2}$

53. $\left(\frac{x}{2} + \frac{3}{x}\right) \cdot \frac{x+1}{x}$

$= \left(\frac{x(x)}{2x} + \frac{3(2)}{2x}\right) \cdot \frac{x+1}{x}$

$= \frac{x(x) + 3(2)}{2x} \cdot \frac{x+1}{x}$

$= \frac{x^2 + 6}{2x} \cdot \frac{x+1}{x}$

$= \frac{(x^2+6)(x+1)}{2x^2}$

$= \frac{x^3 + x^2 + 6x + 6}{2x^2}$

55. $\dfrac{\frac{3x^2y}{2ab}}{\frac{9x}{16a^2}}$

$= \frac{3x^2y}{2ab} \div \frac{9x}{16a^2}$

$= \frac{3x^2y}{2ab} \cdot \frac{16a^2}{9x}$

$= \frac{8xya}{3b}$

57. $\dfrac{\frac{3}{a} - \frac{2}{a}}{\frac{5}{a}}$

$= \dfrac{\frac{3}{a} - \frac{2}{a}}{\frac{5}{a}} \cdot \frac{a}{a}$

$= \dfrac{\frac{3}{a} \cdot \frac{a}{1} - \frac{2}{a} \cdot \frac{a}{1}}{\frac{5}{a} \cdot \frac{a}{1}}$

$= \frac{3-2}{5}$

$= \frac{1}{5}$

59. $\dfrac{\frac{3}{b+1} + 2}{\frac{2}{b-1} + b}$

$= \dfrac{\frac{3}{b+1} + 2}{\frac{2}{b-1} + b} \cdot \frac{(b+1)(b-1)}{(b+1)(b-1)}$

$= \dfrac{\frac{3}{b+1} \cdot \frac{(b+1)(b-1)}{1} + 2(b+1)(b-1)}{\frac{2}{b-1} \cdot \frac{(b+1)(b-1)}{1} + b(b+1)(b-1)}$

$= \frac{3(b-1) + 2(b^2-1)}{2(b+1) + b(b^2-1)}$

$= \frac{3b - 3 + 2b^2 - 2}{2b + 2 + b^3 - b}$

$= \frac{2b^2 + 3b - 5}{b^3 + b + 2}$

61. $f(x) - g(x)$

$= \frac{x+3}{x-2} - \frac{1}{x-4}$

$= \frac{(x+3)(x-4)}{(x-2)(x-4)} - \frac{x-2}{(x-2)(x-4)}$

$= \frac{(x+3)(x-4) - (x-2)}{(x-2)(x-4)}$

$= \frac{x^2 - x - 12 - x + 2}{(x-2)(x-4)}$

$= \frac{x^2 - 2x - 10}{(x-2)(x-4)}$

63. $g(x) = \frac{1}{x-4}$

$g\left(\frac{2}{3}\right) = \dfrac{1}{\frac{2}{3} - 4}$

$= \dfrac{1}{\frac{-10}{3}}$

$= -\frac{3}{10}$

65.
$$\frac{x}{3} + \frac{x-1}{2} = \frac{7}{6}$$
$$6\left(\frac{x}{3} + \frac{x-1}{2}\right) = 6\left(\frac{7}{6}\right)$$
$$\frac{6}{1} \cdot \frac{x}{3} + \frac{6}{1} \cdot \frac{x-1}{2} = \frac{6}{1} \cdot \frac{7}{6}$$
$$\begin{aligned} 2x + 3(x-1) &= 7 \\ 2x + 3x - 3 &= 7 \\ 5x - 3 &= 7 \\ 5x &= 10 \\ x &= 2 \end{aligned}$$

67.
$$\frac{x}{5} - \frac{x+1}{3} < \frac{1}{3}$$
$$15\left(\frac{x}{5} - \frac{x+1}{3}\right) < 15\left(\frac{1}{3}\right)$$
$$\frac{15}{1} \cdot \frac{x}{5} - \frac{15}{1} \cdot \frac{x+1}{3} < \frac{15}{1} \cdot \frac{1}{3}$$
$$\begin{aligned} 3x - 5(x+1) &< 5 \\ 3x - 5x - 5 &< 5 \\ -2x - 5 &< 5 \\ -2x &< 10 \\ x &> -5 \end{aligned}$$

69.
$$\frac{5}{x} - \frac{1}{3} = \frac{11}{3x}$$
$$3x\left(\frac{5}{x} - \frac{1}{3}\right) = 3x\left(\frac{11}{3x}\right)$$
$$\frac{3x}{1} \cdot \frac{5}{x} - \frac{3x}{1} \cdot \frac{1}{3} = \frac{3x}{1} \cdot \frac{11}{3x}$$
$$\begin{aligned} 15 - x &= 11 \\ -x &= -4 \\ x &= 4 \end{aligned}$$

71.
$$\frac{x+1}{3} - \frac{x}{2} > 4$$
$$6\left(\frac{x+1}{3} - \frac{x}{2}\right) > 6(4)$$
$$\frac{6}{1} \cdot \frac{x+1}{3} - \frac{6}{1} \cdot \frac{x}{2} > 24$$
$$\begin{aligned} 2(x+1) - 3x &> 24 \\ 2x + 2 - 3x &> 24 \\ -x + 2 &> 24 \\ -x &> 22 \\ x &< -22 \end{aligned}$$

73.
$$\begin{aligned} -\frac{7}{x} + 1 &= -13 \\ x\left(-\frac{7}{x} + 1\right) &= x(-13) \\ \frac{x}{1}\left(-\frac{7}{x}\right) + x &= -13x \\ -7 + x &= -13x \\ -7 &= -14x \\ \frac{1}{2} &= x \end{aligned}$$

75.
$$\frac{5}{x-2} - 1 = 0$$
$$(x-2)\left(\frac{5}{x-2} - 1\right) = (x-2)(0)$$
$$\frac{x-2}{1} \cdot \frac{5}{x-2} - (x-2)(1) = 0$$
$$\begin{aligned} 5 - x + 2 &= 0 \\ -x + 7 &= 0 \\ 7 &= x \end{aligned}$$

77.
$$\frac{x-2}{5} - \frac{3-x}{15} > \frac{1}{9}$$
$$45\left(\frac{x-2}{5} - \frac{3-x}{15}\right) > 45\left(\frac{1}{9}\right)$$
$$\frac{45}{1} \cdot \frac{x-2}{5} - \frac{45}{1} \cdot \frac{3-x}{15} > \frac{45}{1} \cdot \frac{1}{9}$$
$$\begin{aligned} 9(x-2) - 3(3-x) &> 5 \\ 9x - 18 - 9 + 3x &> 5 \\ 12x - 27 &> 5 \\ 12x &> 32 \\ x &> \frac{8}{3} \end{aligned}$$

79.
$$\frac{7}{x-1} + 4 = \frac{x+6}{x-1}$$
$$(x-1)\left(\frac{7}{x-1} + 4\right) = (x-1)\left(\frac{x+6}{x-1}\right)$$
$$\frac{x-1}{1} \cdot \frac{7}{x-1} + (x-1)(4) = \frac{x-1}{1} \cdot \frac{x+6}{x-1}$$
$$\begin{aligned} 7 + 4x - 4 &= x + 6 \\ 4x + 3 &= x + 6 \\ 3x + 3 &= 6 \\ 3x &= 3 \\ x &= 1 \end{aligned}$$

No solution, since $x = 1$ causes a denominator to equal 0 in the original equation.

81.
$$\frac{4x+1}{x^2-x-6}=\frac{2}{x-3}+\frac{5}{x+2}$$
$$\frac{4x+1}{(x-3)(x+2)}=\frac{2}{x-3}+\frac{5}{x+2}$$
$$(x-3)(x+2)\left[\frac{4x+1}{(x-3)(x+2)}\right]=(x-3)(x+2)\left(\frac{2}{x-3}+\frac{5}{x+2}\right)$$
$$\frac{(x-3)(x+2)}{1}\cdot\frac{4x+1}{(x-3(x+2)}=\frac{(x-3)(x+2)}{1}\cdot\frac{2}{x-3}+\frac{(x-3)(x+2)}{1}\cdot\frac{5}{x+2}$$
$$\begin{aligned}4x+1&=2(x+2)+5(x-3)\\4x+1&=2x+4+5x-15\\4x+1&=7x-11\\1&=3x-11\\12&=3x\\4&=x\end{aligned}$$

83.
$$\begin{aligned}5x-3y&=2x+7y\\3x-3y&=7y\\3x&=10y\\x&=\frac{10}{3}y\end{aligned}$$

85.
$$\begin{aligned}3xy&=2xy+4\\xy&=4\\y&=\frac{4}{x}\end{aligned}$$

87.
$$\frac{2x+1}{y}=x$$
$$\frac{y}{1}\cdot\frac{2x+1}{y}=yx$$
$$2x+1=yx$$
$$\frac{2x+1}{x}=\frac{yx}{x}$$
$$\frac{2x+1}{x}=y$$

89.
$$\frac{ax+b}{cx+d}=y$$
$$\frac{cx+d}{1}\cdot\frac{ax+b}{cx+d}=(cx+d)y$$
$$\begin{aligned}ax+b&=cxy+dy\\ax-cxy&=dy-b\\x(a-cy)&=dy-b\\x&=\frac{dy-b}{a-cy}\end{aligned}$$

91.
$$\frac{1}{a}+\frac{1}{b}+\frac{1}{c}=\frac{1}{d}$$
$$abcd\left(\frac{1}{a}+\frac{1}{b}+\frac{1}{c}\right)=abcd\left(\frac{1}{d}\right)$$
$$\frac{abcd}{1}\cdot\frac{1}{a}+\frac{abcd}{1}\cdot\frac{1}{b}+\frac{abcd}{1}\cdot\frac{1}{c}=\frac{abcd}{1}\cdot\frac{1}{d}$$
$$\begin{aligned}bcd+acd+abd&=abc\\acd&=abc-bcd-abd\\acd&=b(ac-cd-ad)\\\frac{acd}{ac-cd-ad}&=b\end{aligned}$$

93. Let x = number of inches in 1 cm
$$\frac{1\text{ inch}}{2.54\text{ cm}}=\frac{x\text{ inches}}{1\text{ cm}}$$
$$\frac{1}{2.54}=\frac{x}{1}$$
$$0.3937=x$$
There is 0.3937 inches in 1 c.m.

95.

x = Total distance

1/2 x 1/3(1/2x)   1 1/2 miles

$$\frac{1}{2}x + \frac{1}{3}\left(\frac{1}{2}x\right) + 1\frac{1}{2} = x$$

$$\frac{1}{2}x + \frac{1}{6}x + \frac{3}{2} = x$$

$$6\left(\frac{1}{2}x + \frac{1}{6}x + \frac{3}{2}\right) = 6(x)$$

$$\begin{aligned} 3x + x + 9 &= 6x \\ 4x + 9 &= 6x \\ 9 &= 2x \\ \frac{9}{2} &= x \end{aligned}$$

Carol walked $\frac{9}{2} = 4\frac{1}{2}$ miles.

97. Let x = number of hours Charles and Ellen work together.

Portion of job completed by Charles	+	Portion of job completed by Ellen	=	1 whole job

$$\frac{x}{2\frac{1}{2}} + \frac{x}{2\frac{1}{3}} = 1$$

$$\frac{2}{5}x + \frac{3}{7}x = 1$$

$$35\left(\frac{2}{5}x + \frac{3}{7}x\right) = 35(1)$$

$$\begin{aligned} 14x + 15x &= 35 \\ 29x &= 35 \\ x &= \frac{35}{29} \end{aligned}$$

It would take them $\frac{35}{29} = 1\frac{6}{29}$ days working together.

99. Let x = number of hours for John to complete the job alone.

Tali and John together	+	John alone	=	whole job

$$2\left(\frac{1}{6}\right) + 2\left(\frac{1}{x}\right) + 1\left(\frac{1}{x}\right) = 1$$

$$6x\left(\frac{1}{3} + \frac{2}{x} + \frac{1}{x}\right) = 6x(1)$$

$$\begin{aligned} 2x + 12 + 6 &= 6x \\ 2x + 18 &= 6x \\ 18 &= 4x \\ \frac{18}{4} &= x \end{aligned}$$

It would take John $\frac{18}{4} = 4\frac{1}{2}$ hours working alone.

101. Let x = amount of 35% solution

$$\begin{aligned} 0.35x + 0.70(5) &= 0.60(x + 5) \\ 100[0.35x + 0.70(5)] &= 100[0.60(x + 5)] \\ 35x + 350 &= 60x + 300 \\ 35x + 50 &= 60x \\ 50 &= 25x \\ 2 &= x \end{aligned}$$

He should use 2 liters of the 35% solution.

103. number of children's tickets: x
number of adult tickets: $980 - x$

$$\begin{aligned} 3.50(x) + 6.25(980 - x) &= 4970 \\ 3.50x + 6125 - 6.25x &= 4970 \\ 6125 - 2.75x &= 4970 \\ -2.75x &= -1155 \\ x &= 420 \\ 980 - x &= 980 - 420 = 560 \end{aligned}$$

They sold 420 children's tickets and 560 adult tickets.

105. $c(h) = \dfrac{30.4h}{h^2 + 2}$

(a) $c(0.25) = \dfrac{30.4(0.25)}{(0.25)^2 + 2} = 3.7$

After 0.25 hour there is 3.7 mg/l concentration of the druhg in the bloodstream.

$c(0.5) = \dfrac{30.4(0.5)}{(0.5)^2 + 2} = 6.8$

After 0.5 hour there is 6.8 mg/l concentration of the drug in the bloodstream.

$c(1) = \dfrac{30.4(1)}{1^2 + 2} = 10.1$

After 1 hour there is 10.1 mg/l concentration of the drug in the bloodstream.

$$c(2) = \frac{30.4(2)}{2^2 + 2} = 10.1$$

After 2 hours there is 10.1 mg/l concentration of the drug in the bloodstream.

$$c(4) = \frac{30.4(4)}{4^2 + 2} = 6.8$$

After 4 hours there is 6.8 mg/l concentration of the drug in the bloodstream.

$$c(10) = \frac{30.4(10)}{10^2 + 2} = 3.0$$

After 10 hours there is 3.0 mg/l concentration of the drug in the bloodstream.

(b)

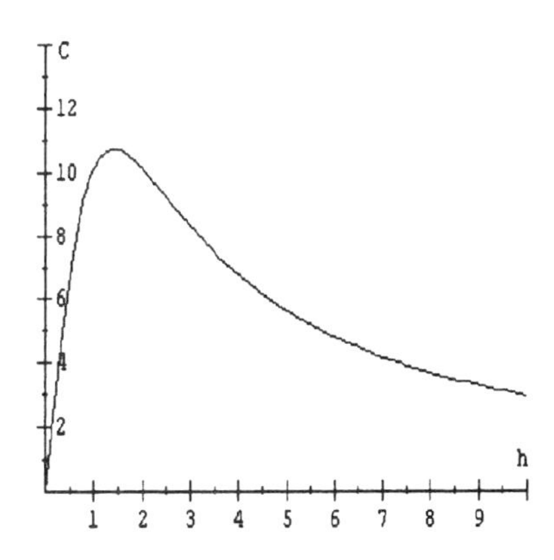

The concentration level increases until between 1 and 2 hours it begins to decrease.

CHAPTER 6 PRACTICE TEST

1. $2x + 5 \neq 0$

$2x \neq -5$

$x \neq -\frac{5}{2}$

Domain: $\left\{x \mid x \neq -\frac{5}{2}\right\}$

3. (a) $\frac{4xy^3}{5ab^4} \cdot \frac{15}{16x^4y^5}$

$$= \frac{4xy^3 \cdot 5 \cdot 3}{5ab^4 \cdot 4 \cdot 4x \cdot x^3y^3 \cdot y^2}$$

$$= \frac{3}{4ab^4x^3y^2}$$

(b) $\frac{3x}{18y^2} + \frac{5}{8x^2y}$

$$= \frac{x}{6y^2} + \frac{5}{8x^2y}$$

$$= \frac{x(4x^2)}{24x^2y^2} + \frac{5(3y)}{24x^2y^2}$$

$$= \frac{x(4x^2) + 5(3y)}{24x^2y^2}$$

$$= \frac{4x^3 + 15y}{24x^2y^2}$$

(c) $\frac{r^2 - rs - 2s^2}{2s^2 + 4rs} \div \frac{r - 2s}{4s^2 + 8rs}$

$$= \frac{r^2 - rs - 2s^2}{2s^2 + 4rs} \cdot \frac{4s^2 + 8rs}{r - 2s}$$

$$= \frac{(r - 2s)(r + s)}{2s(s + 2r)} \cdot \frac{4s(s + 2r)}{r - 2s}$$

$$= 2(r + s)$$

(d) $\frac{9x - 2}{4x - 3} + \frac{x + 4}{3 - 4x}$

$$= \frac{9x - 2}{4x - 3} + \frac{x + 4}{(-1)(4x - 3)}$$

$$= \frac{9x - 2}{4x - 3} + \frac{(-1)(x + 4)}{4x - 3}$$

$$= \frac{9x - 2 + (-1)(x + 4)}{4x - 3}$$

$$= \frac{9x - 2 - x - 4}{4x - 3}$$

$$= \frac{8x - 6}{4x - 3}$$

$$= \frac{2(4x - 3)}{4x - 3}$$

$$= 2$$

(e) $\frac{3x}{x^2 - 4} + \frac{4}{x^2 - 5x + 6} - \frac{2x}{x^2 - x - 6}$

$$= \frac{3x}{(x-2)(x+2)} + \frac{4}{(x-2)(x-3)} - \frac{2x}{(x-3)(x+2)}$$

$$= \frac{3x(x-3)}{(x-2)(x+2)(x-3)} + \frac{4(x+2)}{(x-2)(x+2)(x-3)} - \frac{2x(x-2)}{(x-2)(x+2)(x-3)}$$

$$= \frac{3x(x-3) + 4(x+2) - 2x(x-2)}{(x-2)(x+2)(x-3)}$$

$$= \frac{3x^2 - 9x + 4x + 8 - 2x^2 + 4x}{(x-2)(x+2)(x-3)}$$

$$= \frac{x^2 - x + 8}{(x-2)(x+2)(x-3)}$$

(f) $\left(\frac{3}{x} - \frac{2}{x+1}\right) \div \frac{1}{x+1}$

$$= \left[\frac{3(x+1)}{x(x+1)} - \frac{2(x)}{x(x+1)}\right] \div \frac{1}{x+1}$$

$$= \frac{3(x+1) - 2x}{x(x+1)} \div \frac{1}{x+1}$$

$$= \frac{3x + 3 - 2x}{x(x+1)} \div \frac{1}{x+1}$$

$$= \frac{x+3}{x(x+1)} \cdot \frac{x+1}{1}$$

$$= \frac{x+3}{x}$$

5. $f(x) = \frac{3x+1}{x+4}$

$$f\left(-\frac{3}{5}\right) = \frac{3\left(-\frac{3}{5}\right) + 1}{-\frac{3}{5} + 4}$$

$$= \frac{-\frac{9}{5} + 1}{-\frac{3}{5} + 4}$$

$$= \frac{-\frac{4}{5}}{\frac{17}{5}}$$

$$= -\frac{4}{17}$$

7. $$y = \frac{x - 2}{2x + 1}$$

$$(2x + 1)(y) = \frac{2x + 1}{1} \cdot \frac{x - 2}{2x + 1}$$

$$2xy + y = x - 2$$
$$y + 2 = x - 2xy$$
$$y + 2 = x(1 - 2y)$$
$$\frac{y + 2}{1 - 2y} = x$$

9. Let x = number of hours to complete the job together

Portion of job completed by Jackie	+	Portion of job completed by Eleanor	=	1 whole job
$\frac{x}{3\frac{1}{2}}$	+	$\frac{x}{2}$	=	1

$$\frac{2x}{7} + \frac{x}{2} = 1$$

$$14\left(\frac{2x}{7} + \frac{x}{2}\right) = 14(1)$$

$$2(2x) + 7x = 14$$
$$4x + 7x = 14$$
$$11x = 14$$
$$x = \frac{14}{11}$$

It will take them $\frac{14}{11} = 1\frac{3}{11}$ hours working together.

CHAPTERS 4 - 6 CUMULATIVE REVIEW

1. $3y - 5x + 9 = 0$

x	y
0	-3
9/5	0

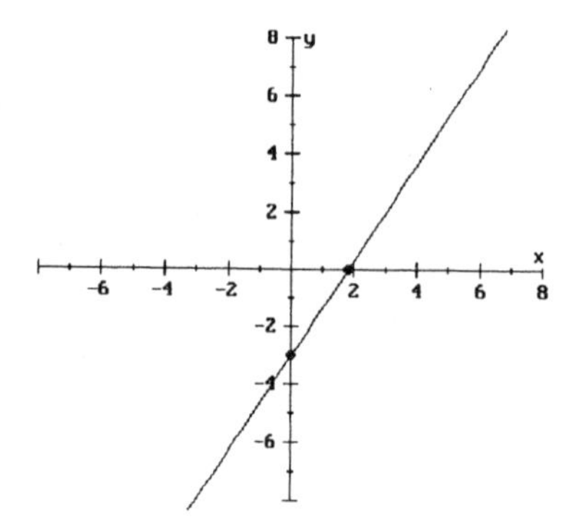

3. $y = -5$
This is a horizontal line passing through (0, -5).

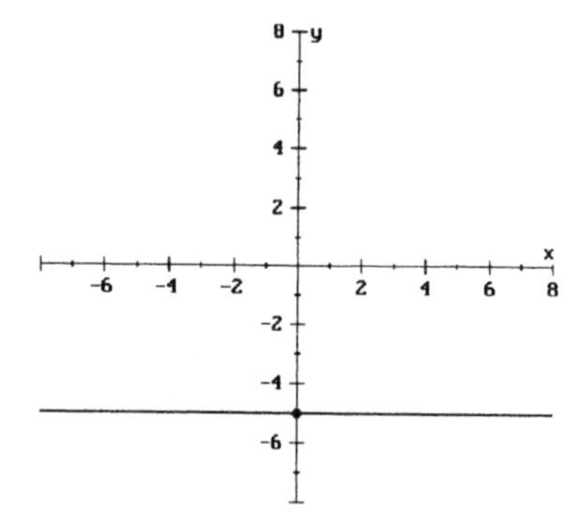

5. $m = 2 = \frac{2}{1}$

From (2, 3), move up 2 then right 1 to locate (3, 5) on the line.

$m = -2 = \frac{-2}{1}$

From (2, 3), move down 2 then right 1 to locate (3, 1) on the line.

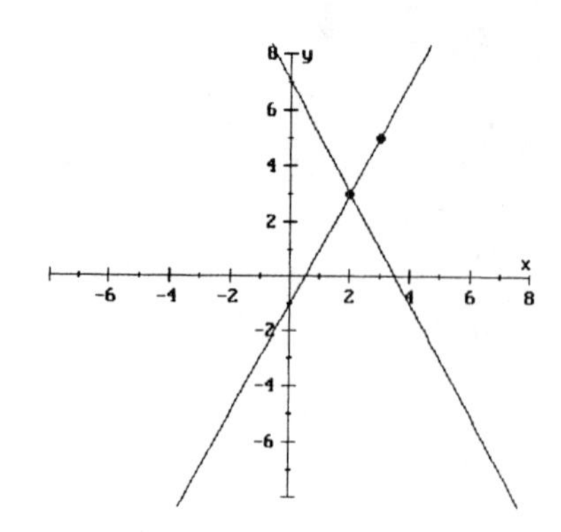

7. $m = \frac{y_2 - y_1}{x_2 - x_1}$

$= \frac{-3 - 5}{2 - 3}$

$= \frac{-8}{-1}$

$= 8$

9. $y = 5x - 8$
$m = 5$

11. $m = \frac{y_2 - y_1}{x_2 - x_1}$

$= \frac{-1 - 4}{2 - 6}$

$= \frac{-5}{-4}$

$= \frac{5}{4}$

Parallel lines have equal slopes.

13. $m = \frac{y_2 - y_1}{x_2 - x_1}$

$3 = \frac{a - (-2)}{2 - a}$

$3 = \frac{a + 2}{2 - a}$

$$(2 - a)(3) = \frac{(2 - a)}{1} \cdot \frac{(a + 2)}{(2 - a)}$$

$$\begin{aligned} 6 - 3a &= a + 2 \\ 6 &= 4a + 2 \\ 4 &= 4a \\ 1 &= a \end{aligned}$$

15. $$\begin{aligned} y - y_1 &= m(x - x_1) \\ y - 7 &= 3[x - (-2)] \\ y - 7 &= 3(x + 2) \\ y - 7 &= 3x + 6 \\ y &= 3x + 13 \end{aligned}$$

17. $$m = \frac{7 - 1}{2 - 3} = -6$$

$$\begin{aligned} y - y_1 &= m(x - x_1) \\ y - 1 &= -6(x - 3) \\ y - 1 &= -6x + 18 \\ y &= -6x + 19 \end{aligned}$$

19. $(0, 2)$; $m = 4$

$$\begin{aligned} y &= mx + b \\ y &= 4x + 2 \end{aligned}$$

21. $$\begin{aligned} 3x + 5y &= 4 \\ 5y &= -3x + 4 \end{aligned}$$

$$y = -\frac{3}{5}x + \frac{4}{5}$$

$$m = -\frac{3}{5}$$

Parallel lines have equal slopes.

$$y - y_1 = m(x - x_1)$$

$$y - (-3) = -\frac{3}{5}(x - 2)$$

$$y + 3 = -\frac{3}{5}x + \frac{6}{5}$$

$$y = -\frac{3}{5}x - \frac{9}{5}$$

23. The line passes through (0, 1) and (2, 4).

$$m = \frac{4 - 1}{2 - 0} = \frac{3}{2}$$

$$y = mx + b$$

$$y = \frac{3}{2}x + 1$$

25. $$\begin{cases} 3x - 2y = 8 \\ 5x + y = 9 \end{cases}$$

Multiply the 2nd equation by 2 and add to the 1st equation:

$$\begin{aligned} 10x + 2y &= 18 \\ 3x - 2y &= 8 \\ \hline 13x &= 26 \\ x &= 2 \end{aligned}$$

$$\begin{aligned} 5(2) + y &= 9 \\ 10 + y &= 9 \\ y &= -1 \end{aligned}$$

$x = 2, \; y = -1$

27. $$\begin{cases} 7u + 5v = 23 \\ 8u + 9v = 23 \end{cases}$$

$$\begin{cases} -8(7u + 5v) = -8(23) \\ 7(8u + 9v) = 7(23) \end{cases}$$

$$\begin{aligned} -56u - 40v &= -184 \\ 56u + 63v &= 161 \\ \hline 23v &= -23 \\ v &= -1 \end{aligned}$$

$$\begin{aligned} 7u + 5(-1) &= 23 \\ 7u - 5 &= 23 \\ 7u &= 28 \\ u &= 4 \end{aligned}$$

$u = 4, \; v = -1$

29. $$\begin{cases} 2m = 3n - 5 \\ 3n = 2m - 5 \end{cases}$$

$$\begin{aligned} 2m - 3n &= -5 \\ -2m + 3n &= -5 \\ \hline 0 &= -10 \end{aligned}$$

Inconsistent

31. $\begin{cases} \dfrac{2}{3}y - \dfrac{1}{2}x = 6 \\ \dfrac{4}{5}y - \dfrac{3}{4}x = 6 \end{cases}$

$$\begin{cases} 6\left(\dfrac{2}{3}y - \dfrac{1}{2}x\right) = 6(6) \\ 20\left(\dfrac{4}{5}y - \dfrac{3}{4}x\right) = 20(6) \end{cases}$$

$$\begin{cases} 4y - 3x = 36 \\ 16y - 15x = 120 \end{cases}$$

$$\begin{cases} -4(4y - 3x) = -4(36) \\ 16y - 15x = 120 \end{cases}$$

$$\begin{aligned} -16y + 12x &= -144 \\ 16y - 15x &= 120 \\ \hline -3x &= -24 \\ x &= 8 \end{aligned}$$

$$\begin{aligned} 4y - 3(8) &= 36 \\ 4y - 24 &= 36 \\ 4y &= 60 \\ y &= 15 \end{aligned}$$

$x = 8, \ y = 15$

33. $2x + 3y \geq 18$ (solid line)

$2x + 3y = 18$

x	y
0	6
9	0

Test point: (0, 0)
$2(0) + 3(0) \geq 18$
$0 \geq 18$
False
Shade the half-plane not containing (0, 0).

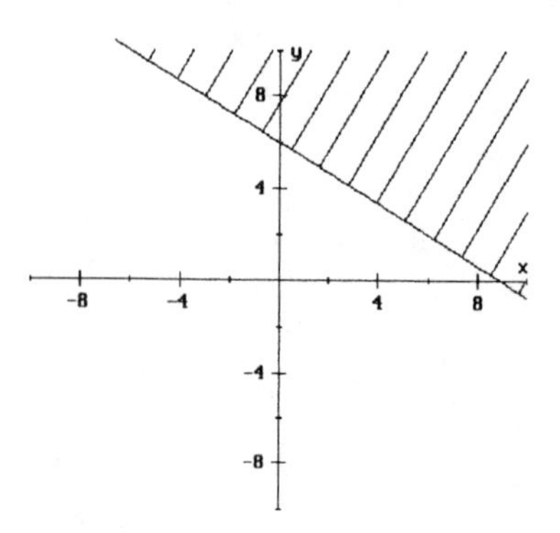

35. $2y < 5x - 20$ (dashed line)

$2y = 5x - 20$

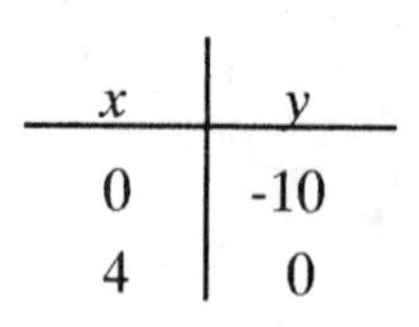

x	y
0	-10
4	0

Test point: (0, 0)
$2(0) < 5(0) - 20$
$0 < -20$
False

Shade the half-plane not containing (0, 0).

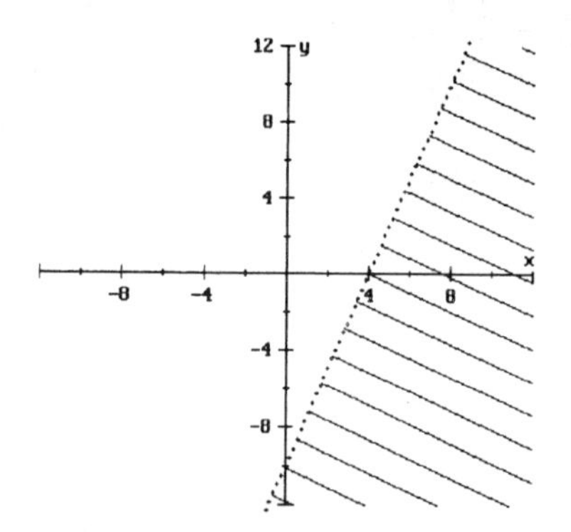

37. degree: $2 + 3 = 5$

39. $(3x^2 + 2x - 4) - (2x^2 - 3x + 5)$
$= 3x^2 + 2x - 4 - 2x^2 + 3x - 5$
$= x^2 + 5x - 9$

41. $(x - y)(x - 3y)$
$= x^2 - 3xy - xy + 3y^2$
$= x^2 - 4xy + 3y^2$

43. $(x + y - 2)(x + y)$
$= x^2 + xy - 2x + xy + y^2 - 2y$
$= x^2 + 2xy + y^2 - 2x - 2y$

45. $(3x - 5y)^2$
$= (3x)^2 - 2(3x)(5y) + (5y)^2$
$= 9x^2 - 30xy + 25y^2$

47. $(3x - 5y)(3x + 5y)$
$= (3x)^2 - (5y)^2$
$= 9x^2 - 25y^2$

49. $(2x + y - 3)(2x + y + 3)$
$= [(2x + y) - 3][(2x + y) + 3]$
$= (2x + y)^2 - 3^2$
$= (2x)^2 + 2(2x)(y) + y^2 - 9$
$= 4x^2 + 4xy + y^2 - 9$

51. $x^2 - 5x - 24$
$= (x - 8)(x + 3)$

53. $y^2 - 12xy + 35x^2$
$= (y - 7x)(y - 5x)$

55. $4y^2 + 16yz + 15z^2$
$= 4y^2 + 10yz + 6yz + 15z^2$
$= 2y(2y + 5z) + 3z(2y + 5z)$
$= (2y + 5z)(2y + 3z)$

57. $25a^2 + 20ab + 4b^2$
$= (5a)^2 + 2(5a)(2b) + (2b)^2$
$= (5a + 2b)^2$

59. $36x^2 - 9$
$= 9(4x^2 - 1)$
$= 9[(2x)^2 - 1^2]$
$= 9(2x - 1)(2x + 1)$

61. Not factorable

63. $3y^3 + 5y^2 - 2y$
$= y(3y^2 + 5y - 2)$
$= y(3y^2 + 6y - y - 2)$
$= y[3y(y + 2) - 1(y + 2)]$
$= y(y + 2)(3y - 1)$

65. $49a^4 - 14a^2z - 3z^2$
$= 49a^4 - 21a^2z + 7a^2z - 3z^2$
$= 7a^2(7a^2 - 3z) + z(7a^2 - 3z)$
$= (7a^2 - 3z)(7a^2 + z)$

67. $(x - y)^2 - 16$
$= (x - y)^2 - 4^2$
$= [(x - y) - 4][(x - y) + 4]$
$= (x - y - 4)(x - y + 4)$

69. $16a^4 - b^4$
$= (4a^2)^2 - (b^2)^2$
$= (4a^2 - b^2)(4a^2 + b^2)$
$= [(2a)^2 - b^2](4a^2 + b^2)$
$= (2a - b)(2a + b)(4a^2 + b^2)$

71. $8a^3 + 125b^3$
$= (2a)^3 + (5b)^3$
$= (2a + 5b)[(2a)^2 - (2a)(5b) + (5b)^2]$
$= (2a + 5b)(4a^2 - 10ab + 25b^2)$

73.
$x^2 - x = 12$
$x^2 - x - 12 = 0$
$(x - 4)(x + 3) = 0$
$x - 4 = 0$ or $x + 3 = 0$
$x = 4$ or $x = -3$

75.
$(a + 3)(a - 5) = 9$
$a^2 - 2a - 15 = 9$
$a^2 - 2a - 24 = 0$
$(a - 6)(a + 4) = 0$
$a - 6 = 0$ or $a + 4 = 0$
$a = 6$ or $a = -4$

77.
$(r + 4)^2 = 36$
$r^2 + 8r + 16 = 36$
$r^2 + 8r - 20 = 0$
$(r + 10)(r - 2) = 0$
$r + 10 = 0$ or $r - 2 = 0$
$r = -10$ or $r = 2$

79.
$f(x) = x^2 - 5x + 4$
$0 = x^2 - 5x + 4$
$0 = (x - 4)(x - 1)$
$x - 4 = 0$ or $x - 1 = 0$
$x = 4$ $\quad x = 1$

81.
$$\begin{array}{r} x - 2 \\ x + 4\overline{)x^2 + 2x + 3} \\ \underline{-(x^2 + 4x)} \\ -2x + 3 \\ \underline{-(-2x - 8)} \\ 11 \end{array}$$

$$x - 2 + \frac{11}{x + 4}$$

83. Long division of $4x^3 + 0x^2 + 3x + 1$ by $2x + 3$:

$$\begin{array}{r} 2x^2 - 3x + 6 \\ 2x + 3 \overline{)4x^3 + 0x^2 + 3x + 1} \\ \underline{-(4x^3 + 6x^2)} \\ -6x^2 + 3x \\ \underline{-(-6x^2 - 9x)} \\ 12x + 1 \\ \underline{-(12x + 18)} \\ -17 \end{array}$$

$$2x^2 - 3x + 6 - \frac{17}{2x + 3}$$

85. $2x + 3 \neq 0$

$2x \neq -3$

$x \neq -\frac{3}{2}$

Domain: $\left\{x \mid x \neq -\frac{3}{2}\right\}$

87. $\frac{18a^3b^2}{16a^5b} = \frac{9b}{8a^2}$

89. $\frac{2a^3 - 5a^2b - 3ab^2}{a^4 - 2a^3b - 3a^2b^2}$

$= \frac{a(2a + b)(a - 3b)}{a^2(a - 3b)(a + b)}$

$= \frac{2a + b}{a(a + b)}$

91. $\frac{3xy^2}{5a^3b} \div \frac{21x^3y}{25ab^3}$

$= \frac{3xy^2}{5a^3b} \cdot \frac{25ab^3}{21x^3y}$

$= \frac{5yb^2}{7x^2a^2}$

93. $\frac{3x}{2y} + \frac{2y}{3x}$

$= \frac{3x(3x)}{6xy} + \frac{2y(2y)}{6xy}$

$= \frac{3x(3x) + 2y(2y)}{6xy}$

$= \frac{9x^2 + 4y^2}{6xy}$

95. $\left(\frac{2x^3 - 2x^2 - 24x}{x + 2}\right)\left(\frac{x + 2}{4x^2 + 12x}\right)$

$= \frac{2x(x - 4)(x + 3)}{x + 2} \cdot \frac{x + 2}{4x(x + 3)}$

$= \frac{x - 4}{2}$

97. $\frac{2}{x - 5} - \frac{3}{x - 2}$

$= \frac{2(x - 2)}{(x - 5)(x - 2)} - \frac{3(x - 5)}{(x - 5)(x - 2)}$

$= \frac{2(x - 2) - 3(x - 5)}{(x - 5)(x - 2)}$

$= \frac{2x - 4 - 3x + 15}{(x - 5)(x - 2)}$

$= \frac{-x + 11}{(x - 5)(x - 2)}$

99. $\frac{x^3 + x^2y}{2x^2 + xy} \div \left[\left(\frac{x^2 + 2xy - 3y^2}{2x^2 - xy - y^2}\right)(x + y)\right]$

$= \frac{x^3 + x^2y}{2x^2 + xy} \div \left[\frac{(x + 3y)(x - y)}{(2x + y)(x - y)} \cdot \frac{(x + y)}{1}\right]$

$= \frac{x^3 + x^2y}{2x^2 + xy} \div \frac{(x + 3y)(x + y)}{2x + y}$

$= \frac{x^2(x + y)}{x(2x + y)} \cdot \frac{2x + y}{(x + 3y)(x + y)}$

$= \frac{x}{x + 3y}$

101. $\frac{x - \frac{2}{x}}{\frac{1}{2} - x}$

$= \frac{x - \frac{2}{x}}{\frac{1}{2} - x} \cdot \frac{2x}{2x}$

$= \frac{2x^2 - 4}{x - 2x^2}$

103. $f(x) = \dfrac{x + 4}{2x + 1}$

$$f\left(\frac{2}{3}\right) = \frac{\frac{2}{3} + 4}{2\left(\frac{2}{3}\right) + 1}$$

$$= \frac{\frac{14}{3}}{\frac{4}{3} + 1}$$

$$= \frac{\frac{14}{3}}{\frac{7}{3}}$$

$$= \frac{14}{7} = 2$$

105. $\dfrac{2}{x} - \dfrac{1}{2} = 2 - \dfrac{1}{x}$

$$2x\left(\frac{2}{x} - \frac{1}{2}\right) = 2x\left(2 - \frac{1}{x}\right)$$

$$\begin{aligned} 4 - x &= 4x - 2 \\ 4 &= 5x - 2 \\ 6 &= 5x \\ \frac{6}{5} &= x \end{aligned}$$

107. $\dfrac{x}{2} - \dfrac{x + 1}{3} > \dfrac{2}{3}$

$$6\left(\frac{x}{2} - \frac{x + 1}{3}\right) > 6\left(\frac{2}{3}\right)$$

$$\begin{aligned} 3x - 2(x + 1) &> 4 \\ 3x - 2x - 2 &> 4 \\ x - 2 &> 4 \\ x &> 6 \end{aligned}$$

109. $\dfrac{3}{x - 2} + \dfrac{5}{x + 1} = \dfrac{1}{x^2 - x - 2}$

$$\frac{3}{x - 2} + \frac{5}{x + 1} = \frac{1}{(x - 2)(x + 1)}$$

$$(x - 2)(x + 1)\left(\frac{3}{x - 2} + \frac{5}{x + 1}\right) = (x - 2)(x + 1)\left[\frac{1}{(x - 2)(x + 1)}\right]$$

$$\begin{aligned} 3(x + 1) + 5(x - 2) &= 1 \\ 3x + 3 + 5x - 10 &= 1 \\ 8x - 7 &= 1 \\ 8x &= 8 \\ x &= 1 \end{aligned}$$

111. $$\begin{aligned} 2a + 3b &= 5b - 4a \\ 6a + 3b &= 5b \\ 6a &= 2b \\ a &= \frac{b}{3} \end{aligned}$$

113. $$\begin{aligned} \frac{x - y}{y} &= x \\ x - y &= xy \\ x &= xy + y \\ x &= y(x + 1) \\ \frac{x}{x + 1} &= y \end{aligned}$$

115. Let x = number of foreign cars

$$\frac{5 \text{ foreign}}{6 \text{ American}} = \frac{x \text{ foreign}}{1200 \text{ American}}$$

$$\frac{5}{6} = \frac{x}{1200}$$

$$1200\left(\frac{5}{6}\right) = x$$

$$1000 = x$$

The total number of cars equals 1000 foreign + 1200 American = 2200 cars.

117. Let x = number of hours to paint the room working together.

$$\frac{x}{4} + \frac{x}{4\frac{1}{2}} = 1$$

$$\frac{x}{4} + \frac{2x}{9} = 1$$

$$36\left(\frac{x}{4} + \frac{2x}{9}\right) = 36(1)$$

$$\begin{aligned} 9x + 8x &= 36 \\ 17x &= 36 \\ x &= \frac{36}{17} \end{aligned}$$

It would take them $\dfrac{36}{17} = 2\dfrac{2}{17}$ hours working together.

119. Let x = number of general admission tickets, then $505 - x$ = number of reserved seats.

$$3.50(x) + 4.25(505 - x) = 1861.25$$
$$3.5x + 2146.25 - 4.25x = 1861.25$$
$$-0.75x + 2146.25 = 1861.25$$
$$-0.75x = -285$$
$$x = 380$$
$$505 - x = 505 - 380 = 125$$

They sold 380 general admission tickets and 125 reserved seat tickets.

CHAPTERS 4 - 6 CUMULATIVE PRACTICE TEST

1. (a) $(2x^2 - 3xy + 4y^2) - (5x^2 - 2xy + y^2)$
$= 2x^2 - 3xy + 4y^2 - 5x^2 + 2xy - y^2$
$= -3x^2 - xy + 3y^2$

(b) $(3a - 2b)(5a + 3b)$
$= 15a^2 + (9 - 10)ab - 6b^2$
$= 15a^2 - ab - 6b^2$

(c) $(2x^2 - y)(2x^2 + y)$
$= (2x^2)^2 - (y)^2$
$= 4x^4 - y^2$

(d) $(3y - 2z)^2$
$= (3y)^2 - 2(3y)(2z) + (2z)^2$
$= 9y^2 - 12yz + 4z^2$

(e) $(x + y - 3)^2$
$= (x + y - 3)(x + y - 3)$
$= x^2 + xy - 3x + xy + y^2 - 3y - 3x - 3y + 9$
$= x^2 + 2xy + y^2 - 6x - 6y + 9$

3. $f(x) = (x - 3)(x + 9)$
$0 = (x - 3)(x + 9)$
$x - 3 = 0$ or $x + 9 = 0$
$x = 3$ $\qquad x = -9$

5. $x + 2 \overline{)2x^3 + 0x^2 + 6x + 5}$, quotient $2x^2 - 4x + 14$

$-(2x^3 + 4x^2)$
$-4x^2 + 6x$
$-(-4x^2 - 8x)$
$14x + 5$
$-(14x + 28)$
-23

$$2x^2 - 4x + 14 - \frac{23}{x + 2}$$

7. (a) $\left(\dfrac{25x^2 - 9y^2}{x - y}\right)\left(\dfrac{2x^2 + xy - 3y^2}{10x^2 + 9xy - 9y^2}\right)$

$= \dfrac{(5x - 3y)(5x + 3y)}{x - y} \cdot \dfrac{(2x + 3y)(x - y)}{(5x - 3y)(2x + 3y)}$

$= 5x + 3y$

(b) $\dfrac{2y}{2x - y} + \dfrac{4x}{y - 2x}$

$= \dfrac{2y}{2x - y} + \dfrac{4x}{(-1)(2x - y)}$

$= \dfrac{2y}{2x - y} + \dfrac{(-1)(4x)}{2x - y}$

$= \dfrac{2y - 4x}{2x - y}$

$= \dfrac{-2(2x - y)}{2x - y}$

$= -2$

(c) $\dfrac{3x}{x^2 - 10x + 21} - \dfrac{2}{x^2 - 8x + 15}$

$= \dfrac{3x}{(x - 7)(x - 3)} - \dfrac{2}{(x - 5)(x - 3)}$

$= \dfrac{3x(x - 5)}{(x - 7)(x - 3)(x - 5)} - \dfrac{2(x - 7)}{(x - 7)(x - 3)(x - 5)}$

$= \dfrac{3x(x - 5) - 2(x - 7)}{(x - 7)(x - 3)(x - 5)}$

$= \dfrac{3x^2 - 15x - 2x + 14}{(x - 7)(x - 3)(x - 5)}$

$= \dfrac{3x^2 - 17x + 14}{(x - 7)(x - 3)(x - 5)}$

9. $f(x) = \dfrac{x}{2x + 1}$

$f(x + h) = \dfrac{x + h}{2(x + h) + 1} = \dfrac{x + h}{2x + 2h + 1}$

$f(x + h) - f(x) = \dfrac{x + h}{2x + 2h + 1} - \dfrac{x}{2x + 1}$

$= \dfrac{(x + h)(2x + 1)}{(2x + 2h + 1)(2x + 1)} - \dfrac{x(2x + 2h + 1)}{(2x + 2h + 1)(2x + 1)}$

$= \dfrac{(x + h)(2x + 1) - x(2x + 2h + 1)}{(2x + 2h + 1)(2x + 1)}$

$= \dfrac{2x^2 + x + 2xh + h - 2x^2 - 2xh - x}{(2x + 2h + 1)(2x + 1)}$

$= \dfrac{h}{(2x + 2h + 1)(2x + 1)}$

11. $y = \dfrac{a}{a + 1}$

$y(a + 1) = \dfrac{a}{a + 1} \cdot \dfrac{a + 1}{1}$

$ya + y = a$

$y = a - ya$

$y = a(1 - y)$

$\dfrac{y}{1 - y} = a$

13. $y = -\dfrac{3}{4}x + 6$

x	y
0	6
8	0

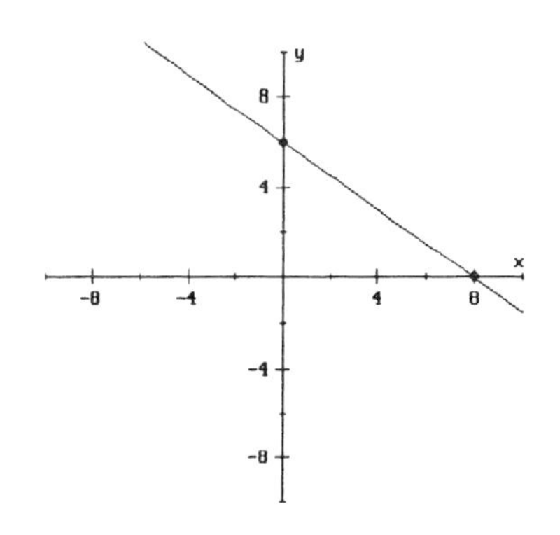

15. (a)

$$3y - 2x = 4$$
$$3y = 2x + 4$$
$$y = \frac{2}{3}x + \frac{4}{3}$$
$$m = \frac{2}{3}$$

Parallel lines have equal slopes.

$$y - y_1 = m(x - x_1)$$
$$y - (-3) = \frac{2}{3}(x - 2)$$
$$y + 3 = \frac{2}{3}x - \frac{4}{3}$$
$$y = \frac{2}{3}x - \frac{13}{3}$$

(b)

$$m_{\perp} = -\frac{3}{2}$$
$$y - y_1 = m(x - x_1)$$
$$y - (-3) = -\frac{3}{2}(x - 2)$$
$$y + 3 = -\frac{3}{2}x + 3$$
$$y = -\frac{3}{2}x$$

17.

$$\begin{cases} 5x + 2y = 4 \\ 2x - 3y = 13 \end{cases}$$

Multiply 1st equation by 3 and 2nd equation by 2, then add:

$$\begin{array}{r} 15x + 6y = 12 \\ \underline{4x - 6y = 26} \\ 19x = 38 \\ x = 2 \end{array}$$

$$5(2) + 2y = 4$$
$$10 + 2y = 4$$
$$2y = -6$$
$$y = -3$$

$x = 2, \quad y = -3$

CHAPTER 7

Exercises 7.1

1. $(x^2x^5)(x^3x) = (x^{2+5})(x^{3+1})$
$= (x^7)(x^4)$
$= x^{7+4}$
$= x^{11}$

3. $(-2a^2b^3)(3a^5b^7) = -6a^{2+5}b^{3+7}$
$= -6a^7b^{10}$

5. $(a^2)^5 = a^{2\cdot 5}$
$= a^{10}$

7. $\left(\frac{2}{3}\right)^2 = \frac{2^2}{3^2}$
$= \frac{4}{9}$

9. $(x + y^3)^2 = x^2 + 2(x)(y^3) + (y^3)^2$
$= x^2 + 2xy^3 + y^6$

11. $(xy^3)^2 = x^2y^{3\cdot 2}$
$= x^2y^6$

13. $(2^3 \cdot 3^2)^2 = 2^{3\cdot 2} \cdot 3^{2\cdot 2}$
$= 2^6 \cdot 3^4$
$= 64 \cdot 81$
$= 5184$

15. $(x^4y^3)^5(x^3y^2)^2$
$= x^{4\cdot 5}y^{3\cdot 5} \cdot x^{3\cdot 2}y^{2\cdot 2}$
$= x^{20}y^{15} \cdot x^6y^4$
$= x^{20+6}y^{15+4}$
$= x^{26}y^{19}$

17. $(-2a^2)^3(ab^2)^4 = (-2)^3a^{2\cdot 3} \cdot a^4b^{2\cdot 4}$
$= -8a^6 \cdot a^4b^8$
$= -8a^{6+4}b^8$
$= -8a^{10}b^8$

19. $(r^2st)^3(-2rs^2t)^4$
$= r^{2\cdot 3}s^3t^3 \cdot (-2)^4r^4s^{2\cdot 4}t^4$
$= r^6s^3t^3 \cdot 16r^4s^8t^4$
$= 16r^{6+4}s^{3+8}t^{3+4}$
$= 16r^{10}s^{11}t^7$

21. $\frac{x^5}{x^2} = x^{5-2}$
$= x^3$

23. $\frac{x^3y^2}{xy^4} = x^{3-1}y^{2-4}$
$= x^2y^{-2}$
$= \frac{x^2}{y^2}$

25. $\frac{5^4 \cdot 2^2}{25^2 \cdot 4^2} = \frac{5^4 \cdot 2^2}{(5^2)^2(2^2)^2}$
$= \frac{5^42^2}{5^42^4}$
$= 5^{4-4}2^{2-4}$
$= 5^02^{-2}$
$= \frac{1}{2^2}$
$= \frac{1}{4}$

27. $\frac{a^5b^9c}{a^4bc^5} = a^{5-4}b^{9-1}c^{1-5}$
$= ab^8c^{-4}$
$= \frac{ab^8}{c^4}$

29. $\frac{(-3)^2xy^4}{-3^2xy^5} = \frac{9}{-9}x^{1-1}y^{4-5}$
$= -x^0y^{-1}$
$= -\frac{1}{y}$

31. $\frac{3^2(-2)^3}{(-9^2)(-4)^2} = \frac{9(-8)}{-81(16)}$
$= \frac{1}{18}$

33. $\left(\frac{y^5}{y^8}\right)^3 = \frac{y^{5\cdot 3}}{y^{8\cdot 3}}$

$= \frac{y^{15}}{y^{24}}$

$= y^{15-24}$
$= y^{-9}$
$= \frac{1}{y^9}$

35. $\left(\frac{y^2y^7}{y^4}\right)^3 = \left(\frac{y^{2+7}}{y^4}\right)^3$

$= \left(\frac{y^9}{y^4}\right)^3$

$= (y^{9-4})^3$
$= (y^5)^3$
$= y^{5\cdot 3}$
$= y^{15}$

37. $\frac{(3r^2s)^3(-2rs^2)^4}{(-18rs)^2}$

$= \frac{3^3r^{2\cdot 3}s^3 \cdot (-2)^4r^4s^{2\cdot 4}}{(-18)^2r^2s^2}$

$= \frac{27r^6s^3 \cdot 16r^4s^8}{324r^2s^2}$

$= \frac{4r^{6+4}s^{3+8}}{3r^2s^2}$

$= \frac{4r^{10}s^{11}}{3r^2s^2}$

$= \frac{4}{3}r^{10-2}s^{11-2}$

$= \frac{4}{3}r^8s^9$

39. $\left(\frac{2x^2y^3}{xy^4}\right)^2\left(\frac{3xy^2}{6}\right)^3$

$= (2x^{2-1}y^{3-4})^2\left(\frac{xy^2}{2}\right)^3$

$= (2xy^{-1})^2\left(\frac{xy^2}{2}\right)^3$

$= 2^2x^2y^{-1\cdot 2} \cdot \frac{x^3y^{2\cdot 3}}{2^3}$

$= 2^2x^2y^{-2} \cdot \frac{x^3y^6}{2^3}$

$= 2^{2-3}x^{2+3}y^{-2+6}$

$= 2^{-1}x^5y^4$

$= \frac{x^5y^4}{2}$

41. $\left(\frac{(6ab^2)^2}{-3ab}\right)^3$

$= \left(\frac{6^2a^2b^{2\cdot 2}}{-3ab}\right)^3$

$= \left(\frac{36a^2b^4}{-3ab}\right)^3$

$= (-12a^{2-1}b^{4-1})^3$

$= (-12ab^3)^3$
$= (-12)^3a^3b^{3\cdot 3}$
$= -1728a^3b^9$

43. $\left(\frac{-2a^2b^3}{ab}\right)^3(-3xy^2)^3$

$= (-2a^{2-1}b^{3-1})^3(-3xy^2)^3$
$= (-2ab^2)^3(-3xy^2)^3$
$= (-2)^3a^3b^{2\cdot 3} \cdot (-3)^3x^3y^{2\cdot 3}$
$= -8a^3b^6 \cdot (-27)x^3y^6$
$= 216a^3b^6x^3y^6$

45. $[(r^3s^2)^3(rs^2)^4]^2$
$= (r^{3\cdot 3}s^{2\cdot 3} \cdot r^4s^{2\cdot 4})^2$
$= (r^9s^6 \cdot r^4s^8)^2$
$= (r^{9+4}s^{6+8})^2$
$= (r^{13}s^{14})^2$
$= r^{13\cdot 2}s^{14\cdot 2}$
$= r^{26}s^{28}$

47. $5^{-8} = 0.000003$

49. $-2 \cdot 5^{-3} + 8 = 7.984000$

51. $x^{-2}x^{4}x^{-3} = x^{-2+4-3}$
$= x^{-1}$
$= \dfrac{1}{x}$

53. $(x^{5}y^{-4})(x^{-3}y^{2}x^{0})$
$= x^{5-3+0}y^{-4+2}$
$= x^{2}y^{-2}$
$= \dfrac{x^{2}}{y^{2}}$

55. $(3^{-2})^{-3}$
$= 3^{(-2)(-3)}$
$= 3^{6}$

57. $(a^{-2}b^{-3})^{2}$
$= a^{-2\cdot 2}b^{-3\cdot 2}$
$= a^{-4}b^{-6}$
$= \dfrac{1}{a^{4}b^{6}}$

59. $(r^{-3}s^{2})^{-4}(r^{2})^{-3}$
$= r^{(-3)(-4)}s^{2(-4)}r^{2(-3)}$
$= r^{12}s^{-8}r^{-6}$
$= r^{12-6}s^{-8}$
$= r^{6}s^{-8}$
$= \dfrac{r^{6}}{s^{8}}$

61. $-2^{-2} = -\dfrac{1}{2^{2}}$
$= -\dfrac{1}{4}$

63. $(-3)^{-2} = \dfrac{1}{(-3)^{2}}$
$= \dfrac{1}{9}$

65. $(2^{-2})^{-3}(3^{-3})^{2}$
$= 2^{(-2)(-3)}3^{(-3)(2)}$
$= 2^{6}3^{-6}$
$= \dfrac{2^{6}}{3^{6}}$

67. $(3^{-2}s^{3})^{4}(9s^{-3})^{-2}$
$= (3^{-2}s^{3})^{4}(3^{2}s^{-3})^{-2}$
$= 3^{(-2)(4)}s^{3\cdot 4} \cdot 3^{(2)(-2)}s^{(-3)(-2)}$
$= 3^{-8}s^{12} \cdot 3^{-4}s^{6}$
$= 3^{-8-4}s^{12+6}$
$= 3^{-12}s^{18}$
$= \dfrac{s^{18}}{3^{12}}$

69. $\dfrac{x^{4}}{x^{-2}} = x^{4-(-2)}$
$= x^{6}$

71. $\dfrac{x^{-3}y^{2}}{x^{-5}y^{0}} = x^{-3-(-5)}y^{2-0}$
$= x^{2}y^{2}$

73. $\left(\dfrac{1}{2}\right)^{-1} = \left(\dfrac{2}{1}\right)^{1}$
$= 2$

75. $\left(-\dfrac{3}{5}\right)^{-3} = \left(-\dfrac{5}{3}\right)^{3}$
$= \dfrac{(-5)^{3}}{3^{3}}$
$= -\dfrac{125}{27}$

77. $\dfrac{x^{-1}xy^{-2}}{x^{4}y^{-3}y} = \dfrac{x^{-1+1}y^{-2}}{x^{4}y^{-3+1}}$
$= \dfrac{x^{0}y^{-2}}{x^{4}y^{-2}}$
$= x^{0-4}y^{-2-(-2)}$
$= x^{-4}y^{0}$
$= \dfrac{1}{x^{4}}$

79. $\dfrac{(a^{-2}b^2)^{-3}}{ab^{-2}}$

$= \dfrac{a^{(-2)(-3)}b^{2(-3)}}{ab^{-2}}$

$= \dfrac{a^6b^{-6}}{ab^{-2}}$

$= a^{6-1}b^{-6-(-2)}$
$= a^5b^{-4}$
$= \dfrac{a^5}{b^4}$

81. $\left(\dfrac{x^{-1}x^{-3}}{x^{-2}}\right)^{-3}$

$= \left(\dfrac{x^{-1-3}}{x^{-2}}\right)^{-3}$

$= \left(\dfrac{x^{-4}}{x^{-2}}\right)^{-3}$

$= (x^{-4-(-2)})^{-3}$
$= (x^{-2})^{-3}$
$= (x^{-2})^{-3}$
$= x^{(-2)(-3)}$
$= x^6$

83. $\dfrac{2^{-2} \cdot 3^2}{6^{-2}} = \dfrac{2^{-2} \cdot 3^2}{(2 \cdot 3)^{-2}}$

$= \dfrac{2^{-2} \cdot 3^2}{2^{-2} \cdot 3^{-2}}$

$= 2^{-2-(-2)} \cdot 3^{2-(-2)}$

$= 2^0 \cdot 3^4$
$= 3^4 = 81$

85. $\dfrac{(3x)^{-2}(2xy^{-1})^0}{(2x^{-2}y^3)^{-2}}$

$= \dfrac{3^{-2}x^{-2} \cdot 1}{2^{-2}x^{(-2)(-2)}y^{3(-2)}}$

$= \dfrac{3^{-2}x^{-2}}{2^{-2}x^4y^{-6}}$

$= \dfrac{2^2y^6}{3^2x^2x^4}$
$= \dfrac{4y^6}{9x^{2+4}}$
$= \dfrac{4y^6}{9x^6}$

87. $\left(\dfrac{x^{-3}y^{-4}}{x^{-5}y^{-7}}\right)^{-3}$
$= \left(x^{-3-(-5)}y^{-4-(-7)}\right)^{-3}$
$= (x^2y^3)^{-3}$
$= x^{2(-3)}y^{3(-3)}$
$= x^{-6}y^{-9}$
$= \dfrac{1}{x^6y^9}$

89. $\dfrac{(-3a^{-4}b^{-2})(-4ab^{-3})^{-1}}{(12ab^2)^{-1}}$

$= \dfrac{(-3a^{-4}b^{-2})(12ab^2)}{-4ab^{-3}}$

$= \dfrac{-36a^{-4+1}b^{-2+2}}{-4ab^{-3}}$

$= \dfrac{-36a^{-3}b^0}{-4ab^{-3}}$
$= 9a^{-3-1}b^{0-(-3)}$
$= 9a^{-4}b^3$
$= \dfrac{9b^3}{a^4}$

91. $(x^2y^{-3})^{-1} = x^{2(-1)}y^{(-3)(-1)}$
$= x^{-2}y^3$
$= \dfrac{y^3}{x^2}$

93. $(x^2 + y^{-3})^{-1} = \left(x^2 + \dfrac{1}{y^3}\right)^{-1}$

$= \left(\dfrac{x^2(y^3)}{y^3} + \dfrac{1}{y^3}\right)^{-1}$

$= \left(\dfrac{x^2y^3 + 1}{y^3}\right)^{-1} = \dfrac{y^3}{x^2y^3 + 1}$

95. $(x^{-2} + y^{-2})^{-2}$

$$= \left(\frac{1}{x^2} + \frac{1}{y^2}\right)^{-2}$$

$$= \left(\frac{y^2}{x^2y^2} + \frac{x^2}{x^2y^2}\right)^{-2}$$

$$= \left(\frac{y^2 + x^2}{x^2y^2}\right)^{-2}$$

$$= \left(\frac{x^2y^2}{y^2 + x^2}\right)^{2}$$

$$= \frac{x^{2(2)}y^{2(2)}}{(y^2 + x^2)^2}$$

$$= \frac{x^4y^4}{(y^2 + x^2)^2}$$

97. $\left(\frac{x^{-4}y^{-7}z^{-6}}{x^{-24}y^{-16}}\right)^0 = 1$

99. $\frac{x^{-1} + y^{-1}}{xy^{-1}}$

$$= \frac{\frac{1}{x} + \frac{1}{y}}{\frac{x}{y}}$$

$$= \frac{\frac{1}{x} + \frac{1}{y}}{\frac{x}{y}} \cdot \frac{xy}{xy}$$

$$= \frac{\frac{1}{x} \cdot \frac{xy}{1} + \frac{1}{y} \cdot \frac{xy}{1}}{\frac{x}{y} \cdot \frac{xy}{1}}$$

$$= \frac{y + x}{x^2}$$

101. $\frac{r^{-2} + s^{-1}}{r^{-1} + s^{-2}}$

$$= \frac{\frac{1}{r^2} + \frac{1}{s}}{\frac{1}{r} + \frac{1}{s^2}}$$

$$= \frac{\frac{1}{r^2} + \frac{1}{s}}{\frac{1}{r} + \frac{1}{s^2}} \cdot \frac{r^2s^2}{r^2s^2}$$

$$= \frac{\frac{1}{r^2} \cdot \frac{r^2s^2}{1} + \frac{1}{s} \cdot \frac{r^2s^2}{1}}{\frac{1}{r} \cdot \frac{r^2s^2}{1} + \frac{1}{s^2} \cdot \frac{r^2s^2}{1}}$$

$$= \frac{s^2 + r^2s}{rs^2 + r^2}$$

103. $\frac{2a^{-1} + b^{-2}}{a^{-2} + b}$

$$= \frac{\frac{2}{a} + \frac{1}{b^2}}{\frac{1}{a^2} + b}$$

$$= \frac{\frac{2}{a} + \frac{1}{b^2}}{\frac{1}{a^2} + b} \cdot \frac{a^2b^2}{a^2b^2}$$

$$= \frac{\frac{2}{a} \cdot \frac{a^2b^2}{1} + \frac{1}{b^2} \cdot \frac{a^2b^2}{1}}{\frac{1}{a^2} \cdot \frac{a^2b^2}{1} + b(a^2b^2)}$$

$$= \frac{2ab^2 + a^2}{b^2 + a^2b^3}$$

111.

$$-2 \le 5 - 2x < 11$$
$$-7 \le -2x < 6$$
$$\frac{7}{2} \ge x > -3$$
or $-3 < x \le \frac{7}{2}$

113. $m = \dfrac{-4 - 0}{0 - 6} = \dfrac{2}{3}$

$y = mx + b$

$y = \dfrac{2}{3}x - 4$

Exercises 7.2

1. $10^{-4} \cdot 10^{7} = 10^{-4+7}$
$= 10^{3}$
$= 1000$

3. $\dfrac{10^{4}}{10^{-5}} = 10^{4-(-5)}$
$= 10^{9}$
$= 1{,}000{,}000{,}000$

5. $\dfrac{10^{-4} \cdot 10^{2}}{10^{-3}} = \dfrac{10^{-4+2}}{10^{-3}}$
$= \dfrac{10^{-2}}{10^{-3}}$
$= 10^{-2-(-3)}$
$= 10$

7. $\dfrac{10^{-4} \cdot 10^{2} \cdot 10^{-3}}{10^{4} \cdot 10^{-3}} = \dfrac{10^{-4+2-3}}{10^{4-3}}$
$= \dfrac{10^{-5}}{10^{1}}$
$= 10^{-5-1}$
$= 10^{-6}$
$= 0.000001$

9. $\dfrac{10^{-4} \cdot 10^{-5} \cdot 10^{7}}{10^{-6} \cdot 10^{-2} \cdot 10^{0}} = \dfrac{10^{-4-5+7}}{10^{-6-2+0}}$
$= \dfrac{10^{-2}}{10^{-8}}$
$= 10^{-2-(-8)}$
$= 10^{6}$
$= 1{,}000{,}000$

11. move 1 place right
$1.62 \times 10^{1} = 16.2$

13. move 8 places right
$7.6 \times 10^{8} = 760{,}000{,}000$

15. move 7 places left
$8.51 \times 10^{-7} = 0.000000851$

17. move 3 places right
$6.0 \times 10^{3} = 6000$

19. $824 = 8.24 \times 10^{?}$
$= 8.24 \times 10^{2}$

21. $5 = 5.0 \times 10^{?}$
$= 5.0 \times 10^{0}$

23. $0.0093 = 9.3 \times 10^{?}$
$= 9.3 \times 10^{-3}$

25. $827{,}546{,}000 = 8.27546 \times 10^{?}$
$= 8.27546 \times 10^{8}$

27. $0.00000072 = 7.2 \times 10^{?}$
$= 7.2 \times 10^{-7}$

29. $79.32 = 7.932 \times 10^{?}$
$= 7.932 \times 10^{1}$

31. $\dfrac{(6000)(0.007)}{(0.021)(12{,}000)}$
$= \dfrac{(6 \times 10^{3})(7.0 \times 10^{-3})}{(2.1 \times 10^{-2})(1.2 \times 10^{4})}$
$= \dfrac{6 \cdot 7}{2.1 \cdot 1.2} \times \dfrac{10^{3} \cdot 10^{-3}}{10^{-2} \cdot 10^{4}}$
$= 16.6\overline{66} \times 10^{-2}$
$= 0.1\overline{66}$

33. $\dfrac{(120)(0.005)}{(10{,}000)(60)}$

$= \dfrac{(1.2 \times 10^2)(5.0 \times 10^{-3})}{(1.0 \times 10^4)(6.0 \times 10^1)}$

$= \dfrac{1.2 \cdot 5.0}{1.0 \cdot 6.0} \times \dfrac{10^{2-3}}{10^{4+1}}$

$= 1 \times 10^{-6}$
$= 0.000001$

35. $5{,}000{,}000 = 5 \times 10^6$
$7500 = 7.5 \times 10^3$

37. $\dfrac{\text{red cells}}{\text{white cells}}: \dfrac{5 \times 10^6}{7.5 \times 10^3} = \dfrac{5}{7.5} \times 10^{6-3}$

$= 0.67 \times 10^3$
$= 6.7 \times 10^{-1} \times 10^3$
$= 6.7 \times 10^2$

The ratio is $\dfrac{6.7 \times 10^2}{1}$ or $\dfrac{670}{1}$.

39. $t = \dfrac{D}{r}$

$= \dfrac{3{,}670{,}000{,}000}{186{,}000}$

$= \dfrac{3.67 \times 10^9}{1.86 \times 10^5}$

$= 1.973118 \times 10^4$
$= 19731.18$ sec
$= 5$ hr 28 min 51 sec

41. 1 day $= 24$ hr
$= (24)(60) = 1440$ min
$= (1440)(60) = 86400$ sec

Hydrogen used = rate · time in 1 day
$= (700{,}000{,}000)(86400)$
$= (70 \times 10^8)(8.64 \times 10^4)$
$= 60.48 \times 10^{12}$
$= 6.048 \times 10^{13}$

6.048×10^{13} tons are used in 1 day

Amount used in 1 yr = # of days in a yr · amount used per day

$= (365)(6.048 \times 10^{13})$
$= 2207.52 \times 10^{13}$
$= 2.20752 \times 10^{16}$

2.20752×10^{16} tons are used in 1 year

43. $\dfrac{1 \text{ Å}}{10^{-8} \text{ cm}} = \dfrac{x \text{ Å}}{10^{-4} \text{ cm}}$

$\dfrac{1}{10^{-8}} = \dfrac{x}{10^{-4}}$

$10^{-4}\left(\dfrac{1}{10^{-8}}\right) = x$

$10^4 = x$

There are $10^4 = 10{,}000$ Å in 10^{-4} cm which is 1 micron. Hence there are 10,000 Å in 1 micron.

45. $\dfrac{10^{-8} \text{ cm}}{1 \text{ angstrom}} = \dfrac{x \text{ cm}}{0.66 \text{ angstrom}}$

$\dfrac{10^{-8}}{1} = \dfrac{x}{0.66}$

$0.66(10^{-8}) = x$
$6.6 \times 10^{-9} = x$

Its radius is 6.6×10^{-9} cm.

47. $\dfrac{1.6 \text{ km}}{1 \text{ mile}} = \dfrac{x \text{ km}}{5.86 \times 10^{12} \text{ miles}}$

$\dfrac{1.6}{1} = \dfrac{x}{5.86 \times 10^{12}}$

$(5.86 \times 10^{12})(1.6) = x$
$9.376 \times 10^{12} = x$

There are 9.376×10^{12} km in 1 light-year.

49. $\dfrac{10^{-4}\text{ cm}}{1\ \mu m} = \dfrac{x\text{ cm}}{60\ \mu m}$

$$\frac{10^{-4}}{1} = \frac{x}{60}$$

$$60 \times 10^{-4} = x$$
$$6.0 \times 10^{-3} = x$$

The diameter is 6.0×10^{-3} or 0.006 cm.

51. $\dfrac{1\text{ atom}}{10^{-23}\text{ gm}} = \dfrac{x\text{ atoms}}{1\text{ gm}}$

$$\frac{1}{10^{-23}} = \frac{x}{1}$$

$$x = \frac{1}{10^{-23}} = 10^{23}$$

There are 10^{23} atoms in 1 gram.

53. distance = number of cheek cells · diameter in cm
= (40000)(0.006)
= 240

They would stretch 240 cm.

55.
$$1\text{ light-year} = 5.86 \times 10^{12}\text{ miles}$$
$$93\text{ million miles} = 9.3 \times 10^{7}\text{ miles}$$

$$\frac{1\ Au}{9.3 \times 10^{7}\text{ miles}} = \frac{x\ Au}{5.86 \times 10^{12}\text{ miles}}$$

$$\frac{1}{9.3 \times 10^{7}} = \frac{x}{5.86 \times 10^{12}}$$

$$(5.86 \times 10^{12})\left(\frac{1}{9.3 \times 10^{7}}\right) = x$$

$$x = 0.6301 \times 10^{5}$$
$$= 6.301 \times 10^{4}$$

There are 6.301×10^{4} AU in a light-year.

59. $(3x - 2)(x - 8) - (x - 4)^2$
$= 3x^2 - 24x - 2x + 16 - (x^2 - 8x + 16)$
$= 3x^2 - 26x + 16 - x^2 + 8x - 16$
$= 2x^2 - 18x$

61. $(x - y)^3 - x^2 = [-3 - (-1)]^3 - (-3)^2$
$= (-2)^3 - (-3)^2$
$= -8 - 9$
$= -17$

Exercises 7.3

1. $8^{1/3} = 2$

3. $(-32)^{1/5} = -2$

5. $-100^{1/2} = -(100^{1/2})$
$= -(10)$
$= -10$

7. $\sqrt[3]{64} = 4$

9. $\sqrt[4]{81} = 3$

11. Not a real number

13. $-\sqrt[9]{-1} = -(-1)$
$= 1$

15. $\sqrt[3]{-343} = -7$

17. $-\sqrt[4]{1296} = -6$

19. $\sqrt[8]{256} = 2$

21. $\sqrt[7]{78{,}125} = 5$

23. $(-32)^{3/5} = [(-32)^{1/5}]^3$
$= (-2)^3$
$= -8$

25. $32^{-1/5} = (32^{1/5})^{-1}$
$= (2)^{-1}$
$= \dfrac{1}{2}$

27. $(-32)^{-1/5} = [(-32)^{1/5}]^{-1}$
$= (-2)^{-1}$
$= -\dfrac{1}{2}$

29. $-(81)^{-1/2} = -[81^{1/2}]^{-1}$
$= -(9)^{-1}$
$= -\dfrac{1}{9}$

31. $(-64)^{-2/3} = \left[(-64)^{1/3}\right]^{-2}$
$= (-4)^{-2}$
$= \dfrac{1}{(-4)^2}$
$= \dfrac{1}{16}$

33. $\sqrt{(-16)^2} = |-16|$
$= 16$

35. $\sqrt{-16}$ is not a real number

37. $-(\sqrt{16})^2 = -(4)^2$
$= -16$

39. $\sqrt[n]{3^{2n}} = (3^{2n})^{1/n}$
$= 3^{(2n)(1/n)}$
$= 3^2$
$= 9$

41. $\left(\dfrac{64}{27}\right)^{1/3} = \dfrac{64^{1/3}}{27^{1/3}}$
$= \dfrac{4}{3}$

43. $\left(\dfrac{81}{16}\right)^{-1/4} = \left(\dfrac{16}{81}\right)^{1/4}$
$= \dfrac{16^{1/4}}{81^{1/4}}$
$= \dfrac{2}{3}$

45. $\left(-\dfrac{1}{32}\right)^{-4/5} = (-32)^{4/5}$
$= \left[(-32)^{1/5}\right]^4$
$= (-2)^4$
$= 16$

47. $x^{1/2}x^{2/3} = x^{1/2+2/3}$
$= x^{7/6}$

49. $(a^{-1/2})^{-3/4} = a^{(-1/2)((-3/4)}$
$= a^{3/8}$

51. $(2^{-1} \cdot 4^{1/2})^{-2} = (2^{-1} \cdot 2)^{-2}$
$= (2^0)^{-2}$
$= 2^{(0)(-2)}$
$= 2^0$
$= 1$

53. $(r^{1/2}r^{-2/3}s^{1/2})^{-2}$
$= (r^{1/2-2/3}s^{1/2})^{-2}$
$= (r^{-1/6}s^{1/2})^{-2}$
$= r^{(-1/6)(-2)}s^{(1/2)(-2)}$
$= r^{1/3}s^{-1}$
$= \dfrac{r^{1/3}}{s}$

55. $(r^{-1}s^{1/2})^{-2}(r^{-1/2}s^{1/3})^2$
$= r^{(-1)(-2)}s^{(1/2)(-2)}r^{(-1/2)(2)}s^{(1/3)(2)}$
$= r^2s^{-1}r^{-1}s^{2/3}$
$= r^{2-1}s^{-1+2/3}$
$= rs^{-1/3}$
$= \dfrac{r}{s^{1/3}}$

57. $\dfrac{x^{-1/2}}{x^{-1/3}} = x^{-1/2-(-1/3)}$
$= x^{-1/6}$
$= \dfrac{1}{x^{1/6}}$

59. $\dfrac{a^{-1/2}b^{1/3}}{a^{1/4}b^{1/5}} = a^{-1/2-1/4}b^{1/3-1/5}$
$= a^{-3/4}b^{2/15}$
$= \dfrac{b^{2/15}}{a^{3/4}}$

61. $\left(\dfrac{x^{1/2}x^{-1}}{x^{1/3}}\right)^{-6} = \left(\dfrac{x^{1/2-1}}{x^{1/3}}\right)^{-6}$
$= \left(\dfrac{x^{-1/2}}{x^{1/3}}\right)^{-6}$
$= (x^{-1/2-1/3})^{-6}$
$= (x^{-5/6})^{-6}$
$= x^{(-5/6)(-6)}$
$= x^5$

63. $\dfrac{(4^{-1/2} \cdot 16^{3/4})^{-2}(64^{5/6})}{(-64)^{1/3}}$

$= \dfrac{\left(\dfrac{1}{4^{1/2}} \cdot 8\right)^{-2}(32)}{-4}$

$= \dfrac{\left(\dfrac{8}{2}\right)^{-2}(32)}{-4}$

$= \dfrac{4^{-2} \cdot 32}{-4}$

$= \dfrac{\dfrac{1}{4^2} \cdot 32}{-4}$

$= \dfrac{\dfrac{1}{16} \cdot 32}{-4}$

$= \dfrac{2}{-4}$

$= -\dfrac{1}{2}$

65. $\dfrac{(x^{1/2}y^{1/3})^{-2}(x^{1/3}y^{1/4})^{-12}}{xy^{1/4}}$

$= \dfrac{x^{(1/2)(-2)}y^{(1/3)(-2)}x^{(1/3)(-12)}y^{(1/4)(-12)}}{xy^{1/4}}$

$= \dfrac{x^{-1}y^{-2/3}x^{-4}y^{-3}}{xy^{1/4}}$

$= \dfrac{x^{-1-4}y^{-2/3-3}}{xy^{1/4}}$

$= \dfrac{x^{-5}y^{-11/3}}{xy^{1/4}}$

$= x^{-5-1}y^{-11/3-1/4}$

$= x^{-6}y^{-47/12}$

$= \dfrac{1}{x^6y^{47/12}}$

67. $(x^{1/2} + x)x^{1/2}$
$= x^{1/2} \cdot x^{1/2} + x \cdot x^{1/2}$
$= x^{1/2+1/2} + x^{1+1/2}$
$= x + x^{3/2}$

69. $(x^{1/2} - 2x^{-1/2})^2$
$= (x^{1/2})^2 - 2(x^{1/2})(2x^{-1/2}) + (2x^{-1/2})^2$
$= x^{(1/2)(2)} - 4x^{1/2-1/2} + 2^2x^{(-1/2)(2)}$
$= x - 4x^0 + 4x^{-1}$
$= x - 4 + \dfrac{4}{x} = \dfrac{x^2 - 4x + 4}{x}$

71. $\sqrt[3]{xy} = (xy)^{1/3}$

73. $\sqrt{x^2 + y^2} = (x^2 + y^2)^{1/2}$

75. $\sqrt[5]{5a^2b^3} = (5a^2b^3)^{1/5}$

77. $2\sqrt[3]{3xyz^4} = 2(3xyz^4)^{1/3}$

79. $5\sqrt[3]{(x - y)^2} = 5(x - y)^{2/3}$

81. $\sqrt[n]{x^n - y^n} = (x^n - y^n)^{1/n}$

83. $\sqrt[n]{x^{5n+1}y^{2n-1}} = (x^{5n+1}y^{2n-1})^{1/n}$

85. $x^{1/3} = \sqrt[3]{x}$

87. $mn^{1/3} = m\sqrt[3]{n}$

89. $(-a)^{2/3} = \sqrt[3]{(-a)^2}$

91. $-a^{2/3} = -\sqrt[3]{a^2}$

93. $(a^2b)^{1/3} = \sqrt[3]{a^2b}$

95. $(x^2 + y^2)^{1/2} = \sqrt{x^2 + y^2}$

97. $(x^n - y^n)^{1/2} = \sqrt{x^n - y^n}$

99. $36^{-1/2} = 0.1667$

101. $3 - 18^{-3/4} = 2.8856$

111. (a) $f(4) = -3$
(b) $f(-3) = -6$
(c) $f(0) = 3$
(d) $f(2) = 0$
(e) $x = -1, 2, 5$

113.
$$\frac{x}{3} - \frac{x}{2} = 4$$
$$6\left(\frac{x}{3} - \frac{x}{2}\right) = 6(4)$$
$$\frac{6}{1} \cdot \frac{x}{3} - \frac{6}{1} \cdot \frac{x}{2} = 24$$
$$2x - 3x = 24$$
$$-x = 24$$
$$x = -24$$

Exercises 7.4

1. $\sqrt{56} = \sqrt{2^3 \cdot 7}$
$= \sqrt{2^2 \cdot 2 \cdot 7}$
$= \sqrt{2^2} \cdot \sqrt{2 \cdot 7}$
$= 2\sqrt{14}$

3. $\sqrt{48} = \sqrt{2^4 \cdot 3}$
$= \sqrt{2^4} \cdot \sqrt{3}$
$= 2^2 \cdot \sqrt{3}$
$= 4\sqrt{3}$

5. $\sqrt[5]{64} = \sqrt[5]{2^6}$
$= \sqrt[5]{2^5 \cdot 2}$
$= \sqrt[5]{2^5} \cdot \sqrt[5]{2}$
$= 2\sqrt[5]{2}$

7. $\sqrt{8}\sqrt{18} = \sqrt{8 \cdot 18}$
$= \sqrt{144}$
$= 12$

9. $\sqrt{64x^8} = \sqrt{64}\sqrt{x^8}$
$= 8x^4$

11. $\sqrt[4]{81x^{12}} = \sqrt[4]{3^4x^{12}}$
$= \sqrt[4]{3^4}\sqrt[4]{x^{12}}$
$= 3x^3$

13. $\sqrt{128x^{60}} = \sqrt{2^7x^{60}}$
$= \sqrt{2^6 \cdot 2x^{60}}$
$= \sqrt{2^6}\sqrt{2}\sqrt{x^{60}}$
$= 2^3\sqrt{2}\,x^{30}$
$= 8x^{30}\sqrt{2}$

15. $\sqrt[4]{128x^{60}} = \sqrt[4]{2^7x^{60}}$
$= \sqrt[4]{2^4 \cdot 2^3x^{60}}$
$= \sqrt[4]{2^4}\sqrt[4]{2^3}\sqrt[4]{x^{60}}$
$= 2\sqrt[4]{8}x^{15}$
$= 2x^{15}\sqrt[4]{8}$

17. $\sqrt[5]{128x^{60}} = \sqrt[5]{2^7x^{60}}$
$= \sqrt[5]{2^5 \cdot 2^2 \cdot x^{60}}$
$= \sqrt[5]{2^5}\sqrt[5]{2^2}\sqrt[5]{x^{60}}$
$= 2\sqrt[5]{4}x^{12}$
$= 2x^{12}\sqrt[5]{4}$

19. $\sqrt[3]{x^3y^6} = \sqrt[3]{x^3}\sqrt[3]{y^6}$
$= xy^2$

21. $\sqrt{32a^2b^4} = \sqrt{2^5a^2b^4}$
$= \sqrt{2^4 \cdot 2a^2b^4}$
$= \sqrt{2^4}\sqrt{2}\sqrt{a^2}\sqrt{b^4}$
$= 2^2\sqrt{2}ab^2$
$= 4ab^2\sqrt{2}$

23. $\sqrt[5]{a^{35}b^{75}} = \sqrt[5]{a^{35}}\sqrt[5]{b^{75}}$
$= a^7b^{15}$

25. $\sqrt{x^3y}\sqrt{xy^3} = \sqrt{x^3y \cdot xy^3}$
$= \sqrt{x^4y^4}$
$= \sqrt{x^4}\sqrt{y^4}$
$= x^2y^2$

27. $\sqrt{\frac{1}{2}} = \frac{\sqrt{1}}{\sqrt{2}}$
$= \frac{1}{\sqrt{2}}$

$$= \frac{1 \cdot \sqrt{2}}{\sqrt{2} \cdot \sqrt{2}}$$

$$= \frac{\sqrt{2}}{\sqrt{2^2}}$$

$$= \frac{\sqrt{2}}{2}$$

29. $\dfrac{\sqrt{x}}{\sqrt{5}} = \dfrac{\sqrt{x} \cdot \sqrt{5}}{\sqrt{5} \cdot \sqrt{5}}$

$$= \frac{\sqrt{x \cdot 5}}{\sqrt{5^2}}$$

$$= \frac{\sqrt{5x}}{5}$$

31. $\sqrt{\dfrac{45}{4}} = \dfrac{\sqrt{45}}{\sqrt{4}}$

$$= \frac{\sqrt{3^2 \cdot 5}}{2}$$

$$= \frac{3\sqrt{5}}{2}$$

33. $\dfrac{1}{\sqrt{75}} = \dfrac{1}{\sqrt{3 \cdot 5^2}}$

$$= \frac{1}{\sqrt{3} \cdot 5}$$

$$= \frac{1 \cdot \sqrt{3}}{5\sqrt{3} \cdot \sqrt{3}}$$

$$= \frac{\sqrt{3}}{5\sqrt{3^2}}$$

$$= \frac{\sqrt{3}}{5 \cdot 3}$$

$$= \frac{\sqrt{3}}{15}$$

35. $\sqrt{64x^5y^8} = \sqrt{2^6x^4 \cdot xy^8}$

$$= \sqrt{2^6}\sqrt{x^4}\sqrt{x}\sqrt{y^8}$$

$$= 2^3x^2\sqrt{x}y^4$$

$$= 8x^2y^4\sqrt{x}$$

37. $\sqrt[3]{81x^8y^7} = \sqrt[3]{3^3 \cdot 3x^6 \cdot x^2y^6 \cdot y}$

$$= \sqrt[3]{3^3} \cdot \sqrt[3]{3} \cdot \sqrt[3]{x^6} \cdot \sqrt[3]{x^2} \cdot \sqrt[3]{y^6} \cdot \sqrt[3]{y}$$

$$= 3\sqrt[3]{3}x^2\sqrt[3]{x^2}y^2\sqrt[3]{y}$$

$$= 3x^2y^2\sqrt[3]{3x^2y}$$

39. $\sqrt[4]{54y^2}\sqrt[4]{48y^4} = \sqrt[4]{54y^2 \cdot 48y^4}$

$$= \sqrt[4]{2592y^6}$$

$$= \sqrt[4]{2^5 \cdot 3^4y^6}$$

$$= \sqrt[4]{2^4 \cdot 2 \cdot 3^4 \cdot y^4 \cdot y^2}$$

$$= \sqrt[4]{2^4} \cdot \sqrt[4]{2} \cdot \sqrt[4]{3^4} \cdot \sqrt[4]{y^4} \cdot \sqrt[4]{y^2}$$

$$= 2\sqrt[4]{2} \cdot 3y \cdot \sqrt[4]{y^2}$$

$$= 6y\sqrt[4]{2y^2}$$

41. $\sqrt[6]{(x + y^2)^6} = x + y^2$

43. Doesn't simplify further

45. $(2s\sqrt{6t})(5t\sqrt{3s}) = 10st\sqrt{6t \cdot 3s}$

$$= 10st\sqrt{18ts}$$

$$= 10st\sqrt{2 \cdot 3^2ts}$$

$$= 10st\sqrt{3^2} \cdot \sqrt{2ts}$$

$$= 10st \cdot 3\sqrt{2ts}$$

$$= 30st\sqrt{2ts}$$

47. $\left(3a\sqrt[3]{2b^4}\right)\left(2a^2\sqrt[3]{4b^2}\right)$

$$= 6a^3\sqrt[3]{2b^4 \cdot 4b^2}$$

$$= 6a^3\sqrt[3]{8b^6}$$

$$= 6a^3\sqrt[3]{8}\sqrt[3]{b^6}$$

$$= 6a^3 \cdot 2 \cdot b^2$$

$$= 12a^3b^2$$

49. $\sqrt[3]{\dfrac{x^3y^6}{8}} = \dfrac{\sqrt[3]{x^3y^6}}{\sqrt[3]{8}}$

$= \dfrac{\sqrt[3]{x^3}\,\sqrt[3]{y^6}}{\sqrt[3]{2^3}}$

$= \dfrac{xy^2}{2}$

51. $\sqrt[4]{\dfrac{32x^9}{y^{12}}} = \dfrac{\sqrt[4]{32x^9}}{\sqrt[4]{y^{12}}}$

$= \dfrac{\sqrt[4]{2^4 \cdot 2x^8 \cdot x}}{\sqrt[4]{y^{12}}}$

$= \dfrac{\sqrt[4]{2^4}\,\sqrt[4]{2}\,\sqrt[4]{x^8}\,\sqrt[4]{x}}{\sqrt[4]{y^{12}}}$

$= \dfrac{2\sqrt[4]{2}x^2\sqrt[4]{x}}{y^3}$

$= \dfrac{2x^2\sqrt[4]{2x}}{y^3}$

53. $\dfrac{\sqrt{54xy}}{\sqrt{2xy}} = \sqrt{\dfrac{54xy}{2xy}}$

$= \sqrt{27}$

$= \sqrt{3^2 \cdot 3}$

$= \sqrt{3^2}\sqrt{3}$

$= 3\sqrt{3}$

55. $\dfrac{\sqrt[4]{x^2y^{17}}}{\sqrt[4]{x^{14}y}} = \sqrt[4]{\dfrac{x^2y^{17}}{x^{14}y}}$

$= \sqrt[4]{\dfrac{y^{16}}{x^{12}}}$

$= \dfrac{\sqrt[4]{y^{16}}}{\sqrt[4]{x^{12}}}$

$= \dfrac{y^4}{x^3}$

57. $\sqrt{\dfrac{3xy}{5x^2y}} = \sqrt{\dfrac{3}{5x}}$

$= \dfrac{\sqrt{3}}{\sqrt{5x}}$

$= \dfrac{\sqrt{3} \cdot \sqrt{5x}}{\sqrt{5x} \cdot \sqrt{5x}}$

$= \dfrac{\sqrt{3 \cdot 5x}}{\sqrt{(5x)^2}}$

$= \dfrac{\sqrt{15x}}{5x}$

59. $\sqrt{\dfrac{3x^2y}{x^3y^4}} = \sqrt{\dfrac{3}{xy^3}}$

$= \dfrac{\sqrt{3}}{\sqrt{x \cdot y^2 \cdot y}}$

$= \dfrac{\sqrt{3}}{y\sqrt{xy}}$

$= \dfrac{\sqrt{3} \cdot \sqrt{xy}}{y\sqrt{xy} \cdot \sqrt{xy}}$

$= \dfrac{\sqrt{3xy}}{y\sqrt{(xy)^2}}$

$= \dfrac{\sqrt{3xy}}{y(xy)}$

$= \dfrac{\sqrt{3xy}}{xy^2}$

61. $\sqrt[3]{\dfrac{3}{2}} = \dfrac{\sqrt[3]{3}}{\sqrt[3]{2}}$

$= \dfrac{\sqrt[3]{3} \cdot \sqrt[3]{2^2}}{\sqrt[3]{2} \cdot \sqrt[3]{2^2}}$

$= \dfrac{\sqrt[3]{3 \cdot 2^2}}{\sqrt[3]{2^3}}$

$= \dfrac{\sqrt[3]{12}}{2}$

63. $\sqrt[3]{\frac{9}{4}} = \frac{\sqrt[3]{9}}{\sqrt[3]{4}}$

$= \frac{\sqrt[3]{9} \cdot \sqrt[3]{2}}{\sqrt[3]{4} \cdot \sqrt[3]{2}}$

$= \frac{\sqrt[3]{9 \cdot 2}}{\sqrt[3]{8}}$

$= \frac{\sqrt[3]{18}}{2}$

65. $\sqrt[4]{\frac{9}{4}} = \frac{\sqrt[4]{9}}{\sqrt[4]{4}}$

$= \frac{\sqrt[4]{9} \cdot \sqrt[4]{4}}{\sqrt[4]{4} \cdot \sqrt[4]{4}}$

$= \frac{\sqrt[4]{9 \cdot 4}}{\sqrt[4]{16}}$

$= \frac{\sqrt[4]{36}}{2} = \frac{\sqrt[4]{6^2}}{2} = \frac{\sqrt{6}}{2}$

67. $\sqrt[3]{\frac{81x^2y^4}{2x^3y}} = \sqrt[3]{\frac{81y^3}{2x}}$

$= \frac{\sqrt[3]{81y^3}}{\sqrt[3]{2x}}$

$= \frac{\sqrt[3]{3^3 \cdot 3y^3}}{\sqrt[3]{2x}}$

$= \frac{3y\sqrt[3]{3}}{\sqrt[3]{2x}}$

$= \frac{3y\sqrt[3]{3} \cdot \sqrt[3]{4x^2}}{\sqrt[3]{2x} \cdot \sqrt[3]{4x^2}}$

$= \frac{3y\sqrt[3]{3 \cdot 4x^2}}{\sqrt[3]{8x^3}}$

$= \frac{3y\sqrt[3]{12x^2}}{2x}$

69. $\frac{3a^2\sqrt{a^2x^5}}{9a^5\sqrt{a^6x}} = \frac{3a^2}{9a^5}\sqrt{\frac{a^2x^5}{a^6x}}$

$= \frac{1}{3a^3}\sqrt{\frac{x^4}{a^4}}$

$= \frac{1}{3a^3} \cdot \frac{\sqrt{x^4}}{\sqrt{a^4}}$

$= \frac{1}{3a^3} \cdot \frac{x^2}{a^2}$

$= \frac{x^2}{3a^5}$

71. $\frac{-3r^2s\sqrt{32r^2s^5}}{2r\sqrt{2r^5}}$

$= \frac{-3r^2s}{2r}\sqrt{\frac{32r^2s^5}{2r^5}}$

$= \frac{-3rs}{2}\sqrt{\frac{16s^5}{r^3}}$

$= \frac{-3rs}{2} \cdot \frac{\sqrt{16s^4 \cdot s}}{\sqrt{r^2 \cdot r}}$

$= \frac{-3rs}{2} \cdot \frac{4s^2\sqrt{s}}{r\sqrt{r}}$

$= \frac{-12s^3\sqrt{s}}{2\sqrt{r}}$

$= \frac{-12s^3\sqrt{s} \cdot \sqrt{r}}{2\sqrt{r} \cdot \sqrt{r}}$

$= \frac{-12s^3\sqrt{sr}}{2\sqrt{r^2}}$

$= -\frac{6s^3\sqrt{sr}}{r}$

73. $\sqrt[12]{a^6} = \sqrt{a}$

75. $\left(\sqrt[3]{x}\right)\left(\sqrt[4]{x^3}\right) = (x^{1/3})(x^{3/4}) = x^{1/3+3/4}$

$= x^{13/12} = \sqrt[12]{x^{13}} = x\sqrt[12]{x}$

77. $\dfrac{\sqrt[3]{a^2}}{\sqrt{a}} = \dfrac{a^{2/3}}{a^{1/2}} = a^{2/3-1/2} = a^{1/6} = \sqrt[6]{a}$

79. $\sqrt[n]{x^{5n}y^{3n}} = \sqrt[n]{(x^5)^n(y^3)^n} = x^5y^3$

83.
$$\begin{array}{r}
x^2 - 4x + 14 \\
x + 4\overline{)x^3 + 0x^2 - 2x + 3} \\
\underline{-(x^3 + 4x^2)} \\
-4x^2 - 2x + 3 \\
\underline{-(-4x^2 - 16x)} \\
14x + 3 \\
\underline{-(14x + 56)} \\
-53
\end{array}$$

$x^2 - 4x + 14$, R -53

Exercises 7.5

1. $5\sqrt{3} - \sqrt{3} = (5 - 1)\sqrt{3}$
$= 4\sqrt{3}$

3. $2\sqrt{5} - 4\sqrt{5} - \sqrt{5} = (2 - 4 - 1)\sqrt{5}$
$= -3\sqrt{5}$

5. $8\sqrt{3} - \left(4\sqrt{3} - 2\sqrt{6}\right)$
$= 8\sqrt{3} - 4\sqrt{3} + 2\sqrt{6}$
$= (8 - 4)\sqrt{3} + 2\sqrt{6}$
$= 4\sqrt{3} + 2\sqrt{6}$

7. $2\sqrt{3} - 2\sqrt{5} - \left(\sqrt{3} - \sqrt{5}\right)$
$= 2\sqrt{3} - 2\sqrt{5} - \sqrt{3} + \sqrt{5}$
$= (2 - 1)\sqrt{3} + (-2 + 1)\sqrt{5}$
$= \sqrt{3} - \sqrt{5}$

9. $2\sqrt{x} - 5\sqrt{x} + 3\sqrt{x}$
$= (2 - 5 + 3)\sqrt{x}$
$= 0$

11. $5a\sqrt{b} - 3a^3\sqrt{b} + 2a\sqrt{b}$
$= (5a - 3a^3 + 2a)\sqrt{b}$
$= (7a - 3a^3)\sqrt{b}$

13. $3x\sqrt[3]{x^2} - 2\sqrt[3]{x^2} + 6\sqrt[3]{x^2}$
$= (3x - 2 + 6)\sqrt[3]{x^2}$
$= (3x + 4)\sqrt[3]{x^2}$

15. $\left(7 - 3\sqrt[3]{a}\right) - \left(6 - \sqrt[3]{a}\right)$
$= 7 - 3\sqrt[3]{a} - 6 + \sqrt[3]{a}$
$= 1 + (-3 + 1)\sqrt[3]{a}$
$= 1 - 2\sqrt[3]{a}$

17. $\sqrt{12} - \sqrt{27}$
$= 2\sqrt{3} - 3\sqrt{3}$
$= (2 - 3)\sqrt{3}$
$= -\sqrt{3}$

19. $\sqrt{24} - \sqrt{27} + \sqrt{54}$
$= 2\sqrt{6} - 3\sqrt{3} + 3\sqrt{6}$
$= (2 + 3)\sqrt{6} - 3\sqrt{3}$
$= 5\sqrt{6} - 3\sqrt{3}$

21. $6\sqrt{3} - 4\sqrt{81}$
$= 6\sqrt{3} - 36$

23. $3\sqrt{24} - 5\sqrt{48} - \sqrt{6}$
$= 6\sqrt{6} - 20\sqrt{3} - \sqrt{6}$
$= (6 - 1)\sqrt{6} - 20\sqrt{3}$
$= 5\sqrt{6} - 20\sqrt{3}$

25. $3\sqrt[3]{24} - 5\sqrt[3]{48} - \sqrt[3]{6}$
$= 6\sqrt[3]{3} - 10\sqrt[3]{6} - \sqrt[3]{6}$
$= 6\sqrt[3]{3} + (-10 - 1)\sqrt[3]{6}$
$= 6\sqrt[3]{3} - 11\sqrt[3]{6}$

27. $2a\sqrt{ab^2} - 3b\sqrt{a^2b} - ab\sqrt{ab}$
$= 2ab\sqrt{a} - 3ab\sqrt{b} - ab\sqrt{ab}$

29. $\sqrt[3]{x^4} - x\sqrt[3]{x}$
$= x\sqrt[3]{x} - x\sqrt[3]{x}$
$= 0$

31. $\sqrt{20x^9y^8} + 2xy\sqrt{5x^7y^6}$
$= 2x^4y^4\sqrt{5x} + 2x^4y^4\sqrt{5x}$
$= (2x^4y^4 + 2x^4y^4)\sqrt{5x}$
$= 4x^4y^4\sqrt{5x}$

33. $5\sqrt[3]{9x^5} - 3x\sqrt[3]{x^2} + 2x\sqrt[3]{72x^2}$
$= 5x\sqrt[3]{9x^2} - 3x\sqrt[3]{x^2} + 4x\sqrt[3]{9x^2}$
$= (5x + 4x)\sqrt[3]{9x^2} - 3x\sqrt[3]{x^2}$
$= 9x\sqrt[3]{9x^2} - 3x\sqrt[3]{x^2}$

35. $4\sqrt[4]{16x} - 7\sqrt[4]{x^5} + x\sqrt[4]{81x}$
$= 8\sqrt[4]{x} - 7x\sqrt[4]{x} + 3x\sqrt[4]{x}$
$= (8 - 7x + 3x)\sqrt[4]{x}$
$= (8 - 4x)\sqrt[4]{x}$

37. $\dfrac{1}{\sqrt{5}} + 2$

$= \dfrac{1 \cdot \sqrt{5}}{\sqrt{5} \cdot \sqrt{5}} + 2$

$= \dfrac{\sqrt{5}}{5} + 2$

$= \dfrac{\sqrt{5}}{5} + \dfrac{10}{5}$

$= \dfrac{\sqrt{5} + 10}{5}$

39. $\dfrac{12}{\sqrt{6}} - 2\sqrt{6}$

$= \dfrac{12\sqrt{6}}{\sqrt{6}\sqrt{6}} - 2\sqrt{6}$

$= \dfrac{12\sqrt{6}}{6} - 2\sqrt{6}$

$= 2\sqrt{6} - 2\sqrt{6}$
$= 0$

41. $\sqrt{\dfrac{1}{2}} + \sqrt{2}$

$= \dfrac{1}{\sqrt{2}} + \sqrt{2}$

$= \dfrac{1 \cdot \sqrt{2}}{\sqrt{2}\sqrt{2}} + \sqrt{2}$

$= \dfrac{\sqrt{2}}{2} + \sqrt{2}$

$= \left(\dfrac{1}{2} + 1\right)\sqrt{2}$

$= \dfrac{3\sqrt{2}}{2}$

43. $\sqrt{\dfrac{5}{2}} + \sqrt{\dfrac{2}{5}}$

$= \dfrac{\sqrt{5}}{\sqrt{2}} + \dfrac{\sqrt{2}}{\sqrt{5}}$

$= \dfrac{\sqrt{5}\sqrt{2}}{\sqrt{2}\sqrt{2}} + \dfrac{\sqrt{2}\sqrt{5}}{\sqrt{5}\sqrt{5}}$

$= \dfrac{\sqrt{10}}{2} + \dfrac{\sqrt{10}}{5}$

$= \dfrac{5\sqrt{10}}{10} + \dfrac{2\sqrt{10}}{10}$

$= \dfrac{5\sqrt{10} + 2\sqrt{10}}{10}$

$= \dfrac{7\sqrt{10}}{10}$

45. $\sqrt{\dfrac{1}{7}} - 3\sqrt{\dfrac{1}{5}}$

$= \dfrac{1}{\sqrt{7}} - \dfrac{3}{\sqrt{5}}$

$= \dfrac{1 \cdot \sqrt{7}}{\sqrt{7}\sqrt{7}} - \dfrac{3\sqrt{5}}{\sqrt{5}\sqrt{5}}$

$= \dfrac{\sqrt{7}}{7} - \dfrac{3\sqrt{5}}{5}$

$= \dfrac{5\sqrt{7}}{35} - \dfrac{21\sqrt{5}}{35}$

$= \dfrac{5\sqrt{7} - 21\sqrt{5}}{35}$

47. $\dfrac{1}{\sqrt[3]{2}} - 6\sqrt[3]{4}$

$= \dfrac{1 \cdot \sqrt[3]{4}}{\sqrt[3]{2} \cdot \sqrt[3]{4}} - 6\sqrt[3]{4}$

$= \dfrac{\sqrt[3]{4}}{\sqrt[3]{8}} - 6\sqrt[3]{4}$

$= \dfrac{\sqrt[3]{4}}{2} - 6\sqrt[3]{4}$

$= \left(\dfrac{1}{2} - 6\right)\sqrt[3]{4}$

$= -\dfrac{11}{2}\sqrt[3]{4}$

$= \dfrac{-11\sqrt[3]{4}}{2}$

49. $\sqrt{\dfrac{1}{x}} + \sqrt{\dfrac{1}{y}}$

$= \dfrac{1}{\sqrt{x}} + \dfrac{1}{\sqrt{y}}$

$= \dfrac{1\sqrt{x}}{\sqrt{x}\sqrt{x}} + \dfrac{1\sqrt{y}}{\sqrt{y}\sqrt{y}}$

$= \dfrac{\sqrt{x}}{x} + \dfrac{\sqrt{y}}{y}$

$= \dfrac{y\sqrt{x}}{xy} + \dfrac{x\sqrt{y}}{xy}$

$= \dfrac{y\sqrt{x} + x\sqrt{y}}{xy}$

51. $\dfrac{1}{\sqrt[3]{9}} - \dfrac{3}{\sqrt[3]{3}}$

$= \dfrac{1 \cdot \sqrt[3]{3}}{\sqrt[3]{9}\sqrt[3]{3}} - \dfrac{3 \cdot \sqrt[3]{9}}{\sqrt[3]{3} \cdot \sqrt[3]{9}}$

$= \dfrac{\sqrt[3]{3}}{\sqrt[3]{27}} - \dfrac{3\sqrt[3]{9}}{\sqrt[3]{27}}$

$= \dfrac{\sqrt[3]{3}}{3} - \dfrac{3\sqrt[3]{9}}{3}$

$= \dfrac{\sqrt[3]{3} - 3\sqrt[3]{9}}{3}$

53. $3\sqrt{\dfrac{2}{49}} + 3\sqrt{7}$

$= \dfrac{3\sqrt{2}}{\sqrt{49}} + 3\sqrt{7}$

$= \dfrac{3\sqrt{2}}{7} + 3\sqrt{7}$

$= \dfrac{3\sqrt{2}}{7} + \dfrac{21\sqrt{7}}{7}$

$= \dfrac{3\sqrt{2} + 21\sqrt{7}}{7}$

55. $3\sqrt{10} - \dfrac{4}{\sqrt{10}} + \dfrac{2}{\sqrt{10}}$

$= 3\sqrt{10} + \dfrac{-4 + 2}{\sqrt{10}}$

$= 3\sqrt{10} - \dfrac{2}{\sqrt{10}}$

$= 3\sqrt{10} - \dfrac{2\sqrt{10}}{\sqrt{10}\sqrt{10}}$

$= 3\sqrt{10} - \dfrac{2\sqrt{10}}{10}$

$= \dfrac{30\sqrt{10}}{10} - \dfrac{2\sqrt{10}}{10}$

$= \dfrac{30\sqrt{10} - 2\sqrt{10}}{10}$

$= \dfrac{28\sqrt{10}}{10}$

$= \dfrac{14\sqrt{10}}{5}$

57. $6\sqrt[3]{25} - \dfrac{15}{\sqrt[3]{5}} + 5\sqrt[3]{\dfrac{1}{5}}$

$= 6\sqrt[3]{25} - \dfrac{15}{\sqrt[3]{5}} + \dfrac{5 \cdot 1}{\sqrt[3]{5}}$

$= 6\sqrt[3]{25} + \dfrac{-15 + 5}{\sqrt[3]{5}}$

$= 6\sqrt[3]{25} - \dfrac{10}{\sqrt[3]{5}}$

$= 6\sqrt[3]{25} - \dfrac{10\sqrt[3]{25}}{\sqrt[3]{5}\sqrt[3]{25}}$

$= 6\sqrt[3]{25} - \dfrac{10\sqrt[3]{25}}{\sqrt[3]{125}}$

$= 6\sqrt[3]{25} - \dfrac{10\sqrt[3]{25}}{5}$

$= 6\sqrt[3]{25} - 2\sqrt[3]{25}$

$= 4\sqrt[3]{25}$

59. $\dfrac{1}{2\sqrt{x - 1}} + \sqrt{x - 1}$

$= \dfrac{1}{2\sqrt{x - 1}} \cdot \dfrac{\sqrt{x - 1}}{\sqrt{x - 1}} + \sqrt{x - 1}$

$= \dfrac{\sqrt{x - 1}}{2(x - 1)} + \sqrt{x - 1}$

$= \dfrac{\sqrt{x - 1}}{2(x - 1)} + \dfrac{\sqrt{x - 1}\,[2(x - 1)]}{2(x - 1)}$

$= \dfrac{\sqrt{x - 1}\,(1 + 2x - 2)}{2(x - 1)}$

$= \dfrac{\sqrt{x - 1}\,(2x - 1)}{2(x - 1)}$

61. (a) $\{x \mid -4 \le x \le 6\}$

(b) $\{y \mid -1 \le y \le 5\}$

(c) the function is increasing (graph is rising) for $\{x \mid -4 \le x \le -2$ and $3 \le x \le 6\}$

(d) the function is decreasing (graph is falling) for $\{x \mid -2 \le x \le 3\}$

63. $y = \dfrac{2x - 3}{5}$

$y = \dfrac{2}{5}x - \dfrac{3}{5}$

$\left(0, -\dfrac{3}{5}\right), \left(\dfrac{3}{2}, 0\right)$

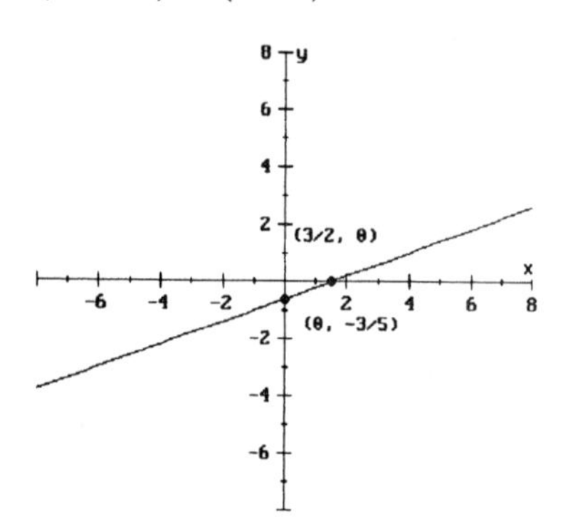

Exercises 7.6

1. $5(\sqrt{5} - 3)$
$= 5\sqrt{5} - 15$

3. $2(\sqrt{3} - \sqrt{5}) - 4(\sqrt{3} + \sqrt{5})$
$= 2\sqrt{3} - 2\sqrt{5} - 4\sqrt{3} - 4\sqrt{5}$
$= -2\sqrt{3} - 6\sqrt{5}$

5. $\sqrt{a}(\sqrt{a} + \sqrt{b})$
$= \sqrt{a}\sqrt{a} + \sqrt{a}\sqrt{b}$
$= a + \sqrt{ab}$

7. $\sqrt{2}(\sqrt{5} + \sqrt{2})$
$= \sqrt{2}\sqrt{5} + \sqrt{2}\sqrt{2}$
$= \sqrt{10} + 2$

9. $3\sqrt{5}(2\sqrt{3} - 4\sqrt{5})$
$= 3\sqrt{5} \cdot 2\sqrt{3} - 3\sqrt{5} \cdot 4\sqrt{5}$
$= 6\sqrt{15} - 12 \cdot 5$
$= 6\sqrt{15} - 60$

11. $\sqrt{2}(\sqrt{3} + \sqrt{2}) - 3(2\sqrt{6} - 4)$
$= \sqrt{2}\sqrt{3} + \sqrt{2}\sqrt{2} - 6\sqrt{6} + 12$
$= \sqrt{6} + 2 - 6\sqrt{6} + 12$
$= 14 - 5\sqrt{6}$

13. $(\sqrt{5} - 2)(\sqrt{3} + 1)$
$= \sqrt{5}\sqrt{3} + \sqrt{5} - 2\sqrt{3} - 2$
$= \sqrt{15} + \sqrt{5} - 2\sqrt{3} - 2$

15. $(\sqrt{5} - \sqrt{3})(\sqrt{5} + \sqrt{3})$
$= \sqrt{5}\sqrt{5} - \sqrt{3}\sqrt{3}$
$= 5 - 3$
$= 2$

17. $(\sqrt{5} - \sqrt{3})^2$
$= (\sqrt{5})^2 - 2\sqrt{5}\sqrt{3} + (\sqrt{3})^2$
$= 5 - 2\sqrt{15} + 3$
$= 8 - 2\sqrt{15}$

19. $(2\sqrt{7} - 5)(2\sqrt{7} + 5)$
$= (2\sqrt{7})^2 - 5^2$
$= 4 \cdot 7 - 25$
$= 3$

21. $(5\sqrt{2} - 3\sqrt{5})(5\sqrt{2} + 3\sqrt{5})$
$= (5\sqrt{2})^2 - (3\sqrt{5})^2$
$= 5^2(\sqrt{2})^2 - 3^2(\sqrt{5})^2$
$= 25 \cdot 2 - 9 \cdot 5$
$= 5$

23. $(2\sqrt{a} - \sqrt{b})^2$
$= (2\sqrt{a})^2 - 2(2\sqrt{a})(\sqrt{b}) + (\sqrt{b})^2$
$= 2^2(\sqrt{a})^2 - 4\sqrt{ab} + b$
$= 4a - 4\sqrt{ab} + b$

25. $(3\sqrt{x} - 2\sqrt{y})(3\sqrt{x} + 2\sqrt{y})$
$= (3\sqrt{x})^2 - (2\sqrt{y})^2$
$= 3^2(\sqrt{x})^2 - 2^2(\sqrt{y})^2$
$= 9x - 4y$

27. $(\sqrt{x} - 3)^2$
$= (\sqrt{x})^2 - 2\sqrt{x} \cdot 3 + 3^2$
$= x - 6\sqrt{x} + 9$

29. $\left(\sqrt{x-3}\right)^2 = x - 3$

31. $\left(\sqrt{x+1}\right)^2 - \left(\sqrt{x}+1\right)^2$
$= x + 1 - \left[\left(\sqrt{x}\right)^2 + 2\sqrt{x}\cdot 1 + 1^2\right]$
$= x + 1 - (x + 2\sqrt{x} + 1)$
$= x + 1 - x - 2\sqrt{x} - 1$
$= -2\sqrt{x}$

33. $\left(\sqrt[3]{2} - \sqrt[3]{3}\right)\left(\sqrt[3]{4} + \sqrt[3]{6} + \sqrt[3]{9}\right)$
$= \sqrt[3]{8} + \sqrt[3]{12} + \sqrt[3]{18} - \sqrt[3]{12} - \sqrt[3]{18} - \sqrt[3]{27}$
$= \sqrt[3]{8} - \sqrt[3]{27}$
$= 2 - 3$
$= -1$

35. $\dfrac{4\sqrt{2} - 6\sqrt{3}}{2}$

$= \dfrac{2\left(2\sqrt{2} - 3\sqrt{3}\right)}{2}$

$= 2\sqrt{2} - 3\sqrt{3}$

37. $\dfrac{5\sqrt{8} - 2\sqrt{7}}{8}$

$= \dfrac{10\sqrt{2} - 2\sqrt{7}}{8}$

$= \dfrac{2\left(5\sqrt{2} - \sqrt{7}\right)}{8}$

$= \dfrac{5\sqrt{2} - \sqrt{7}}{4}$

39. $\dfrac{3\sqrt{50} + 5\sqrt{5}}{5}$

$= \dfrac{15\sqrt{2} + 5\sqrt{5}}{5}$

$= \dfrac{5\left(3\sqrt{2} + \sqrt{5}\right)}{5}$

$= 3\sqrt{2} + \sqrt{5}$

41. $\dfrac{4 + \sqrt{28}}{4}$

$= \dfrac{4 + 2\sqrt{7}}{4}$

$= \dfrac{2\left(2 + \sqrt{7}\right)}{4}$

$= \dfrac{2 + \sqrt{7}}{2}$

43. $\dfrac{1}{\sqrt{2} - 3}$

$= \dfrac{1\left(\sqrt{2} + 3\right)}{\left(\sqrt{2} - 3\right)\left(\sqrt{2} + 3\right)}$

$= \dfrac{\sqrt{2} + 3}{\left(\sqrt{2}\right)^2 - 3^2}$

$= \dfrac{\sqrt{2} + 3}{2 - 9}$

$= \dfrac{\sqrt{2} + 3}{-7}$

$= -\dfrac{\sqrt{2} + 3}{7}$

45. $\dfrac{10}{\sqrt{5} + 1}$

$= \dfrac{10\left(\sqrt{5} - 1\right)}{\left(\sqrt{5} + 1\right)\left(\sqrt{5} - 1\right)}$

$= \dfrac{10\left(\sqrt{5} - 1\right)}{\left(\sqrt{5}\right)^2 - 1^2}$

$= \dfrac{10\left(\sqrt{5} - 1\right)}{5 - 1}$

$= \dfrac{10\left(\sqrt{5} - 1\right)}{4}$

$= \dfrac{5\left(\sqrt{5} - 1\right)}{2}$

$= \dfrac{5\sqrt{5} - 5}{2}$

47. $\dfrac{2}{\sqrt{3} - \sqrt{a}}$

$= \dfrac{2(\sqrt{3} + \sqrt{a})}{(\sqrt{3} - \sqrt{a})(\sqrt{3} + \sqrt{a})}$

$= \dfrac{2(\sqrt{3} + \sqrt{a})}{(\sqrt{3})^2 - (\sqrt{a})^2}$

$= \dfrac{2(\sqrt{3} + \sqrt{a})}{3 - a}$

$= \dfrac{2\sqrt{3} + 2\sqrt{a}}{3 - a}$

49. $\dfrac{\sqrt{x}}{\sqrt{x} - \sqrt{y}}$

$= \dfrac{\sqrt{x}(\sqrt{x} + \sqrt{y})}{(\sqrt{x} - \sqrt{y})(\sqrt{x} + \sqrt{y})}$

$= \dfrac{\sqrt{x}(\sqrt{x} + \sqrt{y})}{(\sqrt{x})^2 - (\sqrt{y})^2}$

$= \dfrac{\sqrt{x}(\sqrt{x} + \sqrt{y})}{x - y}$

$= \dfrac{x + \sqrt{xy}}{x - y}$

51. $\dfrac{\sqrt{2}}{\sqrt{5} - \sqrt{2}}$

$= \dfrac{\sqrt{2}(\sqrt{5} + \sqrt{2})}{(\sqrt{5} - \sqrt{2})(\sqrt{5} + \sqrt{2})}$

$= \dfrac{\sqrt{2}(\sqrt{5} + \sqrt{2})}{(\sqrt{5})^2 - (\sqrt{2})^2}$

$= \dfrac{\sqrt{2}(\sqrt{5} + \sqrt{2})}{5 - 2}$

$= \dfrac{\sqrt{10} + 2}{3}$

53. $\dfrac{2\sqrt{2}}{2\sqrt{5} - \sqrt{2}}$

$= \dfrac{2\sqrt{2}(2\sqrt{5} + \sqrt{2})}{(2\sqrt{5} - \sqrt{2})(2\sqrt{5} + \sqrt{2})}$

$= \dfrac{2\sqrt{2}(2\sqrt{5} + \sqrt{2})}{(2\sqrt{5})^2 - (\sqrt{2})^2}$

$= \dfrac{2\sqrt{2}(2\sqrt{5} + \sqrt{2})}{20 - 2}$

$= \dfrac{2\sqrt{2}(2\sqrt{5} + \sqrt{2})}{18}$

$= \dfrac{\sqrt{2}(2\sqrt{5} + \sqrt{2})}{9}$

$= \dfrac{2\sqrt{10} + 2}{9}$

55. $\dfrac{2\sqrt{5} - \sqrt{2}}{2\sqrt{2}}$

$= \dfrac{(2\sqrt{5} - \sqrt{2})\sqrt{2}}{2\sqrt{2}\sqrt{2}}$

$= \dfrac{2\sqrt{10} - 2}{4}$

$= \dfrac{2(\sqrt{10} - 1)}{4}$

$= \dfrac{\sqrt{10} - 1}{2}$

57. $\dfrac{\sqrt{3} + \sqrt{2}}{\sqrt{3} - \sqrt{2}}$

$= \dfrac{(\sqrt{3} + \sqrt{2})(\sqrt{3} + \sqrt{2})}{(\sqrt{3} - \sqrt{2})(\sqrt{3} + \sqrt{2})}$

$= \dfrac{(\sqrt{3})^2 + 2(\sqrt{3})(\sqrt{2}) + (\sqrt{2})^2}{(\sqrt{3})^2 - (\sqrt{2})^2}$

$= \dfrac{3 + 2\sqrt{6} + 2}{3 - 2}$

$= 5 + 2\sqrt{6}$

59. $\dfrac{3\sqrt{5} - 2\sqrt{2}}{2\sqrt{5} - 3\sqrt{2}}$

$= \dfrac{(3\sqrt{5} - 2\sqrt{2})(2\sqrt{5} + 3\sqrt{2})}{(2\sqrt{5} - 3\sqrt{2})(2\sqrt{5} + 3\sqrt{2})}$

$= \dfrac{6 \cdot 5 + 9\sqrt{10} - 4\sqrt{10} - 6}{(2\sqrt{5})^2 - (3\sqrt{2})^2}$

$= \dfrac{30 + 5\sqrt{10} - 12}{20 - 18}$

$= \dfrac{18 + 5\sqrt{10}}{2}$

61. $\dfrac{x - y}{\sqrt{x} - \sqrt{y}}$

$= \dfrac{(x - y)(\sqrt{x} + \sqrt{y})}{(\sqrt{x} - \sqrt{y})(\sqrt{x} + \sqrt{y})}$

$= \dfrac{(x - y)(\sqrt{x} + \sqrt{y})}{(\sqrt{x})^2 - (\sqrt{y})^2}$

$= \dfrac{(x - y)(\sqrt{x} + \sqrt{y})}{x - y}$

$= \sqrt{x} + \sqrt{y}$

63. $\dfrac{x^2 - x - 2}{\sqrt{x} - \sqrt{2}}$

$= \dfrac{(x^2 - x - 2)(\sqrt{x} + \sqrt{2})}{(\sqrt{x} - \sqrt{2})(\sqrt{x} + \sqrt{2})}$

$= \dfrac{(x^2 - x - 2)(\sqrt{x} + \sqrt{2})}{(\sqrt{x})^2 - (\sqrt{2})^2}$

$= \dfrac{(x - 2)(x + 1)(\sqrt{x} + \sqrt{2})}{x - 2}$

$= (x + 1)(\sqrt{x} + \sqrt{2})$

65. $\dfrac{12}{\sqrt{6}-2} - \dfrac{36}{\sqrt{6}}$

$= \dfrac{12(\sqrt{6}+2)}{(\sqrt{6}-2)(\sqrt{6}+2)} - \dfrac{36\sqrt{6}}{\sqrt{6}\sqrt{6}}$

$= \dfrac{12(\sqrt{6}+2)}{(\sqrt{6})^2 - 2^2} - \dfrac{36\sqrt{6}}{6}$

$= \dfrac{12(\sqrt{6}+2)}{6-4} - 6\sqrt{6}$

$= \dfrac{12(\sqrt{6}+2)}{2} - 6\sqrt{6}$

$= 6(\sqrt{6}+2) - 6\sqrt{6}$

$= 6\sqrt{6} + 12 - 6\sqrt{6}$

$= 12$

67. $\dfrac{20}{\sqrt{7}+\sqrt{3}} + \dfrac{28}{\sqrt{7}}$

$= \dfrac{20(\sqrt{7}-\sqrt{3})}{(\sqrt{7}+\sqrt{3})(\sqrt{7}-\sqrt{3})} + \dfrac{28\sqrt{7}}{\sqrt{7}\sqrt{7}}$

$= \dfrac{20(\sqrt{7}-\sqrt{3})}{(\sqrt{7})^2 - (\sqrt{3})^2} + \dfrac{28\sqrt{7}}{7}$

$= \dfrac{20(\sqrt{7}-\sqrt{3})}{7-3} + 4\sqrt{7}$

$= \dfrac{20(\sqrt{7}-\sqrt{3})}{4} + 4\sqrt{7}$

$= 5(\sqrt{7}-\sqrt{3}) + 4\sqrt{7}$

$= 5\sqrt{7} - 5\sqrt{3} + 4\sqrt{7}$

$= 9\sqrt{7} - 5\sqrt{3}$

69. $x = \dfrac{-3 + \sqrt{3^2 - 4(1)(2)}}{2(1)}$

$= \dfrac{-3 + \sqrt{1}}{2}$

$= \dfrac{-3+1}{2} = \dfrac{-2}{2} = -1$

71. $x = \dfrac{-6 + \sqrt{6^2 - 4(2)(3)}}{2(2)}$

$= \dfrac{-6 + \sqrt{12}}{4}$

$= \dfrac{-6 + 2\sqrt{3}}{4}$

$= \dfrac{2(-3 + \sqrt{3})}{4}$

$= \dfrac{-3 + \sqrt{3}}{2}$

73. $x = \dfrac{-6 + \sqrt{6^2 - 4(2)(-3)}}{2(2)}$

$= \dfrac{-6 + \sqrt{60}}{4}$

$= \dfrac{-6 + 2\sqrt{15}}{4}$

$= \dfrac{2(-3 + \sqrt{15})}{4}$

$= \dfrac{-3 + \sqrt{15}}{2}$

75. On the calculator the answer to #65 is 12. On the calculator the answer to #66 is 12. The answers are the same before and after simplification.

77. $(2x - 1)(2x + 3) = 12$

$4x^2 + 4x - 3 = 12$

$4x^2 + 4x - 15 = 0$

$(2x - 3)(2x + 5) = 0$

$2x - 3 = 0$ or $2x + 5 = 0$

$2x = 3$ $\quad$ $2x = -5$

$x = \dfrac{3}{2}$ or $x = \dfrac{-5}{2}$

79. $12x^2 - 46x + 40$
$= 2(6x^2 - 23x + 20)$
$= 2(6x^2 - 8x - 15x + 20)$
$= 2[2x(3x - 4) - 5(3x - 4)]$
$= 2(3x - 4)(2x - 5)$

Exercises 7.7

1. $\sqrt{a} = 7$
$(\sqrt{a})^2 = 7^2$
$a = 49$

3. $6 = \sqrt{a + 5}$
$6^2 = (\sqrt{a + 5})^2$
$36 = a + 5$
$31 = a$

5. $\sqrt{3x - 1} = 5$
$(\sqrt{3x - 1})^2 = 5^2$
$3x - 1 = 25$
$3x = 26$
$x = \frac{26}{3}$

7. $\sqrt{y + 5} - 4 = 7$
$\sqrt{y + 5} = 11$
$(\sqrt{y + 5})^2 = 11^2$
$y + 5 = 121$
$y = 116$

9. $2 = 5 - \sqrt{3x - 1}$
$-3 = -\sqrt{3x - 1}$
$(-3)^2 = (-\sqrt{3x - 1})^2$
$9 = 3x - 1$
$10 = 3x$
$\frac{10}{3} = x$

11. $3\sqrt{x} = 12$
$\sqrt{x} = 4$
$(\sqrt{x})^2 = 4^2$
$x = 16$

13. $6\sqrt{a} + 3.4 = 21.3$
$6\sqrt{a} = 17.9$
$(6\sqrt{a})^2 = (17.9)^2$
$36a = 320.41$
$a = 8.90$

15. $7 + \sqrt{x - 1} = 3$
$\sqrt{x - 1} = -4$

Since square root is nonnegative, there is no solution.

17. $\sqrt{2x - 3} - \sqrt{x + 5} = 0$
$\sqrt{2x - 3} = \sqrt{x + 5}$
$(\sqrt{2x - 3})^2 = (\sqrt{x + 5})^2$
$2x - 3 = x + 5$
$x - 3 = 5$
$x = 8$

19. $\sqrt{5 - 4x} - \sqrt{6 - 3x} = 0$
$\sqrt{5 - 4x} = \sqrt{6 - 3x}$
$(\sqrt{5 - 4x})^2 = (\sqrt{6 - 3x})^2$
$5 - 4x = 6 - 3x$
$5 = 6 + x$
$-1 = x$

21. $4 = \sqrt[3]{x}$
$4^3 = (\sqrt[3]{x})^3$
$64 = x$

23. $\sqrt[5]{s} = -3$
$(\sqrt[5]{s})^5 = (-3)^5$
$s = -243$

25. $-3 = \sqrt[3]{y + 5}$
$(-3)^3 = (\sqrt[3]{y + 5})^3$
$-27 = y + 5$
$-32 = y$

27. $x^{1/3} = 5$
$(x^{1/3})^3 = 5^3$
$x = 125$

29. $$\begin{aligned} x^{1/3} + 7 &= 5 \\ x^{1/3} &= -2 \\ (x^{1/3})^3 &= (-2)^3 \\ x &= -8 \end{aligned}$$

31. $$\begin{aligned} (x + 7)^{1/3} &= 5 \\ \left[(x + 7)^{1/3}\right]^3 &= 5^3 \\ x + 7 &= 125 \\ x &= 118 \end{aligned}$$

33. $$\begin{aligned} x^{-1/4} &= 4 \\ (x^{-1/4})^{-4} &= 4^{-4} \\ x &= \frac{1}{4^4} \\ x &= \frac{1}{256} \end{aligned}$$

35. $$\begin{aligned} (x + 3)^{1/3} + 4 &= 2 \\ (x + 3)^{1/3} &= -2 \\ \left[(x + 3)^{1/3}\right]^3 &= (-2)^3 \\ x + 3 &= -8 \\ x &= -11 \end{aligned}$$

37. $$\begin{aligned} f(x) &= \sqrt{x} \\ 7.8 &= \sqrt{x} \\ (7.8)^2 &= \left(\sqrt{x}\right)^2 \\ 60.84 &= x \end{aligned}$$

39. $$\begin{aligned} h(x) &= \sqrt{x - 4.3} \\ 9.38 &= \sqrt{x - 4.3} \\ (9.38)^2 &= \left(\sqrt{x - 4.3}\right)^2 \\ 87.98 &= x - 4.3 \\ 92.28 &= x \end{aligned}$$

41. $$\begin{aligned} g(x) &= (3x - 1)^{1/2} \\ 5.24 &= (3x - 1)^{1/2} \\ (5.24)^2 &= \left[(3x - 1)^{1/2}\right]^2 \\ 27.46 &= 3x - 1 \\ 28.46 &= 3x \\ 9.49 &= x \end{aligned}$$

43. $$\begin{aligned} f(x) &= 6\sqrt{x} + 3.4 \\ 21 &= 6\sqrt{x} + 3.4 \\ 17.6 &= 6\sqrt{x} \\ 2.93 &= \sqrt{x} \\ (2.93)^2 &= \left(\sqrt{x}\right)^2 \\ 8.58 &= x \end{aligned}$$

45. $$\begin{aligned} f(x) &= 7 - \sqrt{x - 2} \\ 4 &= 7 - \sqrt{x - 2} \\ -3 &= -\sqrt{x - 2} \\ (-3)^2 &= \left(-\sqrt{x - 2}\right)^2 \\ 9 &= x - 2 \\ 11 &= x \end{aligned}$$

47. $$\begin{aligned} f(x) &= x^{1/3} - 3 \\ 5 &= x^{1/3} - 3 \\ 8 &= x^{1/3} \\ 8^3 &= (x^{1/3})^3 \\ 512 &= x \end{aligned}$$

49. $$\begin{aligned} \sqrt{3a} &= 5 \\ \left(\sqrt{3a}\right)^2 &= 5^2 \\ 3a &= 25 \\ a &= \frac{25}{3} \end{aligned}$$

51. $$\begin{aligned} 8 &= \sqrt{3x - 2} \\ 8^2 &= \left(\sqrt{3x - 2}\right)^2 \\ 64 &= 3x - 2 \\ 66 &= 3x \\ 22 &= x \end{aligned}$$

53. $$\begin{aligned} \sqrt{y + 4} + 8 &= 15.6 \\ \sqrt{y + 4} &= 7.6 \\ \left(\sqrt{y + 4}\right)^2 &= (7.6)^2 \\ y + 4 &= 57.76 \\ y &= 53.76 \end{aligned}$$

55. $$\begin{aligned} 3.5\sqrt{2x} &= 8.4 \\ \sqrt{2x} &= 2.4 \\ \left(\sqrt{2x}\right)^2 &= (2.4)^2 \\ 2x &= 5.76 \\ x &= 2.88 \end{aligned}$$

57. $$\begin{aligned} \sqrt{7 - 9x} - \sqrt{13 - 6x} &= 0 \\ \sqrt{7 - 9x} &= \sqrt{13 - 6x} \\ \left(\sqrt{7 - 9x}\right)^2 &= \left(\sqrt{13 - 6x}\right)^2 \\ 7 - 9x &= 13 - 6x \\ 7 &= 13 + 3x \\ -6 &= 3x \\ -2 &= x \end{aligned}$$

59.
$$\begin{aligned} v &= 2\sqrt{5s} \\ 55 &= 2\sqrt{5s} \\ 27.5 &= \sqrt{5s} \\ (27.5)^2 &= \left(\sqrt{5s}\right)^2 \\ 756.25 &= 5s \\ 151.25 &= s \end{aligned}$$
The skid marks are 151.25 ft.

61.
$$\begin{aligned} T &= 2\pi\sqrt{\frac{L}{980}} \\ 8 &= 2\pi\sqrt{\frac{L}{980}} \\ \frac{4}{\pi} &= \sqrt{\frac{L}{980}} \\ \left(\frac{4}{\pi}\right)^2 &= \left(\sqrt{\frac{L}{980}}\right)^2 \\ 1.6211 &= \frac{L}{980} \\ 1588.68 &= L \end{aligned}$$
It is 1588.68 cm.

63.
$$r = \sqrt[3]{\frac{3v}{4\pi}}$$
$$r = \sqrt[3]{\frac{3(100)}{4\pi}} \approx 2.9$$
The radius is approximately 2.9 inches.

67.
$$\begin{aligned} \frac{x-3}{4} - \frac{x+2}{5} &= \frac{x-5}{2} \\ 20\left(\frac{x-3}{4} - \frac{x+2}{5}\right) &= 20\left(\frac{x-5}{2}\right) \\ 5(x-3) - 4(x+2) &= 10(x-5) \\ 5x - 15 - 4x - 8 &= 10x - 50 \\ x - 23 &= 10x - 50 \\ -23 &= 9x - 50 \\ 27 &= 9x \\ 3 &= x \end{aligned}$$

Exercises 7.8

1. $$\begin{aligned} i^3 &= i^2 \cdot i \\ &= -1 \cdot i \\ &= -i \end{aligned}$$

3. $$\begin{aligned} -i^{21} &= -(i^4)^5 \cdot i \\ &= -(1)^5 \cdot i \\ &= -i \end{aligned}$$

5. $$\begin{aligned} (-i)^{29} &= (-1)^{29} i^{29} \\ &= -(i^4)^7 \cdot i \\ &= -(1)^7 \cdot i \\ &= -i \end{aligned}$$

7. $$\begin{aligned} i^{72} &= (i^4)^{18} \\ &= 1^{18} \\ &= 1 \end{aligned}$$

9. $$\begin{aligned} i^{67} &= (i^4)^{16} \cdot i^3 \\ &= 1^{16} \cdot (-i) \\ &= -i \end{aligned}$$

11. $$\begin{aligned} -i^{16} &= -(i^4)^4 \\ &= -(1)^4 \\ &= -1 \end{aligned}$$

13. $$\begin{aligned} \sqrt{-5} &= \sqrt{5(-1)} \\ &= \sqrt{5i^2} \\ &= \sqrt{5}\sqrt{i^2} \\ &= \sqrt{5}i \text{ or } i\sqrt{5} \end{aligned}$$

15. $$\begin{aligned} 3 - \sqrt{-2} &= 3 - \sqrt{2(-1)} \\ &= 3 - \sqrt{2i^2} \\ &= 3 - \sqrt{2}\sqrt{i^2} \\ &= 3 - \sqrt{2}i \text{ or } 3 - i\sqrt{2} \end{aligned}$$

17. $$\begin{aligned} &2\sqrt{-4} - \sqrt{28} \\ &= 2\sqrt{4(-1)} - \sqrt{4 \cdot 7} \\ &= 2\sqrt{4i^2} - 2\sqrt{7} \\ &= 2\sqrt{4}\sqrt{i^2} - 2\sqrt{7} \\ &= 2(2)i - 2\sqrt{7} \\ &= 4i - 2\sqrt{7} \\ &= -2\sqrt{7} + 4i \end{aligned}$$

19. $\dfrac{3 - \sqrt{-2}}{5} = \dfrac{3 - i\sqrt{2}}{5}$

$= \dfrac{3}{5} - \dfrac{\sqrt{2}}{5}i$

21. $\dfrac{6 - \sqrt{-3}}{3} = \dfrac{6 - i\sqrt{3}}{3}$

$= \dfrac{6}{3} - \dfrac{\sqrt{3}}{3}i$

$= 2 - \dfrac{\sqrt{3}}{3}i$

23. $(3 + 2i) - (2 + 3i)$
$= (3 - 2) + (2 - 3)i$
$= 1 - i$

25. $(2 - i) + (2 + i)$
$= (2 + 2) + (-1 + 1)i$
$= 4$

27. $5(3 - 7i) = 5(3) - 5(7i)$
$= 15 - 35i$

29. $(2 - i)(-2) = 2(-2) - i(-2)$
$= -4 + 2i$

31. $(5i)(3i) = 15i^2$
$= -15$

33. $(-3i)^2 = 9i^2$
$= -9$

35. $2i(3 + 2i) = 6i + 4i^2$
$= 6i - 4$
$= -4 + 6i$

37. $-2i(-3 + 5i) = 6i - 10i^2$
$= 6i + 10$
$= 10 + 6i$

39. $(3 + 2i)(0) = 0$

41. $(7 - 4i)(3 + i)$
$= 21 + 7i - 12i - 4i^2$
$= 21 + 7i - 12i + 4$
$= 25 - 5i$

43. $(3 + i)(3 - i)$
$= 3^2 - i^2$
$= 9 + 1$
$= 10$

45. $(2 - 7i)(2 + 7i)$
$= 2^2 - (7i)^2$
$= 4 - 49i^2$
$= 4 + 49$
$= 53$

47. $(i + 1)^2$
$= i^2 + 2(i)(1) + 1^2$
$= -1 + 2i + 1$
$= 2i$

49. $(2 - i)^2 - 4(2 - i)$
$= 2^2 - 2(2)(i) + i^2 - 8 + 4i$
$= 4 - 4i - 1 - 8 + 4i$
$= -5$

51. $(1 + i)^2 - 2(1 + i) + 1$
$= 1^2 + 2(1)(i) + i^2 - 2 - 2i + 1$
$= 1 + 2i - 1 - 2 - 2i + 1$
$= -1$

53. $\dfrac{6 - 5i}{3} = \dfrac{6}{3} - \dfrac{5i}{3}$
$= 2 - \dfrac{5}{3}i$

55. $\dfrac{3 - i}{i}$

$= \dfrac{(3 - i) \cdot i}{i \cdot i}$

$= \dfrac{3i - i^2}{i^2}$

$= \dfrac{3i + 1}{-1}$

$= \dfrac{1 + 3i}{-1}$

$= -1 - 3i$

57. $\dfrac{5-2i}{2i}$

$= \dfrac{(5-2i)\cdot i}{2i\cdot i}$

$= \dfrac{5i-2i^2}{2i^2}$

$= \dfrac{5i+2}{-2}$

$= \dfrac{2+5i}{-2}$

$= -1 - \dfrac{5}{2}i$

59. $\dfrac{2i}{5-2i}$

$= \dfrac{2i(5+2i)}{(5-2i)(5+2i)}$

$= \dfrac{10i+4i^2}{5^2-(2i)^2}$

$= \dfrac{10i+4i^2}{5^2-(2i)^2}$

$= \dfrac{10i-4}{25+4}$

$= \dfrac{-4+10i}{29}$

$= -\dfrac{4}{29} + \dfrac{10}{29}i$

61. $\dfrac{2}{5-2i}$

$= \dfrac{2(5+2i)}{(5-2i)(5+2i)}$

$= \dfrac{10+4i}{5^2-(2i)^2}$

$= \dfrac{10+4i}{25+4}$

$= \dfrac{10+4i}{29}$

$= \dfrac{10}{29} + \dfrac{4}{29}i$

63. $\dfrac{2+i}{2-i}$

$= \dfrac{(2+i)(2+i)}{(2-i)(2+i)}$

$= \dfrac{2^2+2(2)i+i^2}{2^2-i^2}$

$= \dfrac{4+4i-1}{4+1}$

$= \dfrac{3+4i}{5}$

$= \dfrac{3}{5} + \dfrac{4}{5}i$

65. $\dfrac{2-5i}{2+5i}$

$= \dfrac{(2-5i)(2-5i)}{(2+5i)(2-5i)}$

$= \dfrac{2^2-2(2)(5i)+(5i)^2}{2^2-(5i)^2}$

$= \dfrac{4-20i+25i^2}{4-25i^2}$

$= \dfrac{4-20i-25}{4+25}$

$= \dfrac{-21-20i}{29}$

$= -\dfrac{21}{29} - \dfrac{20}{29}i$

67. $\dfrac{3-7i}{5+2i}$

$= \dfrac{(3-7i)(5-2i)}{(5+2i)(5-2i)}$

$= \dfrac{15-6i-35i+14i^2}{5^2-(2i)^2}$

$= \dfrac{15-6i-35i-14}{25-4i^2}$

$$= \frac{1 - 41i}{25 + 4}$$

$$= \frac{1 - 41i}{29}$$

$$= \frac{1}{29} - \frac{41}{29}i$$

69. $\dfrac{5 - \sqrt{-4}}{3 + \sqrt{-25}}$

$$= \frac{5 - 2i}{3 + 5i}$$

$$= \frac{(5 - 2i)(3 - 5i)}{(3 + 5i)(3 - 5i)}$$

$$= \frac{16 - 25i - 6i + 10i^2}{3^2 - (5i)^2}$$

$$= \frac{15 - 25i - 6i - 10}{9 - 25i^2}$$

$$= \frac{5 - 31i}{9 + 25}$$

$$= \frac{5 - 31i}{34}$$

$$= \frac{5}{34} - \frac{31}{34}i$$

71. $x = \dfrac{-4 + \sqrt{4^2 - 4(1)(5)}}{2(1)}$

$$= \frac{-4 + \sqrt{-4}}{2}$$

$$= \frac{-4 + 2i}{2}$$

$$= \frac{2(-2 + i)}{2}$$

$$= -2 + i$$

73. $x = \dfrac{-4 + \sqrt{4^2 - 4(2)(5)}}{2(2)}$

$$= \frac{-4 + \sqrt{-24}}{4}$$

$$= \frac{-4 + 2i\sqrt{6}}{4}$$

$$= \frac{2\left(-2 + i\sqrt{6}\right)}{4}$$

$$= \frac{-2 + i\sqrt{6}}{2}$$

75.
$$\begin{aligned} x^2 + 4 &= 0 \\ (2i)^2 + 4 &= 0 \\ 4i^2 + 4 &= 0 \\ -4 + 4 &= 0 \\ 0 &= 0 \end{aligned}$$

True

$x = 2i$ is a solution.

77.
$$\begin{aligned} x^2 - 4x + 5 &= 0 \\ (2 - i)^2 - 4(2 - i) + 5 &= 0 \\ 2^2 - 2(2)(i) + i^2 - 8 + 4i + 5 &= 0 \\ 4 - 4i - 1 - 8 + 4i + 5 &= 0 \\ 0 &= 0 \end{aligned}$$

True

$x = 2 - i$ is a solution.

83. $\dfrac{x + 3}{x - 6} \div (x^2 - 3x - 18)$

$$= \frac{x + 3}{x - 6} \cdot \frac{1}{x^2 - 3x - 18}$$

$$= \frac{x + 3}{x - 6} \cdot \frac{1}{(x - 6)(x + 3)}$$

$$= \frac{1}{(x - 6)^2}$$

85.
$$\begin{aligned} x^2 - 25 &\neq 0 \\ (x - 5)(x + 5) &\neq 0 \end{aligned}$$

$$x - 5 \neq 0 \qquad x + 5 \neq 0$$
$$x \neq 5 \qquad x \neq -5$$

Domain: $\{x \mid x \neq \pm 5\}$

CHAPTER 7 REVIEW EXERCISES

1. $(x^2x^5)(x^4x)$
$= (x^{2+5})(x^{4+1})$
$= (x^7)(x^5)$
$= x^{7+5}$
$= x^{12}$

3. $(-3x^2y)(-2xy^4)(-x)$
$= -6x^{2+1+1}y^{1+4}$
$= -6x^4y^5$

5. $(a^3)^4 = a^{(3)(4)}$
$= a^{12}$

7. $(a^2b^3)^7 = a^{(2)(7)}b^{(3)(7)}$
$= a^{14}b^{21}$

9. $(a^2b^3)^2(a^2b)^3$
$= a^{(2)(2)}b^{(3)(2)}a^{(2)(3)}b^3$
$= a^4b^6a^6b^3$
$= a^{4+6}b^{6+3}$
$= a^{10}b^9$

11. $(a^2bc^2)^2(ab^2c)^3$
$= a^{(2)(2)}b^2c^{(2)(2)}a^3b^{(2)(3)}c^3$
$= a^4b^2c^4a^3b^6c^3$
$= a^{4+3}b^{2+6}c^{4+3}$
$= a^7b^8c^7$

13. $\frac{a^5}{a^6} = a^{5-6}$
$= a^{-1}$
$= \frac{1}{a}$

15. $\frac{x^2x^5}{x^4x^3} = \frac{x^{2+5}}{x^{4+3}}$
$= \frac{x^7}{x^7}$
$= x^{7-7}$
$= x^0$
$= 1$

17. $\frac{(x^3y^2)^3}{(x^5y^4)^5} = \frac{x^9y^6}{x^{25}y^{20}}$
$= x^{9-25}y^{6-20}$
$= x^{-16}y^{-14}$
$= \frac{1}{x^{16}y^{14}}$

19. $\left(\frac{a^2b}{ab}\right)^4 = (a^{2-1}b^{1-1})^4$
$= (ab^0)^4$
$= a^4$

21. $\frac{(2ax^2)^2(3ax)^2}{(-2x)^2}$
$= \frac{2^2a^2x^43^2a^2x^2}{(-1)^2(2)^2x^2}$
$= \frac{a^2x^43^2a^2}{1}$
$= 9a^4x^4$

23. $\left(\frac{-3xy}{x^2}\right)^2\left(\frac{-2xy^2}{x}\right)^3$
$= \left(\frac{-3y}{x}\right)^2(-2y^2)^3$
$= \frac{(-3)^2y^2}{x^2} \cdot (-2)^3y^6$
$= \frac{9y^2}{x^2} \cdot \frac{-8y^6}{1}$
$= \frac{-72y^8}{x^2}$

25. $a^{-3}a^{-4}a^5 = a^{-3-4+5}$
$= a^{-2}$
$= \frac{1}{a^2}$

27. $(x^{-2}y^5)^{-4} = x^{(-2)(-4)}y^{5(-4)}$

$= x^8y^{-20}$

$= \dfrac{x^8}{y^{20}}$

29. $(-3)^{-4}(-2)^{-1}$

$= \dfrac{1}{(-3)^4} \cdot \dfrac{1}{(-2)}$

$= \dfrac{1}{81} \cdot \dfrac{1}{(-2)}$

$= -\dfrac{1}{162}$

31. $\left(\dfrac{3x^{-5}y^2z^{-4}}{2x^{-7}y^{-4}}\right)^0 = 1$

33. $\dfrac{x^{-3}x^{-6}}{x^{-5}x^0} = \dfrac{x^{-3-6}}{x^{-5+0}}$

$= \dfrac{x^{-9}}{x^{-5}}$

$= x^{-9-(-5)}$

$= x^{-4}$

$= \dfrac{1}{x^4}$

35. $\left(\dfrac{x^{-2}y^{-3}}{y^{-3}x^2}\right)^{-2}$

$= (x^{-2-2}y^{-3-(-3)})^{-2}$

$= (x^{-4}y^0)^{-2}$

x $= (x^{-4})^{-2}$

$= x^{(-4)(-2)}$

$= x^8$

37. $\left(\dfrac{r^{-2}s^{-3}r^{-2}}{s^{-4}}\right)^{-2}\left(\dfrac{r^{-1}}{s^{-1}}\right)^{-3}$

$= \left(r^{-2-2}s^{-3-(-4)}\right)^{-2}\left(\dfrac{r^{(-1)(-3)}}{s^{(-1)(-3)}}\right)$

$= (r^{-4}s)^{-2}\left(\dfrac{r^3}{s^3}\right)$

$= r^{(-4)(-2)}s^{-2}\left(\dfrac{r^3}{s^3}\right)$

$= \dfrac{r^8s^{-2}r^3}{s^3}$

$= r^{8+3}s^{-2-3}$

$= r^{11}s^{-5}$

$= \dfrac{r^{11}}{s^5}$

39. $\left(\dfrac{2}{5}\right)^{-2} = \left(\dfrac{5}{2}\right)^2$

$= \dfrac{25}{4}$

41. $\dfrac{(2x^2y^{-1}z)^{-2}}{(3xy^2)^{-3}}$

$= \dfrac{2^{-2}x^{-4}y^2z^{-2}}{3^{-3}x^{-3}y^{-6}}$

$= \dfrac{3^3}{2^2}x^{-4-(-3)}y^{2-(-6)}z^{-2}$

$= \dfrac{27}{4}x^{-1}y^8z^{-2}$

$= \dfrac{27y^8}{4xz^2}$

43. $(x^{-1} + y^{-1})(x - y)$

$= x^{-1+1} - x^{-1}y + xy^{-1} - y^{-1+1}$

$= x^0 - \dfrac{y}{x} + \dfrac{x}{y} - y^0$

$= 1 - \dfrac{y}{x} + \dfrac{x}{y} - 1$

$= \dfrac{x}{y} - \dfrac{y}{x}$

$= \dfrac{x \cdot x}{xy} - \dfrac{y \cdot y}{xy}$

$= \dfrac{x^2 - y^2}{xy}$

45. $\dfrac{x^{-1} + y^{-3}}{x^{-1}y^2}$

$= \dfrac{\dfrac{1}{x} + \dfrac{1}{y^3}}{\dfrac{y^2}{x}}$

$= \dfrac{\left(\dfrac{1}{x} + \dfrac{1}{y^3}\right)xy^3}{\dfrac{y^2}{x} \cdot xy^3}$

$= \dfrac{y^3 + x}{y^5}$

47. move 4 places right
$2.83 \times 10^4 = 28{,}300$

49. move 5 places left
$7.96 \times 10^{-5} = 0.0000796$

51. $92.59 = 9.259 \times 10^?$
$= 9.259 \times 10^1$

53. $625{,}897 = 6.25897 \times 10^?$
$= 6.25897 \times 10^5$

55. $\dfrac{(0.0014)(9{,}000)}{(20{,}000)(63{,}000)}$

$= \dfrac{(1.4 \times 10^{-3})(9 \times 10^3)}{(2 \times 10^4)(6.3 \times 10^4)}$

$= \dfrac{(1.4)(9)}{(2)(6.3)} \times \dfrac{(10^{-3})(10^3)}{(10^4)(10^4)}$

$= 1 \times \dfrac{10^0}{10^8}$

$= 1 \times 10^{-8}$
$= 0.00000001$

57. $x^{1/2}x^{1/3} = x^{1/2 + 1/3}$
$= x^{5/6}$

59. $(x^{-1/2}x^{1/3})^{-6}$
$= (x^{-1/2 + 1/3})^{-6}$
$= (x^{-1/6})^{-6}$
$= x^{(-1/6)(-6)}$
$= x^1 = x$

61. $\dfrac{x^{1/2}x^{1/3}}{x^{2/5}} = \dfrac{x^{1/2 + 1/3}}{x^{2/5}}$

$= \dfrac{x^{5/6}}{x^{2/5}}$

$= x^{5/6 - 2/5}$

$= x^{13/30}$

63. $\left(\dfrac{a^{1/2}a^{-1/3}}{a^{1/2}b^{1/5}}\right)^{-15}$

$= \left(\dfrac{a^{1/2 - 1/3}}{a^{1/2}b^{1/5}}\right)^{-15}$

$= \left(\dfrac{a^{1/6}}{a^{1/2}b^{1/5}}\right)^{-15}$

$= \left(\dfrac{a^{1/6 - 1/2}}{b^{1/5}}\right)^{-15}$

$= \left(\dfrac{a^{-1/3}}{b^{1/5}}\right)^{-15}$

$= \dfrac{a^{(-1/3)(-15)}}{b^{(1/5)(-15)}}$

$= \dfrac{a^5}{b^{-3}}$

$= a^5b^3$

65. $\dfrac{(x^{-1/2}y^{1/2})^{-2}}{(x^{-1/3}y^{-1/3})^{-1/2}}$

$= \dfrac{x^{(-1/2)(-2)}y^{(1/2)(-2)}}{x^{(-1/3)(-1/2)}y^{(-1/3)(-1/2)}}$

$= \dfrac{xy^{-1}}{x^{1/6}y^{1/6}}$

$= x^{1-1/6}y^{-1-1/6}$
$= x^{5/6}y^{-7/6}$
$= \dfrac{x^{5/6}}{y^{7/6}}$

67. $\left(\dfrac{4^{-1/2} \cdot 16^{-3/4}}{8^{1/3}}\right)^{-2}$

$= \left[\dfrac{(2^2)^{-1/2}(2^4)^{-3/4}}{(2^3)^{1/3}}\right]^{-2}$

$= \left(\dfrac{2^{-1}2^{-3}}{2}\right)^{-2}$

$= \left(\dfrac{2^{-4}}{2}\right)^{-2}$

$= (2^{-5})^{-2}$
$= 2^{10}$

69. $(a^{1/3} + b^{1/3})(a^{1/3} - b^{1/3})$
$= (a^{1/3})^2 - (b^{1/3})^2$
$= a^{2/3} - b^{2/3}$

71. $x^{1/2} = \sqrt{x}$

73. $xy^{1/2} = x\sqrt{y}$

75. $m^{2/3} = \sqrt[3]{m^2}$

77. $(5x)^{3/4} = \sqrt[4]{(5x)^3}$

79. $\sqrt[3]{a} = a^{1/3}$

81. $-\sqrt[5]{n^4} = -n^{4/5}$

83. $\dfrac{1}{\sqrt[5]{t^7}} = \dfrac{1}{t^{7/5}}$

$= t^{-7/5}$

85. $\sqrt{54} = \sqrt{9 \cdot 6}$
$= \sqrt{9}\sqrt{6}$
$= 3\sqrt{6}$

87. $\sqrt{x^{60}} = x^{30}$

89. $\sqrt[3]{48x^4y^8} = \sqrt[3]{8 \cdot 6x^3xy^6y^2}$
$= 2xy^2\sqrt[3]{6xy^2}$

91. $\sqrt{75xy}\sqrt{3x} = \sqrt{(75xy)(3x)}$
$= \sqrt{225x^2y}$
$= 15x\sqrt{y}$

93. $\left(x\sqrt{xy}\right)\left(2x^2y\sqrt{xy^2}\right)$
$= \left(x\sqrt{xy}\right)\left(2x^2y^2\sqrt{x}\right)$
$= 2x^3y^2\sqrt{x^2y}$
$= 2x^4y^2\sqrt{y}$

95. $\dfrac{\sqrt{28}}{\sqrt{63}} = \sqrt{\dfrac{28}{63}}$

$= \sqrt{\dfrac{4}{9}}$

$= \dfrac{\sqrt{4}}{\sqrt{9}}$

$= \dfrac{2}{3}$

97. $\dfrac{y}{x\sqrt{y}}$

$= \dfrac{y\sqrt{y}}{x\sqrt{y}\sqrt{y}}$

$= \dfrac{y\sqrt{y}}{x\sqrt{y^2}}$

$= \dfrac{y\sqrt{y}}{xy} = \dfrac{\sqrt{y}}{x}$

99. $\sqrt{\dfrac{48a^2b}{3a^5b^2}}$

$= \sqrt{\dfrac{16}{a^3b}}$

$= \dfrac{\sqrt{16}}{\sqrt{a^3b}}$

$= \dfrac{4}{a\sqrt{ab}}$

$= \dfrac{4\sqrt{ab}}{a\sqrt{ab}\sqrt{ab}}$

$= \dfrac{4\sqrt{ab}}{a\sqrt{(ab)^2}}$

$= \dfrac{4\sqrt{ab}}{a(ab)}$

$= \dfrac{4\sqrt{ab}}{a^2b}$

101. $\sqrt{\dfrac{5}{a}}$

$= \dfrac{\sqrt{5}}{\sqrt{a}}$

$= \dfrac{\sqrt{5}\sqrt{a}}{\sqrt{a}\sqrt{a}}$

$= \dfrac{\sqrt{5a}}{\sqrt{a^2}}$

$= \dfrac{\sqrt{5a}}{a}$

103. $\dfrac{4}{\sqrt[3]{2a}}$

$= \dfrac{4 \cdot \sqrt[3]{4a^2}}{\sqrt[3]{2a} \cdot \sqrt[3]{4a^2}}$

$= \dfrac{4\sqrt[3]{4a^2}}{\sqrt[3]{8a^3}}$

$= \dfrac{4\sqrt[3]{4a^2}}{2a}$

$= \dfrac{2\sqrt[3]{4a^2}}{a}$

105. $\sqrt[6]{x^4} = \sqrt[3]{x^2}$

107. $\sqrt{2}\sqrt[3]{2}$
$= 2^{1/2} \cdot 2^{1/3}$
$= 2^{1/2 + 1/3}$
$= 2^{5/6}$
$= \sqrt[6]{2^5}$

109. $4\sqrt{2} + \sqrt{2} - 5\sqrt{2}$
$= (4 + 1 - 5)\sqrt{2}$
$= 0$

111. $6\sqrt{12} - 4\sqrt{27}$
$= 12\sqrt{3} - 12\sqrt{3}$
$= 0$

113. $\sqrt{\dfrac{3}{2}} + \sqrt{\dfrac{5}{3}}$

$= \dfrac{\sqrt{3}}{\sqrt{2}} + \dfrac{\sqrt{5}}{\sqrt{3}}$

$= \dfrac{\sqrt{3}\sqrt{2}}{\sqrt{2}\sqrt{2}} + \dfrac{\sqrt{5}\sqrt{3}}{\sqrt{3}\sqrt{3}}$

$= \dfrac{\sqrt{6}}{2} + \dfrac{\sqrt{15}}{3}$

$= \dfrac{3\sqrt{6}}{6} + \dfrac{2\sqrt{15}}{6}$

$= \dfrac{3\sqrt{6} + 2\sqrt{15}}{6}$

115. $\sqrt{3}(\sqrt{6} - \sqrt{2}) + \sqrt{2}(\sqrt{3} - 3)$
$= \sqrt{18} - \sqrt{6} + \sqrt{6} - 3\sqrt{2}$
$= 3\sqrt{2} - \sqrt{6} + \sqrt{6} - 3\sqrt{2}$
$= (3 - 3)\sqrt{2} + (-1 + 1)\sqrt{6}$
$= 0$

117. $(3\sqrt{7} - \sqrt{3})(\sqrt{7} - 2\sqrt{3})$
$= 3(7) - 6\sqrt{21} - \sqrt{21} + 2(3)$
$= 21 - 6\sqrt{21} - \sqrt{21} + 6$
$= 27 - 7\sqrt{21}$

119. $(\sqrt{x} - 5)^2$
$= (\sqrt{x})^2 - 2\sqrt{x}(5) + 5^2$
$= x - 10\sqrt{x} + 25$

121. $(\sqrt{a + 7})^2 - (\sqrt{a} + 7)^2$
$= a + 7 - [(\sqrt{a})^2 + 2\sqrt{a}(7) + 7^2]$
$= a + 7 - (a + 14\sqrt{a} + 49)$
$= a + 7 - a - 14\sqrt{a} - 49$
$= -42 - 14\sqrt{a}$

123. $\dfrac{12}{\sqrt{5} + \sqrt{3}}$

$= \dfrac{12(\sqrt{5} - \sqrt{3})}{(\sqrt{5} + \sqrt{3})(\sqrt{5} - \sqrt{3})}$

$= \dfrac{12(\sqrt{5} - \sqrt{3})}{5 - 3}$

$= \dfrac{12(\sqrt{5} - \sqrt{3})}{2}$

$= 6(\sqrt{5} - \sqrt{3})$
$= 6\sqrt{5} - 6\sqrt{3}$

125. $\dfrac{8x - 20y}{\sqrt{2x} - \sqrt{5y}}$

$= \dfrac{(8x - 20y)(\sqrt{2x} + \sqrt{5y})}{(\sqrt{2x} - \sqrt{5y})(\sqrt{2x} + \sqrt{5y})}$

$= \dfrac{4(2x - 5y)(\sqrt{2x} + \sqrt{5y})}{2x - 5y}$
$= 4(\sqrt{2x} + \sqrt{5y})$
$= 4\sqrt{2x} + 4\sqrt{5y}$

127. $\sqrt{2x} - 5 = 7$
$$\begin{aligned} \sqrt{2x} &= 12 \\ \left(\sqrt{2x}\right)^2 &= 12^2 \\ 2x &= 144 \\ x &= 72 \end{aligned}$$

129. $\sqrt{2x - 5} = 7$
$$\begin{aligned} \left(\sqrt{2x - 5}\right)^2 &= 7^2 \\ 2x - 5 &= 49 \\ 2x &= 54 \\ x &= 27 \end{aligned}$$

131. $\sqrt[4]{x - 4} = 3$
$$\begin{aligned} \left(\sqrt[4]{x - 4}\right)^4 &= 3^4 \\ x - 4 &= 81 \\ x &= 85 \end{aligned}$$

133. $x^{1/4} + 1 = 3$
$$\begin{aligned} x^{1/4} &= 2 \\ (x^{1/4})^4 &= 2^4 \\ x &= 16 \end{aligned}$$

135. $f(x) = \sqrt{3x}$
$$\begin{aligned} 7.25 &= \sqrt{3x} \\ (7.25)^2 &= \left(\sqrt{3x}\right)^2 \\ 52.5625 &= 3x \\ 17.52 &= x \end{aligned}$$

137. $G = 34.2d^2\sqrt{p}$
$$\begin{aligned} 450 &= 34.2(3)^2\sqrt{p} \\ 450 &= 307.8\sqrt{p} \\ 1.4620 &= \sqrt{p} \\ (1.45620)^2 &= \left(\sqrt{p}\right)^2 \\ 2.1 &= p \end{aligned}$$

The pressure is 2.1 *psi*.

139. $i^{11} = (i^4)^2 \cdot i^3$
$$\begin{aligned} &= 1^2 \cdot (-i) \\ &= -i \end{aligned}$$

141. $-i^{14} = -(i^4)^3 i^2$
$$\begin{aligned} &= -(1)^3(-1) \\ &= 1 \end{aligned}$$

143. $(5 + i) + (4 - 2i)$
$$\begin{aligned} &= (5 + 4) + (1 - 2)i \\ &= 9 - i \end{aligned}$$

145. $(7 - 2i)(2 - 3i)$
$$\begin{aligned} &= 14 - 21i - 4i + 6i^2 \\ &= 14 - 21i - 4i - 6 \\ &= 8 - 25i \end{aligned}$$

147. $2i(3i - 4)$
$$\begin{aligned} &= 6i^2 - 8i \\ &= -6 - 8i \end{aligned}$$

149. $(3 - 2i)^2$
$$\begin{aligned} &= 3^2 - 2(3)(2i) + (2i)^2 \\ &= 9 - 12i + 4i^2 \\ &= 9 - 12i - 4 \\ &= 5 - 12i \end{aligned}$$

151. $(6 - i)^2 - 12(6 - i)$
$$\begin{aligned} &= 6^2 - 2(6)i + i^2 - 72 + 12i \\ &= 36 - 12i - 1 - 72 + 12i \\ &= -37 \end{aligned}$$

153. $\dfrac{4 - 3i}{3 + i}$
$$\begin{aligned} &= \frac{(4 - 3i)(3 - i)}{(3 + i)(3 - i)} \\ &= \frac{12 - 4i - 9i + 3i^2}{9 - i^2} \\ &= \frac{12 - 4i - 9i - 3}{9 + 1} \\ &= \frac{9 - 13i}{10} \\ &= \frac{9}{10} - \frac{13}{10}i \end{aligned}$$

155.
$$\begin{aligned} x^2 - 2x + 5 &= 0 \\ (1 - 2i)^2 - 2(1 - 2i) + 5 &= 0 \\ 1^2 - 2(1)(2i) + (2i)^2 - 2 + 4i + 5 &= 0 \\ 1 - 4i + 4i^2 - 2 + 4i + 5 &= 0 \\ 1 - 4i - 4 - 2 + 4i + 5 &= 0 \\ 0 &= 0 \end{aligned}$$
True

$x = 1 - 2i$ is a solution.

CHAPTER 7 PRACTICE TEST

1. $(a^2b^3)(ab^4)^2$
$= (a^2b^3)(a^2b^8)$
$= a^4b^{11}$

3. $(-3ab^{-2})^{-1}(-2x^{-1}y)^2$
$= (-3)^{-1}a^{-1}b^2(4x^{-2}y^2)$
$= \dfrac{b^2}{-3a} \cdot \dfrac{4y^2}{x^2}$
$= -\dfrac{4b^2y^2}{3ax^2}$

5. $\left(\dfrac{5r^{-1}s^{-3}}{3rs^2}\right)^{-2}$
$= \left(\dfrac{5}{3}r^{-1-1}s^{-3-2}\right)^{-2}$
$= \left(\dfrac{5}{3}r^{-2}s^{-5}\right)^{-2}$
$= \left(\dfrac{5}{3}\right)^{-2}r^4s^{10}$
$= \left(\dfrac{3}{5}\right)^{2}r^4s^{10}$
$= \dfrac{9}{25}r^4s^{10}$

7. $\left(\dfrac{x^{1/4}x^{-2/3}}{x^{-1}}\right)^4$
$= \left(\dfrac{x^{1/4-2/3}}{x^{-1}}\right)^4$
$= \left(\dfrac{x^{-5/12}}{x^{-1}}\right)^4$
$= (x^{-5/12-(-1)})^4$
$= (x^{7/12})^4$
$= x^{(7/12)(4)}$
$= x^{7/3}$

9. $x^{1/3}(x^{2/3} - x)$
$= x^{1/3+2/3} - x^{1/3+1}$
$= x - x^{4/3}$

11. $3a^{2/5} = 3\sqrt[3]{a^2}$

13. 1 hr = 60 min = (60)(60) sec
= 3600 sec

$D = r \cdot t$
$= (186000)(3600)$
$= (1.86 \times 10^5)(3.6 \times 10^3)$
$= (1.86)(3.6) \times (10^5 \cdot 10^3)$
$= 6.696 \times 10^8$

It travels 6.696×10^8 miles.

15. $\sqrt[3]{4x^2y^2}\sqrt[3]{2x}$
$= \sqrt[3]{8x^3y^2}$
$= 2x\sqrt[3]{y^2}$

17. $\sqrt{\dfrac{5}{7}} = \dfrac{\sqrt{5}}{\sqrt{7}}$
$= \dfrac{\sqrt{5}\sqrt{7}}{\sqrt{7}\sqrt{7}}$
$= \dfrac{\sqrt{35}}{7}$

19. $\dfrac{(xy\sqrt{2xy})(3x\sqrt{y})}{\sqrt{4x^3}}$
$= \dfrac{3x^2y\sqrt{2xy^2}}{2x\sqrt{x}}$
$= \dfrac{3x^2y^2\sqrt{2x}}{2x\sqrt{x}}$
$= \dfrac{3xy^2}{2}\sqrt{\dfrac{2x}{x}}$
$= \dfrac{3xy^2\sqrt{2}}{2}$

21. $\sqrt{50} - 3\sqrt{8} + 2\sqrt{18}$

$= 5\sqrt{2} - 6\sqrt{2} + 6\sqrt{2}$

$= (5 - 6 + 6)\sqrt{2}$

$= 5\sqrt{2}$

23. $\dfrac{\sqrt{6}}{\sqrt{6} - 2}$

$= \dfrac{\sqrt{6}(\sqrt{6} + 2)}{(\sqrt{6} - 2)(\sqrt{6} + 2)}$

$= \dfrac{\sqrt{6}(\sqrt{6} + 2)}{6 - 4}$

$= \dfrac{6 + 2\sqrt{6}}{2}$

$= \dfrac{2(3 + \sqrt{6})}{2}$

$= 3 + \sqrt{6}$

25. $\sqrt{x} - 3 = 8$

$\sqrt{x} = 11$

$(\sqrt{x})^2 = 11^2$

$x = 121$

27. $i^{51} = (i^4)^{12} \cdot i^3$

$= 1^{12} \cdot (-i)$

$= -i$

29. $\dfrac{2i + 3}{i - 2}$

$= \dfrac{(2i + 3)(i + 2)}{(i - 2)(i + 2)}$

$= \dfrac{2i^2 + 4i + 3i + 6}{i^2 - 4}$

$= \dfrac{-2 + 4i + 3i + 6}{-1 - 4}$

$= \dfrac{4 + 7i}{-5}$

$= -\dfrac{4}{5} - \dfrac{7}{5}i$

CHAPTER 8

Exercises 8.1

1. $P(x) = -3.5x^2 + 600.4x$

(a)

x	$-3.5x^2 + 600.4x = P$
30	$-3.5(30)^2 + 600.4(30) = 14{,}862$
50	$-3.5(50)^2 + 600.4(50) = 21{,}270$
70	$-3.5(70)^2 + 600.4(70) = 24{,}878$
100	$-3.5(100)^2 + 600.4(100) = 25{,}040$
120	$-3.5(120)^2 + 600.4(120) = 21{,}648$

As the number of items x increases, the profit rises for a while and when the number of items exceeds 100, the profit begins to decrease.

(b)

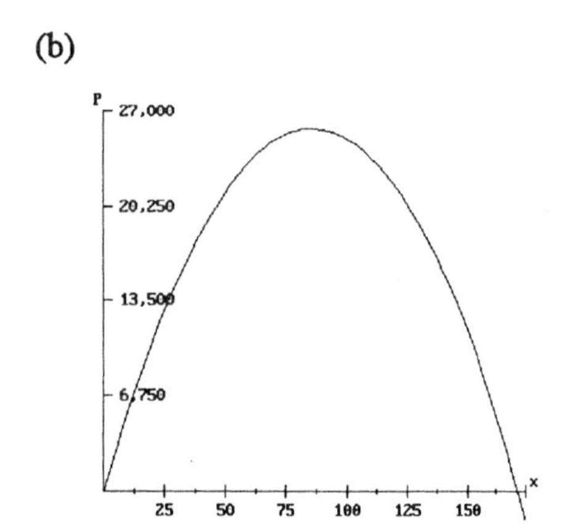

The profit seems to reach a maximum of $25,700 occurring at $x = 85$ items.

3. $s(t) = -16t^2 + 725t$

(a)

t	$-16t^2 + 725t = S$
5	$-16(5)^2 + 725(5) = 3225$
10	$-16(10)^2 + 725(10) = 5650$
20	$-16(20)^2 + 725(20) = 8100$
30	$-16(30)^2 + 725(30) = 7350$
40	$-16(40)^2 + 725(40) = 3400$
45	$-16(45)^2 + 725(45) = 225$

(b)

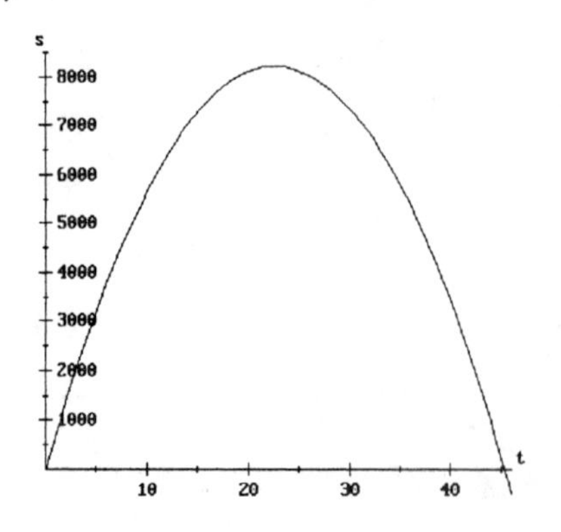

The maximum height is about 8200 feet, occurring at 20–25 seconds.

(c) The rocket hits the ground between 45 and 46 seconds.

5. (a)

$$2x + 2y = 200$$
$$2y = -2x + 200$$
$$y = -x + 100$$

$$\text{Area} = xy$$
$$A(x) = x(-x + 100)$$
$$A(x) = -x^2 + 100x$$

(b)

x	$-x^2 + 100x = A$
10	$-(10)^2 + 100(10) = 900$
40	$-(40)^2 + 100(40) = 2400$
60	$-(60)^2 + 100(60) = 2400$
70	$-(70)^2 + 100(70) = 2100$
90	$-(90)^2 + 100(90) = 900$

As the length x increases, the area increases until the length x is between 40 and 60 feet, then the area begins to decrease.

(c)

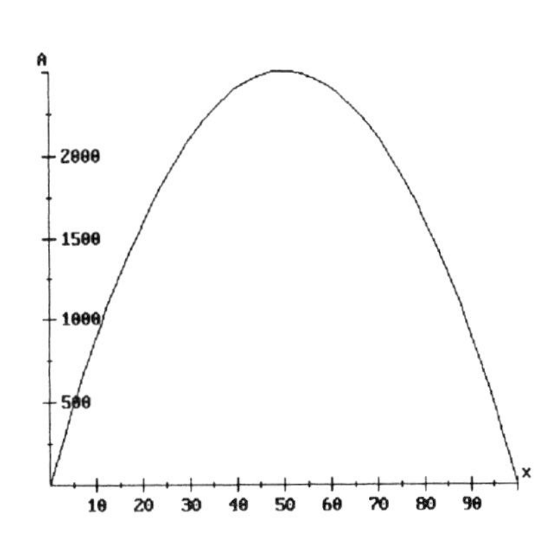

The maximum area is approximately 2500 sq. feet when $x = 50$ feet.

7. Let $x = 1^{st}$ number
then $225 - x = 2^{nd}$ number

$$P(x) = x(225 - x)$$
$$P(x) = 225x - x^2$$

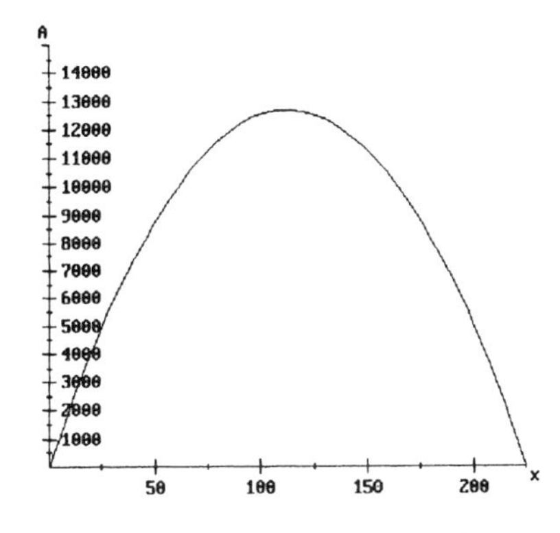

A will be a maximum when x is between 112 and 113, and hence the other number $(225 - x)$ will also be between 112 and 113.

9. (a) $4x + 3y = 1000$

$$3y = 1000 - 4x$$
$$y = \frac{1000 - 4x}{3}$$

$$A(x) = (2x)(y)$$

$$A(x) = (2x)\left(\frac{1000 - 4x}{3}\right)$$

$$A(x) = \frac{2000x - 8x^2}{3}$$

(b)

x	$\frac{2000x - 8x^2}{3}$
50	26,667
100	40,000
120	41,600
150	40,000
200	26,667

As the length x increases, the area increases until x is approximately 120 feet, then the area decreases.

(c)

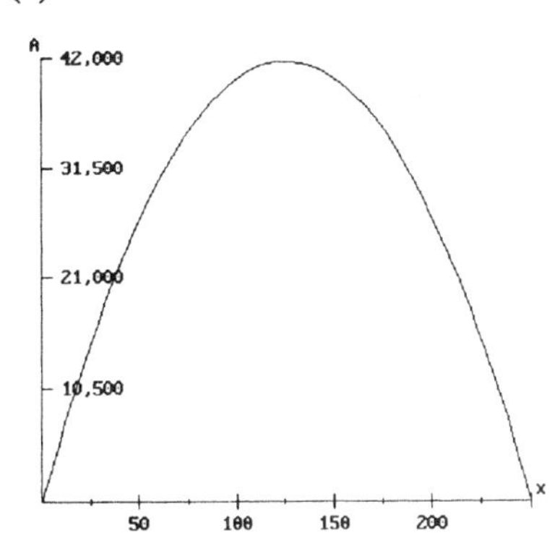

The maximum area is approximately 41,667 sq. feet when x is 125 feet and y is 167 feet.

11. $p = 10 - \dfrac{x}{6}$

(a)
$$6 = 10 - \frac{x}{6}$$
$$6(6) = 6\left(10 - \frac{x}{6}\right)$$
$$36 = 60 - x$$
$$-24 = -x$$
$$24 = x$$

24 items would be sold.

(b)
$$p = 10 - \frac{30}{6}$$
$$p = 5$$

The price should be \$5 per item.

(c)

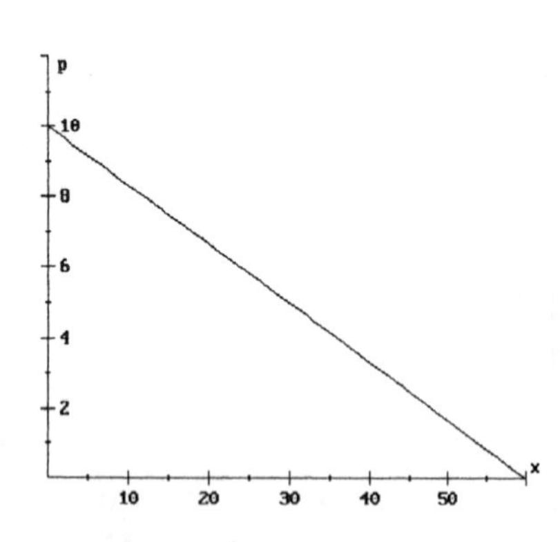

(d) $R = xp$

$$R = x\left(10 - \frac{x}{6}\right)$$

$$R = 10x - \frac{x^2}{6}$$

(e)

x	$10x - (x^2)/6$
5	45.83
15	112.5
25	145.83
35	145.83
40	133.33

As the number of units sold, x increases, the revenue increases until x is between 25 and 35, then it begins to decrease.

(f)

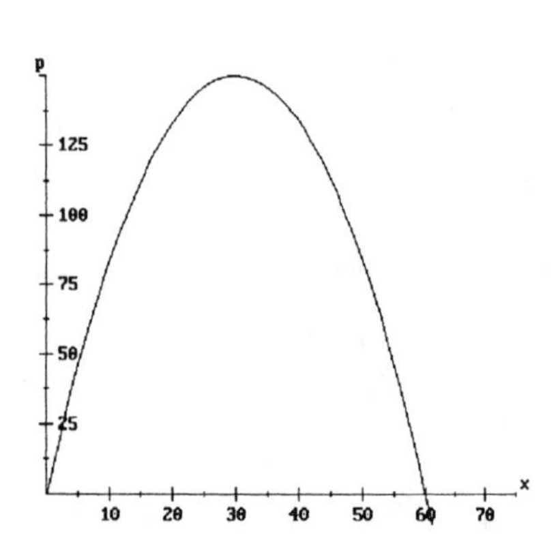

The maximum revenue is approximately \$150 when 30 units are sold.

13. (a) $\text{Revenue} = \begin{pmatrix}\text{number}\\ \text{of cubes}\\ \text{sold}\end{pmatrix} \cdot \begin{pmatrix}\text{price}\\ \text{per}\\ \text{cube}\end{pmatrix}$

$$R(x) = (200000 - 7250x)(3 + 0.25x)$$

(b)

x	$(200000 - 7250x)(3 + 0.25x)$
3	668,438
6	704,250
9	707,438
12	678,000

As the number of 25 cent price increases rises, the revenue increases until x is approximately 8, then it begins to decrease.

(c)

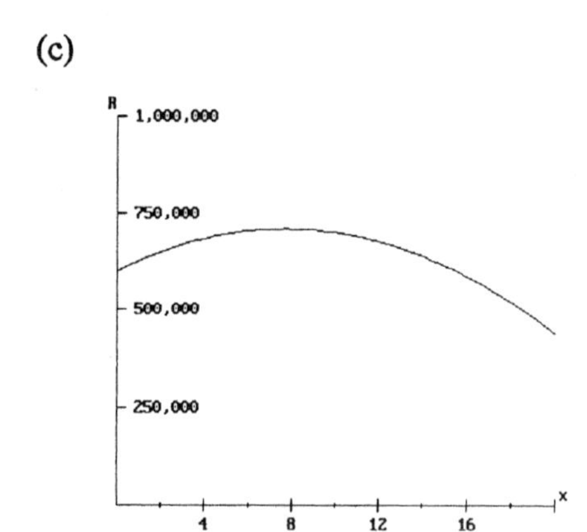

The maximum revenue is approximately \$710,000 and it occurs when $x \approx 8$. At this time the price is $3 + 0.25(8) = \$5$.

15. $3x^3y^2 - x^2 + y^3$
$= 3(-2)^3(-3)^2 - (-2)^2 + (-3)^3$
$= 3(-8)(9) - 4 - 27$
$= -247$

17. $|5 - 2x| \geq 6$
$5 - 2x \geq 6$ or $5 - 2x \leq -6$
$-2x \geq 1$ $\qquad -2x \leq -11$
$x \leq -\frac{1}{2}$ or $x \geq \frac{11}{2}$

Exercises 8.2

1. $(x + 2)(x - 3) = 0$
   $x + 2 = 0$ or $x - 3 = 0$
   $x = -2$ or $x = 3$

3. $x^2 = 25$
   $x = \pm\sqrt{25}$
   $x = \pm 5$

5. $x(x - 4) = 0$
   $x = 0$ or $x - 4 = 0$
   $x = 0$ or $x = 4$

7. $12 = x(x - 4)$
   $12 = x^2 - 4x$
   $0 = x^2 - 4x - 12$
   $0 = (x - 6)(x + 2)$
   $x - 6 = 0$ or $x + 2 = 0$
   $x = 6$ or $x = -2$

9. $5y(y - 7) = 0$
   $5y = 0$ or $y - 7 = 0$
   $y = 0$ or $y = 7$

11. $x^2 - 16 = 0$
    $x^2 = 16$
    $x = \pm\sqrt{16}$
    $x = \pm 4$

13. $0 = 9c^2 - 16$
    $16 = 9c^2$
    $\frac{16}{9} = c^2$
    $\pm\sqrt{\frac{16}{9}} = c$
    $\pm\frac{4}{3} = c$

15. $8a^2 - 18 = 0$
    $8a^2 = 18$
    $a^2 = \frac{9}{4}$
    $a = \pm\sqrt{\frac{9}{4}}$
    $a = \pm\frac{3}{2}$

17. $0 = x^2 - x - 6$
    $0 = (x - 3)(x + 2)$
    $x - 3 = 0$ or $x + 2 = 0$
    $x = 3$ or $x = -2$

19. $2y^2 - 3y + 1 = 0$
    $(2y - 1)(y - 1) = 0$
    $2y - 1 = 0$ or $y - 1 = 0$
    $2y = 1$ $\quad$ $y = 1$
    $y = \frac{1}{2}$ or $y = 1$

21. $f(x) = 4x^2 - 24$
    $0 = 4x^2 - 24$
    $24 = 4x^2$
    $6 = x^2$
    $\pm\sqrt{6} = x$

23. $g(x) = 7x^2$
    $19 = 7x^2$
    $\frac{19}{7} = x^2$
    $\pm\sqrt{\frac{19}{7}} = x$
    $\pm\frac{\sqrt{19}\cdot\sqrt{7}}{\sqrt{7}\cdot\sqrt{7}} = x$
    $\pm\frac{\sqrt{133}}{7} = x$

25. $10 = x^2 - 3x$
    $0 = x^2 - 3x - 10$
    $0 = (x - 5)(x + 2)$
    $x - 5 = 0$ or $x + 2 = 0$
    $x = 5$ or $x = -2$

27. $6a^2 + 3a - 1 = 4a$
    $6a^2 - a - 1 = 0$
    $(3a + 1)(2a - 1) = 0$
    $3a + 1 = 0$ or $2a - 1 = 0$
    $3a = -1$ $\quad$ $2a = 1$
    $a = -\frac{1}{3}$ or $a = \frac{1}{2}$

29. $3y^2 - 5y + 8 = 9y^2 - 10y + 2$

$0 = 6y^2 - 5y - 6$

$0 = (2y - 3)(3y + 2)$

$2y - 3 = 0$ or $3y + 2 = 0$

$2y = 3$ $\quad 3y = -2$

$y = \frac{3}{2}$ or $y = -\frac{2}{3}$

31. $0 = 3x^2 + 5$

$-5 = 3x^2$

$-\frac{5}{3} = x^2$

$\pm\sqrt{-\frac{5}{3}} = x$

$\pm i\frac{\sqrt{5}}{\sqrt{3}} = x$

$\pm\frac{i\sqrt{5}\sqrt{3}}{\sqrt{3}\sqrt{3}} = x$

$\pm\frac{\sqrt{15}}{3}i = x$

33. $(2s - 3)(3s + 1) = 7$

$6s^2 - 7s - 3 = 7$

$6s^2 - 7s - 10 = 0$

$(6s + 5)(s - 2) = 0$

$6s + 5 = 0$ or $s - 2 = 0$

$6s = -5$ $\quad s = 2$

$s = -\frac{5}{6}$ or $s = 2$

35. $(x + 2)^2 = 25$

$x + 2 = \pm\sqrt{25}$

$x + 2 = \pm 5$

$x + 2 = 5$ or $x + 2 = -5$

$x = 3$ or $x = -7$

37. $(x - 2)(3x - 1) = (2x - 3)(x + 1)$

$3x^2 - 7x + 2 = 2x^2 - x - 3$

$x^2 - 6x + 5 = 0$

$(x - 5)(x - 1) = 0$

$x - 5 = 0$ or $x - 1 = 0$

$x = 5$ or $x = 1$

39. $(x + 3)^2 = x(x + 5)$

$x^2 + 6x + 9 = x^2 + 5x$

$6x + 9 = 5x$

$9 = -x$

$-9 = x$

41. $x^2 + 3 = 1$

$x^2 = -2$

$x = \pm\sqrt{-2}$

$x = \pm i\sqrt{2}$

43. $8 = (x - 8)^2$

$\pm\sqrt{8} = x - 8$

$\pm 2\sqrt{2} = x - 8$

$8 \pm 2\sqrt{2} = x$

45. $(y - 5)^2 = -16$

$y - 5 = \pm\sqrt{-16}$

$y - 5 = \pm 4i$

$y = 5 \pm 4i$

47. $\frac{2}{x - 1} + x = 4$

$(x - 1)\left(\frac{2}{x - 1} + x\right) = (x - 1)4$

$2 + x(x - 1) = 4x - 4$

$2 + x^2 - x = 4x - 4$

$x^2 - 5x + 6 = 0$

$(x - 2)(x - 3) = 0$

$x - 2 = 0$ or $x - 3 = 0$

$x = 2$ or $x = 3$

49. $2a - 5 = \frac{3(a + 2)}{a + 4}$

$(a + 4)(2a - 5) = (a + 4)\left[\frac{3(a + 2)}{a + 4}\right]$

$2a^2 + 3a - 20 = 3a + 6$

$2a^2 = 26$

$a^2 = 13$

$a = \pm\sqrt{13}$

51. $\frac{x}{x + 2} - \frac{3}{x} = \frac{x + 1}{x}$

$x(x + 2)\left(\frac{x}{x + 2} - \frac{3}{x}\right) = x(x + 2)\left(\frac{x + 1}{x}\right)$

$x^2 - 3(x + 2) = (x + 2)(x + 1)$

$x^2 - 3x - 6 = x^2 + 3x + 2$

$-3x - 6 = 3x + 2$

$-6 = 6x + 2$

$-8 = 6x$

$-\frac{4}{3} = x$

53. $\frac{1}{x} + x = 2$

$$x\left(\frac{1}{x} + x\right) = x(2)$$

$$\begin{aligned} 1 + x^2 &= 2x \\ x^2 - 2x + 1 &= 0 \\ (x - 1)^2 &= 0 \\ x - 1 &= \pm\sqrt{0} \\ x - 1 &= 0 \\ x &= 1 \end{aligned}$$

55. $\frac{3}{x - 2} + \frac{7}{x + 2} = \frac{x + 1}{x - 2}$

$$(x - 2)(x + 2)\left(\frac{3}{x - 2} + \frac{7}{x + 2}\right) = (x - 2)(x + 2)\left(\frac{x + 1}{x - 2}\right)$$

$$\begin{aligned} 3(x + 2) + 7(x - 2) &= (x + 2)(x + 1) \\ 3x + 6 + 7x - 14 &= x^2 + 3x + 2 \\ 10x - 8 &= x^2 + 3x + 2 \\ 0 &= x^2 - 7x + 10 \\ 0 &= (x - 5)(x - 2) \end{aligned}$$

$$x - 5 = 0 \quad \text{or} \quad x - 2 = 0$$
$$x = 5 \quad \text{or} \quad x = 2$$

The only solution is $x = 5$, since $x = 2$ causes a denominator to equal 0 in the original equation.

57. $\frac{2}{x - 1} + \frac{3x}{x + 2} = \frac{2(5x + 9)}{x^2 + x - 2}$

$$\frac{2}{x - 1} + \frac{3x}{x + 2} = \frac{2(5x + 9)}{(x + 2)(x - 1)}$$

$$(x - 1)(x + 2)\left(\frac{2}{x - 1} + \frac{3x}{x + 2}\right) = (x - 1)(x + 2)\left[\frac{2(5x + 9)}{(x + 2)(x - 1)}\right]$$

$$\begin{aligned} 2(x + 2) + 3x(x - 1) &= 2(5x + 9) \\ 2x + 4 + 3x^2 - 3x &= 10x + 18 \\ 3x^2 - x + 4 &= 10x + 18 \\ 3x^2 - 11x - 14 &= 0 \\ (3x - 14)(x + 1) &= 0 \end{aligned}$$

$$3x - 14 = 0 \quad \text{or} \quad x + 1 = 0$$
$$3x = 14 \qquad x = -1$$
$$x = \frac{14}{3} \quad \text{or} \quad x = -1$$

59. $$\begin{aligned} 8a^2 + 3b &= 5b \\ 8a^2 &= 2b \\ a^2 &= \frac{b}{4} \\ a &= \pm\sqrt{\frac{b}{4}} \\ a &= \pm\frac{\sqrt{b}}{2} \end{aligned}$$

61. $$\begin{aligned} 5x^2 + 7y^2 &= 9 \\ 7y^2 &= 9 - 5x^2 \\ y^2 &= \frac{9 - 5x^2}{7} \\ y &= \pm\sqrt{\frac{9 - 5x^2}{7}} \\ y &= \pm\frac{\sqrt{9 - 5x^2}}{\sqrt{7}} \\ y &= \pm\frac{\sqrt{9 - 5x^2} \cdot \sqrt{7}}{\sqrt{7} \cdot \sqrt{7}} \\ y &= \pm\frac{\sqrt{7(9 - 5x^2)}}{7} \end{aligned}$$

63. $$\begin{aligned} V &= \frac{2}{3}\pi r^2 \\ 3V &= 2\pi r^2 \\ \frac{3V}{2\pi} &= r^2 \\ \sqrt{\frac{3V}{2\pi}} &= r \\ \frac{\sqrt{3V}}{\sqrt{2\pi}} &= r \\ \frac{\sqrt{3V}\sqrt{2\pi}}{\sqrt{2\pi}\sqrt{2\pi}} &= r \\ \frac{\sqrt{6V\pi}}{2\pi} &= r \end{aligned}$$

65. $a^2 - 4b^2 = 0$
$$a^2 = 4b^2$$
$$a = \pm\sqrt{4b^2}$$
$$a = \pm 2b$$

67. $x^2 - xy - 6y^2 = 0$
$$(x - 3y)(x + 2y) = 0$$
$$x - 3y = 0 \quad \text{or} \quad x + 2y = 0$$
$$x = 3y \quad \text{or} \quad x = -2y$$

69. Let x = the number

$$x^2 - 5 = 1 + 5x$$
$$x^2 - 5x - 6 = 0$$
$$(x - 6)(x + 1) = 0$$
$$x - 6 = 0 \quad \text{or} \quad x + 1 = 0$$
$$x = 6 \quad \text{or} \quad x = -1$$

$x = 6$, since the number is positive.

71. Let x = other number

$$x^2 + 8^2 = 68$$
$$x^2 + 64 = 68$$
$$x^2 = 4$$
$$x = \pm\sqrt{4}$$
$$x = \pm 2$$

$x = 2$, since the number is positive.

73. Let x = the number

$$(x + 6)^2 = 169$$
$$x + 6 = \pm\sqrt{169}$$
$$x + 6 = \pm 13$$
$$x + 6 = 13 \quad \text{or} \quad x + 6 = -13$$
$$x = 7 \quad \text{or} \quad x = -19$$

The numbers are 7 and -19.

75. Let x = the number

$$x + \frac{1}{x} = \frac{13}{6}$$

$$6x\left(x + \frac{1}{x}\right) = 6x\left(\frac{13}{6}\right)$$

$$6x^2 + 6 = 13x$$
$$6x^2 - 13x + 6 = 0$$
$$(3x - 2)(2x - 3) = 0$$

$$3x - 2 = 0 \quad \text{or} \quad 2x - 3 = 0$$
$$3x = 2 \qquad 2x = 3$$
$$x = \frac{2}{3} \quad \text{or} \quad x = \frac{3}{2}$$

The numbers are $\frac{2}{3}$ and $\frac{3}{2}$.

77. $P(d) = 10000(-d^2 + 12d - 35)$

(a) $P(5) = 10000\left[-5^2 + 12(5) - 35\right]$
$= 0$

The profit is \$0.

(b) $10000 = 10000(-d^2 + 12d - 35)$
$$1 = -d^2 + 12d - 35$$
$$d^2 - 12d + 36 = 0$$
$$(d - 6)^2 = 0$$
$$d - 6 = \pm\sqrt{0}$$
$$d - 6 = 0$$
$$d = 6$$

The price must be \$6.

79. $h(t) = -16t^2 + 40$

(a) $h(1) = -16(1)^{22} + 40$
$= 24$

He is 24 feet above the pool.

(b) $0 = -16t^2 + 40$
$$16t^2 = 40$$
$$t^2 = \frac{5}{2}$$

$$t = \sqrt{\frac{5}{2}}$$

$$t = \frac{\sqrt{10}}{2} \approx 1.58$$

It will take 1.58 sec.

(c) $h(0) = -16(0)^2 + 40$
$= 40$

It is 40 feet high.

81. Let x = width
then $x + 2$ = length

$$\begin{aligned} \text{Area} &= wl \\ 80 &= x(x+2) \\ 80 &= x^2 + 2x \\ 0 &= x^2 + 2x - 80 \\ 0 &= (x+10)(x-8) \end{aligned}$$

$x + 10 = 0$ or $x - 8 = 0$
$x = -10$ or $x = 8$

Since x is positive, $x = 8$ and
$x + 2 = 8 + 2 = 10$.
The dimensions are 8 ft by 10 ft.

83. Let x = side length

$$\begin{aligned} \text{Area} &= s^2 \\ 60 &= x^2 \\ \pm\sqrt{60} &= x \\ \pm 2\sqrt{15} &= x \end{aligned}$$

Since x is positive, $x = 2\sqrt{15}$.
The square is $2\sqrt{15}$ in. by $2\sqrt{15}$ in.

85. Let x = 1st number,
then $x - 2$ = 2nd number

$$\begin{aligned} x(x-2) &= 120 \\ x^2 - 2x &= 120 \\ x^2 - 2x - 120 &= 0 \\ (x-12)(x+10) &= 0 \end{aligned}$$

$x - 12 = 0$ or $x + 10 = 0$
$x = 12$ or $x = -10$
$x - 2 = 10$ $\quad x - 2 = -12$

The numbers are 12 and 10 or −10 and −12.

87. Let x = 1st number,
then $3x + 2$ = 2nd number

$$\begin{aligned} x(3x+2) &= 85 \\ 3x^2 + 2x &= 85 \\ 3x^2 + 2x - 85 &= 0 \\ (3x+17)(x-5) &= 0 \end{aligned}$$

$3x + 17 = 0$ or $x - 5 = 0$
$3x = -17$ $\quad x = 5$
$x = -\frac{17}{3}$ or $x = 5$
$3x + 2 = -15$ $\quad 3x + 2 = 17$

The numbers are $-\frac{17}{3}$ and −15 or 5 and 17.

89.
$$\begin{aligned} a^2 + b^2 &= c^2 \\ 4^2 + b^2 &= 7^2 \\ 16 + b^2 &= 49 \\ b^2 &= 33 \\ b &= \sqrt{33} \end{aligned}$$

The other leg is $\sqrt{33}$ in.

91.
$$\begin{aligned} a^2 + b^2 &= c^2 \\ 8^2 + (x-2)^2 &= (x+4)^2 \\ 64 + x^2 - 4x + 4 &= x^2 + 8x + 16 \\ x^2 - 4x + 68 &= x^2 + 8x + 16 \\ -4x + 68 &= 8x + 16 \\ 68 &= 12x + 16 \\ 52 &= 12x \\ \frac{13}{3} &= x \end{aligned}$$

$$x - 2 = \frac{13}{3} - 2 = \frac{7}{3}$$

$$x + 4 = \frac{13}{3} + 4 = \frac{25}{3}$$

They are 8, $\frac{7}{3}$ and $\frac{25}{3}$ in.

93.

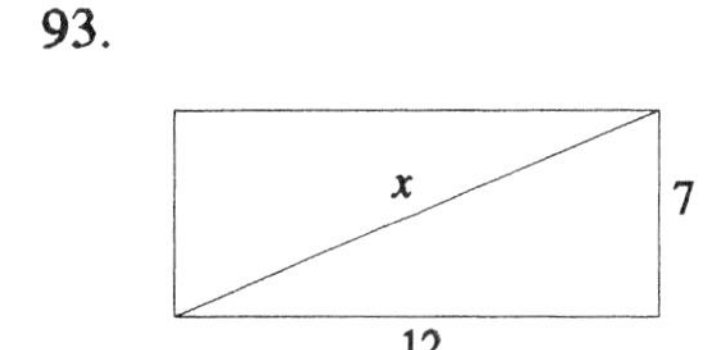

$$\begin{aligned} 7^2 + 12^2 &= x^2 \\ 49 + 144 &= x^2 \\ 193 &= x^2 \\ \sqrt{193} &= x \end{aligned}$$

The diagonal is $\sqrt{193}$ in.

95. Let x = height

$$8^2 + x^2 = 30^2$$
$$64 + x^2 = 900$$
$$x^2 = 836$$
$$x = \sqrt{836}$$
$$x = 2\sqrt{209}$$

It reaches $2\sqrt{209}$ ft.

97.

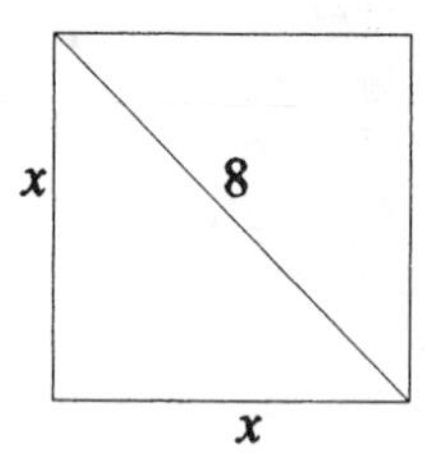

$$x^2 + x^2 = 8^2$$
$$2x^2 = 64$$
$$x^2 = 32$$
$$x = 4\sqrt{2}$$

$$\text{Area} = x^2$$
$$= (4\sqrt{2})^2$$
$$= 32$$

The area is 32 sq in.

99.

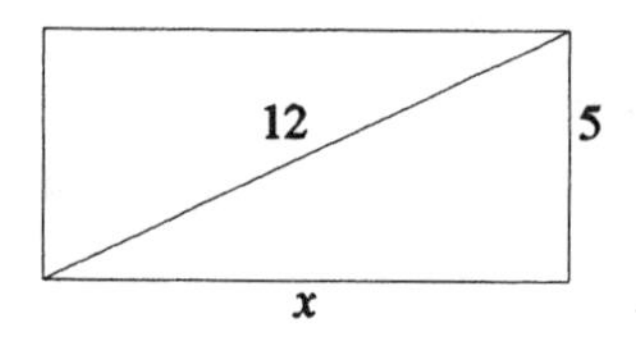

$$x^2 + 5^2 = 12^2$$
$$x^2 + 25 = 144$$
$$x^2 = 119$$
$$x = \sqrt{119}$$

$$\text{Area} = wl$$
$$= 5\sqrt{119}$$

The area is $5\sqrt{119}$ sq. in.

101. $(1, n)$ and $(3, n^2)$

$$m_1 = \frac{n^2 - n}{3 - 1} = \frac{n^2 - n}{2}$$

$(-6, 0)$ and $(-5, 6)$

$$m_2 = \frac{6 - 0}{-5 - (-6)} = 6$$

$$m_1 = m_2$$

$$\frac{n^2 - n}{2} = 6$$

$$n^2 - n = 12$$
$$n^2 - n - 12 = 0$$
$$(n - 4)(n + 3) = 0$$
$$n - 4 = 0 \quad \text{or} \quad n + 3 = 0$$
$$n = 4 \quad \text{or} \quad n = -3$$

103.

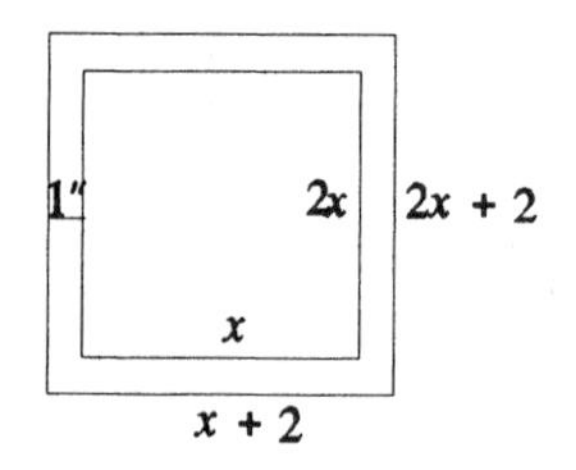

$$(x + 2)(2x + 2) = 60$$
$$2x^2 + 6x + 4 = 60$$
$$2x^2 + 6x - 56 = 0$$
$$x^2 + 3x - 28 = 0$$
$$(x + 7)(x - 4) = 0$$
$$x + 7 = 0 \quad \text{or} \quad x - 4 = 0$$
$$x = -7 \quad \text{or} \quad x = 4$$

Since x is positive, $x = 4$ and $2x = 8$. The dimensions are 4" by 8".

105. Let x = boat's rate in still water,

then rate upstream = $x - 5$
and rate downstream = $x + 5$

(time upstream) + (time downstream) = $\frac{4}{3}$

$$\frac{20}{x-5} + \frac{10}{x+5} = \frac{4}{3}$$

$$3(x-5)(x+5)\left(\frac{20}{x-5} + \frac{10}{x+5}\right) = 3(x-5)(x+5)\left(\frac{4}{3}\right)$$

$$\begin{aligned} 60(x+5) + 30(x-5) &= 4(x-5)(x+5) \\ 60x + 300 + 30x - 150 &= 4(x^2 - 25) \\ 90x + 150 &= 4x^2 - 100 \\ 0 &= 4x^2 - 90x - 250 \\ 0 &= 2x^2 - 45x - 125 \\ 0 &= (2x+5)(x-25) \\ 2x + 5 = 0 \text{ or } x - 25 &= 0 \\ 2x = -5 \qquad x &= 25 \\ x &= -\frac{5}{2} \end{aligned}$$

Since x is positive, $x = 25$.
The boat's rate is 25 kph.

Exercises 8.3

1. $$\begin{aligned} x^2 - 6x - 1 &= 0 \\ x^2 - 6x &= 1 \\ x^2 - 6x + 9 &= 1 + 9 \\ (x-3)^2 &= 10 \\ x - 3 &= \pm\sqrt{10} \\ x &= 3 \pm \sqrt{10} \end{aligned}$$

3. $$\begin{aligned} 0 &= c^2 - 2c - 5 \\ 5 &= c^2 - 2c \\ 5 + 1 &= c^2 - 2c + 1 \\ 6 &= (c-1)^2 \\ \pm\sqrt{6} &= c - 1 \\ 1 \pm \sqrt{6} &= c \end{aligned}$$

5. $$\begin{aligned} y^2 + 5y - 2 &= 0 \\ y^2 + 5y &= 2 \\ y^2 + 5y + \frac{25}{4} &= 2 + \frac{25}{4} \\ \left(y + \frac{5}{2}\right)^2 &= \frac{33}{4} \\ y + \frac{5}{2} &= \pm\sqrt{\frac{33}{4}} \\ y + \frac{5}{2} &= \pm\frac{\sqrt{33}}{2} \\ y &= -\frac{5}{2} \pm \frac{\sqrt{33}}{2} \\ \text{or } y &= \frac{-5 \pm \sqrt{33}}{2} \end{aligned}$$

7. $$\begin{aligned} 2x^2 + 3x - 1 &= x^2 - 2 \\ x^2 + 3x &= -1 \\ x^2 + 3x + \frac{9}{4} &= -1 + \frac{9}{4} \\ \left(x + \frac{3}{2}\right)^2 &= \frac{5}{4} \\ x + \frac{3}{2} &= \pm\sqrt{\frac{5}{4}} \\ x + \frac{3}{2} &= \pm\frac{\sqrt{5}}{2} \\ x &= -\frac{3}{2} \pm \frac{\sqrt{5}}{2} \\ \text{or } x &= \frac{-3 \pm \sqrt{5}}{2} \end{aligned}$$

9. $(a - 2)(a + 1) = 2$
$$a^2 - a - 2 = 2$$
$$a^2 - a = 4$$
$$a^2 - a + \frac{1}{4} = 4 + \frac{1}{4}$$
$$\left(a - \frac{1}{2}\right)^2 = \frac{17}{4}$$
$$a - \frac{1}{2} = \pm\sqrt{\frac{17}{4}}$$
$$a - \frac{1}{2} = \pm\frac{\sqrt{17}}{2}$$
$$a = \frac{1}{2} \pm \frac{\sqrt{17}}{2}$$
$$\text{or } a = \frac{1 \pm \sqrt{17}}{2}$$

11. $10 = 5a^2 + 10a + 20$
$$-10 = 5a^2 + 10a$$
$$-2 = a^2 + 2a$$
$$-2 + 1 = a^2 + 2a + 1$$
$$-1 = (a + 1)^2$$
$$\pm\sqrt{-1} = a + 1$$
$$\pm i = a + 1$$
$$-1 \pm i = a$$

13. $3x^2 + 3x = x^2 - 5x + 4$
$$2x^2 + 8x = 4$$
$$x^2 + 4x = 2$$
$$x^2 + 4x + 4 = 2 + 4$$
$$(x + 2)^2 = 6$$
$$x + 2 = \pm\sqrt{6}$$
$$x = -2 \pm \sqrt{6}$$

15. $(a - 2)(a + 1) = 6$
$$a^2 - a - 2 = 6$$
$$a^2 - a = 8$$
$$a^2 - a + \frac{1}{4} = 8 + \frac{1}{4}$$
$$\left(a - \frac{1}{2}\right)^2 = \frac{33}{4}$$
$$a - \frac{1}{2} = \pm\sqrt{\frac{33}{4}}$$
$$a - \frac{1}{2} = \pm\frac{\sqrt{33}}{2}$$
$$a = \frac{1}{2} \pm \frac{\sqrt{33}}{2}$$
$$\text{or } a = \frac{1 \pm \sqrt{33}}{2}$$

17. $2x - 7 = x^2 - 3x + 4$
$$-11 = x^2 - 5x$$
$$-11 + \frac{25}{4} = x^2 - 5x + \frac{25}{4}$$
$$-\frac{19}{4} = \left(x - \frac{5}{2}\right)^2$$
$$\pm\sqrt{-\frac{19}{4}} = x - \frac{5}{2}$$
$$\pm\frac{\sqrt{19}}{2}i = x - \frac{5}{2}$$
$$\frac{5}{2} \pm \frac{\sqrt{19}}{2}i = x$$
$$\text{or } \frac{5 \pm i\sqrt{19}}{2} = x$$

19. $5x^2 + 10x - 14 = 20$
$$5x^2 + 10x = 34$$
$$x^2 + 2x = \frac{34}{5}$$
$$x^2 + 2x + 1 = \frac{34}{5} + 1$$

$$(x+1)^2 = \frac{39}{5}$$

$$x + 1 = \pm\sqrt{\frac{39}{5}}$$

$$x + 1 = \pm\frac{\sqrt{195}}{5}$$

$$x = -1 \pm \frac{\sqrt{195}}{5}$$

$$\text{or } x = \frac{-5 \pm \sqrt{195}}{5}$$

21. $$2t^2 + 3t - 4 = 2t - 1$$

$$2t^2 + t = 3$$

$$t^2 + \frac{1}{2}t = \frac{3}{2}$$

$$t^2 + \frac{1}{2}t + \frac{1}{16} = \frac{3}{2} + \frac{1}{16}$$

$$\left(t + \frac{1}{4}\right)^2 = \frac{25}{16}$$

$$t + \frac{1}{4} = \pm\sqrt{\frac{25}{16}}$$

$$t + \frac{1}{4} = \pm\frac{5}{4}$$

$$t + \frac{1}{4} = \frac{5}{4} \quad \text{or} \quad t + \frac{1}{4} = -\frac{5}{4}$$

$$t = 1 \quad \text{or} \quad t = -\frac{3}{2}$$

23. $$(2x+5)(x-3) = (x+4)(x-1)$$

$$2x^2 - x - 15 = x^2 + 3x - 4$$

$$x^2 - 4x = 11$$

$$x^2 - 4x + 4 = 11 + 4$$

$$(x-2)^2 = 15$$

$$x - 2 = \pm\sqrt{15}$$

$$x = 2 \pm \sqrt{15}$$

25. $$\frac{2x}{2x-3} = \frac{3x-1}{x+1}$$

$$(2x-3)(x+1)\left(\frac{2x}{2x-3}\right) = (2x-3)(x+1)\left(\frac{3x-1}{x+1}\right)$$

$$2x(x+1) = (2x-3)(3x-1)$$

$$2x^2 + 2x = 6x^2 - 11x + 3$$

$$-4x^2 + 13x = 3$$

$$x^2 - \frac{13}{4}x = -\frac{3}{4}$$

$$x^2 - \frac{13}{4}x + \frac{169}{64} = -\frac{3}{4} + \frac{169}{64}$$

$$\left(x - \frac{13}{8}\right)^2 = \frac{121}{64}$$

$$x - \frac{13}{8} = \pm\sqrt{\frac{121}{64}}$$

$$x - \frac{13}{8} = \pm\frac{11}{8}$$

$$x - \frac{13}{8} = \frac{11}{8} \quad \text{or} \quad x - \frac{13}{8} = -\frac{11}{8}$$

$$x = 3 \quad \text{or} \quad x = \frac{1}{4}$$

27. $$a^2 - a + 1 = 0$$

$$a^2 - a = -1$$

$$a^2 - a + \frac{1}{4} = -1 + \frac{1}{4}$$

$$\left(a - \frac{1}{2}\right)^2 = -\frac{3}{4}$$

$$a - \frac{1}{2} = \pm\sqrt{-\frac{3}{4}}$$

$$a - \frac{1}{2} = \pm\frac{\sqrt{3}}{2}i$$

$$a = \frac{1}{2} \pm \frac{\sqrt{3}}{2}i$$

$$\text{or } a = \frac{1 \pm i\sqrt{3}}{2}$$

29.
$$0 = 2y^2 + 2y + 5$$
$$-5 = 2y^2 + 2y$$
$$-\frac{5}{2} = y^2 + y$$
$$-\frac{5}{2} + \frac{1}{4} = y^2 + y + \frac{1}{4}$$
$$-\frac{9}{4} = \left(y + \frac{1}{2}\right)^2$$
$$\pm\sqrt{-\frac{9}{4}} = y + \frac{1}{2}$$
$$\pm\frac{3}{2}i = y + \frac{1}{2}$$
$$-\frac{1}{2} \pm \frac{3}{2}i = y$$
$$\text{or } y = \frac{-1 \pm 3i}{2}$$

31.
$$5n^2 - 3n = 2n^2 - 6$$
$$3n^2 - 3n = -6$$
$$n^2 - n = -2$$
$$n^2 - n + \frac{1}{4} = -2 + \frac{1}{4}$$
$$\left(n - \frac{1}{2}\right)^2 = -\frac{7}{4}$$
$$n - \frac{1}{2} = \pm\sqrt{-\frac{7}{4}}$$
$$n - \frac{1}{2} = \pm\frac{\sqrt{7}}{2}i$$
$$n = \frac{1}{2} \pm \frac{\sqrt{7}}{2}i$$
$$\text{or } n = \frac{1 \pm i\sqrt{7}}{2}$$

33.
$$(3t + 5)(t + 1) = (t + 4)(t + 2)$$
$$3t^2 + 8t + 5 = t^2 + 6t + 8$$
$$2t^2 + 2t = 3$$
$$t^2 + t = \frac{3}{2}$$
$$t^2 + t + \frac{1}{4} = \frac{3}{2} + \frac{1}{4}$$
$$\left(t + \frac{1}{2}\right)^2 = \frac{7}{4}$$
$$t + \frac{1}{2} = \pm\sqrt{\frac{7}{4}}$$
$$t + \frac{1}{2} = \pm\frac{\sqrt{7}}{2}$$
$$t = -\frac{1}{2} \pm \frac{\sqrt{7}}{2}$$
$$\text{or } t = \frac{-1 \pm \sqrt{7}}{2}$$

35.
$$\frac{3}{x + 2} - \frac{2}{x - 1} = 5$$
$$(x + 2)(x - 1)\left(\frac{3}{x + 2} - \frac{2}{x - 1}\right) = (x + 2)(x - 1)(5)$$
$$3(x - 1) - 2(x + 2) = 5(x^2 + x - 2)$$
$$3x - 3 - 2x - 4 = 5x^2 + 5x - 10$$
$$x - 7 = 5x^2 + 5x - 10$$
$$-7 = 5x^2 + 4x - 10$$
$$3 = 5x^2 + 4x$$
$$\frac{3}{5} + \frac{4}{25} = x^2 + \frac{4}{5}x + \frac{4}{25}$$
$$\frac{19}{25} = \left(x + \frac{2}{5}\right)^2$$
$$\pm\sqrt{\frac{19}{25}} = x + \frac{2}{5}$$
$$\pm\frac{\sqrt{19}}{5} = x + \frac{2}{5}$$
$$-\frac{2}{5} \pm \frac{\sqrt{19}}{5} = x$$
$$\frac{-2 \pm \sqrt{19}}{5} = x$$

41.
$$\begin{cases} 0.3x + 0.4y = 15 \\ 0.5x + 0.6y = 24 \end{cases}$$

Multiply both equations by 10:

$$\begin{cases} 3x + 4y = 150 \\ 5x + 6y = 240 \end{cases}$$

Multiply 1st equation by -3 and 2nd equation by 2, then add:

$$\begin{aligned} -9x - 12y &= -450 \\ 10x + 12y &= 480 \\ \hline x \quad &= 30 \end{aligned}$$

$$\begin{aligned} 3(30) + 4y &= 150 \\ 90 + 4y &= 150 \\ 4y &= 60 \\ y &= 15 \end{aligned}$$

$x = 30, \quad y = 15$

43. $5n^0 + n^{-1} + 6n^{-2}$

$$= 5(1) + \frac{1}{n} + \frac{6}{n^2}$$

$$= 5 + \frac{1}{n} + \frac{6}{n^2}$$

When $n = 3$:

$$5 + \frac{1}{3} + \frac{6}{3^2}$$

$$= 5 + \frac{1}{3} + \frac{2}{3}$$

$$= 6$$

Exercises 8.4

1. $x^2 - 4x - 5 = 0$
   $A = 1, \quad B = -4, \quad C = -5$

$$x = \frac{-B \pm \sqrt{B^2 - 4AC}}{2A}$$

$$= \frac{-(-4) \pm \sqrt{(-4)^2 - 4(1)(-5)}}{2(1)}$$

$$= \frac{4 \pm \sqrt{16 + 20}}{2}$$

$$= \frac{4 \pm \sqrt{36}}{2}$$

$$= \frac{4 \pm 6}{2}$$

$$x = \frac{4 + 6}{2} = 5 \quad \text{or} \quad x = \frac{4 - 6}{2} = -1$$

3. $2a^2 - 3a - 4 = 0$
   $A = 2, \quad B = -3, \quad C = -4$

$$a = \frac{-B \pm \sqrt{B^2 - 4AC}}{2A}$$

$$= \frac{-(-3) \pm \sqrt{(-3)^2 - 4(2)(-4)}}{2(2)}$$

$$= \frac{3 \pm \sqrt{9 + 32}}{4}$$

$$= \frac{3 \pm \sqrt{41}}{4}$$

5. $(3y - 1)(2y - 3) = y$

$$\begin{aligned} 6y^2 - 11y + 3 &= y \\ 6y^2 - 12y + 3 &= 0 \\ 2y^2 - 4y + 1 &= 0 \end{aligned}$$

$A = 2, \quad B = -4, \quad C = 1$

$$y = \frac{-B \pm \sqrt{B^2 - 4AC}}{2A}$$

$$= \frac{-(-4) \pm \sqrt{(-4)^2 - 4(2)(1)}}{2(2)}$$

$$= \frac{4 \pm \sqrt{16 - 8}}{4}$$

$$= \frac{4 \pm \sqrt{8}}{4}$$

$$= \frac{4 \pm 2\sqrt{2}}{4}$$

$$= \frac{2(2 \pm \sqrt{2})}{4}$$

$$= \frac{2 \pm \sqrt{2}}{2}$$

7. $y^2 - 3y + 4 = 2y^2 + 4y - 3$

$$0 = y^2 + 7y - 7$$

$A = 1, \quad B = 7, \quad C = -7$

$$y = \frac{-B \pm \sqrt{B^2 - 4AC}}{2A}$$

$$= \frac{-7 \pm \sqrt{7^2 - 4(1)(-7)}}{2(1)}$$

$$= \frac{-7 \pm \sqrt{49 + 28}}{2}$$

$$= \frac{-7 \pm \sqrt{77}}{2}$$

9. $3a^2 + a + 2 = 0$
$A = 3, \quad B = 1, \quad C = 2$

$$a = \frac{-B \pm\sqrt{B^2 - 4AC}}{2A}$$

$$= \frac{-1 \pm \sqrt{1^2 - 4(3)(2)}}{2(3)}$$

$$= \frac{-1 \pm \sqrt{1 - 24}}{6}$$

$$= \frac{-1 \pm \sqrt{-23}}{6}$$

$$= \frac{-1 \pm i\sqrt{23}}{6}$$

11. $(s - 3)(s + 4) = (2s - 1)(s + 2)$
$s^2 + s - 12 = 2s^2 + 3s - 2$
$0 = s^2 + 2s + 10$
$A = 1, \quad B = 2, \quad C = 10$

$$s = \frac{-B \pm \sqrt{B^2 - 4AC}}{2A}$$

$$= \frac{-2 \pm \sqrt{2^2 - 4(1)(10)}}{2(1)}$$

$$= \frac{-2 \pm \sqrt{4 - 40}}{2}$$

$$= \frac{-2 \pm \sqrt{-36}}{2}$$

$$= \frac{-2 \pm 6i}{2}$$

$$= \frac{2(-1 \pm 3i)}{2}$$

$$= -1 \pm 3i$$

13. $3a^2 - 2a + 5 = 0$
$A = 3, \quad B = -2, \quad C = 5$

$B^2 - 4AC$
$= (-2)^2 - 4(3)(5)$
$= 4 - 60$
$= -56 < 0$
The roots are not real.

15. $(3y + 5)(2y - 8) = (y - 4)(y + 1)$
$6y^2 - 14y - 40 = y^2 - 3y - 4$
$5y^2 - 11y - 36 = 0$
$A = 5, \quad B = -11, \quad C = -36$

$B^2 - 4AC$
$= (-11)^2 - 4(5)(-36)$
$= 121 + 720$
$= 841 > 0$
The roots are real and distinct.

17. $2a^2 + 4a = 0$
$A = 2, \quad B = 4, \quad C = 0$

$B^2 - 4AC$
$= 4^2 - 4(2)(0)$
$= 16 > 0$
The roots are real and distinct.

19. $(2y + 3)(y - 1) = y + 5$
$2y^2 + y - 3 = y + 5$
$2y^2 - 8 = 0$
$y^2 - 4 = 0$
$A = 1, \quad B = 0, \quad C = -4$

$B^2 - 4AC$
$= 0^2 - 4(1)(-4)$
$= 16 > 0$
The roots are real and distinct.

21. $a^2 - 3a - 4 = 0$
$(a - 4)(a + 1) = 0$

$a - 4 = 0$ or $a + 1 = 0$
$a = 4$ or $a = -1$

23. $8y^2 = 3$

$$y^2 = \frac{3}{8}$$

$$y = \pm\sqrt{\frac{3}{8}}$$

$$y = \pm\frac{\sqrt{3}}{2\sqrt{2}}$$

$$y = \pm\frac{\sqrt{6}}{4}$$

25. $(5x - 4)(2x - 3) = 0$

$5x - 4 = 0$ or $2x - 3 = 0$

$5x = 4$ $\quad$ $2x = 3$

$x = \frac{4}{5}$ or $x = \frac{3}{2}$

27. $(5x - 4) + (2x - 3) = 0$

$$7x - 7 = 0$$

$$7x = 7$$

$$x = 1$$

29. $(5x - 4)(2x - 3) = 17$

$10x^2 - 23x + 12 = 17$

$10x^2 - 23x - 5 = 0$

$(5x + 1)(2x - 5) = 0$

$5x + 1 = 0$ or $2x - 5 = 0$

$5x = -1$ $\quad$ $2x = 5$

$x = -\frac{1}{5}$ or $x = \frac{5}{2}$

31. $(a - 1)(a + 2) = -2$

$a^2 + a - 2 = -2$

$a^2 + a = 0$

$a(a + 1) = 0$

$a = 0$ or $a + 1 = 0$

$a = 0$ or $a = -1$

33. $2.4x^2 - 12.72x + 3.6 = 0$

$A = 2.4$, $B = -12.72$, $C = 3.6$

$$x = \frac{-B \pm \sqrt{B^2 - 4AC}}{2A}$$

$$= \frac{-(-12.72) \pm \sqrt{(-12.72)^2 - 4(2.4)(3.6)}}{2(2.4)}$$

$$= \frac{12.72 \pm \sqrt{127.2384}}{4.8}$$

$x = 0.3$ or $x = 5$

35. $x^2 - 3x + 5 = 0$

$A = 1$, $B = -3$, $C = 5$

$$x = \frac{-B \pm \sqrt{B^2 - 4AC}}{2A}$$

$$= \frac{-(-3) \pm \sqrt{(-3)^2 - 4(1)(5)}}{2(1)}$$

$$= \frac{3 \pm \sqrt{9 - 20}}{2}$$

$$= \frac{3 \pm \sqrt{-11}}{2}$$

$$= \frac{3 \pm i\sqrt{11}}{2}$$

37. $2s^2 - 5s - 12 = -5s$

$$2s^2 - 12 = 0$$

$$2s^2 = 12$$

$$s^2 = 6$$

$$s = \pm\sqrt{6}$$

39. $3x^2 - 2x + 9 = 2x^2 - 3x - 1$

$x^2 + x + 10 = 0$

$A = 1$, $B = 1$, $C = 10$

$$x = \frac{-B \pm \sqrt{B^2 - 4AC}}{2A}$$

$$= \frac{-1 \pm \sqrt{1^2 - 4(1)(10)}}{2(1)}$$

$$= \frac{-1 \pm \sqrt{1 - 40}}{2}$$

$$= \frac{-1 \pm \sqrt{-39}}{2}$$

$$= \frac{-1 \pm i\sqrt{39}}{2}$$

41. $x^2 + 3x - 8 = x^2 - x + 11$

$$3x - 8 = -x + 11$$

$$4x = 19$$

$$x = \frac{19}{4}$$

43. $3x^2 + 2.7x - 14.58 = 0$
$A = 3,\ B = 2.7,\ C = -14.58$

$$x = \frac{-B \pm \sqrt{B^2 - 4AC}}{2A}$$
$$= \frac{-2.7 \pm \sqrt{(2.7)^2 - 4(3)(-14.58)}}{2(3)}$$
$$= \frac{-2.7 \pm \sqrt{182.25}}{6}$$
$x = -2.7$ or $x = 1.8$

45. $3a^2 - 4a + 2 = 0$
$A = 3,\ B = -4,\ C = 2$

$$a = \frac{-B \pm \sqrt{B^2 - 4AC}}{2A}$$
$$= \frac{-(-4) \pm \sqrt{(-4)^2 - 4(3)(2)}}{2(3)}$$
$$= \frac{4 \pm \sqrt{16 - 24}}{6}$$
$$= \frac{4 \pm \sqrt{-8}}{6}$$
$$= \frac{4 \pm 2i\sqrt{2}}{6}$$
$$= \frac{2(2 \pm i\sqrt{2})}{6}$$
$$= \frac{2 \pm i\sqrt{2}}{3}$$

47. $(x - 4)(2x + 3) = x^2 - 4$
$2x^2 - 5x - 12 = x^2 - 4$
$x^2 - 5x - 8 = 0$
$A = 1,\ B = -5,\ C = -8$

$$x = \frac{-B \pm \sqrt{B^2 - 4AC}}{2A}$$
$$= \frac{-(-5) \pm \sqrt{(-5)^2 - 4(1)(-8)}}{2(1)}$$
$$= \frac{5 \pm \sqrt{25 + 32}}{2}$$
$$= \frac{5 \pm \sqrt{57}}{2}$$

49. $(a + 1)(a - 3) = (3a + 1)(a - 2)$
$a^2 - 2a - 3 = 3a^2 - 5a - 2$
$0 = 2a^2 - 3a + 1$
$0 = (2a - 1)(a - 1)$
$2a - 1 = 0$ or $a - 1 = 0$
$2a = 1$ $\quad a = 1$
$a = \frac{1}{2}$ or $a = 1$

51. $(z + 3)(z - 1) = (z - 2)^2$
$z^2 + 2z - 3 = z^2 - 4z + 4$
$2z - 3 = -4z + 4$
$6z - 3 = 4$
$6z = 7$
$z = \frac{7}{6}$

53. $$\frac{y}{y - 2} = \frac{y - 3}{y}$$
$$y(y - 2)\left(\frac{y}{y - 2}\right) = y(y - 2)\left(\frac{y - 3}{y}\right)$$
$y^2 = (y - 2)(y - 3)$
$y^2 = y^2 - 5y + 6$
$0 = -5y + 6$
$5y = 6$
$y = \frac{6}{5}$

55. $0.001x^2 - 2x = 0.1$
$0.001x^2 - 2x - 0.1 = 0$

$A = 0.001,\ B = -2,\ C = -0.1$

$$x = \frac{-B \pm \sqrt{B^2 - 4AC}}{2A}$$
$$= \frac{-(-2) \pm \sqrt{(-2)^2 - 4(0.001)(-0.1)}}{2(0.001)}$$
$$= \frac{2 \pm \sqrt{4.0004}}{0.002}$$

$x = -0.05$ or $x = 2000.05$

57.
$$\frac{3a}{a+1} + \frac{2}{a-2} = 5$$
$$(a+1)(a-2)\left(\frac{3a}{a+1} + \frac{2}{a-2}\right) = (a+1)(a-2)(5)$$
$$3a(a-2) + 2(a+1) = 5(a^2 - a - 2)$$
$$3a^2 - 6a + 2a + 2 = 5a^2 - 5a - 10$$
$$3a^2 - 4a + 2 = 5a^2 - 5a - 10$$
$$0 = 2a^2 - a - 12$$
$A = 2, \quad B = -1, \quad C = -12$
$$a = \frac{-B \pm \sqrt{B^2 - 4AC}}{2A}$$
$$= \frac{-(-1) \pm \sqrt{(-1)^2 - 4(2)(-12)}}{2(2)}$$
$$= \frac{1 \pm \sqrt{1 + 96}}{4}$$
$$= \frac{1 \pm \sqrt{97}}{4}$$

59.
$$\frac{3}{a+2} - \frac{5}{a-2} = 2$$
$$(a+2)(a-2)\left(\frac{3}{a+2} - \frac{5}{a-2}\right) = (a+2)(a-2)(2)$$
$$3(a-2) - 5(a+2) = 2(a^2 - 4)$$
$$3a - 6 - 5a - 10 = 2a^2 - 8$$
$$-2a - 16 = 2a^2 - 8$$
$$0 = 2a^2 + 2a + 8$$
$$0 = a^2 + a + 4$$
$A = 1, \quad B = 1, \quad C = 4$
$$a = \frac{-B \pm \sqrt{B^2 - 4AC}}{2A}$$
$$= \frac{-1 \pm \sqrt{1^2 - 4(1)(4)}}{2(1)}$$
$$= \frac{-1 \pm \sqrt{1 - 16}}{2}$$
$$= \frac{-1 \pm \sqrt{-15}}{2}$$
$$= \frac{-1 \pm i\sqrt{15}}{2}$$

61. Let x = 1st number,

then $\frac{40}{x}$ = 2nd number
$$x - \frac{40}{x} = 3$$
$$x\left(x - \frac{40}{x}\right) = x(3)$$
$$x^2 - 40 = 3x$$
$$x^2 - 3x - 40 = 0$$
$$(x - 8)(x + 5) = 0$$
$$x - 8 = 0 \quad \text{or} \quad x + 5 = 0$$
$$x = 8 \quad \text{or} \quad x = -5$$
$$\frac{40}{x} = 5 \qquad \frac{40}{x} = -8$$

The numbers are 8 and 5 or −5 and −8.

63. $a^2 + b^2 = c^2$
$$3^2 + 8^2 = c^2$$
$$9 + 64 = c^2$$
$$73 = c^2$$
$$\sqrt{73} = c$$

The hypotenuse is $\sqrt{73} \approx 8.54''$.

65.

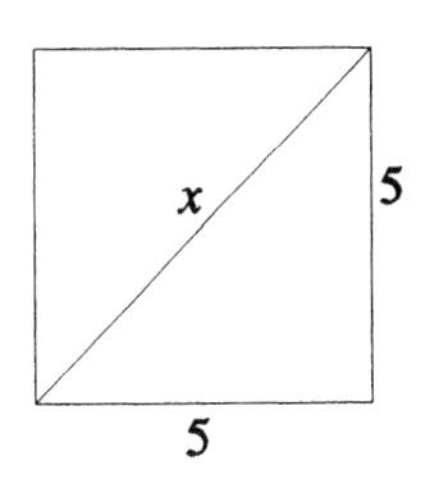

$$5^2 + 5^2 = x^2$$
$$25 + 25 = x^2$$
$$50 = x^2$$
$$\sqrt{50} = x$$
$$5\sqrt{2} = x$$

The diagonals are $5\sqrt{2} \approx 7.07''$.

67.

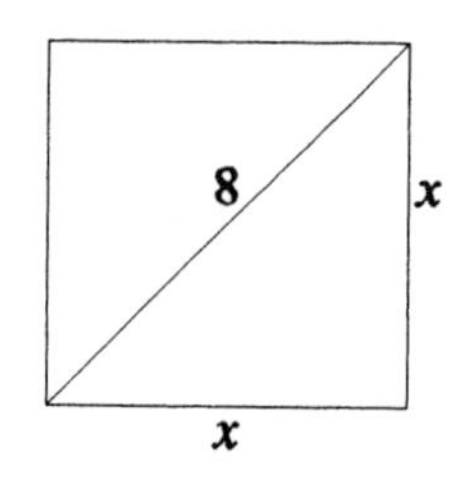

$$x^2 + x^2 = 8^2$$
$$2x^2 = 64$$
$$x^2 = 32$$
$$x = \sqrt{32}$$
$$x = 4\sqrt{2}$$

The sides are $4\sqrt{2} \approx 5.66''$.

69. (0, 1) and (c, c)

$$m_1 = \frac{c - 1}{c - 0} = \frac{c - 1}{c}$$

(0, 2) and (c, c)

$$m_2 = \frac{c - 2}{c - 0} = \frac{c - 2}{c}$$

$$m_1 = -\frac{1}{m_2}$$

$$\frac{c - 1}{c} = -\frac{c}{c - 2}$$

$$c(c - 2)\left(\frac{c - 1}{c}\right) = c(c - 2)\left(-\frac{c}{c - 2}\right)$$

$$(c - 2)(c - 1) = -c^2$$
$$c^2 - 3c + 2 = -c^2$$
$$2c^2 - 3c + 2 = 0$$

$$c = \frac{-(-3) \pm \sqrt{(-3)^2 - 4(2)(2)}}{2(2)}$$

$$= \frac{3 \pm \sqrt{9 - 16}}{4}$$

$$= \frac{3 \pm \sqrt{-7}}{4} = \frac{3 \pm i\sqrt{7}}{4}$$

Since c is not real, no solution.

71.

$$P(w) = 20w - w^2$$
$$-2400 = 20w - w^2$$
$$w^2 - 20w - 2400 = 0$$
$$(w - 60)(w + 40) = 0$$
$$w - 60 = 0 \quad \text{or} \quad w + 40 = 0$$
$$w = 60 \quad \text{or} \quad w = -40$$

Since w is positive, $w = 60$.

They sold 60 items.

73.

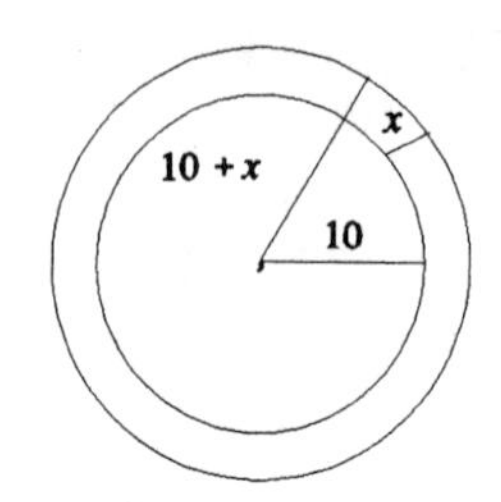

Area of path = area of outer circle − area of inner circle

$$44\pi = \pi(10 + x)^2 - \pi(10)^2$$
$$44\pi = \pi(100 + 20x + x^2) - 100\pi$$
$$44\pi = 100\pi + 20\pi x + \pi x^2 - 100\pi$$
$$44\pi = 20\pi x + \pi x^2$$
$$44 = 20x + x^2$$
$$0 = x^2 + 20x - 44$$
$$0 = (x + 22)(x - 2)$$
$$x + 22 = 0 \quad \text{or} \quad x - 2 = 0$$
$$x = -22 \quad \text{or} \quad x = 2$$

The path is 2 ft wide.

75. Let x = price per roll during 1st week,
then $x - 2$ = price per roll during 2nd week

$$(x - 2)\left(2250 - \frac{10000}{x}\right) = 10000$$

$$x\left[(x - 2)\left(2250 - \frac{10000}{x}\right)\right] = x(10000)$$

$$2250x(x - 2) - 10000(x - 2) = 10000x$$
$$2250x^2 - 4500x - 10000x + 20000 = 10000x$$
$$2250x^2 - 24500x + 20000 = 0$$
$$9x^2 - 98x + 80 = 0$$
$$(9x - 8)(x - 10) = 0$$

$$9x - 8 = 0 \quad \text{or} \quad x - 10 = 0$$
$$9x = 8 \qquad x = 10$$
$$x = \frac{8}{9} \quad \text{or} \quad x = 10$$

$$x - 2 = -\frac{10}{9} \qquad x - 2 = 8$$

Since both prices are positive, $x = 10$ and $x - 2 = 8$. The price during the 1st week was \$10 and during the 2nd week, \$8.

81. $(-27)^{-2/3} = \left[(-27)^{1/3}\right]^{-2}$
$$= (-3)^{-2}$$
$$= \frac{1}{(-3)^2}$$
$$= \frac{1}{9}$$

83. Let x = amount invested at 8%
and y = amount invested at 10%

$$\begin{cases} 0.08x + 0.10y = 730 \\ 0.08y + 0.10x = 710 \end{cases}$$

$$\begin{cases} 8x + 10y = 73000 \\ 10x + 8y = 71000 \end{cases}$$

$$-40x - 50y = -365000$$
$$\underline{40x + 32y = 284000}$$
$$-18y = -81000$$
$$y = 4500$$
$$8x + 10(4500) = 73000$$
$$8x + 45000 = 73000$$
$$8x = 28000$$
$$x = 3500$$
$$x + y = 3500 + 4500$$
$$= 8000$$

\$8000 was invested.

Exercises 8.5

Note: Be sure to check all solutions. Only the checks of those numbers leading to extraneous solutions have been shown in this manual.

1. $\sqrt{x} + 3 = 2x$
$$\sqrt{x} = 2x - 3$$
$$\left(\sqrt{x}\right)^2 = (2x - 3)^2$$
$$x = 4x^2 - 12x + 9$$
$$0 = 4x^2 - 13x + 9$$
$$0 = (4x - 9)(x - 1)$$
$$4x - 9 = 0 \quad \text{or} \quad x - 1 = 0$$
$$4x = 9 \qquad x = 1$$
$$x = \frac{9}{4}$$

When you check $x = 1$, you find it to be extraneous.

$$\sqrt{1} + 3 = 2(1)$$
$$1 + 3 = 2$$
$$4 = 2$$
False

The only solution is $\frac{9}{4}$.

3. $\sqrt{x + 5} = 7 - x$
$$\left(\sqrt{x + 5}\right)^2 = (7 - x)^2$$
$$x + 5 = 49 - 14x + x^2$$
$$0 = x^2 - 15x + 44$$
$$0 = (x - 4)(x - 11)$$
$$x - 4 = 0 \quad \text{or} \quad x - 11 = 0$$
$$x = 4 \quad \text{or} \quad x = 11$$

When you check $x = 11$, you find it to be extraneous.

$$\sqrt{11 + 5} = 7 - 11$$
$$\sqrt{16} = -4$$
$$4 = -4$$
False

The only solution is $x = 4$.

5. $\sqrt{5a - 1} + 5 = a$
$$\sqrt{5a - 1} = a - 5$$
$$\left(\sqrt{5a - 1}\right)^2 = (a - 5)^2$$
$$5a - 1 = a^2 - 10a + 25$$
$$0 = a^2 - 15a + 26$$
$$0 = (a - 2)(a - 13)$$
$$a - 2 = 0 \quad \text{or} \quad a - 13 = 0$$
$$a = 2 \quad \text{or} \quad a = 13$$

When you check $a = 2$, you find it to be extraneous.

$$\sqrt{5(2) - 1} + 5 = 2$$
$$\sqrt{9} + 5 = 2$$
$$3 + 5 = 2$$
$$8 = 2$$
False

The only solution is $a = 13$.

7. $\sqrt{a + 1} + a = 11$
$$\sqrt{a + 1} = 11 - a$$
$$\left(\sqrt{a + 1}\right)^2 = (11 - a)^2$$
$$a + 1 = 121 - 22a + a^2$$
$$0 = a^2 - 23a + 120$$
$$0 = (a - 8)(a - 15)$$
$$a - 8 = 0 \text{ or } a - 15 = 0$$
$$a = 8 \text{ or } a = 15$$

When you check $a = 15$, you find it to be extraneous.

$$\sqrt{15 + 1} + 15 = 11$$
$$\sqrt{16} + 15 = 11$$
$$4 + 15 = 11$$
$$19 = 11$$
False

The only solution is $a = 8$.

9. $\sqrt{3x + 1} + 3 = x$
$$\sqrt{3x + 1} = x - 3$$
$$\left(\sqrt{3x + 1}\right)^2 = (x - 3)^2$$
$$3x + 1 = x^2 - 6x + 9$$
$$0 = x^2 - 9x + 8$$
$$0 = (x - 8)(x - 1)$$
$$x - 8 = 0 \text{ or } x - 1 = 0$$
$$x = 8 \text{ or } x = 1$$

When you check $x = 1$, you find it to be extraneous.

$$\sqrt{3(1) + 1} + 3 = 1$$
$$\sqrt{4} + 3 = 1$$
$$2 + 3 = 1$$
$$5 = 1$$
False

The only solution is $x = 8$.

11. $5a - 2\sqrt{a + 3} = 2a - 1$
$$-2\sqrt{a + 3} = -3a - 1$$
$$2\sqrt{a + 3} = 3a + 1$$
$$\left(2\sqrt{a + 3}\right)^2 = (3a + 1)^2$$
$$4(a + 3) = 9a^2 + 6a + 1$$
$$4a + 12 = 9a^2 + 6a + 1$$
$$0 = 9a^2 + 2a - 11$$
$$0 = (9a + 11)(a - 1)$$
$$9a + 11 = 0 \text{ or } a - 1 = 0$$
$$a = -\frac{11}{9} \text{ or } a = 1$$

When you check $a = -\frac{11}{9}$, you find it to be extraneous.

$$5\left(-\frac{11}{9}\right) - 2\sqrt{-\frac{11}{9} + 3} = 2\left(-\frac{11}{9}\right) - 1$$

$$-\frac{55}{4} - 2\sqrt{-\frac{2}{9}} = -\frac{22}{9} - 1$$

$\sqrt{-\frac{2}{9}}$ is not a real number.

The only solution is $a = 1$.

13. $\sqrt{y + 3} = 1 + \sqrt{y}$
$$\left(\sqrt{y + 3}\right)^2 = \left(1 + \sqrt{y}\right)^2$$
$$y + 3 = 1 + 2\sqrt{y} + y$$
$$2 = 2\sqrt{y}$$
$$1 = \sqrt{y}$$
$$1^2 = \left(\sqrt{y}\right)^2$$
$$1 = y$$

15. $\sqrt{a + 7} = 1 + \sqrt{2a}$
$$\left(\sqrt{a + 7}\right)^2 = \left(1 + \sqrt{2a}\right)^2$$
$$a + 7 = 1 + 2\sqrt{2a} + 2a$$
$$-a + 6 = 2\sqrt{2a}$$
$$(-a + 6)^2 = \left(2\sqrt{2a}\right)^2$$
$$a^2 - 12a + 36 = 4(2a)$$
$$a^2 - 12a + 36 = 8a$$
$$a^2 - 20a + 36 = 0$$
$$(a - 2)(a - 18) = 0$$

$$a - 2 = 0 \quad \text{or} \quad a - 18 = 0$$
$$a = 2 \quad \text{or} \quad a = 18$$

When you check $a = 18$, you find it to be extraneous.

$$\sqrt{18 + 7} = 1 + \sqrt{2(18)}$$
$$\sqrt{25} = 1 + \sqrt{36}$$
$$5 = 1 + 6$$
$$5 = 7$$
False

The only solution is $a = 2$.

17. $\sqrt{7s + 1} - 2\sqrt{s} = 2$
$$\sqrt{7s + 1} = 2 + 2\sqrt{s}$$
$$\left(\sqrt{7s + 1}\right)^2 = \left(2 + 2\sqrt{s}\right)^2$$
$$7s + 1 = 4 + 8\sqrt{s} + 4s$$
$$3s - 3 = 8\sqrt{s}$$
$$(3s - 3)^2 = \left(8\sqrt{s}\right)^2$$
$$9s^2 - 18s + 9 = 64s$$
$$9s^2 - 82s + 9 = 0$$
$$(9s - 1)(s - 9) = 0$$
$$9s - 1 = 0 \quad \text{or} \quad s - 9 = 0$$
$$s = \frac{1}{9} \quad \text{or} \quad s = 9$$

When you check $s = \frac{1}{9}$, you find it to be extraneous.

$$\sqrt{7\left(\frac{1}{9}\right) + 1} - 2\sqrt{\frac{1}{9}} = 2$$
$$\frac{4}{3} - 2\left(\frac{1}{3}\right) = 2$$
$$\frac{2}{3} = 2$$
False

The only solution is $s = 9$.

19. $\sqrt{7 - a} - \sqrt{3 + a} = 2$
$$\sqrt{7 - a} = 2 + \sqrt{3 + a}$$
$$\left(\sqrt{7 - a}\right)^2 = \left(2 + \sqrt{3 + a}\right)^2$$
$$7 - a = 4 + 4\sqrt{3 + a} + 3 + a$$
$$7 - a = 7 + a + 4\sqrt{3 + a}$$
$$-2a = 4\sqrt{3 + a}$$
$$(-2a)^2 = \left(4\sqrt{3 + a}\right)^2$$
$$4a^2 = 16(3 + a)$$
$$4a^2 = 48 + 16a$$
$$4a^2 - 16a - 48 = 0$$
$$a^2 - 4a - 12 = 0$$
$$(a - 6)(a + 2) = 0$$
$$a - 6 = 0 \quad \text{or} \quad a + 2 = 0$$
$$a = 6 \quad \text{or} \quad a = -2$$

When you check $a = 6$, you find it to be extraneous.

$$\sqrt{7 - 6} - \sqrt{3 + 6} = 2$$
$$\sqrt{1} - \sqrt{9} = 2$$
$$1 - 3 = 2$$
$$-2 = 2$$
False

The only solution is $a = -2$.

21. $\sqrt{x} + a = b$
$$\sqrt{x} = b - a$$
$$\left(\sqrt{x}\right)^2 = (b - a)^2$$
$$x = (b - a)^2$$

23. $\dfrac{\sqrt{\pi L}}{g} = T$
$$\sqrt{\pi L} = gT$$
$$\left(\sqrt{\pi L}\right)^2 = (gT)^2$$
$$\pi L = g^2T^2$$
$$L = \frac{g^2T^2}{\pi}$$

25. $\sqrt{5x + b} = 6 + b$
$$\left(\sqrt{5x + b}\right)^2 = (6 + b)^2$$
$$5x + b = 36 + 12b + b^2$$
$$5x = b^2 + 11b + 36$$
$$x = \frac{b^2 + 11b + 36}{5}$$

27. $$t = \frac{\overline{X} - a}{\frac{s}{\sqrt{n}}}$$

$$\frac{s}{\sqrt{n}}(t) = \frac{s}{\sqrt{n}}\left(\frac{\overline{X} - a}{\frac{s}{\sqrt{n}}}\right)$$

$$\frac{st}{\sqrt{n}} = \overline{X} - a$$

$$st = \sqrt{n}(\overline{X} - a)$$

$$\frac{st}{\overline{X} - a} = \sqrt{n}$$

$$\left(\frac{st}{\overline{X} - a}\right)^2 = (\sqrt{n})^2$$

$$\left(\frac{st}{\overline{X} - a}\right)^2 = n$$

29. $x^3 - 2x^2 - 15x = 0$
$x(x - 5)(x + 3) = 0$
$x = 0$ or $x - 5 = 0$ or $x + 3 = 0$
$x = 0$ or $x = 5$ or $x = -3$

31. $6a^3 - a^2 - 2a = 0$
$a(3a - 2)(2a + 1) = 0$
$a = 0$ or $3a - 2 = 0$ or $2a + 1 = 0$
$a = 0$ or $a = \frac{2}{3}$ or $a = -\frac{1}{2}$

33. $y^4 - 17y^2 + 16 = 0$
$(y^2 - 16)(y^2 - 1) = 0$
$y^2 - 16 = 0$ or $y^2 - 1 = 0$
$y^2 = 16$ $\quad y^2 = 1$
$y = \pm 4$ or $y = \pm 1$

35. $3a^4 + 24 = 18a^2$
$3a^4 - 18a^2 + 24 = 0$
$a^4 - 6a^2 + 8 = 0$
$(a^2 - 4)(a^2 - 2) = 0$
$a^2 - 4 = 0$ or $a^2 - 2 = 0$
$a^2 = 4$ $\quad a^2 = 2$
$a = \pm 2$ or $a = \pm\sqrt{2}$

37. $b^4 + 112 = 23b^2$
$b^4 - 23b^2 + 112 = 0$
$(b^2 - 7)(b^2 - 16) = 0$
$b^2 - 7 = 0$ or $b^2 - 16 = 0$
$b^2 = 7$ $\quad b^2 = 16$
$b = \pm\sqrt{7}$ or $b = \pm 4$

39. $$9 - \frac{8}{x^2} = x^2$$

$$x^2\left(9 - \frac{8}{x^2}\right) = x^2(x^2)$$

$9x^2 - 8 = x^4$
$0 = x^4 - 9x^2 + 8$
$0 = (x^2 - 8)(x^2 - 1)$
$x^2 - 8 = 0$ or $x^2 - 1 = 0$
$x^2 = 8$ $\quad x^2 = 1$
$x = \pm 2\sqrt{2}$ or $x = \pm 1$

41. $x^3 + x^2 - x - 1 = 0$
$(x^2 - 1)(x + 1) = 0$
$x^2 - 1 = 0$ or $x + 1 = 0$
$x^2 = 1$ $\quad x = -1$
$x = \pm 1$

43. $x^{2/3} - 4 = 0$
$x^{2/3} = 4$
$(x^{2/3})^3 = 4^3$
$x^2 = 64$
$x = \pm 8$

45. $x + x^{1/2} - 6 = 0$
$(x^{1/2})^2 + x^{1/2} - 6 = 0$
Let $u = x^{1/2}$.

$u^2 + u - 6 = 0$
$(u + 3)(u - 2) = 0$
$u + 3 = 0$ or $u - 2 = 0$
$u = -3$ or $u = 2$
$x^{1/2} = -3$ $\quad x^{1/2} = 2$
No solution $\quad (x^{1/2})^2 = 2^2$
$x = 4$

The solution is $x = 4$.

47. $y^{2/3} - 4y^{1/3} - 5 = 0$
$(y^{1/3})^2 - 4y^{1/3} - 5 = 0$
Let $u = y^{1/3}$.

$u^2 - 4u - 5 = 0$
$(u - 5)(u + 1) = 0$
$u - 5 = 0$ or $u + 1 = 0$
$u = 5$ $\quad u = -1$
$y^{1/3} = 5$ $\quad y^{1/3} = -1$
$y = 5^3$ $\quad y = (-1)^3$
$y = 125$ or $y = -1$

49. $x^{1/2} + 8x^{1/4} + 7 = 0$
$(x^{1/4})^2 + 8x^{1/4} + 7 = 0$
Let $u = x^{1/4}$.

$u^2 + 8u + 7 = 0$
$(u + 7)(u + 1) = 0$
$u + 7 = 0$ or $u + 1 = 0$
$u = -7$ $\quad u = -1$
$x^{1/4} = -7$ $\quad x^{1/4} = -1$
No solution $\quad$ No solution

No solution

51. $x^{-2} - 5x^{-1} + 6 = 0$
$(x^{-1})^2 - 5x^{-1} + 6 = 0$
Let $u = x^{-1}$.

$u^2 - 5u + 6 = 0$
$(u - 2)(u - 3) = 0$
$u - 2 = 0$ or $u - 3 = 0$
$u = 2$ $\quad u = 3$
$x^{-1} = 2$ $\quad x^{-1} = 3$
$(x^{-1})^{-1} = 2^{-1}$ $\quad (x^{-1})^{-1} = 3^{-1}$
$x = \frac{1}{2}$ or $x = \frac{1}{3}$

53. $6x^{-2} + x^{-1} - 1 = 0$
$6(x^{-1})^2 + x^{-1} - 1 = 0$
Let $u = x^{-1}$.
$6u^2 + u - 1 = 0$
$(3u - 1)(2u + 1) = 0$
$3u - 1 = 0$ or $2u + 1 = 0$
$u = \frac{1}{3}$ $\quad u = -\frac{1}{2}$
$x^{-1} = \frac{1}{3}$ $\quad x^{-1} = -\frac{1}{2}$
$(x^{-1})^{-1} = \left(\frac{1}{3}\right)^{-1}$ $\quad (x^{-1})^{-1} = \left(-\frac{1}{2}\right)^{-1}$
$x = 3$ $\quad x = -2$

55. $x^{-4} - 13x^{-2} + 36 = 0$
$(x^{-2})^2 - 13x^{-2} + 36 = 0$
Let $u = x^{-2}$.

$u^2 - 13u + 36 = 0$
$(u - 9)(u - 4) = 0$
$u - 9 = 0$ or $u - 4 = 0$
$u = 9$ $\quad u = 4$
$x^{-2} = 9$ $\quad x^{-2} = 4$
$(x^{-2})^{-1/2} = \pm 9^{-1/2}$ $\quad (x^{-2})^{-1/2} = \pm 4^{-1/2}$

$x = \pm\frac{1}{3}$ $\quad x = \pm\frac{1}{2}$

57. $\sqrt{a} - \sqrt[4]{a} - 6 = 0$
$a^{1/2} - a^{1/4} - 6 = 0$
$(a^{1/4})^2 - a^{1/4} - 6 = 0$
Let $u = a^{1/4}$.

$u^2 - u - 6 = 0$
$(u - 3)(u + 2) = 0$
$u - 3 = 0$ or $u + 2 = 0$
$u = 3$ $\quad u = -2$
$a^{1/4} = 3$ $\quad a^{1/4} = -2$
$(a^{1/4})^4 = 3^4$ $\quad$ No solution
$a = 81$

The only solution is $a = 81$.

59. $\sqrt{x} - 4\sqrt[4]{x} = 5$
$x^{1/2} - 4x^{1/4} = 5$
$(x^{1/4})^2 - 4x^{1/4} - 5 = 0$
Let $u = x^{1/4}$

$u^2 - 4u - 5 = 0$
$(u - 5)(u + 1) = 0$
$u - 5 = 0$ or $u + 1 = 0$
$u = 5$ $\quad u = -1$
$(x^{1/4})^4 = 5^4$ $\quad$ No solution
$x = 625$

The only solution is 625.

61. $(a + 4)^2 + 6(a + 4) + 9 = 0$
Let $u = a + 4$.

$u^2 + 6u + 9 = 0$
$(u + 3)^2 = 0$
$u + 3 = \pm\sqrt{0}$
$u + 3 = 0$
$u = -3$
$a + 4 = -3$
$a = -7$

63. $2(3x + 1)^2 - 5(3x + 1) - 3 = 0$
Let $u = 3x + 1$.

$$2u^2 - 5u - 3 = 0$$
$$(2u + 1)(u - 3) = 0$$
$$2u + 1 = 0 \quad \text{or} \quad u - 3 = 0$$
$$u = -\frac{1}{2} \qquad u = 3$$
$$3x + 1 = -\frac{1}{2} \qquad 3x + 1 = 3$$
$$3x = -\frac{3}{2} \qquad 3x = 2$$
$$x = -\frac{1}{2} \qquad x = \frac{2}{3}$$

65. $(x^2 + x)^2 - 4 = 0$
$$(x^2 + x)^2 = 4$$
$$x^2 + x = \pm 2$$

$$x^2 + x = 2$$
$$x^2 + x - 2 = 0$$
$$(x + 2)(x - 1) = 0$$
$$x + 2 = 0 \quad \text{or} \quad x - 1 = 0$$
$$x = -2 \quad \text{or} \quad x = 1$$

or
$$x^2 + x = -2$$
$$x^2 + x + 2 = 0$$
$$x = \frac{-1 \pm \sqrt{1^2 - 4(1)(2)}}{2(1)}$$
$$= \frac{-1 \pm \sqrt{1 - 8}}{2}$$
$$= \frac{-1 \pm \sqrt{-7}}{2}$$
$$= \frac{-1 \pm i\sqrt{7}}{2}$$

The solutions are -2, 1, $\frac{-1 \pm i\sqrt{7}}{2}$.

67. $\left(a - \frac{10}{a}\right)^2 - 12\left(a - \frac{10}{a}\right) + 27 = 0$

Let $u = a - \frac{10}{a}$.

$$u^2 - 12u + 27 = 0$$
$$(u - 9)(u - 3) = 0$$
$$u - 9 = 0$$
$$u = 9$$
$$a - \frac{10}{a} = 9$$
$$a^2 - 10 = 9a$$
$$a^2 - 9a - 10 = 0$$
$$(a - 10)(a + 1) = 0$$
$$a - 10 = 0 \quad \text{or} \quad a + 1 = 0$$
$$a = 10 \quad \text{or} \quad a = -1$$

or
$$u - 3 = 0$$
$$u = 3$$
$$a - \frac{10}{a} = 3$$
$$a^2 - 10 = 3a$$
$$a^2 - 3a - 10 = 0$$
$$(a - 5)(a + 2) = 0$$
$$a - 5 = 0 \quad \text{or} \quad a + 2 = 0$$
$$a = 5 \quad \text{or} \quad a = -2$$

$a = 10$ or $a = -1$ or $a = 5$ or $a = -2$

69.
$$s_e = s_y\sqrt{1 - r^2}$$
$$1.4 = 2.2\sqrt{1 - r^2}$$
$$(1.4)^2 = \left(2.2\sqrt{1 - r^2}\right)^2$$
$$1.96 = 4.84(1 - r^2)$$
$$0.4050 = 1 - r^2$$
$$-0.595 = -r^2$$
$$0.595 = r^2$$
$$\pm 0.77 = r$$

73. $2x + 5y - 8 = 0$
$$5y = -2x + 8$$
$$y = -\frac{2}{5}x + \frac{8}{5}$$
$$m = -\frac{2}{5}$$

75. $\frac{\text{winner}}{\text{loser}} = \frac{8}{5} = \frac{875{,}400}{x}$

$$8x = 5(875{,}400)$$
$$8x = 4377000$$
$$x = 547125$$

Total number of votes = 875,400 + 547,125
= 1,422,525 votes

Exercises 8.6

1. $y = 3x^2$

x	y
-2	12
-1	3
0	0
1	3
2	12

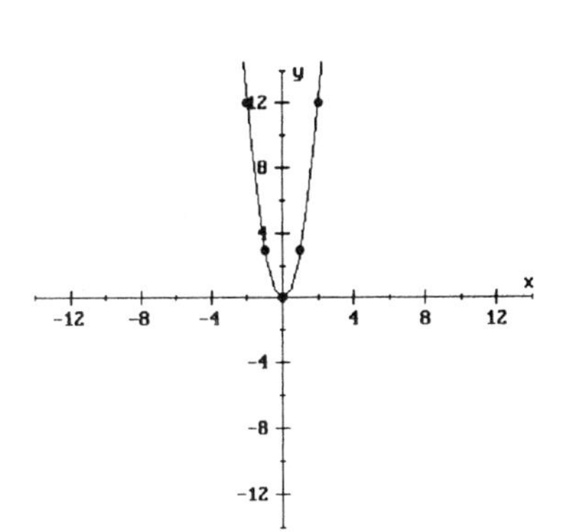

3. $y = \frac{1}{2}x^2$

x	y
-2	2
-1	1/2
0	0
1	1/2
2	2

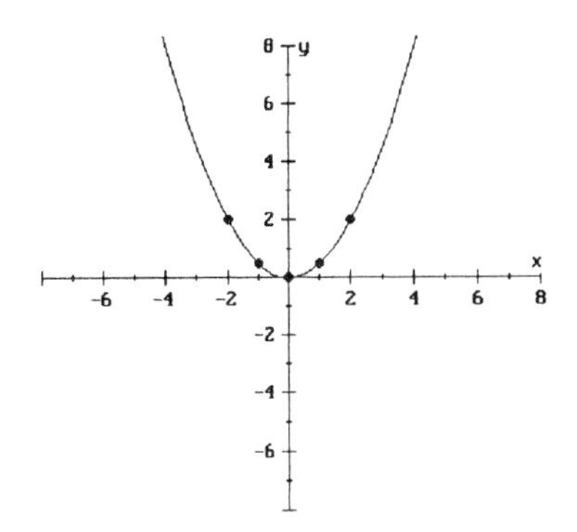

5. $y = -4x^2$

x	y
-2	-16
-1	-4
0	0
1	-4
2	-16

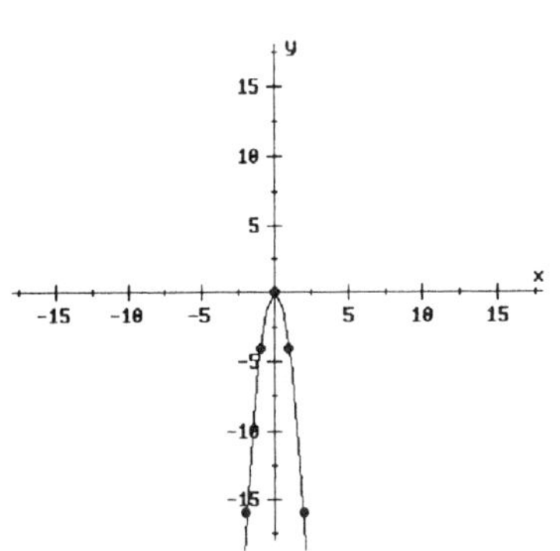

7. $6x^2 - y = 0$
$6x^2 = y$

x	y
-2	24
-1	6
0	0
1	6
2	24

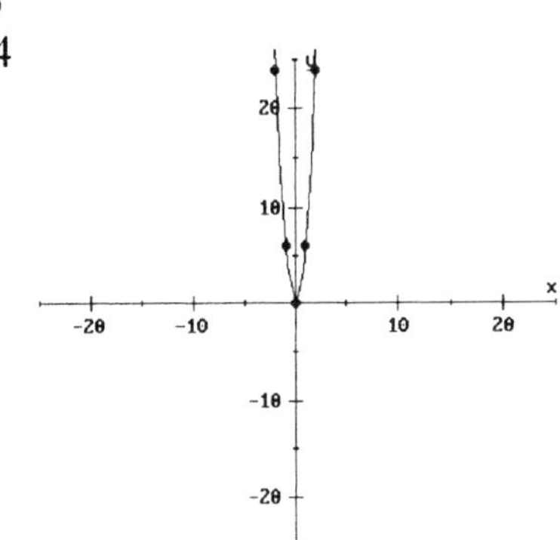

9. $y = x^2 + 4$

Shift the graph of $y = x^2$ up 4 units.

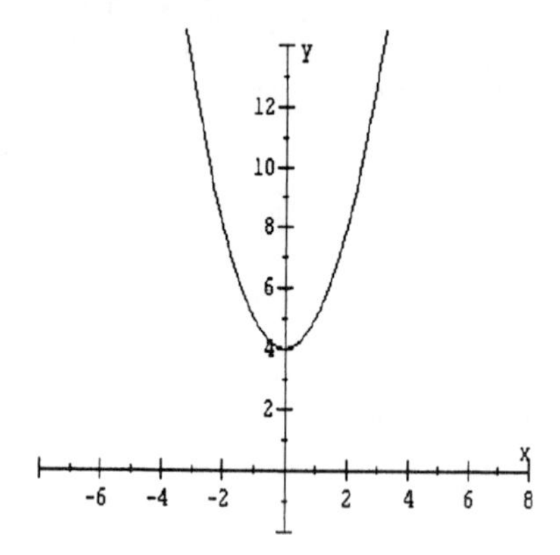

11. $y = -2x^2 + 1$

Shift the graph of $y = -2x^2$ up 1 unit.

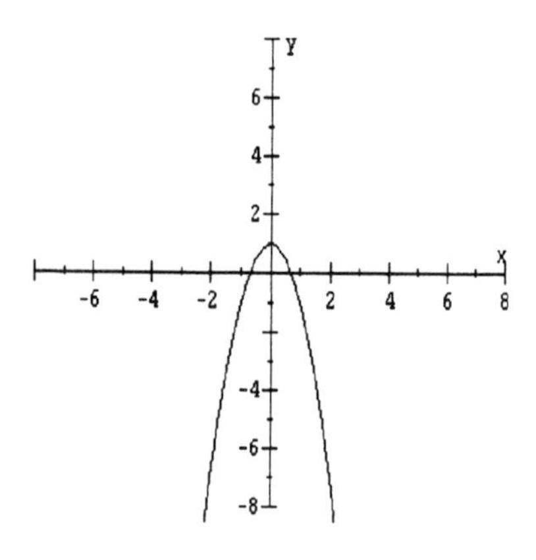

13. $y = (x - 4)^2$

Shift the graph of $y = x^2$ to the right 4 units.

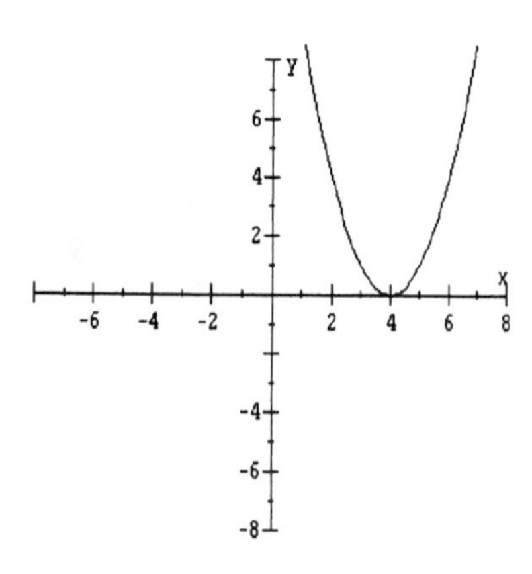

15. $y = -(x + 2)^2$

Shift the graph of $y = -x^2$ to the left 2 units.

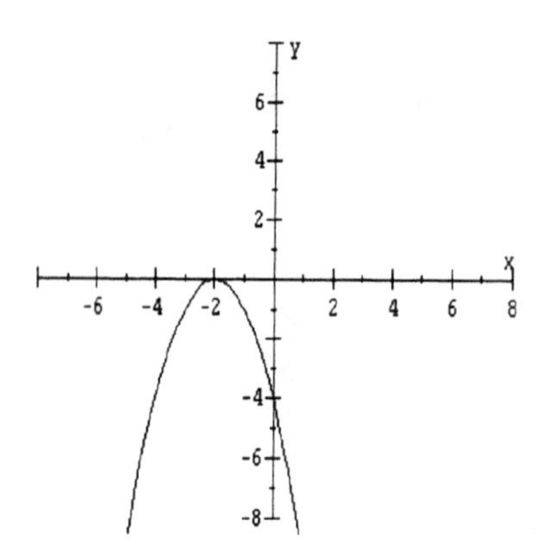

17. $y = (x - 3)^2 + 4$
Vertex: $(3, 4)$
Axis of symmetry: $x = 3$

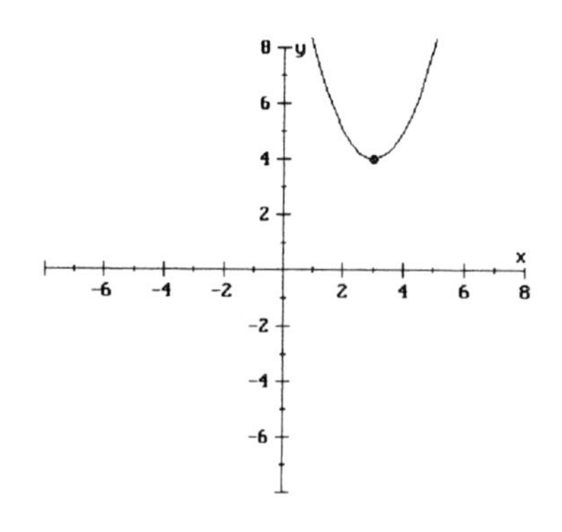

19. $y = -3(x - 1)^2 - 8$
Vertex: $(1, -8)$
Axis of symmetry: $x = 1$

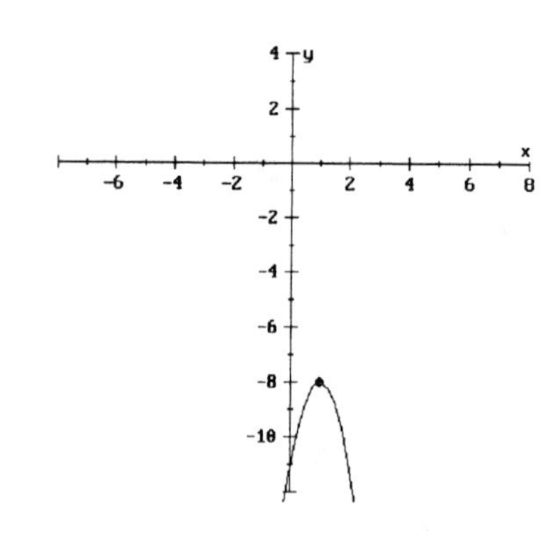

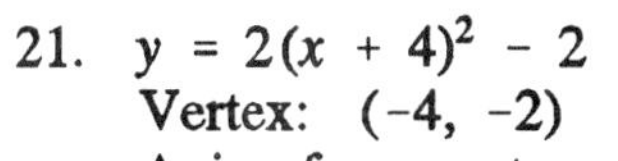
21. $y = 2(x + 4)^2 - 2$
Vertex: $(-4, -2)$
Axis of symmetry: $x = -4$

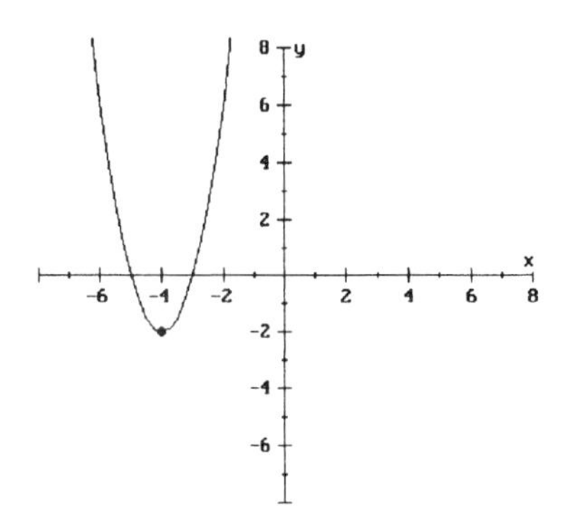

23. $y = 2x^2 - 12x + 19$
$y = 2(x^2 - 6x) + 19$
$y = 2(x^2 - 6x + 9) + 19 - 18$
$y = 2(x - 3)^2 + 1$
Vertex: $(3, 1)$
Axis of symmetry: $x = 3$

25. $y = -x^2 + 4x - 1$
$y = -(x^2 - 4x) - 1$
$y = -(x^2 - 4x + 4) - 1 + 4$
$y = -(x - 2)^2 + 3$
Vertex: $(2, 3)$
Axis of symmetry: $x = 2$

27. $y = -3x^2 + 30x - 70$
$y = -3(x^2 - 10x) - 70$
$y = -3(x^2 - 10x + 25) - 70 + 75$
$y = -3(x - 5)^2 + 5$
Vertex: $(5, 5)$
Axis of symmetry: $x = 5$

29. $y = x^2 - 4$
$y = (x - 0)^2 - 4$

Vertex: $(0, -4)$
Axis of symmetry: $x = 0$
x-intercepts: Let $y = 0$
$0 = x^2 - 4$
$4 = x^2$
$\pm 2 = x$

y-intercept: Let $x = 0$
$y = 0^2 - 4$
$y = -4$

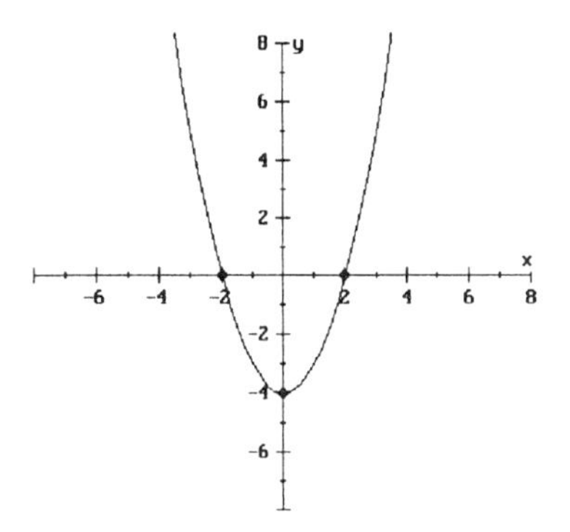

31. $y = 2x^2 + 8$
$y = 2(x - 0)^2 + 8$

Vertex: $(0, 8)$
Axis of symmetry: $x = 0$
x-intercepts: Let $y = 0$
$0 = 2x^2 + 8$
$-8 = 2x^2$
$-4 = x^2$
none

y-intercept: Let $x = 0$
$y = 2(0)^2 + 8$
$y = 8$

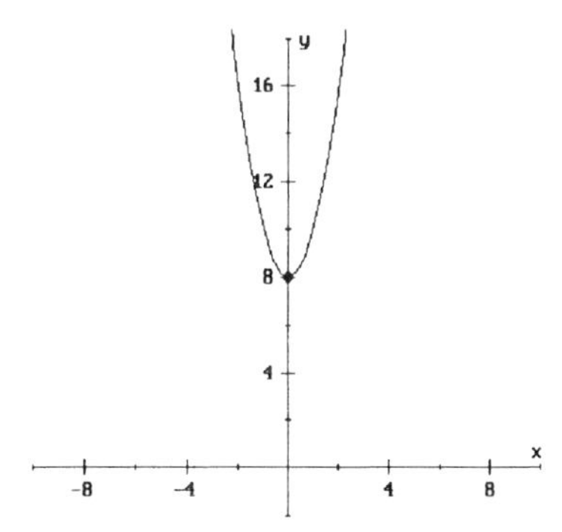

33. $y = -x^2 - 10x - 25$
$y = -(x^2 + 10x) - 25$
$y = -(x^2 + 10x + 25) - 25 + 25$
$y = -(x + 5)^2$

Vertex: $(-5, 0)$
Axis of symmetry: $x = -5$
x-intercepts: Let $y = 0$
$0 = -(x + 5)^2$
$0 = (x + 5)^2$
$0 = x + 5$
$-5 = x$

y-intercept: Let $x = 0$
$y = -0^2 - 10(0) - 25$
$y = -25$

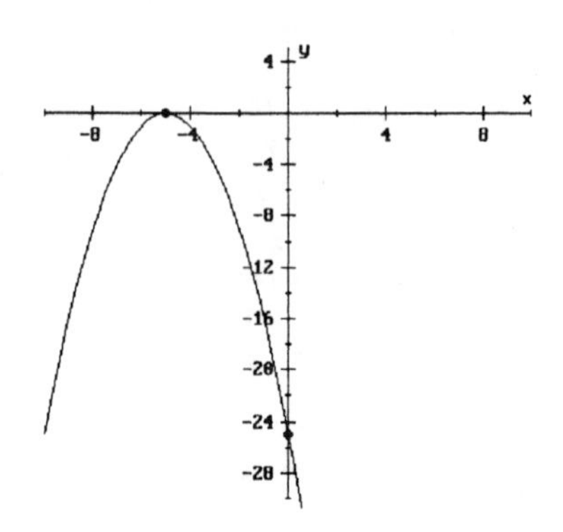

35. $y = x^2 - 2x - 35$
$y = (x^2 - 2x + 1) - 35 - 1$
$y = (x - 1)^2 - 36$

Vertex: $(1, -36)$
Axis of symmetry: $x = 1$
x-intercepts: Let $y = 0$
$0 = (x - 1)^2 - 36$
$36 = (x - 1)^2$
$\pm 6 = x - 1$
$x - 1 = 6$ or $x - 1 = -6$
$x = 7$ or $x = -5$

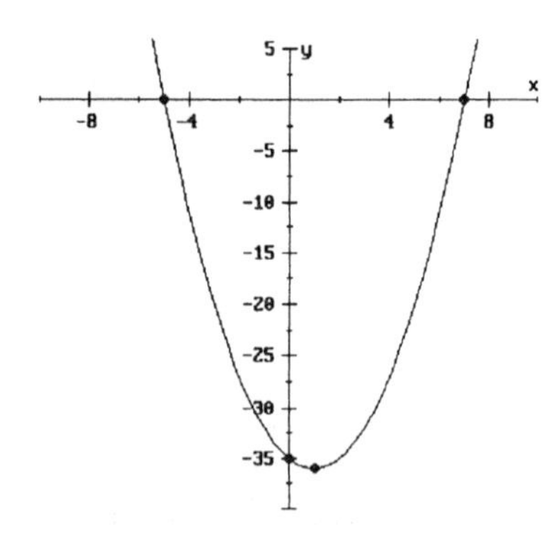

y-intercept: Let $x = 0$
$y = 0^2 - 2(0) - 35$
$y = -35$

37. $y = 2x^2 - 4x + 4$
$y = 2(x^2 - 2x) + 4$
$y = 2(x^2 - 2x + 1) + 4 - 2$
$y = 2(x - 1)^2 + 2$

Vertex: $(1, 2)$
Axis of symmetry: $x = 1$
x-intercepts: Let $y = 0$
$0 = 2(x - 1)^2 + 2$
$-2 = 2(x - 1)^2$
$-1 = (x - 1)^2$
none

y-intercept: Let $x = 0$
$y = 2(0)^2 - 4(0) + 4$
$y = 4$

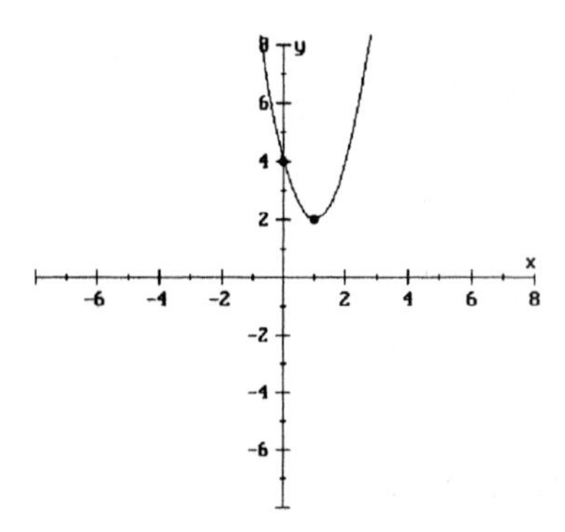

39. $y = -x^2 - 3x - 4$
$y = -(x^2 + 3x) - 4$

$$y = -\left(x^2 + 3x + \frac{9}{4}\right) - 4 + \frac{9}{4}$$

$$y = -\left(x + \frac{3}{2}\right)^2 - \frac{7}{4}$$

Vertex: $\left(-\frac{3}{2}, -\frac{7}{4}\right)$

Axis of symmetry: $x = -\frac{3}{2}$

x-intercepts: Let $y = 0$

$$0 = -\left(x + \frac{3}{2}\right)^2 - \frac{7}{4}$$

$$\left(x + \frac{3}{2}\right)^2 = -\frac{7}{4}$$

none

y-intercept: Let $x = 0$
$y = -0^2 - 3(0) - 4$
$y = -4$

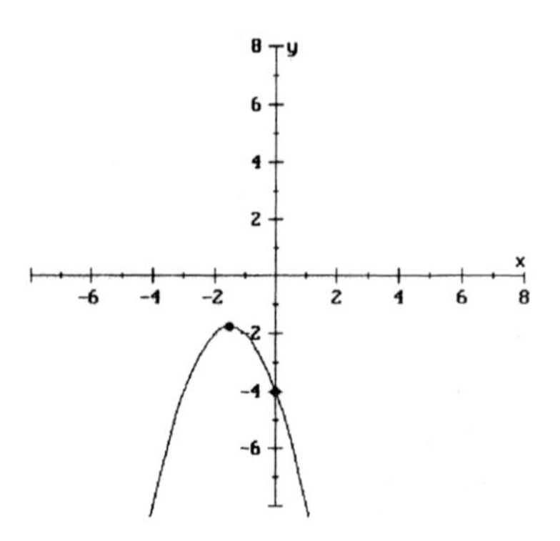

41. $y = 2x^2 - 3x + 2$

$y = 2\left(x^2 - \frac{3}{2}x\right) + 2$

$y = 2\left(x^2 - \frac{3}{2}x + \frac{9}{16}\right) + 2 - \frac{9}{8}$

$y = 2\left(x - \frac{3}{4}\right)^2 + \frac{7}{8}$

Vertex: $\left(\frac{3}{4}, \frac{7}{8}\right)$

Axis of symmetry: $x = \frac{3}{4}$

x-intercepts: Let $y = 0$

$0 = 2\left(x - \frac{3}{4}\right)^2 + \frac{7}{8}$

$-\frac{7}{8} = 2\left(x - \frac{3}{4}\right)^2$

$-\frac{7}{16} = \left(x - \frac{3}{4}\right)^2$

none

y-intercept: Let $x = 0$

$y = 2(0)^2 - 3(0) + 2$

$y = 2$

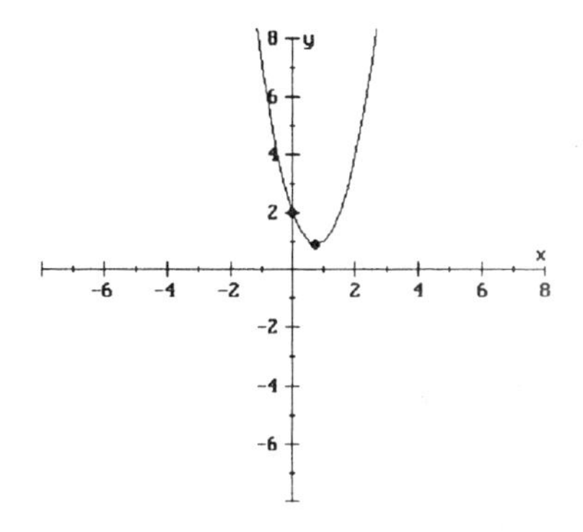

43. $y = -2x^2 + 4x - 1$

$y = -2(x^2 - 2x) - 1$

$y = -2(x^2 - 2x + 1) - 1 + 2$

$y = -2(x - 1)^2 + 1$

Vertex: $(1, 1)$

Axis of symmetry: $x = 1$

x-intercepts: Let $y = 0$

$0 = -2(x - 1)^2 + 1$

$2(x - 1)^2 = 1$

$(x - 1)^2 = \frac{1}{2}$

$x - 1 = \pm\sqrt{\frac{1}{2}}$

$x = 1 \pm \frac{\sqrt{2}}{2}$

y-intercept: Let $x = 0$

$y = -2(0)^2 + 4(0) - 1$

$y = -1$

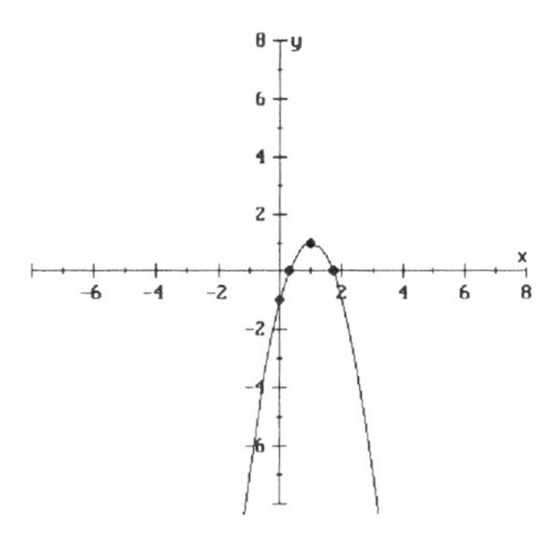

45. $P(x) = -x^2 + 70x$

$P(x) = -(x^2 - 70x)$

$P(x) = -(x^2 - 70x + 1225) + 1225$

$P(x) = -(x - 35)^2 + 1225$

Vertex: (35, 1225), maximum

3500 bagels must be produced and the maximum profit is $1225.

47. $M(t) = -4.185t^2 + 703.65t - 27132.38$

$t = \frac{-B}{2A} = \frac{-703.65}{2(-4.185)} \approx 84$

The number of marriages was at a maximum in 1984.

49. $P(x) = -x^2 + 112x - 535$

$P(x) = -(x^2 - 112x) - 535$

$P(x) = -(x^2 - 112x + 3136) - 535 + 3136$

$P(x) = -(x - 56)^2 + 2601$

Vertex: (56, 2601), maximum

56 cases of candy canes made daily will yield a maximum profit of $2601.

51. $y = -16t^2 + 400t$

$y = -16(t^2 - 25t)$

$y = -16\left(t^2 - 25t + \frac{625}{4}\right) + 2500$

$y = -16\left(t - \frac{25}{2}\right)^2 + 2500$

Vertex: $\left(\frac{25}{2}, 2500\right)$, maximum

It will take $\frac{25}{2} = 12.5$ sec to reach a maximum height of 2500 feet.

53.

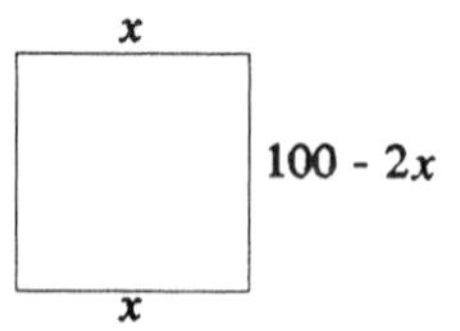

$A = x(100 - 2x)$

$A = 100x - 2x^2$

$A = -2(x^2 - 50x)$

$A = -2(x^2 - 50x + 625) + 1250$

$A = -2(x - 25)^2 + 1250$

Vertex: (25, 1250), maximum

$x = 25$ and $100 - 2x = 100 - 2(25) = 50$

The dimensions are 25 ft and 50 ft.

55. $R = xp$

$R = x\left(10 - \frac{x}{6}\right)$

$R = 10x - \frac{x^2}{6}$

$R = -\frac{1}{6}(x^2 - 60x)$

$R = -\frac{1}{6}(x^2 - 60x + 900) + 150$

$R = -\frac{1}{6}(x - 30)^2 + 150$

Vertex: (30, 150), maximum
The maximum revenue is $150 when 30 items are sold.

57. (a) $R(x) = (200000 - 7250x)(3 + 0.25x)$

(b) $R(x) = 600000 + 28250x - 1812.5x^2$

$R(x) = -1812.5(x^2 - 15.59x) + 600000$

$R(x) = -1812.5(x^2 - 15.59x + 60.76) + 600000 + 1101275$

$R(x) = -1812.5(x - 7.8)^2 + 710127.5$

Vertex: (7.8, 710128), maximum

The maximum revenue is $710,000 when $x = 8$.

price $= 3 + (0.25)8 = \$5$

65. $$\frac{3}{2x - 4} = \frac{x - 2}{x + 1}$$

$$(2x - 4)(x + 1)\left(\frac{3}{2x - 4}\right) = (2x - 4)(x + 1)\left(\frac{x - 2}{x + 1}\right)$$

$$3(x + 1) = (2x - 4)(x - 2)$$
$$3x + 3 = 2x^2 - 8x + 8$$
$$0 = 2x^2 - 11x + 5$$
$$0 = (2x - 1)(x - 5)$$

$2x - 1 = 0$ or $x - 5 = 0$

$2x = 1$ or $x = 5$

$x = \frac{1}{2}$ or $x = 5$

67. Let x = number of ounces removed
= number of ounces of water added

$$0.05(60 - x) + 0(x) = 0.04(60)$$
$$3 - 0.05x = 2.4$$
$$-0.05x = -0.6$$
$$x = 12$$

12 oz. must be removed and replaced with water.

Exercises 8.7

1. $(x + 4)(x - 2) < 0$

$x + 4 = 0$ $\quad$ $x - 2 = 0$

$x = -4$ $\quad$ $x = 2$

The intervals are $x < -4$, $-4 < x < 2$, and $x > 2$.

Test points:

	$(x + 4)(x - 2)$	
-5	$(-5 + 4)(-5 - 2)$	pos
0	$(0 + 4)(0 - 2)$	neg
3	$(3 + 4)(3 - 2)$	pos

We want $(x + 4)(x - 2) < 0$: neg

Solution: $-4 < x < 2$

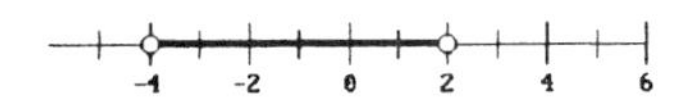

3. $(x + 4)(x - 2) > 0$: positive
   See exercise #1.
   $x < -4$ or $x > 2$

5. $(x + 2)(x - 5) \le 0$
   $x + 2 = 0 \qquad x - 5 = 0$
   $x = -2 \qquad x = 5$

   The intervals are $x < -2$, $-2 < x < 5$, and $x > 5$

   Test points:

	$(x + 2)(x - 5)$	
-3	$(-3 + 2)(-3 - 5)$	pos
0	$(0 + 2)(0 - 5)$	neg
6	$(6 + 2)(6 - 5)$	pos

We want $(x + 2)(x - 5) \le 0$: neg or 0

Solution: $-2 \le x \le 5$

7. $(x - 3)(2x - 1) \ge 0$
   $x - 3 = 0 \qquad 2x - 1 = 0$
   $x = 3 \qquad x = \frac{1}{2}$

   The intervals are $x < \frac{1}{2}$, $\frac{1}{2} < x < 3$, $x > 3$

   Test points:

	$(x - 3)(2x - 1)$	
0	$(0 - 3)(2 \cdot 0 - 1)$	pos
1	$(1 - 3)(2 \cdot 1 - 1)$	neg
4	$(4 - 3)(2 \cdot 4 - 1)$	pos

We want $(x - 3)(2x - 1) \ge 0$; pos or 0

Solution: $x \le \frac{1}{2}$ or $x \ge 3$

9. $a^2 - a - 20 < 0$
   $(a - 5)(a + 4) = 0$
   $a - 5 = 0 \qquad a + 4 = 0$
   $a = 5 \qquad a = -4$

   The intervals are $a < -4$, $-4 < a < 5$, $a > 5$

Test points:

	$(a - 5)(a + 4)$	
-5	$(-5 - 5)(-5 + 4)$	pos
0	$(0 - 5)(0 + 4)$	neg
6	$(6 - 5)(6 + 4)$	pos

We want $(a - 5)(a + 4) < 0$: neg

Solution $-4 < a < 5$

11. $$\begin{aligned} x^2 + x - 12 &\geq 0 \\ (x + 4)(x - 3) &\geq 0 \\ x + 4 = 0 \qquad & x - 3 = 0 \\ x = -4 \qquad & x = 3 \end{aligned}$$

The intervals are $x < -4$, $-4 < x < 3$, $x > 3$

Test points:

	$(x + 4)(x - 3)$	
-5	$(-5 + 4)(-5 - 3)$	pos
0	$(0 + 4)(0 - 3)$	neg
4	$(4 + 4)(4 - 3)$	pos

We want $(x + 4)(x - 3) \geq 0$: pos or 0

Solution: $x \leq -4$ or $x \geq 3$

13. $$\begin{aligned} 2a^2 - 9a &\leq 5 \\ 2a^2 - 9a - 5 &\leq 0 \\ (2a + 1)(a - 5) &\leq 0 \\ (2a + 1) = 0 \qquad & a - 5 = 0 \\ a = -\frac{1}{2} \qquad & a = 5 \end{aligned}$$

The intervals are $a < -\frac{1}{2}$, $-\frac{1}{2} < a < 5$, $a > 5$

Test points:

	$(2a + 1)(a - 5)$	
-1	$(2(-1) + 1) + (-1 - 5)$	pos
0	$(2 \cdot 0 + 1)(0 - 5)$	neg
6	$(2 \cdot 6 + 1)(6 - 5)$	pos

We want $(2a + 1)(a - 5) \leq 0$; neg or 0

Solution: $-\frac{1}{2} \leq a \leq 5$

15. $$\begin{aligned} 6y^2 - y &> 1 \\ 6y^2 - y - 1 &> 0 \\ (3y + 1)(2y - 1) &> 0 \\ 3y + 1 = 0 \qquad & 2y - 1 = 0 \\ y = -\frac{1}{3} \qquad & y = \frac{1}{2} \end{aligned}$$

The intervals are $y < -\frac{1}{3}$, $-\frac{1}{3} < y < \frac{1}{2}$, $y > \frac{1}{2}$

Test points:

	$(3y + 1)(2y - 1)$	
-1	$[3(-1) + 1][2(-1) - 1]$	pos
0	$(3 \cdot 0 + 1)(2 \cdot 0 - 1)$	neg
1	$(3 \cdot 1 + 1)(2 \cdot 1 - 1)$	pos

We want $(3y + 1)(2y - 1) > 0$; pos

Solution: $y < -\frac{1}{3}$ or $y > \frac{1}{2}$

17. $$\begin{aligned} 3x^2 &\leq 10 - 13x \\ 3x^2 + 13x - 10 &\leq 0 \\ (3x - 2)(x + 5) &\leq 0 \\ 3x - 2 = 0 \qquad & x + 5 = 0 \\ x = \frac{2}{3} \qquad & x = -5 \end{aligned}$$

The intervals are $x < -5$, $-5 < x < \frac{2}{3}$, $x > \frac{2}{3}$

Test points:

	$(3x - 2)(x + 5)$	
-6	$[3(-6) - 2](-6 + 5)$	pos
0	$(3 \cdot 0 - 2)(0 + 5)$	neg
1	$(3 \cdot 1 - 2)(1 + 5)$	pos

We want $(3x - 2)(x + 5) \le 0$; neg or 0

Solution: $-5 \le x \le \frac{2}{3}$

19. $x^2 + 2x + 1 \ge 0$

$(x + 1)^2 \ge 0$

A quantity squared is always ≥ 0.
Solution: All real numbers

21. $x^2 - 6x + 9 < 0$

$(x - 3)^2 < 0$

A quantity squared is never < 0.
No solution

23. $2x^2 - 13x > -15$

$2x^2 - 13x + 15 > 0$

$(2x - 3)(x - 5) > 0$

$2x - 3 = 0 \qquad x - 5 = 0$

$x = \frac{3}{2} \qquad x = 5$

The intervals are $x < \frac{3}{2}, \frac{3}{2} < x < 5, x > 5$

Test points:

	$(2x - 3)(x - 5)$	
0	$(2 \cdot 0 - 3)(0 - 5)$	pos
4	$(2 \cdot 4 - 3)(4 - 5)$	neg
6	$(2 \cdot 6 - 3)(6 - 5)$	pos

We want $(2x - 3)(x - 5) > 0$; pos

Solution: $x < \frac{3}{2}$ or $x > 5$

25. $3y^2 \ge 5y + 2$

$3y^2 - 5y - 2 \ge 0$

$(3y + 1)(y - 2) \ge 0$

$3y + 1 = 0 \qquad y - 2 = 0$

$y = -\frac{1}{3} \qquad y = 2$

The intervals are $y < -\frac{1}{3}, -\frac{1}{3} < y < 2, y > 2$

Test points:

	$(3y + 1)(y - 2)$	
-1	$[3(-1) + 1](-1 - 2)$	pos
0	$(3 \cdot 0 + 1)(0 - 2)$	neg
3	$(3 \cdot 3 + 1)(3 - 2)$	pos

We want $(3y + 1)(y - 2) \ge 0$; pos or 0

Solution: $y \le -\frac{1}{3}$ or $y \ge 2$

27. $x^2 + 2x \le -1$

$x^2 + 2x + 1 \le 0$

$(x + 1)^2 \le 0$

A quantity squared cannot be < 0.
Hence $x + 1 = 0$

$x = -1$

29. $\frac{x - 2}{x + 1} < 0$

$x - 2 = 0 \qquad x + 1 = 0$

$x = 2 \qquad x = -1$

The intervals are $x < -1, -1 < x < 2, x > 2$

Test points:

	$(x - 2)/(x + 1)$	
-2	$(-2 - 2)/(-2 + 1)$	pos
0	$(0 - 2)/(0 + 1)$	neg
3	$(3 - 2)/(3 + 1)$	pos

We want $\frac{x - 2}{x + 1} < 0$; neg

Solution: $-1 < x < 2$

31. $\dfrac{y + 3}{y - 5} \ge 0$

$y + 3 = 0 \qquad y - 5 = 0$
$y = -3 \qquad y = 5$

The intervals are $y < -3$, $-3 < y < 5$, $y > 5$

Test points:

	$(y + 3)/(y - 5)$	
-4	(-4 + 3)/(-4 - 5)	pos
0	(0 + 3)/(0 - 5)	neg
6	(6 + 3)/(6 - 5)	pos

We want $\dfrac{y + 3}{y - 5} \ge 0$; pos or 0

Solution: $y \le -3$ or $y > 5$

33. $\dfrac{a - 6}{a + 4} < 0$

$a - 6 = 0 \qquad a + 4 = 0$
$a = 6 \qquad a = -4$

The intervals are $a < -4$, $-4 < a < 6$, $a > 6$

Test points:

	$(a - 6)/(a + 4)$	
-5	(-5 - 6)/(-5 + 4)	pos
0	(0 - 6)/(0 + 4)	neg
7	(7 - 6)/(7 + 4)	pos

We want $\dfrac{a - 6}{a + 4} < 0$; neg

Solution: $-4 < a < 6$

35. $\dfrac{5}{y - 4} > 0$

$y - 4 = 0$
$y = 4$

The intervals are $y < 4$, $y > 4$

Test points:

	$(5)/(y - 4)$	
0	(5)/(0 - 4)	neg
5	(5)/(5 - 4)	pos

We want $\dfrac{5}{y - 4} > 0$; pos

Solution: $y > 4$

37. $\dfrac{3}{y - 1} < 1$

$\dfrac{3}{y - 1} - 1 < 0$

$\dfrac{3 - (y - 1)}{y - 1} < 0$

$\dfrac{-y + 4}{y - 1} < 0$

$-y + 4 = 0 \qquad y - 1 = 0$
$4 = y \qquad y = 1$

The intervals are $y < 1$, $1 < y < 4$, $y > 4$

Test points:

	$(-y + 4)/(y - 1)$	
0	(-0 + 4)/(0 - 1)	neg
2	(-2 + 4)/(2 - 1)	pos
5	(-5 + 4)/(5 - 1)	neg

We want $\dfrac{-y + 4}{y - 1} < 0$; neg

Solution: $y < 1$ or $y > 4$

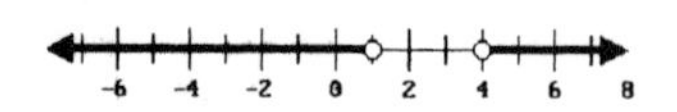

39. $$\frac{y+1}{y+2} > 3$$

$$\frac{y+1}{y+2} - 3 > 0$$

$$\frac{y+1-3(y+2)}{y+2} > 0$$

$$\frac{-2y-5}{y+2} > 0$$

$$-2y - 5 = 0 \qquad y + 2 = 0$$

$$y = -\frac{5}{2} \qquad y = -2$$

The intervals are $y < -\frac{5}{2}$, $-\frac{5}{2} < y < -2$, $y > -2$

Test points:

	$(-2y - 5)/(y + 2)$	
-3	$(-2(-3) - 5)/(-3 + 2)$	neg
-9/4	$(-2(-9/4) - 5)/(-9/4 + 2)$	pos
0	$(-2(0) - 5)/(0 + 2)$	neg

We want $\frac{-2y-5}{y+2} > 0$; pos

Solution: $-\frac{5}{2} < y < -2$

41. $$\frac{2y+3}{y-1} \le 2$$

$$\frac{2y+3}{y-1} - 2 \le 0$$

$$\frac{2y+3-2(y-1)}{y-1} \le 0$$

$$\frac{5}{y-1} \le 0$$

$$y - 1 = 0$$
$$y = 1$$

The intervals are $y < 1$, $y > 1$

Test points:

	$(5)/(y - 1)$	
0	$(5)/(0 - 1)$	neg
2	$(5)/(2 - 1)$	pos

We want $\frac{5}{y-1} \le 0$; neg or 0

Solution: $y < 1$

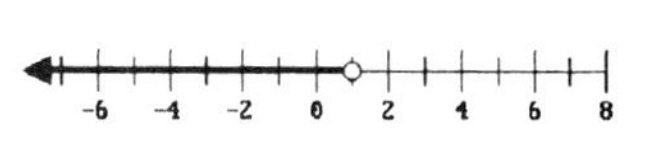

43. $$\frac{y-3}{y+1} > -2$$

$$\frac{y-3}{y+1} + 2 > 0$$

$$\frac{y-3+2(y+1)}{y+1} > 0$$

$$\frac{3y-1}{y+1} > 0$$

$$3y - 1 = 0 \qquad y + 1 = 0$$

$$y = \frac{1}{3} \qquad y = -1$$

The intervals are $y < -1$, $-1 < y < \frac{1}{3}$, $y > \frac{1}{3}$

Test points:

	$(3y - 1)/(y + 1)$	
-2	$(3(-2) - 1)/(-2 + 1)$	pos
0	$(3(0) - 1)/(0 + 1)$	neg
1	$(3(1) - 1)/(1 + 1)$	pos

We want $\frac{3y-1}{y+1} > 0$; pos

Solution: $y < -1$ or $y > \frac{1}{3}$

45. $$\frac{x}{x-1} \le \frac{3}{x-1}$$

$$\frac{x}{x-1} - \frac{3}{x-1} \le 0$$

$$\frac{x-3}{x-1} \le 0$$

$x - 3 = 0$ $\quad$ $x - 1 = 0$

$x = 3$ $\quad$ $x = 1$

The intervals are $x < 1$, $1 < x < 3$, $x > 3$

Test points:

	$(x - 3)/(x - 1)$	
0	$(0 - 3)/(0 - 1)$	pos
2	$(2 - 3)/(2 - 1)$	neg
4	$(4 - 3)/(4 - 1)$	pos

We want $\frac{x-3}{x-1} \le 0$; neg or 0

Solution: $1 < x \le 3$

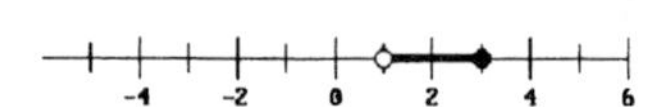

47. $$\frac{x-2}{x+3} > \frac{x+4}{x+3}$$

$$\frac{x-2}{x+3} - \frac{x+4}{x+3} > 0$$

$$\frac{x-2-(x+4)}{x+3} > 0$$

$$\frac{-6}{x+3} > 0$$

$x + 3 = 0$

$x = -3$

The intervals are $x < -3$, $x > -3$

Test points:

	$(-6)/(x + 3)$	
-4	$(-6)/(-4 + 3)$	pos
0	$(-6)/(0 + 3)$	neg

We want $\frac{-6}{x+3} > 0$; pos

Solution: $x < -3$

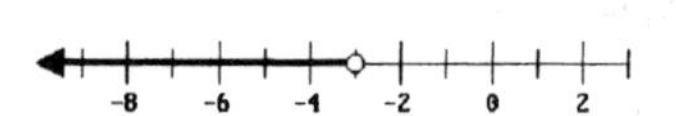

49. $$\frac{1}{x-2} + \frac{2}{x+3} \le \frac{3}{x+3}$$

$$\frac{1}{x-2} + \frac{2}{x+3} - \frac{3}{x+3} \le 0$$

$$\frac{1}{x-2} - \frac{1}{x+3} \le 0$$

$$\frac{x+3-(x-2)}{(x-2)(x+3)} \le 0$$

$$\frac{5}{(x-2)(x+3)} \le 0$$

$x - 2 = 0$ $\quad$ $x + 3 = 0$

$x = 2$ $\quad$ $x = -3$

The intervals are $x < -3$, $-3 < x < 2$, $x > 2$

Test points:

	$(5)/\{(x - 2)(x + 3)\}$	
-4	$(5)/\{(-4 - 2)(-4 + 3)\}$	pos
0	$(5)/\{(0 - 2)(0 + 3)\}$	neg
3	$(5)/\{(3 - 2)(3 + 3)\}$	pos

We want $\frac{5}{(x-2)(x+3)} \le 0$; neg or 0

Solution: $-3 < x < 2$

51. The graph is on or above the x-axis for $0.7 \le x \le 4.3$.

57. (x, P): (50, 600) and (65, 750)

(a) $m = \dfrac{750 - 600}{65 - 50} = 10$

$$P - 600 = 10(x - 50)$$
$$P - 600 = 10x - 500$$
$$P = 10x + 100$$

(b) When $x = 90$:
$P = 10(90) + 100 = 1000$

The profit would be \$1000.

59. $(x^{-2} + y)^{-1} = \left(\dfrac{1}{x^2} + y\right)^{-1} = \left(\dfrac{1 + x^2y}{x^2}\right)^{-1}$
$= \dfrac{x^2}{1 + x^2y}$

Exercises 8.8

1. $d = \sqrt{(6 - 3)^2 + (9 - 5)^2}$
$= \sqrt{3^2 + 4^2}$
$= \sqrt{9 + 16}$
$= \sqrt{25}$
$= 5$

3. $d = \sqrt{(6 - 3)^2 + (3 - 6)^2}$
$= \sqrt{3^2 + (-3)^2}$
$= \sqrt{9 + 9}$
$= \sqrt{18}$
$= 3\sqrt{2}$

5. $d = \sqrt{[6 - (-6)]^2 + (-9 - 9)^2}$
$= \sqrt{12^2 + (-18)^2}$
$= \sqrt{144 + 324}$
$= \sqrt{468}$
$= 6\sqrt{13}$

7. $d = \sqrt{[-8 - (-7)]^2 + [-3 - (-3)]^2}$
$= \sqrt{(-1)^2 + 0^2}$
$= \sqrt{1}$
$= 1$

9. $d = \sqrt{\left(\dfrac{1}{2} - \dfrac{1}{3}\right)^2 + (0 - 2)^2}$
$= \sqrt{\left(\dfrac{1}{6}\right)^2 + (-2)^2}$
$= \sqrt{\dfrac{1}{36} + 4}$
$= \sqrt{\dfrac{145}{36}}$
$= \dfrac{\sqrt{145}}{6}$

11. $d = \sqrt{(1.7 - 1.4)^2 + (1.2 - 0.8)^2}$
$= \sqrt{(0.3)^2 + (0.4)^2}$
$= \sqrt{0.09 + 0.16}$
$= \sqrt{0.25}$
$= 0.5$

13. midpoint $= \left(\dfrac{0 + 0}{2}, \dfrac{5 + 7}{2}\right)$
$= (0, 6)$

15. midpoint $= \left(\dfrac{-3 + 3}{2}, \dfrac{1 + 1}{2}\right)$
$= (0, 1)$

17. midpoint $= \left(\dfrac{-3 + 3}{2}, \dfrac{4 + (-4)}{2}\right)$
$= (0, 0)$

19. $\text{midpoint} = \left(\dfrac{\frac{2}{5} + \frac{1}{3}}{2}, \dfrac{\frac{3}{4} + 2}{2}\right)$

$= \left(\dfrac{\frac{11}{15}}{2}, \dfrac{\frac{11}{4}}{2}\right)$

$= \left(\dfrac{11}{30}, \dfrac{11}{8}\right)$

21. $d(P, Q) = |5 - 8| = 3$

$d(Q, R) = |2 - 6| = 4$

$d(P, R) = \sqrt{(5 - 8)^2 + (2 - 6)^2}$
$= \sqrt{9 + 16}$
$= \sqrt{25}$
$= 5$

$[d(P, Q)]^2 + [d(Q, R)]^2 = [d(P, R)]^2$
$3^2 + 4^2 = 5^2$
$9 + 16 = 25$
$25 = 25$
True

They are vertices of a right triangle.

23. $d(P, Q) = \sqrt{[4 - (-1)]^2 + (2 - 5)^2}$
$= \sqrt{25 + 9}$
$= \sqrt{34}$

$d(Q, R) = \sqrt{(-1 - 3)^2 + (5 - 9)^2}$
$= \sqrt{16 + 16}$
$= \sqrt{32}$
$= 4\sqrt{2}$

$d(P, R) = \sqrt{(4 - 3)^2 + (2 - 9)^2}$
$= \sqrt{1 + 49}$
$= \sqrt{50}$
$= 5\sqrt{2}$

$[d(P, Q)]^2 + [d(Q, R)]^2 = [d(P, R)]^2$

$(\sqrt{34})^2 + (4\sqrt{2})^2 = (5\sqrt{2})^2$
$34 + 32 = 50$
$66 = 50$
False

They are not vertices of a right triangle.

25. $P(5, 3),\ Q(7, 4),\ R(9, 7),\ S(7, 6)$

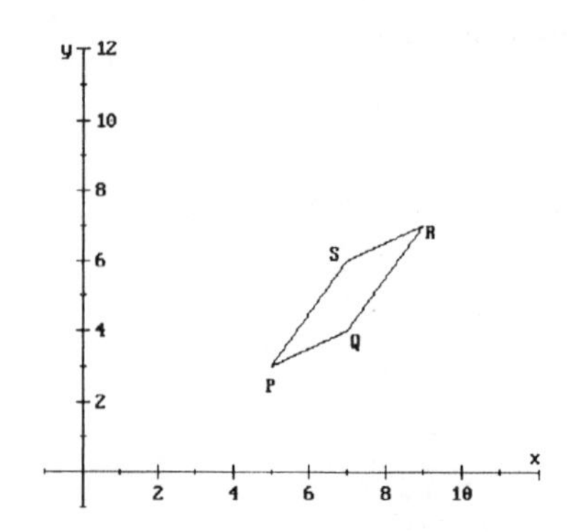

$\text{midpoint}_{PR} = \left(\dfrac{5 + 9}{2}, \dfrac{3 + 7}{2}\right) = (7, 5)$

$\text{midpoint}_{SQ} = \left(\dfrac{7 + 7}{2}, \dfrac{4 + 6}{2}\right) = (7, 5)$

Yes, since both diagonals have the same midpoint (7, 5).

27. $P(-2, 3),\ Q(5, -4),\ R(-6, 5),\ S(3, -4)$

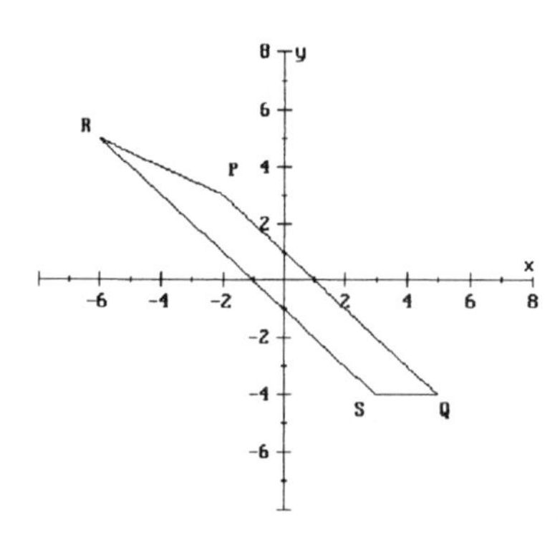

$\text{midpoint}_{RQ} = \left(\dfrac{-6 + 5}{2}, \dfrac{5 - 4}{2}\right) = \left(-\dfrac{1}{2}, \dfrac{1}{2}\right)$

$\text{midpoint}_{PS} = \left(\dfrac{-2 + 3}{2}, \dfrac{3 - 4}{2}\right) = \left(\dfrac{1}{2}, -\dfrac{1}{2}\right)$

No, since the diagonals have different midpoints.

29. $(x - h)^2 + (y - k)^2 = r^2$
$(x - 0)^2 + (y - 0)^2 = 1^2$
$x^2 + y^2 = 1$

31. $(x - h)^2 + (y - k)^2 = r^2$
$(x - 1)^2 + (y - 0)^2 = 7^2$
$(x - 1)^2 + y^2 = 49$

33. $(x - h)^2 + (y - k)^2 = r^2$
$(x - 2)^2 + (y - 5)^2 = 6^2$
$(x - 2)^2 + (y - 5)^2 = 36$

35. $(x - h)^2 + (y - k)^2 = r^2$
$(x - 6)^2 + [y - (-2)]^2 = 5^2$
$(x - 6)^2 + (y + 2)^2 = 25$

37. $(x - h)^2 + (y - k)^2 = r^2$
$[x - (-3)]^2 + [y - (-2)]^2 = 1^2$
$(x + 3)^2 + (y + 2)^2 = 1$

39. $x^2 + y^2 = 16$
$(x - 0)^2 + (y - 0)^2 = 4^2$
Center: $(0, 0)$
$r = 4$

41. $x^2 + y^2 = 24$
$(x - 0)^2 + (y - 0)^2 = \left(\sqrt{24}\right)^2$
Center: $(0, 0)$
$r = \sqrt{24} = 2\sqrt{6}$

43. $(x - 3)^2 + y^2 = 16$
$(x - 3)^2 + (y - 0)^2 = 4^2$
Center: $(3, 0)$
$r = 4$

45. $(x - 2)^2 + (y - 1)^2 = 1$
Center: $(2, 1)$
$r = \sqrt{1} = 1$

47. $(x + 1)^2 + (y - 3)^2 = 25$
Center: $(-1, 3)$
$r = \sqrt{25} = 5$

49. $(x + 2)^2 + (y + 3)^2 = 32$
Center: $(-2, -3)$
$r = \sqrt{32} = 4\sqrt{2}$

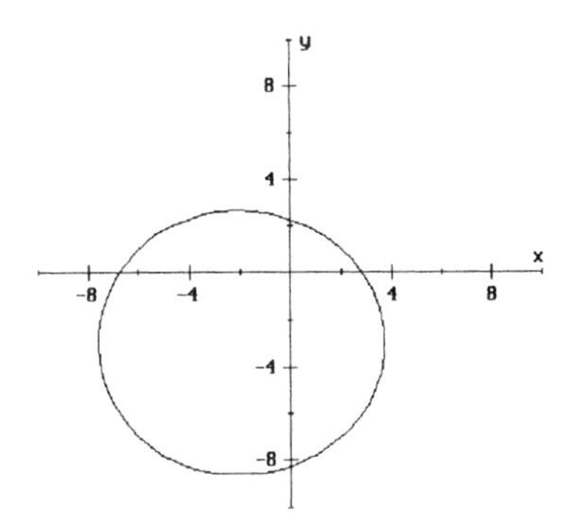

51. $(x + 7)^2 + (y + 1)^2 = 2$
Center: $(-7, -1)$
$r = \sqrt{2}$

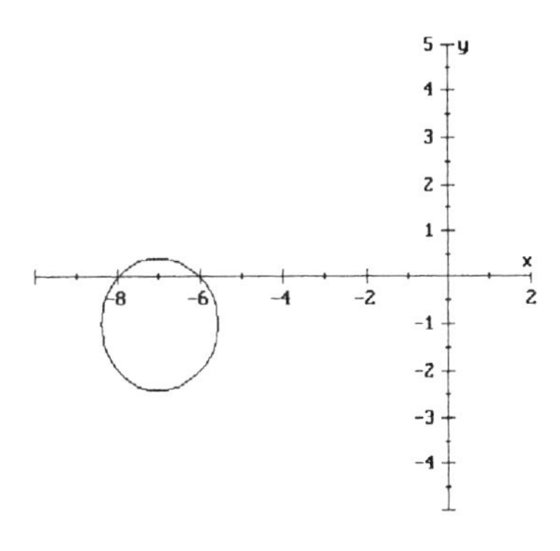

53. $x^2 + y^2 - 2x = 15$
$(x^2 - 2x + 1) + y^2 = 15 + 1$
$(x - 1)^2 + (y - 0)^2 = 16$
Center: $(1, 0)$
$r = 4$

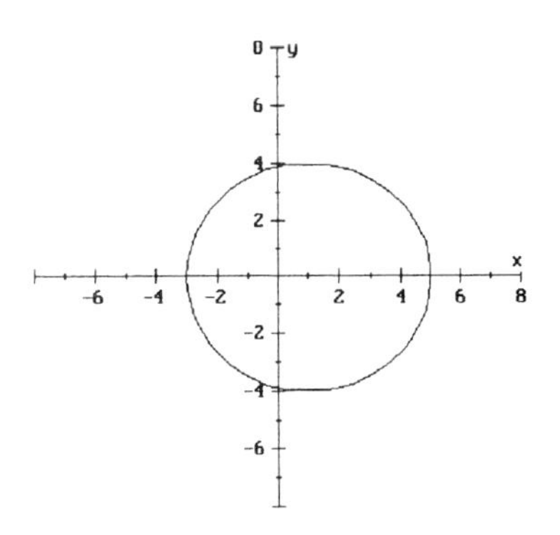

55. $$x^2 + y^2 - 4x - 2y = 20$$
$$(x^2 - 4x + 4) + (y^2 - 2y + 1) = 20 + 4 + 1$$
$$(x - 2)^2 + (y - 1)^2 = 25$$
Center: $(2, 1)$
$r = 5$

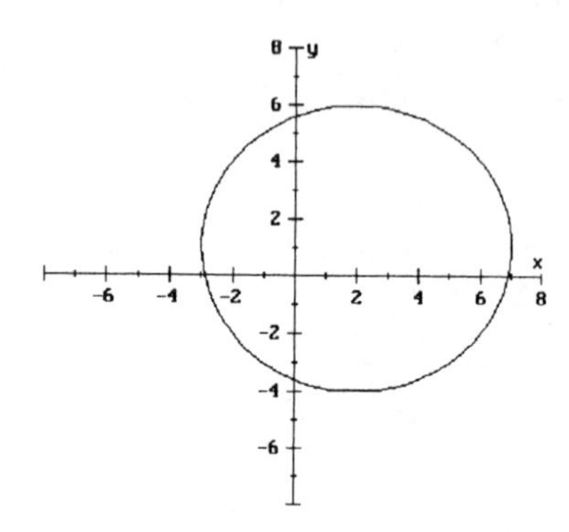

57. $$x^2 + y^2 - 2x = 20 + 4y$$
$$x^2 - 2x + y^2 - 4y = 20$$
$$(x^2 - 2x + 1) + (y^2 - 4y + 4) = 20 + 1 + 4$$
$$(x - 1)^2 + (y - 2)^2 = 25$$
Center: $(1, 2)$
$r = 5$

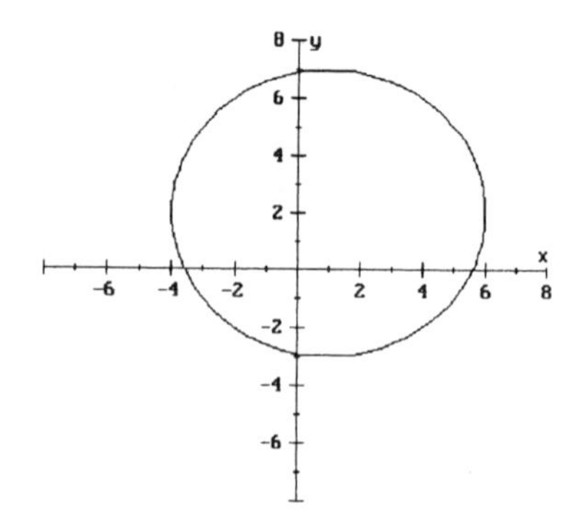

59. $$x^2 + 10y = 71 - y^2 + 4x$$
$$x^2 - 4x + y^2 + 10y = 71$$
$$(x^2 - 4x + 4) + (y^2 + 10y + 25) = 71 + 4 + 25$$
$$(x - 2)^2 + (y + 5)^2 = 100$$
Center: $(2, -5)$
$r = 10$

61. $$x^2 + y^2 = 2y - 6x - 2$$
$$x^2 + 6x + y^2 - 2y = -2$$
$$(x^2 + 6x + 9) + (y^2 - 2y + 1) = -2 + 9 + 1$$
$$(x + 3)^2 + (y - 1)^2 = 8$$
Center: $(-3, 1)$
$r = \sqrt{8} = 2\sqrt{2}$

63. $$2x^2 + 2y^2 - 4x + 4y = 22$$
$$x^2 + y^2 - 2x + 2y = 11$$
$$(x^2 - 2x + 1) + (y^2 + 2y + 1) = 11 + 1 + 1$$
$$(x - 1)^2 + (y + 1)^2 = 13$$
Center: $(1, -1)$
$r = \sqrt{13}$

65. $$x^2 + y^2 - x + 2y = \frac{59}{4}$$
$$\left(x^2 - x + \frac{1}{4}\right) + (y^2 + 2y + 1) = \frac{59}{4} + \frac{1}{4} + 1$$
$$\left(x - \frac{1}{2}\right)^2 + (y + 1)^2 = 16$$
Center: $\left(\frac{1}{2}, -1\right)$
$r = 4$

67. $x^2 + y^2 = 16$
Center: $(0, 0)$
$r = 4$

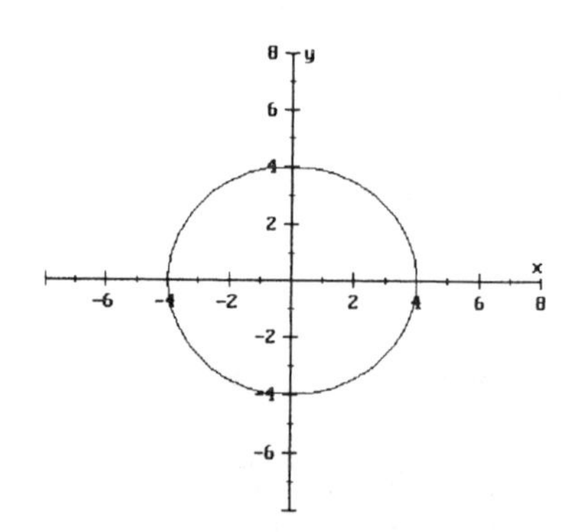

69. $x + y = 4$

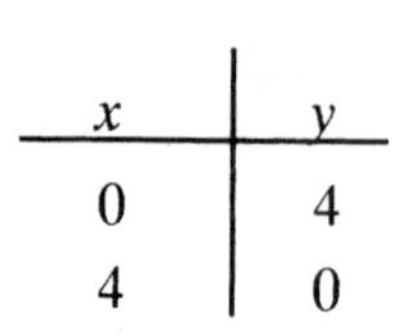

x	y
0	4
4	0

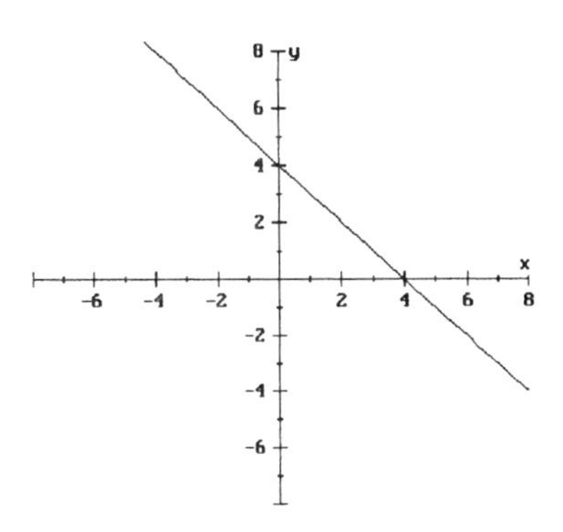

71. Center = midpoint

$$= \left(\frac{-2 + 4}{2}, \frac{8 - 5}{2}\right)$$

$$= \left(1, \frac{3}{2}\right)$$

$$r = \sqrt{(-2 - 1)^2 + \left(8 - \frac{3}{2}\right)^2}$$

$$= \sqrt{9 + \frac{169}{4}}$$

$$= \sqrt{\frac{205}{4}}$$

$$r^2 = \frac{205}{4}$$

$$(x - h)^2 + (y - k)^2 = r^2$$

$$(x - 1)^2 + \left(y - \frac{3}{2}\right)^2 = \frac{205}{4}$$

73. $r = \sqrt{(3 - 2)^2 + (-5 - 6)^2}$
$= \sqrt{1 + 121}$
$= \sqrt{122}$
$r^2 = 122$

$$(x - h)^2 + (y - k)^2 = r^2$$
$$(x - 3)^2 + [y - (-5)]^2 = 122$$
$$(x - 3)^2 + (y + 5)^2 = 122$$

75. $r = \sqrt{(-3 - 5)^2 + (4 - 2)^2}$
$= \sqrt{64 + 4}$
$= \sqrt{68}$
$= 2\sqrt{17}$

$C = 2\pi r$
$= 2\pi(2\sqrt{17})$
$= 4\pi\sqrt{17}$

77. It will touch the x-axis at (3, 0).

$r = |-2 - 0| = 2$

$$(x - h)^2 + (y - k)^2 = r^2$$
$$(x - 3)^2 + [y - (-2)]^2 = 2^2$$
$$(x - 3)^2 + (y + 2)^2 = 4$$

79. The center will be the point of intersection of the horizontal line through (0, −3) with the vertical line through (3, 0).

$$\begin{cases} y = -3 \\ x = 3 \end{cases}$$

$C(3, -3)$
$r = |3 - 0| = 3$

$$(x - h)^2 + (y - k)^2 = r^2$$
$$(x - 3)^2 + [y - (-3)]^2 = 3^2$$
$$(x - 3)^2 + (y + 3)^2 = 9$$

83. $\dfrac{83700}{0.0042} = \dfrac{8.37 \times 10^4}{4.2 \times 10^{-3}}$

$$= \frac{8.37}{4.2} \times \frac{10^4}{10^{-3}}$$

$$= 1.99 \times 10^7$$

It is closest to 10^7.

85. $2x^{1/2} - (5x)^{2/3}$
$= 2\sqrt{x} - \sqrt[3]{(5x)^2}$
$= 2\sqrt{x} - \sqrt[3]{25x^2}$

CHAPTER 8 REVIEW EXERCISES

1. $d = 18v - \left(\frac{v}{2.3}\right)^2 \quad , \quad 10 \le v \le 90$

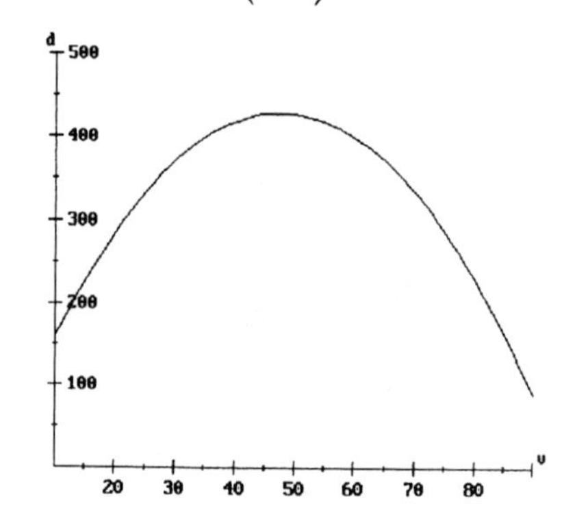

(a)

v	d
30	369.9 miles
40	417.5 miles
50	427.4 miles
60	399.5 miles

(b) When $d = 400$, $v = 35$ mph and $v = 60$ mph

3.
$$\begin{aligned} 2y^2 - y - 1 &= 0 \\ (2y + 1)(y - 1) &= 0 \\ 2y + 1 = 0 \quad &\text{or} \quad y - 1 = 0 \\ 2y = -1 \quad & \quad\quad y = 1 \\ y = -\frac{1}{2} \quad &\text{or} \quad y = 1 \end{aligned}$$

5.
$$\begin{aligned} 3x^2 - 17x &= 28 \\ 3x^2 - 17x - 28 &= 0 \\ (3x + 4)(x - 7) &= 0 \\ 3x + 4 = 0 \quad &\text{or} \quad x - 7 = 0 \\ 3x = -4 \quad & \quad\quad x = 7 \\ x = -\frac{4}{3} \quad &\text{or} \quad x = 7 \end{aligned}$$

7.
$$\begin{aligned} 81 &= a^2 \\ \pm\sqrt{81} &= \sqrt{a^2} \\ \pm 9 &= a \end{aligned}$$

9.
$$\begin{aligned} z^2 + 7 &= 2 \\ z^2 &= -5 \\ \sqrt{z^2} &= \pm\sqrt{-5} \\ z &= \pm i\sqrt{5} \end{aligned}$$

11.
$$\begin{aligned} 4x^2 + 36 &= 24x \\ 4x^2 - 24x + 36 &= 0 \\ x^2 - 6x + 9 &= 0 \\ (x - 3)^2 &= 0 \\ x - 3 &= \pm\sqrt{0} \\ x - 3 &= 0 \\ x &= 3 \end{aligned}$$

13.
$$\begin{aligned} (a + 7)(a + 3) &= (3a + 1)(a + 1) \\ a^2 + 10a + 21 &= 3a^2 + 4a + 1 \\ 0 &= 2a^2 - 6a - 20 \\ 0 &= a^2 - 3a - 10 \\ 0 &= (a - 5)(a + 2) \\ a - 5 = 0 \quad &\text{or} \quad a + 2 = 0 \\ a = 5 \quad &\text{or} \quad a = -2 \end{aligned}$$

15.
$$x - 2 = \frac{1}{x + 2}$$
$$\begin{aligned} (x + 2)(x - 2) &= 1 \\ x^2 - 4 &= 1 \\ x^2 &= 5 \\ x &= \pm\sqrt{5} \end{aligned}$$

17.
$$\frac{2}{x - 2} - \frac{5}{x + 2} = 1$$
$$(x - 2)(x + 2)\left(\frac{2}{x - 2} - \frac{5}{x + 2}\right) = (x - 2)(x + 2)(1)$$
$$\begin{aligned} 2(x + 2) - 5(x - 2) &= x^2 - 4 \\ 2x + 4 - 5x + 10 &= x^2 - 4 \\ -3x + 14 &= x^2 - 4 \\ 0 &= x^2 + 3x - 18 \\ 0 &= (x + 6)(x - 3) \\ x + 6 = 0 \quad &\text{or} \quad x - 3 = 0 \\ x = -6 \quad &\text{or} \quad x = 3 \end{aligned}$$

19.
$$\begin{aligned} x^2 + 2x - 4 &= 0 \\ x^2 + 2x &= 4 \\ x^2 + 2x + 1 &= 4 + 1 \\ (x + 1)^2 &= 5 \\ x + 1 &= \pm\sqrt{5} \\ x &= -1 \pm \sqrt{5} \end{aligned}$$

21. $2y^2 + 4y - 3 = 0$

$$y^2 + 2y - \frac{3}{2} = 0$$

$$y^2 + 2y = \frac{3}{2}$$

$$y^2 + 2y + 1 = \frac{3}{2} + 1$$

$$(y + 1)^2 = \frac{5}{2}$$

$$y + 1 = \pm\sqrt{\frac{5}{2}}$$

$$y = -1 \pm \frac{\sqrt{10}}{2}$$

$$\text{or} \quad y = \frac{-2 \pm \sqrt{10}}{2}$$

23. $3a^2 + 6a - 5 = 0$

$$a^2 + 2a - \frac{5}{3} = 0$$

$$a^2 + 2a = \frac{5}{3}$$

$$a^2 + 2a = \frac{5}{3}$$

$$a^2 + 2a + 1 = \frac{5}{3} + 1$$

$$(a + 1)^2 = \frac{8}{3}$$

$$a + 1 = \pm\sqrt{\frac{8}{3}}$$

$$a = -1 \pm \frac{2\sqrt{6}}{3}$$

$$\text{or} \quad a = \frac{-3 \pm 2\sqrt{6}}{3}$$

25. $\frac{1}{a - 5} + \frac{3}{a + 2} = 4$

$$(a - 5)(a + 2)\left(\frac{1}{a - 5} + \frac{3}{a + 2}\right) = (a - 5)(a + 2)(4)$$

$$a + 2 + 3(a - 5) = 4(a^2 - 3a - 10)$$
$$a + 2 + 3a - 15 = 4a^2 - 12a - 40$$
$$4a - 13 = 4a^2 - 12a - 40$$
$$27 = 4a^2 - 16a$$

$$\frac{27}{4} = a^2 - 4a$$

$$\frac{27}{4} + 4 = a^2 - 4a + 4$$

$$\frac{43}{4} = (a - 2)^2$$

$$\pm\sqrt{\frac{43}{4}} = a - 2$$

$$\frac{\pm\sqrt{43}}{2} = a - 2$$

$$2 \pm \frac{\sqrt{43}}{2} = a$$

$$\text{or} \quad \frac{4 \pm \sqrt{43}}{2} = a$$

27. $6a^2 - 13a = 5$

$$6a^2 - 13a - 5 = 0$$
$$(3a + 1)(2a - 5) = 0$$
$$3a + 1 = 0 \quad \text{or} \quad 2a - 5 = 0$$
$$3a = -1 \qquad 2a = 5$$
$$a = -\frac{1}{3} \quad \text{or} \quad a = \frac{5}{2}$$

29. $3.2a^2 - 5.8a - 5.6 = 0$
$A = 3.2,\ B = -5.8,\ C = -5.6$

$$a = \frac{-B \pm \sqrt{B^2 - 4AC}}{2A}$$

$$= \frac{-(-5.8) \pm \sqrt{(-5.8)^2 - 4(3.2)(-5.6)}}{2(3.2)}$$

$$= \frac{5.8 \pm \sqrt{105.32}}{6.4}$$

$$a = -0.70 \quad \text{or} \quad a = 2.51$$

31. $5a^2 - 3a = 3 - 3a + 2a^2$

$$3a^2 = 3$$
$$a^2 = 1$$
$$a = \pm 1$$

33. $(x - 4)(x + 1) = x - 2$

$$x^2 - 3x - 4 = x - 2$$
$$x^2 - 4x - 2 = 0$$

$$x = \frac{-(-4) \pm \sqrt{(-4)^2 - 4(1)(-2)}}{2(1)}$$

$$= \frac{4 \pm \sqrt{16 + 8}}{2}$$

$$= \frac{4 \pm \sqrt{24}}{2}$$

$$= \frac{4 \pm 2\sqrt{6}}{2}$$

$$= \frac{2\left(2 \pm \sqrt{6}\right)}{2}$$

$$= 2 \pm \sqrt{6}$$

35. $(t + 3)(t - 4) = t(t + 2)$

$$t^2 - t - 12 = t^2 + 2t$$
$$-12 = 3t$$
$$-4 = t$$

37. $8x^2 = 12$

$$x^2 = \frac{3}{2}$$

$$x = \pm\sqrt{\frac{3}{2}}$$

$$x = \pm\frac{\sqrt{6}}{2}$$

39. $3x^2 - 2x + 5 = 7x^2 - 2x + 5$

$$0 = 4x^2$$
$$0 = x^2$$
$$\pm\sqrt{0} = x$$
$$0 = x$$

41. $(x + 2)(x - 4) = 2x - 10$

$$x^2 - 2x - 8 = 2x - 10$$
$$x^2 - 4x + 2 = 0$$

$$x = \frac{-(-4) \pm \sqrt{(-4)^2 - 4(1)(2)}}{2(1)}$$

$$= \frac{4 \pm \sqrt{16 - 8}}{2}$$

$$= \frac{4 \pm \sqrt{8}}{2}$$

$$= \frac{4 \pm 2\sqrt{2}}{2}$$

$$= \frac{2\left(2 \pm \sqrt{2}\right)}{2}$$

$$= 2 \pm \sqrt{2}$$

43. $\dfrac{1}{z + 2} = z - 4$

$$1 = (z + 2)(z - 4)$$
$$1 = z^2 - 2z - 8$$
$$0 = z^2 - 2z - 9$$

$$z = \frac{-(-2) \pm \sqrt{(-2)^2 - 4(1)(-9)}}{2(1)}$$

$$= \frac{2 \pm \sqrt{4 + 36}}{2}$$

$$= \frac{2 \pm \sqrt{40}}{2}$$

$$= \frac{2 \pm 2\sqrt{10}}{2}$$

$$= \frac{2\left(1 \pm \sqrt{10}\right)}{2}$$

$$= 1 \pm \sqrt{10}$$

45. $$\frac{1}{x+4} - \frac{3}{x+2} = 5$$

$$(x+4)(x+2)\left(\frac{1}{x+4} - \frac{3}{x+2}\right) = (x+4)(x+2)(5)$$

$$x + 2 - 3(x+4) = 5(x^2 + 6x + 8)$$
$$x + 2 - 3x - 12 = 5x^2 + 30x + 40$$
$$-2x - 10 = 5x^2 + 30x + 40$$
$$0 = 5x^2 + 32x + 50$$

$$x = \frac{-32 \pm \sqrt{32^2 - 4(5)(50)}}{2(5)}$$

$$= \frac{-32 \pm \sqrt{1024 - 1000}}{10}$$

$$= \frac{-32 \pm \sqrt{24}}{10}$$

$$= \frac{-32 \pm 2\sqrt{6}}{10}$$

$$= \frac{2\left(-16 \pm \sqrt{6}\right)}{10}$$

$$= \frac{-16 \pm \sqrt{6}}{5}$$

47. $$\frac{3}{x-4} + \frac{2x}{x-5} = \frac{3}{x-5}$$

$$(x-4)(x-5)\left(\frac{3}{x-4} + \frac{2x}{x-5}\right) = (x-4)(x-5)\left(\frac{3}{x-5}\right)$$

$$3(x-5) + 2x(x-4) = 3(x-4)$$
$$3x - 15 + 2x^2 - 8x = 3x - 12$$
$$2x^2 - 5x - 15 = 3x - 12$$
$$2x^2 - 8x - 3 = 0$$

$$x = \frac{-(-8) \pm \sqrt{(-8)^2 - 4(2)(-3)}}{2(2)}$$

$$= \frac{8 \pm \sqrt{64 + 24}}{4}$$

$$= \frac{8 \pm \sqrt{88}}{4}$$

$$= \frac{8 \pm 2\sqrt{22}}{4}$$

$$= \frac{2\left(4 \pm \sqrt{22}\right)}{4}$$

$$= \frac{4 \pm \sqrt{22}}{2}$$

49. $A = \pi r^2 h$

$\frac{A}{\pi h} = r^2$

$\sqrt{\frac{A}{\pi h}} = r$

$\frac{\sqrt{A\pi h}}{\pi h} = r$

51. $2x^2 + xy - 3y^2 = 0$
$(2x + 3y)(x - y) = 0$
$2x + 3y = 0$ or $x - y = 0$
$2x = -3y$ $\quad x = y$
$x = -\frac{3}{2}y$ or $x = y$

53. $\sqrt{2a + 3} = a$
$(\sqrt{2a + 3})^2 = a^2$
$2a + 3 = a^2$
$0 = a^2 - 2a - 3$
$0 = (a - 3)(a + 1)$
$a - 3 = 0$ or $a + 1 = 0$
$a = 3$ or $a = -1$
When you check $a = -1$, you find it to be extraneous. The only solution is $a = 3$.

55. $\sqrt{3a + 1} + 1 = a$
$\sqrt{3a + 1} = a - 1$
$(\sqrt{3a + 1})^2 = (a - 1)^2$
$3a + 1 = a^2 - 2a + 1$
$0 = a^2 - 5a$
$0 = a(a - 5)$
$a = 0$ or $a - 5 = 0$
$a = 5$
When you check a = 0, you find it to be extraneous. The only solution is $a = 5$.

57. $\sqrt{2x} + 1 - \sqrt{x - 3} = 4$
$\sqrt{2x} = 3 + \sqrt{x - 3}$
$(\sqrt{2x})^2 = (3 + \sqrt{x - 3})^2$
$2x = 9 + 6\sqrt{x - 3} + x - 3$
$2x = 6 + 6\sqrt{x - 3} + x$
$x - 6 = 6\sqrt{x - 3}$
$(x - 6)^2 = (6\sqrt{x - 3})^2$
$x^2 - 12x + 36 = 36(x - 3)$
$x^2 - 12x + 36 = 36x - 108)$
$x^2 - 48x + 144 = 0$
$x = \frac{-(-48) \pm \sqrt{(-48)^2 - 4(1)(144)}}{2(1)}$
$= \frac{48 \pm \sqrt{1728}}{2}$
$x = 3.2$ or $x = 44.8$
When you check $x = 3.2$, you find it to be extraneous. The only solution is $x = 44.8$.

59. $\sqrt{3x + 4} - \sqrt{x - 3} = 3$
$\sqrt{3x + 4} = 3 + \sqrt{x - 3}$
$(\sqrt{3x + 4})^2 = (3 + \sqrt{x - 3})^2$
$3x + 4 = 9 + 6\sqrt{x - 3} + x - 3$
$3x + 4 = 6 + x + 6\sqrt{x - 3}$
$2x - 2 = 6\sqrt{x - 3}$
$x - 1 = 3\sqrt{x - 3}$
$(x - 1)^2 = (3\sqrt{x - 3})^2$
$x^2 - 2x + 1 = 9(x - 3)$
$x^2 - 2x + 1 = 9x - 27$
$x^2 - 11x + 28 = 0$
$(x - 4)(x - 7) = 0$
$x - 4 =$ or $x - 7 = 0$
$x = 4$ or $x = 7$

61. $\sqrt{3y + z} = x$
$(\sqrt{3y + z})^2 = x^2$
$3y + z = x^2$
$3y = x^2 - z$
$y = \frac{x^2 - z}{3}$

63. $\sqrt{3y} + z = x$
$\sqrt{3y} = x - z$
$(\sqrt{3y})^2 = (x - z)^2$
$3y = (x - z)^2$
$y = \frac{(x - z)^2}{3}$

65. $x^3 - 2x^2 - 15x = 0$
$x(x - 5)(x + 3) = 0$
$x = 0$ or $x - 5 = 0$ or $x + 3 = 0$
$x = 0$ or $x = 5$ or $x = -3$

67. $4x^3 - 10x^2 - 6x = 0$
$2x(2x + 1)(x - 3) = 0$
$2x = 0$ or $2x + 1 = 0$ or $x - 3 = 0$
$x = 0$ or $x = -\frac{1}{2}$ or $x = 3$

69. $$a^4 - 17a^2 = -16$$
$$a^4 - 17a^2 + 16 = 0$$
$$(a^2 - 1)(a^2 - 16) = 0$$
$$a^2 - 1 = 0 \quad \text{or} \quad a^2 - 16 = 0$$
$$a^2 = 1 \qquad a^2 = 16$$
$$a = \pm 1 \quad \text{or} \quad a = \pm 4$$

71. $$y^4 - 3y^2 = 4$$
$$y^4 - 3y^2 - 4 = 0$$
$$(y^2 - 4)(y^2 + 1) = 0$$
$$y^2 - 4 = 0 \quad \text{or} \quad y^2 + 1 = 0$$
$$y^2 = 4 \qquad y^2 = -1$$
$$y = \pm 2 \quad \text{or} \quad y = \pm i$$

73. $$z^4 = 6z^2 - 5$$
$$z^4 - 6z^2 + 5 = 0$$
$$(z^2 - 5)(z^2 - 1) = 0$$
$$z^2 - 5 = 0 \quad \text{or} \quad z^2 - 1 = 0$$
$$z^2 = 5 \qquad z^2 = 1$$
$$z = \pm\sqrt{5} \quad \text{or} \quad z = \pm 1$$

75. $$a^{1/2} - a^{1/4} - 6 = 0$$
$$(a^{1/4})^2 - a^{1/4} - 6 = 0$$
Let $u = a^{1/4}$.
$$u^2 - u - 6 = 0$$
$$(u - 3)(u + 2) = 0$$
$$u - 3 = 0 \quad \text{or} \quad u + 2 = 0$$
$$u = 3 \qquad u = -2$$
$$a^{1/4} = 3 \qquad a^{1/4} = -2$$
$$(a^{1/4})^4 = 3^4 \qquad \text{No solution}$$
$$a = 81$$
The only solution is $a = 81$.

77. $$2x^{2/3} = 5x^{1/3} + 3$$
$$2(x^{1/3})^2 - 5x^{1/3} - 3 = 0$$
Let $u = x^{1/3}$.
$$2u^2 - 5u - 3 = 0$$
$$(2u + 1)(u - 3) = 0$$
$$2u + 1 = 0 \quad \text{or} \quad u - 3 = 0$$
$$u = -\frac{1}{2} \qquad u = 3$$
$$x^{1/3} = -\frac{1}{2} \qquad x^{1/3} = 3$$
$$(x^{1/3})^3 = \left(-\frac{1}{2}\right)^3 \qquad (x^{1/3})^3 = 3^3$$
$$x = -\frac{1}{8} \quad \text{or} \quad x = 27$$

79. $$\sqrt{x} + 2\sqrt[4]{x} - 35 = 0$$
$$x^{1/2} + 2x^{1/4} - 35 = 0$$
$$(x^{1/4})^2 + 2x^{1/4} - 35 = 0$$
Let $u = x^{1/4}$.
$$u^2 + 2u - 35 = 0$$
$$(u + 7)(u - 5) = 0$$
$$u + 7 = 0 \quad \text{or} \quad u - 5 = 0$$
$$u = -7 \qquad u = 5$$
$$x^{1/4} = -7 \qquad x^{1/4} = 5$$
$$\text{No solution} \qquad (x^{1/4})^4 = 5^4$$
$$x = 625$$

The only solution is $x = 625$.

81. $$3x^{-2} + x^{-1} - 2 = 0$$
$$3(x^{-1})^2 + x^{-1} - 2 = 0$$
Let $u = x^{-1}$.
$$3u^2 + u - 2 = 0$$
$$(3u - 2)(u + 1) = 0$$
$$3u - 2 = 0 \quad \text{or} \quad u + 1 = 0$$
$$u = \frac{2}{3} \qquad u = -1$$
$$x^{-1} = \frac{2}{3} \qquad x^{-1} = -1$$
$$(x^{-1})^{-1} = \left(\frac{2}{3}\right)^{-1} \qquad (x^{-1})^{-1} = (-1)^{-1}$$
$$x = \frac{3}{2} \quad \text{or} \quad x = -1$$

83. $y = (x - 2)^2 + 1$
Vertex: (2, 1)
Axis of symmetry: $x = 2$

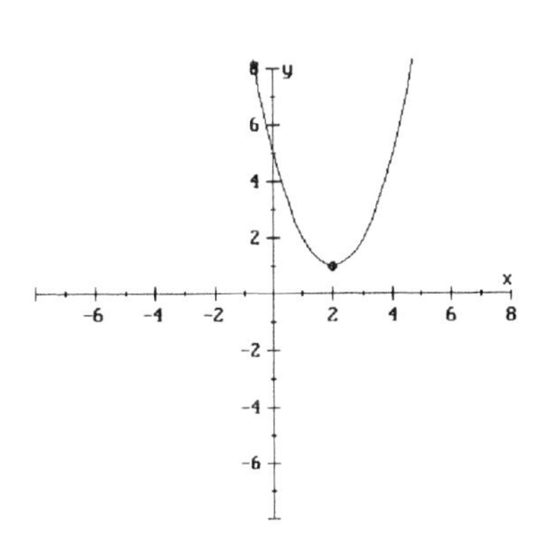

85. $y = -3(x - 2)^2 - 4$
Vertex: $(2, -4)$
Axis of symmetry: $x = 2$

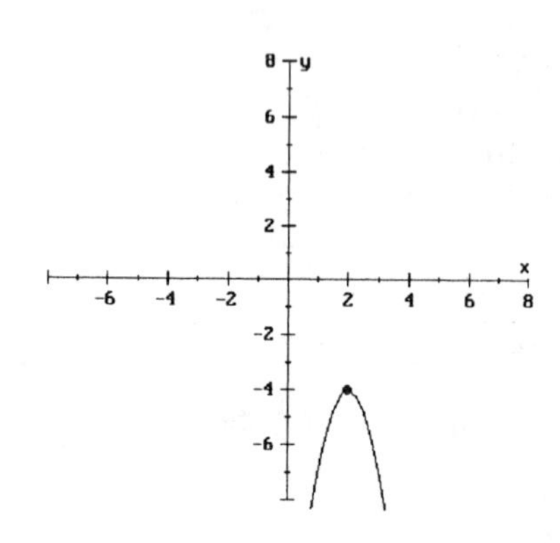

87. $y = 2x^2 - 12x + 4$
$y = 2(x^2 - 6x) + 4$
$y = 2(x^2 - 6x + 9) + 4 - 18$
$y = 2(x - 3)^2 - 14$
Vertex: $(3, -14)$
Axis of symmetry: $x = 3$

89. $y = -x^2 + 4x - 12$
$y = -(x^2 - 4x) - 12$
$y = -(x^2 - 4x + 4) - 12 + 4$
$y = -(x - 2)^2 - 8$
Vertex: $(2, -8)$
Axis of symmetry: $x = 2$

91. $y = 7x^2$
Vertex: $(0, 0)$
Axis of symmetry: $x = 0$
x-intercept: Let $y = 0$
$0 = 7x^2$
$0 = x^2$
$0 = x$

y-intercept: Let $x = 0$
$y = 7(0)^2$
$y = 0$

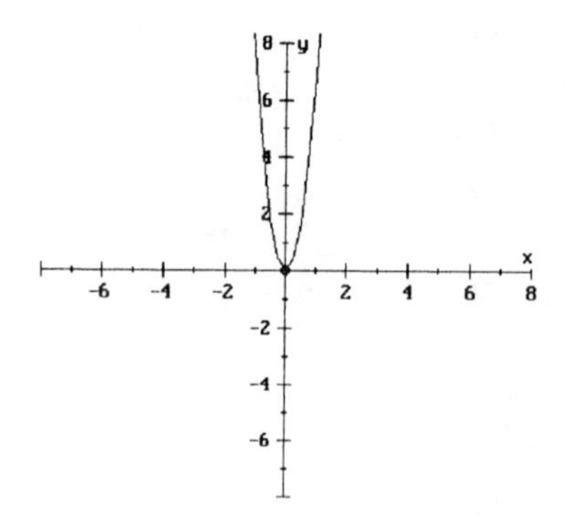

93. $y = -7x^2 + 3$
Vertex: $(0, 3)$
Axis of symmetry: $x = 0$
x-intercepts: Let $y = 0$

$0 = -7x^2 + 3$
$7x^2 = 3$

$x^2 = \frac{3}{7}$

$x = \pm\sqrt{\frac{3}{7}}$

$x = \pm\frac{\sqrt{21}}{7}$

y-intercept: Let $x = 0$
$y = -7(0)^2 + 3$
$= 3$

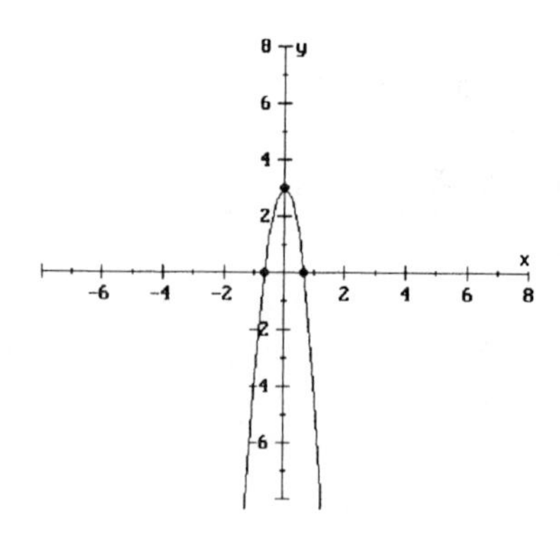

95. $y = x^2 - 6x$
$y = (x^2 - 6x + 9) - 9$
$y = (x - 3)^2 - 9$

Vertex: $(3, -9)$
Axis of symmetry: $x = 3$
x-intercepts: Let $y = 0$

$0 = (x - 3)^2 - 9$
$9 = (x - 3)^2$
$\pm 3 = x - 3$
$x - 3 = 3$ or $x - 3 = -3$
$x = 6$ or $x = 0$

y-intercept: Let $x = 0$

$y = 0^2 - 6(0)$
$= 0$

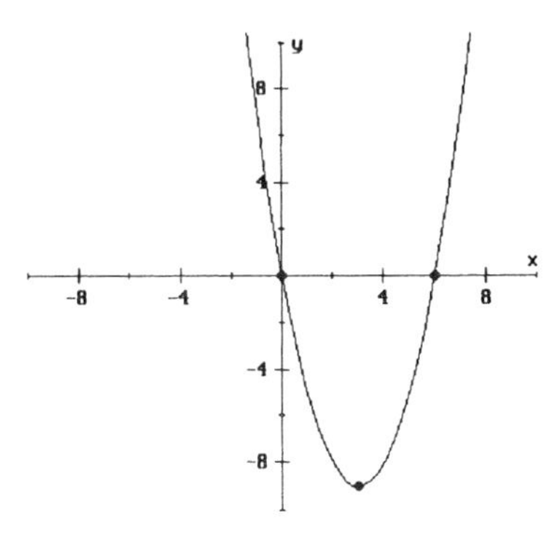

97. $y = x^2 - 2x - 8$
$y = (x^2 - 2x + 1) - 8 - 1$
$y = (x - 1)^2 - 9$
Vertex: $(1, -9)$
Axis of symmetry: $x = 1$
x-intercepts: Let $y = 0$

$0 = (x - 1)^2 - 9$
$9 = (x - 1)^2$
$\pm 3 = x - 1$
$x - 1 = 3$ or $x - 1 = -3$
$x = 4$ or $x = -2$

y-intercept: Let $x = 0$

$y = 0^2 - 2(0) - 8$
$= -8$

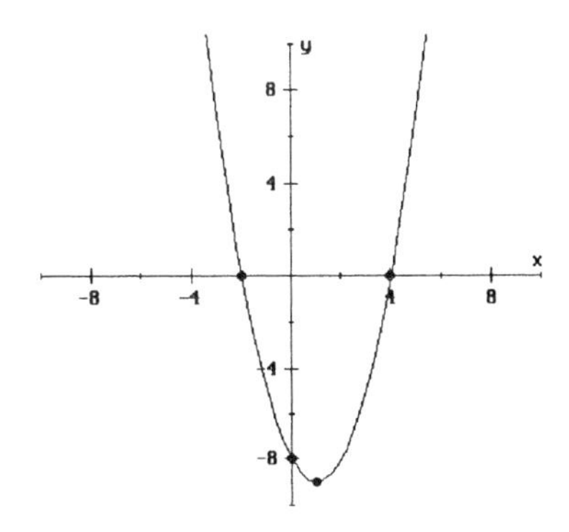

99. $y = -x^2 + 2x - 5$
$y = -(x^2 - 2x + 1) - 5 + 1$
$y = -(x - 1)^2 - 4$
Vertex: $(1, -4)$
Axis of symmetry: $x = 1$
x-intercepts: Let $y = 0$

$0 = -(x - 1)^2 - 4$
$(x - 1)^2 = -4$
none

y-intercept: Let $x = 0$

$y = -0^2 + 2(0) - 5$
$= -5$

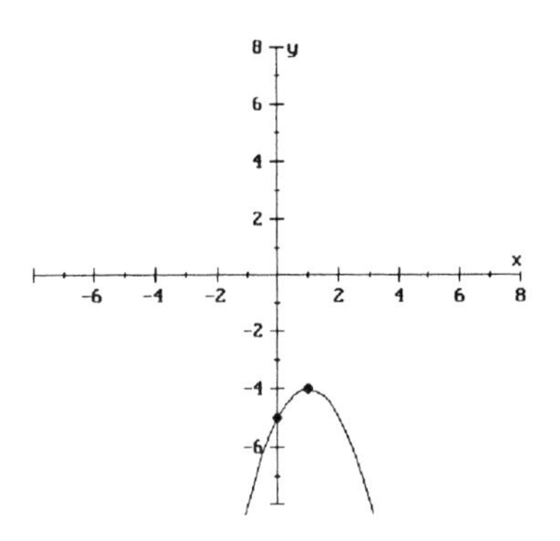

101. $(x - 2)(x + 1) > 0$
$x - 2 = 0 \quad x + 1 = 0$
$x = 2 \quad x = -1$
The intervals are $x < -1$, $-1 < x < 2$, $x > 2$

	$(x - 2)(x + 1)$	
-2	$(-2 - 2)(-2 + 1)$	pos
0	$(0 - 2)(0 + 1)$	neg
3	$(3 - 2)(3 + 1)$	pos

We want $(x - 2)(x + 1) > 0$: pos
Solution: $x < -1$ or $x > 2$

103. $(3x + 1)(x - 2) \le 0$
$3x + 1 = 0 \quad x - 2 = 0$
$x = -\frac{1}{3} \quad x = 2$
The intervals are $x < -\frac{1}{3}$, $-\frac{1}{3} < x < 2$, $x > 2$
Test points:

	$(3x + 1)(x - 2)$	
-1	$[3(-1) + 1](-1 - 2)$	pos
0	$(3 \cdot 0 + 1)(0 - 2)$	neg
3	$(3 \cdot 3 + 1)(3 - 2)$	pos

We want $(3x + 1)(x - 2) \leq 0$ neg or 0

Solution: $-\frac{1}{3} \leq x \leq 2$

105. $y^2 - 5y + 4 > 0$

$(y - 4)(y - 1) > 0$

$y - 4 = 0 \qquad y - 1 = 0$

$y = 4 \qquad y = 1$

The intervals are $y < 1$, $1 < y < 4$, $y > 4$

Test points:

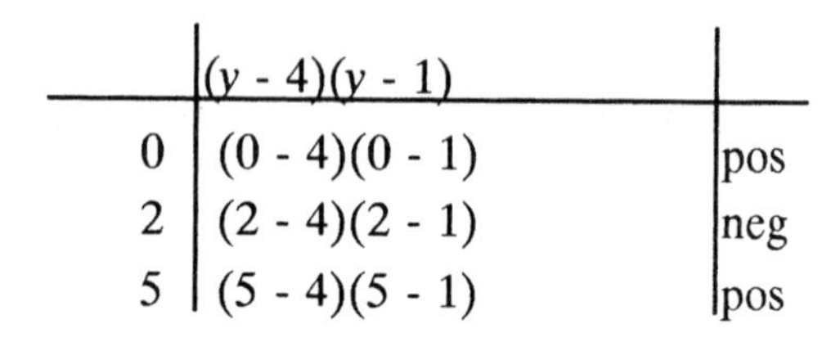

	$(y - 4)(y - 1)$	
0	$(0 - 4)(0 - 1)$	pos
2	$(2 - 4)(2 - 1)$	neg
5	$(5 - 4)(5 - 1)$	pos

We want $(y - 4)(y - 1) > 0$; pos

Solution: $y < 1$ or $y > 4$

107. $a^2 < 81$

$a^2 - 81 < 0$

$(a - 9)(a + 9) < 0$

$a - 9 = 0 \qquad a + 9 = 0$

$a = 9 \qquad a = -9$

The intervals are $a < -9$, $-9 < a < 9$, $a > 9$

Test points:

	$(a - 9)(a + 9)$	
-10	$(-10 - 9)(-10 + 9)$	pos
0	$(0 - 9)(0 + 9)$	neg
10	$(10 - 9)(10 + 9)$	pos

We want $(a - 9)(a + 9) < 0$; neg

$-9 < a < 9$

109. $5s^2 - 18s \geq 8$

$5s^2 - 18s - 8 \geq 0$

$(5s + 2)(s - 4) \geq 0$

$5s + 2 = 0 \qquad s - 4 = 0$

$s = -\frac{2}{5} \qquad s = 4$

The intervals are $s < -\frac{2}{5}$, $-\frac{2}{5} < s < 4$, $s > 4$

Test points:

	$(5s + 2)(s - 4)$	
-1	$[5(-1) + 2](-1 - 4)$	pos
0	$(5 \cdot 0 + 2)(0 - 4)$	neg
5	$(5 \cdot 5 + 2)(5 - 4)$	pos

We want $(5s + 2)(s - y) \geq 0$; pos or 0

Solution: $s \leq -\frac{2}{5}$ or $s \geq 4$

111. $\dfrac{x - 3}{x + 2} < 0$

$x - 3 = 0 \qquad x + 2 = 0$

$x = 3 \qquad x = -2$

The intervals are $x < -2$, $-2 < x < 3$, $x > 3$

Test points:

	$(x - 3)/(x + 2)$	
-3	(-3 - 3)/(-3 + 2)	pos
0	(0 - 3)/(0 + 2)	neg
4	(4 - 3)/(4 + 2)	pos

We want $\dfrac{x - 3}{x + 2} < 0$; neg

Solution: $-2 < x < 3$

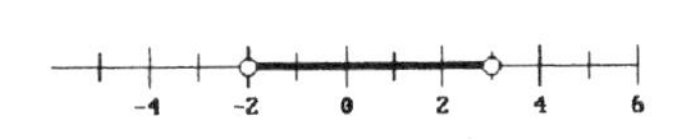

113. $\dfrac{x - 3}{x + 2} \geq 0$; pos or 0

(See exercise #111)

Solution: $x < -2$ or $x \geq 3$

115. $\dfrac{2x + 1}{x - 3} < 2$

$$\frac{2x + 1}{x - 3} - 2 < 0$$

$$\frac{2x + 1 - 2(x - 3)}{x - 3} < 0$$

$$\frac{4}{x - 3} < 0$$

$x - 3 = 0$

$x = 3$

The intervals are $x < 3$, $x > 3$

Test points:

	$4/(x - 3)$	
0	4/(0 - 3)	neg
4	4/(4 - 3)	pos

We want $\dfrac{4}{x - 3} < 0$; neg

Solution: $x < 3$

117. $\dfrac{5}{x + 4} \geq 4$

$$\frac{5}{x + 4} - 4 \geq 0$$

$$\frac{5 - 4(x + 4)}{x + 4} \geq 0$$

$$\frac{-4x - 11}{x + 4} \geq 0$$

$-4x - 11 = 0 \qquad x + 4 = 0$

$x = -\dfrac{11}{4} \qquad x = -4$

The intervals are $x < -4$, $-4 < x < -\dfrac{11}{4}$, $x > -\dfrac{11}{4}$

Test points:

	$(-4x - 11)/(x + 4)$	
-5	(-4(-5) - 11)/(5 + 4)	neg
-3	(-4(-3) - 11)/(-3 + 4)	pos
0	(-4(0) - 11)/(0 + 4)	neg

We want $\dfrac{-4x - 11}{x + 4} \geq 0$; pos or 0

Solution: $-4 < x \leq -\dfrac{11}{4}$

119. The graph appears to be below the x-axis for $x < -3.5$ or $x > 0.5$.

121. $d = \sqrt{(2 - 0)^2 + (6 - 0)^2}$

$= \sqrt{2^2 + 6^2}$

$= \sqrt{4 + 36}$

$= \sqrt{40}$

$= 2\sqrt{10}$

midpoint $= \left(\dfrac{0 + 2}{2}, \dfrac{0 + 6}{2}\right) = (1, 3)$

123. $d = |2 - (-2)| = 4$

$$\text{midpoint} = \left(\frac{2 - 2}{2}, \frac{5 + 5}{2}\right) = (0, 5)$$

125. $d = \sqrt{(6 - 4)^2 + (-4 - (-6))^2}$
$= \sqrt{2^2 + 2^2} = \sqrt{4 + 4} = \sqrt{8} = 2\sqrt{2}$

$$\text{midpoint} = \left(\frac{6 + 4}{2}, \frac{-4 - 6}{2}\right) = (5, -5)$$

127. $d = \sqrt{(-2 - 2)^2 + (-5 - 5)^2}$
$= \sqrt{(-4)^2 + (-10)^2}$
$= \sqrt{16 + 100} = \sqrt{116} = 2\sqrt{29}$

$$\text{midpoint} = \left(\frac{-2 + 2}{2}, \frac{-5 + 5}{2}\right) = (0, 0)$$

129. $d(P, Q) = \sqrt{(5 - 3)^2 + (9 - 6)^2}$
$= \sqrt{4 + 9} = \sqrt{13}$

$d(Q, R) = \sqrt{(8 - 5)^2 + (7 - 9)^2}$
$= \sqrt{9 + 4} = \sqrt{13}$

$d(P, R) = \sqrt{(8 - 3)^2 + (7 - 6)^2}$
$= \sqrt{25 + 1} = \sqrt{26}$

$$[d(P, Q)]^2 + [d(Q, R)]^2 = [d(P, R)]^2$$
$$(\sqrt{13})^2 + (\sqrt{13})^2 = (\sqrt{26})^2$$
$$13 + 13 = 26$$
$$26 = 26$$

True

Yes, they form a right triangle.

131. $x^2 + y^2 = 100$
Center: $(0, 0)$
$r = \sqrt{100} = 10$

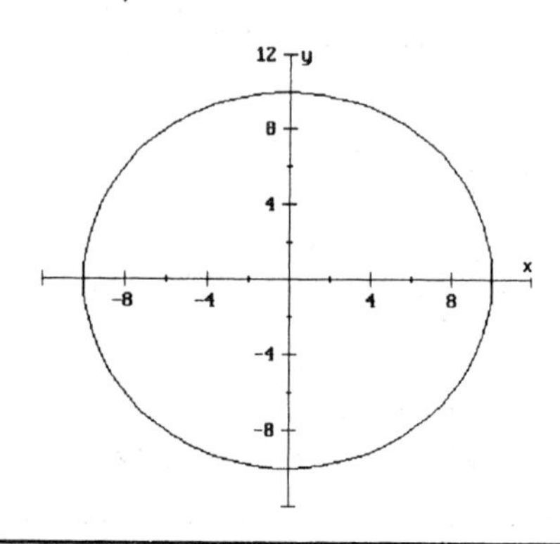

133.
$$x^2 + y^2 - 4x - 14y = -52$$
$$x^2 - 4x + y^2 - 14y = -52$$
$$(x^2 - 4x + 4) + (y^2 - 14y + 49) = -52 + 4 + 49$$
$$(x - 2)^2 + (y - 7)^2 = 1$$
Center: $(2, 7)$
$r = \sqrt{1} = 1$

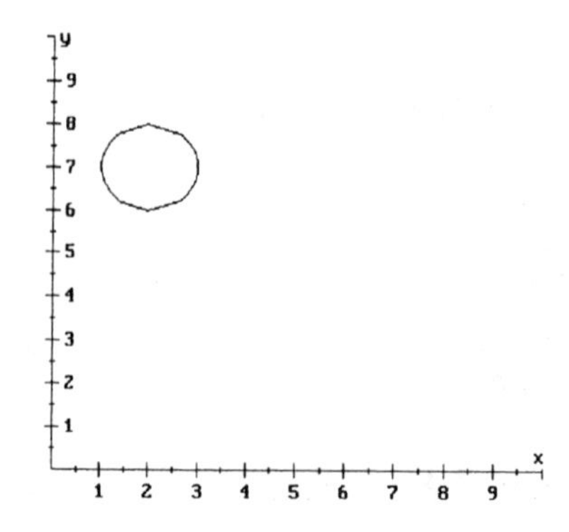

135.
$$x^2 + y^2 - 6x + 4y = 68$$
$$x^2 - 6x + y^2 + 4y = 68$$
$$(x^2 - 6x + 9) + (y^2 + 4y + 4) = 68 + 9 + 4$$
$$(x - 3)^2 + (y + 2)^2 = 81$$
Center: $(3, -2)$
$r = \sqrt{81} = 9$

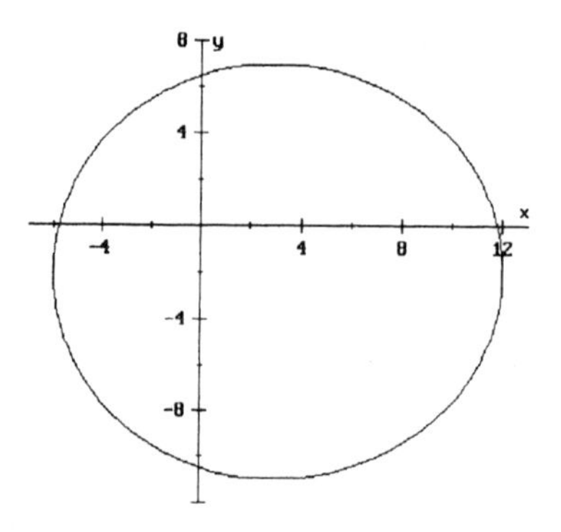

137. Center = midpoint
$$= \left(\frac{-2 + 6}{2}, \frac{4 + 8}{2}\right)$$
$$= (2, 6)$$

$r = \sqrt{(6 - 2)^2 + (8 - 6)^2}$
$= \sqrt{16 + 4}$
$= \sqrt{20}$
$r^2 = (\sqrt{20})^2 = 20$

$$(x - h)^2 + (y - k)^2 = r^2$$
$$(x - 2)^2 + (y - 6)^2 = 20$$

139. $P(x) = 10000(-x^2 + 12x - 35)$

$$-\frac{b}{2a} = -\frac{12}{2(-1)}$$
$$= 6$$

\$6 per ticket would produce the maximum profit.

141. Let x = the number

$$x^2 + 4 = 36$$
$$x^2 = 32$$
$$x = \pm\sqrt{32}$$
$$x = \pm 4\sqrt{2}$$

The numbers are $\pm 4\sqrt{2}$.

143. Let x = the number

$$x + \frac{1}{x} = \frac{53}{14}$$
$$14x\left(x + \frac{1}{x}\right) = 14x\left(\frac{53}{14}\right)$$
$$14x^2 + 14 = 53x$$
$$14x^2 - 53x + 14 = 0$$
$$(2x - 7)(7x - 2) = 0$$
$$2x - 7 = 0 \quad \text{or} \quad 7x - 2 = 0$$
$$2x = 7 \qquad 7x = 2$$
$$x = \frac{7}{2} \qquad x = \frac{2}{7}$$

The numbers are $\frac{7}{2}$ and $\frac{2}{7}$.

145. width: x
length: $2x$

$$A = wl$$
$$50 = x(2x)$$
$$50 = 2x^2$$
$$25 = x^2$$
$$5 = x$$
$$2x = 2(5) = 10$$

The dimensions are 5 ft by 10 ft.

147.

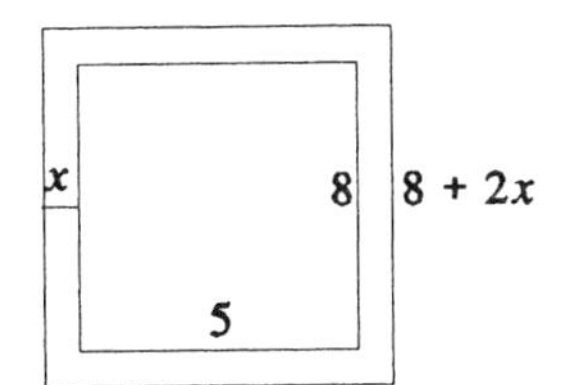

$$(8 + 2x)(5 + 2x) - (8)(5) = 114$$
$$40 + 26x + 4x^2 - 40 = 114$$
$$26x + 4x^2 = 114$$
$$4x^2 + 26x - 114 = 0$$
$$2x^2 + 13x - 57 = 0$$
$$(2x + 19)(x - 3) = 0$$
$$2x + 19 = 0 \quad \text{or} \quad x - 3 = 0$$
$$x = -\frac{19}{2} \quad \text{or} \quad x = 3$$

Since $x > 0$, $x = 3$. The width of the frame is 3″.

149. $a^2 + b^2 = c^2$

$$5^2 + 15^2 = c^2$$
$$25 + 225 = c^2$$
$$250 = c^2$$
$$5\sqrt{10} = c$$

The hypotenuse is $5\sqrt{10}$″.

151. $a^2 + b^2 = c^2$

$$5^2 + 4^2 = c^2$$
$$25 + 16 = c^2$$
$$41 = c^2$$
$$\sqrt{41} = c$$

The length of the diagonal is $\sqrt{41} \approx 6.4$″.

153. Let x = rate of the wind

then $200 - x$ = rate into the wind
and $200 + x$ = rate with the wind

$$(\text{time into the wind}) + (\text{time with the wind}) = \frac{25}{8}$$
$$\frac{300}{200 - x} + \frac{300}{200 + x} = \frac{25}{8}$$
$$8(200 - x)(200 + x)\left(\frac{300}{200 - x} + \frac{300}{200 + x}\right) = 8(200 - x)(200 + x)\left(\frac{25}{8}\right)$$
$$2400(200 + x) + 2400(200 - x) = 25(40000 - x^2)$$
$$480000 + 2400x + 480000 - 2400x = 1000000 - 25x^2$$

$$960000 = 1000000 - 25x^2$$
$$-40000 = -25x^2$$
$$1600 = x^2$$
$$40 = x$$

The wind's rate is 40 mph.

CHAPTER 8 PRACTICE TEST

1.

(a) $(3z - 1)(z - 4) = z^2 - 8z + 7$

$$3z^2 - 13z + 4 = z^2 - 8z + 7$$
$$2z^2 - 5z - 3 = 0$$
$$(2z + 1)(z - 3) = 0$$
$$2z + 1 = 0 \quad \text{or} \quad z - 3 = 0$$
$$2z = -1 \qquad z = 3$$
$$z = -\frac{1}{2} \quad \text{or} \quad z = 3$$

(b) $3 + \dfrac{5}{x^2} = 4$

$$\frac{5}{x^2} = 1$$
$$5 = x^2$$
$$\pm\sqrt{5} = x$$

3. $2x^2 - 3x + 5 = 0$

$$b^2 - 4ac = (-3)^2 - 4(2)(5)$$
$$= -31 < 0$$

The roots are not real.

5. $\sqrt{3x} = 2 + \sqrt{x + 4}$

$$\left(\sqrt{3x}\right)^2 = \left(2 + \sqrt{x + 4}\right)^2$$
$$3x = 4 + 4\sqrt{x + 4} + x + 4$$
$$3x = 8 + x + 4\sqrt{x + 4}$$
$$2x - 8 = 4\sqrt{x + 4}$$
$$x - 4 = 2\sqrt{x + 4}$$
$$(x - 4)^2 = \left(2\sqrt{x + 4}\right)^2$$
$$x^2 - 8x + 16 = 4(x + 4)$$
$$x^2 - 8x + 16 = 4x + 16$$
$$x^2 - 12x = 0$$
$$x(x - 12) = 0$$
$$x = 0 \quad \text{or} \quad x - 12 = 0$$
$$x = 12$$

When you check $x = 0$, you find it to be extraneous. The only solution is $x = 12$.

7. (a) $x^2 - 5x \geq 36$

$$x^2 - 5x - 36 \geq 0$$
$$(x - 9)(x + 4) \geq 0$$
$$x - 9 = 0 \qquad x + 4 = 0$$
$$x = 9 \qquad x = -4$$

The intervals are $x < -4$, $-4 < x < 9$, $x > 9$

Test points:

	$(x - 9)(x + 4)$	
-5	(-5 - 9)(-5 + 4)	pos
0	(0 - 9)(0 + 4)	neg
10	(10 - 9)(10 + 4)	pos

We want $(x - 9)(x + 4) \geq 0$; pos or 0

Solution: $x \leq -4$ or $x \geq 9$

(b) $\dfrac{3x - 2}{x - 6} < 0$

$$3x - 2 = 0 \qquad x - 6 = 0$$
$$x = \frac{2}{3} \qquad x = 6$$

The intervals are $x < \frac{2}{3}$, $\frac{2}{3} < x < 6$, $x > 6$

Test points:

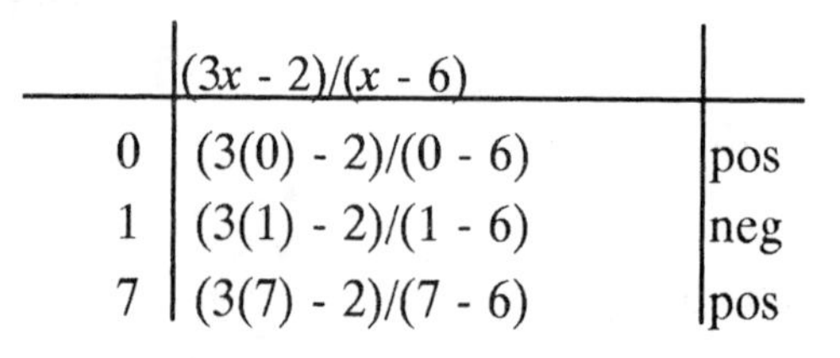

	$(3x - 2)/(x - 6)$	
0	(3(0) - 2)/(0 - 6)	pos
1	(3(1) - 2)/(1 - 6)	neg
7	(3(7) - 2)/(7 - 6)	pos

We want $\dfrac{3x - 2}{x - 6} < 0$; neg

Solution: $\dfrac{2}{3} < x < 6$

9. Let x = rate of the boat in still water,

then $x - 3$ = rate upstream
and $x + 3$ = rate downstream

$$\frac{15}{x-3} + \frac{12}{x+3} = \frac{9}{4}$$

$$4(x-3)(x+3)\left(\frac{15}{x-3} + \frac{12}{x+3}\right) = 4(x-3)(x+3)\left(\frac{9}{4}\right)$$

$$\begin{aligned}
60(x + 3) + 48(x - 3) &= 9(x^2 - 9) \\
60x + 180 + 48x - 144 &= 9x^2 - 81 \\
108x + 36 &= 9x^2 - 81 \\
0 &= 9x^2 - 108x - 117 \\
0 &= x^2 - 12x - 13 \\
0 &= (x - 13)(x + 1) \\
x - 13 = 0 \quad &\text{or} \quad x + 1 = 0 \\
x = 13 \quad & \qquad x = -1
\end{aligned}$$

Since $x > 0$, $x = 13$
The boat's rate is 13 mph.

11. $C(x) = \dfrac{-x^2}{10} + 100x - 24000$

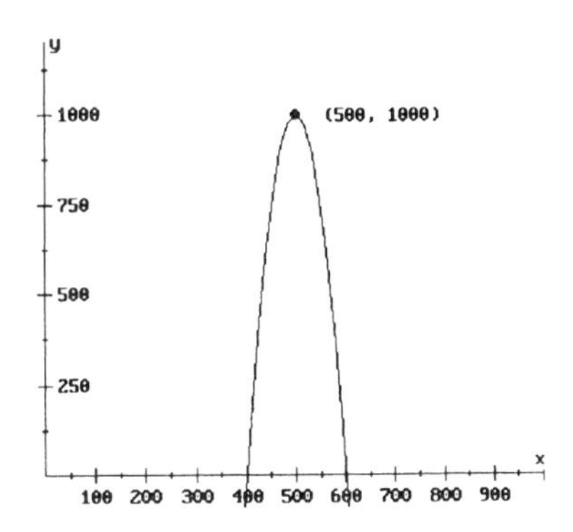

maximum: (500, 1000)

500 widgets would produce the maximum cost of $1000.

13.
$$\begin{aligned}
x^2 + y^2 - 8x + 12y &= -2 \\
x^2 - 8x + y^2 + 12y &= -2 \\
(x^2 - 8x + 16) + (y^2 + 12y + 36) &= -2 + 16 + 36 \\
(x - 4)^2 + (y + 6)^2 &= 50
\end{aligned}$$

Center: $(4, -6)$

$r = \sqrt{50} = 5\sqrt{2}$

CHAPTER 9

Exercises 9.1

1. $f(x) = 7x - 2$
   $f(-2) = 7(-2) - 2 = -16$
   $f(0) = 7(0) - 2 = -2$
   $f(2) = 7(2) - 2 = 12$

3. $f(x) = \sqrt{3x - 2}$
   -2 and 0 are not in the domain of $f(x)$.
   $f(2) = \sqrt{3(2) - 2} = \sqrt{4} = 2$

5. $f(x) = \sqrt{2 - x}$
   $f(-2) = \sqrt{2 - (-2)}$
   $= \sqrt{4}$
   $= 2$
   $f(0) = \sqrt{2 - 0}$
   $= \sqrt{2}$
   $f(2) = \sqrt{2 - 2}$
   $= \sqrt{0}$
   $= 0$

7. $f(x) = \dfrac{2}{x}$
   0 is not in the domain of $f(x)$.
   $f(-2) = \dfrac{2}{-2} = -1$
   $f(2) = \dfrac{2}{2} = 1$

9. $f(x) = \dfrac{x + 2}{x - 2}$
   2 is not in the domain of $f(x)$.
   $f(-2) = \dfrac{-2 + 2}{-2 - 2}$
   $= \dfrac{0}{-4}$
   $= 0$
   $f(0) = \dfrac{0 + 2}{0 - 2}$
   $= -1$

11. all real numbers

13. all real numbers

15. all real numbers

17. $2x - 1 \geq 0$
    $2x \geq 1$
    $x \geq \dfrac{1}{2}$
    $\left\{x \,\middle|\, x \geq \dfrac{1}{2}\right\}$

19. $2 - 3x \geq 0$
    $-3x \geq -2$
    $x \leq \dfrac{2}{3}$
    $\left\{x \,\middle|\, x \leq \dfrac{2}{3}\right\}$

21. $\{x \mid x \neq 0\}$

23. $x - 4 \neq 0$
    $x \neq 4$
    $\{x \mid x \neq 4\}$

25. $x - 1 \neq 0 \qquad x + 2 \neq 0$
    $x \neq 1 \qquad x \neq -2$
    $\{x \mid x \neq 1, -2\}$

27. $x^2 - x - 12 \neq 0$
    $(x - 4)(x + 3) \neq 0$
    $x - 4 \neq 0 \qquad x + 3 \neq 0$
    $x \neq 4 \qquad x \neq -3$
    $\{x \mid x \neq -3, 4\}$

29. $\{x \mid x > 0\}$

31. $2x - 1 > 0$
    $2x > 1$
    $x > \dfrac{1}{2}$
    $\left\{x \,\middle|\, x > \dfrac{1}{2}\right\}$

33. $f(x) = 2x - 3$

    (a) $f(5) = 2(5) - 3 = 7$
    $f(x) + f(5) = 2x - 3 + 7$
    $= 2x + 4$

    (b) $f(x + 5) = 2(x + 5) - 3$
    $= 2x + 10 - 3$
    $= 2x + 7$

(c) $f(2x) = 2(2x) - 3$
$= 4x - 3$

(d) $2f(x) = 2(2x - 3)$
$= 4x - 6$

(e) $f(x + 5) - f(x)$ (see part b)
$= (2x + 7) - (2x - 3)$
$= 2x + 7 - 2x + 3$
$= 10$

(f) $\dfrac{f(x + 5) - f(x)}{5} = \dfrac{10}{5} = 2$

(see part e)

(g) $f(x + h) = 2(x + h) - 3$
$= 2x + 2h - 3$

(h) $f(h) = 2h - 3$
$f(x) + f(h) = (2x - 3) + (2h - 3)$
$= 2x + 2h - 6$

(i) $\dfrac{f(x + h) - f(x)}{h}$

$= \dfrac{2x + 2h - 3 - (2x - 3)}{h}$ (see part g)

$= \dfrac{2x + 2h - 3 - 2x + 3}{h}$

$= \dfrac{2h}{h}$
$= 2$

35. $f(x) = \dfrac{5}{x - 1}$

(a) $f(3) = \dfrac{5}{3 - 1} = \dfrac{5}{2}$

$f(x) + f(3) = \dfrac{5}{x - 1} + \dfrac{5}{2}$

$= \dfrac{5(2) + 5(x - 1)}{2(x - 1)}$

$= \dfrac{10 + 5x - 5}{2x - 2}$

$= \dfrac{5x + 5}{2x - 2}$

(b) $f(x + 3) = \dfrac{5}{x + 3 - 1}$

$= \dfrac{5}{x + 2}$

(c) $f(3x) = \dfrac{5}{3x - 1}$

(d) $3f(x) = 3\left(\dfrac{5}{x - 1}\right)$

$= \dfrac{15}{x - 1}$

(e) $f(x + 3) - f(x)$ (see part b)

$= \dfrac{5}{x + 2} - \dfrac{5}{x - 1}$

$= \dfrac{5(x - 1) - 5(x + 2)}{(x + 2)(x - 1)}$

$= \dfrac{5x - 5 - 5x - 10}{(x + 2)(x - 1)}$

$= -\dfrac{15}{(x + 2)(x - 1)}$

(f) $\dfrac{f(x + 3) - f(x)}{3}$ (see part e)

$= \dfrac{\dfrac{-15}{(x + 2)(x - 2)}}{3}$

$= \dfrac{-15}{3(x + 2)(x - 1)}$

$= \dfrac{-5}{(x + 2)(x - 1)}$

(g) $f(x + h)$

$= \dfrac{5}{x + h - 1}$

(h) $f(h) = \dfrac{5}{h - 1}$

$f(x) + f(h)$

$$= \frac{5}{x - 1} + \frac{5}{h - 1}$$

$$= \frac{5(h - 1) + 5(x - 1)}{(x - 1)(h - 1)}$$

$$= \frac{5h - 5 + 5x - 5}{(x - 1)(h - 1)}$$

$$= \frac{5x + 5h - 10}{(x - 1)(h - 1)}$$

(i) $\dfrac{f(x + h) - f(x)}{h}$ (see part g)

$$= \frac{\dfrac{5}{x + h - 1} - \dfrac{5}{x - 1}}{h}$$

$$= \frac{\dfrac{5(x - 1) - 5(x + h - 1)}{(x - 1)(x + h - 1)}}{h}$$

$$= \frac{5x - 5 - 5x - 5h + 5}{h(x - 1)(x + h - 1)}$$

$$= \frac{-5h}{h(x - 1)(x + h - 1)}$$

$$= \frac{-5}{(x - 1)(x + h - 1)}$$

37. $f(x) = \begin{cases} x - 4 & \text{if } x \le 2 \\ x + 1 & \text{if } x > 2 \end{cases}$

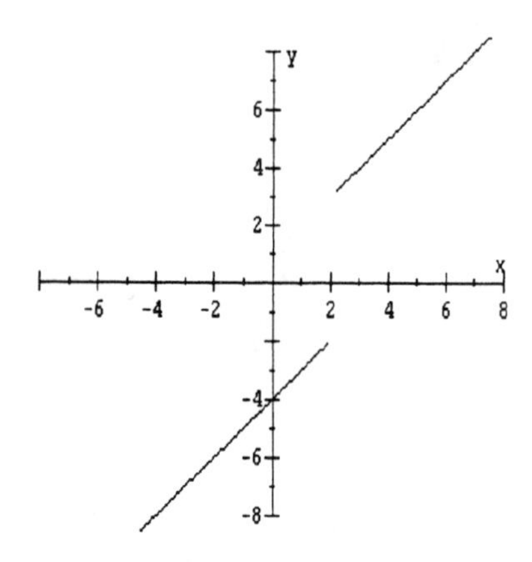

$f(0) = 0 - 4 = -4$
$f(2) = 2 - 4 = -2$
$f(3) = 3 + 1 = 4$

39. $h(x) = \begin{cases} 2x + 3 & \text{if } -4 \le x < 2 \\ 6 - x & \text{if } x \ge 2 \end{cases}$

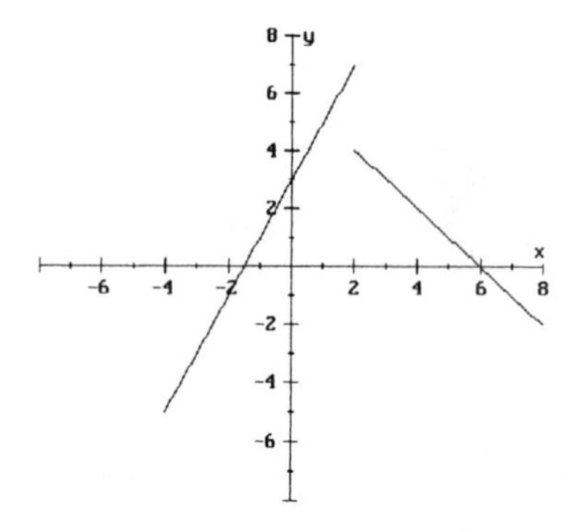

$h(-1) = 2(-1) + 3 = 1$
$h(2) = 6 - 2 = 4$
$h(5) = 6 - 5 = 1$

41. $g(x) = \begin{cases} x^2 - 1 & \text{if } -3 \le x \le 3 \\ 14 - 2x & \text{if } x > 3 \end{cases}$

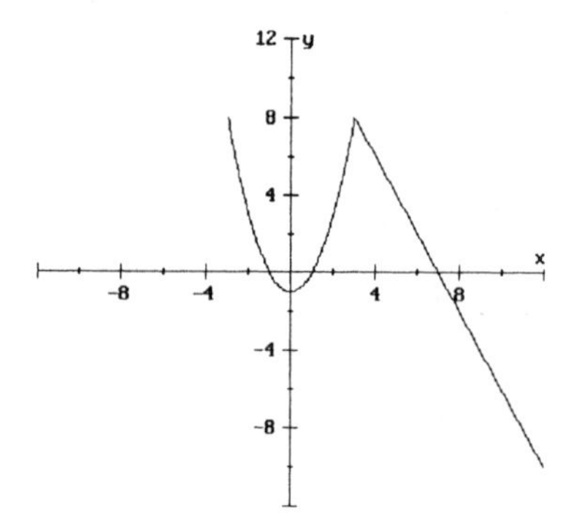

$g(-3) = (-3)^2 - 1 = 8$
$g(3) = 3^2 - 1 = 8$
$g(6) = 14 - 2(6) = 2$

43. $f(x) = \begin{cases} 9 - x^2 & \text{if } -2 \le x < 2 \\ x + 1 & \text{if } x \ge 2 \end{cases}$

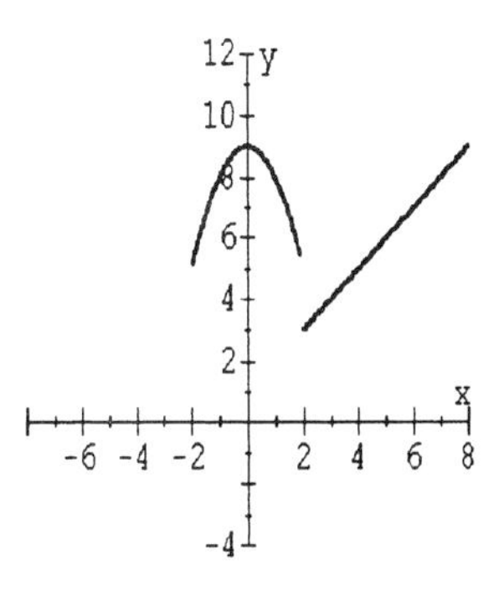

-3 is not in the domain of $f(x)$.

$f(1) = 9 - 1^2 = 8$
$f(5) = 5 + 3 = 8$

45. (a) $$C = \begin{cases} 0.61t, & \text{if } 0 \le t \le 200 \\ 0.61(200) + 0.59(t - 200) & \text{if } t > 200 \end{cases}$$

$$C = \begin{cases} 0.61t, & \text{if } 0 \le t \le 200 \\ 0.59t + 4 & \text{if } t > 200 \end{cases}$$

(b) $C(150) = 0.61(150) = \$91.50$

$C(200) = 0.61(200) = \$122$

$C(300) = 0.59(300) + 4 = \$181$

47. (a) $$C(k) = \begin{cases} 2.50 + 0.125k, \text{ if } 0 \le k \le 560 \\ 0.118(k - 560) + 2.50 + 0.125(560), \text{ if } k > 560 \end{cases}$$

$$= \begin{cases} 2.50 + 0.125k, \text{ if } 0 \le k \le 560 \\ 0.118(k - 560) + 72.50, \text{ if } k > 560 \end{cases}$$

(b)

k	C
300	\$40
600	\$77.22
1000	\$124.42

51. $\frac{x}{x+1} - \frac{3}{x-1} + 1$

$= \frac{x(x-1)}{(x+1)(x-1)} - \frac{3(x+1)}{(x+1)(x-1)} + \frac{(x+1)(x-1)}{(x+1)(x-1)}$

$= \frac{x(x-1) - 3(x+1) + (x+1)(x-1)}{(x+1)(x-1)}$

$= \frac{x^2 - x - 3x - 3 + x^2 - 1}{(x+1)(x-1)}$

$= \frac{2x^2 - 4x - 4}{(x+1)(x-1)}$

53. Let x = rate going
then $x + 12$ = rate returning

$$\frac{60}{x+12} = \frac{60}{x} - \frac{1}{4}$$

$$4x(x+12)\left(\frac{60}{x+12}\right) = 4x(x+12)\left(\frac{60}{x} - \frac{1}{4}\right)$$

$$\begin{aligned} 240x &= 240(x+12) - x(x+12) \\ 240x &= 240x + 2880 - x^2 - 12x \\ x^2 + 12x - 2880 &= 0 \\ (x+60)(x-48) &= 0 \\ x + 60 = 0 \quad &\text{or} \quad x - 48 = 0 \\ x = -60 \quad &\text{or} \quad x = 48 \end{aligned}$$

Since $x > 0$, $x = 48$

$x = 48$ going rate

$x + 12 = 48 + 12 = 60$ returning rate

$\frac{60}{x} = \frac{60}{48} = 1\frac{1}{4}$ going time

$\frac{60}{x+12} = \frac{60}{60} = 1$ returning time

Exercises 9.2

1. $(f + g)(0) = f(0) + g(0)$
$= [0^2 - 2(0) - 3] + (0 - 1)$
$= -4$

3. $(g - f)(2) = g(2) - f(2)$
$= (2 - 1) - [2^2 - 2(2) - 3]$
$= 1 - (-3)$
$= 4$

5. $\left(\frac{f}{g}\right)(3) = \frac{f(3)}{g(3)}$

$= \frac{3^2 - 2(3) - 3}{3 - 1}$

$= \frac{0}{2}$

$= 0$

7. $\left(\frac{f}{g}\right)(1) = \frac{f(1)}{g(1)}$

$= \frac{1^2 - 2(1) - 3}{1 - 1}$

$= \frac{-4}{0}$

undefined

9. $(f + g)(x) = f(x) + g(x)$
$= (x^2 - 2x - 3) + (x - 1)$
$= x^2 - x - 4$

11. $\left(\frac{g}{f}\right) = \frac{g(x)}{f(x)}$

$= \frac{x - 1}{x^2 - 2x - 3}$

13. $(h + g)(2) = h(2) + g(2)$

$= \frac{1}{2(2) - 1} + 2(2) + 1$

$= \frac{1}{3} + 5$

$= \frac{16}{3}$

15. $(g - h)\left(\frac{1}{2}\right) = g\left(\frac{1}{2}\right) - h\left(\frac{1}{2}\right)$

$$= 2\left(\frac{1}{2}\right) + 1 - \frac{1}{2\left(\frac{1}{2}\right) - 1}$$

$$= 2 - \frac{1}{0}$$

undefined

17. $\left(\frac{h}{g}\right)(3) = \frac{h(3)}{g(3)}$

$$= \frac{\frac{1}{2(3) - 1}}{2(3) + 1}$$

$$= \frac{\frac{1}{5}}{7}$$

$$= \frac{1}{35}$$

19. $\left(\frac{h}{g}\right)(x) = \frac{h(x)}{g(x)}$

$$= \frac{\frac{1}{2x - 1}}{2x + 1}$$

$$= \frac{1}{(2x - 1)(2x + 1)}$$

21. $g(3) = \sqrt{3 + 1} = 2$
$f[g(3)] = [g(3)]^2 - 4$
$= 2^2 - 4$
$= 0$

23. $f(3) = 3^2 - 4 = 5$
$g[f(3)] = \sqrt{f(3) + 1}$
$= \sqrt{5 + 1}$
$= \sqrt{6}$

25. $f[g(x)] = [g(x)]^2 - 4$
$= \left(\sqrt{x + 1}\right)^2 - 4$
$= x + 1 - 4$
$= x - 3$

27. $g[f(x)] = \sqrt{f(x) + 1}$
$= \sqrt{x^2 - 4 + 1}$
$= \sqrt{x^2 - 3}$

29. $h(3) = \frac{1}{3}$

$g[h(3)] = \sqrt{h(3) + 1}$

$$= \sqrt{\frac{1}{3} + 1}$$

$$= \sqrt{\frac{4}{3}}$$

$$= \frac{2}{\sqrt{3}}$$

$$= \frac{2\sqrt{3}}{3}$$

31. $g\left(\frac{1}{2}\right) = \sqrt{\frac{1}{2} + 1} = \sqrt{\frac{3}{2}} = \frac{\sqrt{6}}{2}$

$f\left[g\left(\frac{1}{2}\right)\right] = \left[g\left(\frac{1}{2}\right)\right]^2 - 4$

$$= \left(\frac{\sqrt{6}}{2}\right)^2 - 4$$

$$= \frac{3}{2} - 4$$

$$= -\frac{5}{2}$$

33. $g[h(x)] = \sqrt{h(x) + 1}$

$$= \sqrt{\frac{1}{x} + 1}$$

$$= \sqrt{\frac{1 + x}{x}}$$

$$= \frac{\sqrt{x + x^2}}{x}$$

35. $f(1) = 1^2 - 4 = -3$

This is not in the domain of $g(x)$, hence $x = 1$ is not in the domain of $g[f(x)]$.

37. $g(2) = 3(2) - 1 = 5$

$$f[g(2)] = [g(2)]^2 - 3[g(2)] + 5$$
$$= 5^2 - 3(5) + 5$$
$$= 15$$

39. $g(-1) = 3(-1) - 1 = -4$

$$f[g(-1)] = [g(-1)]^2 - 3[g(-1)] + 5$$
$$= (-4)^2 - 3(-4) + 5$$
$$= 33$$

41. $h\left(\frac{1}{3}\right) = \dfrac{\frac{1}{3} - 1}{4 - 2\left(\frac{1}{3}\right)}$

$$= \frac{-\frac{2}{3}}{\frac{10}{3}}$$

$$= -\frac{1}{5}$$

$$f\left[h\left(\frac{1}{3}\right)\right] = \left[h\left(\frac{1}{3}\right)\right]^2 - 3\left[h\left(\frac{1}{3}\right)\right] + 5$$

$$= \left(-\frac{1}{5}\right)^2 - 3\left(-\frac{1}{5}\right) + 5$$

$$= \frac{141}{25}$$

43. $h(2) = \dfrac{2 - 1}{4 - 2(2)} = \dfrac{1}{0}$

2 is not in the domain of $g[h(x)]$.

45. $f\left(\frac{1}{2}\right) = \left(\frac{1}{2}\right)^2 - 3\left(\frac{1}{2}\right) + 5 = \frac{15}{4}$

$$g\left[f\left(\frac{1}{2}\right)\right] = 3\left[f\left(\frac{1}{2}\right)\right] - 1$$

$$= 3\left(\frac{15}{4}\right) - 1$$

$$= \frac{41}{4}$$

47. $f[g(x)] = [g(x)]^2 - 3[g(x)] + 5$

$$= (3x - 1)^2 - 3(3x - 1) + 5$$
$$= 9x^2 - 6x + 1 - 9x + 3 + 5$$
$$= 9x^2 - 15x + 9$$

49. $g[h(x)] = 3[h(x)] - 1$

$$= 3\left(\frac{x - 1}{4 - 2x}\right) - 1$$

$$= \frac{3(x - 1) - (4 - 2x)}{4 - 2x}$$

$$= \frac{3x - 3 - 4 + 2x}{4 - 2x}$$

$$= \frac{5x - 7}{4 - 2x}$$

51. $g[g(x)] = 3[g(x)] - 1$

$$= 3(3x - 1) - 1$$
$$= 9x - 3 - 1$$
$$= 9x - 4$$

53. (a) $r(t) = 3t$

$$r(2) = 3(2) = 6$$

$$A(r) = \pi r^2$$
$$A[r(2)] = \pi[r(2)]^2$$
$$= \pi(6)^2$$
$$= 36\pi$$

The area is $36\pi \approx 113.10$ sq. in.

(b) $A[r(t)] = \pi[r(t)]^2$

$$= \pi(3t)^2$$
$$= 9\pi t^2$$

The area is $9\pi t^2$ sq. in.

55. (*a*) $r(t) = 3t$
$r(5) = 3(5) = 15$

$$V[r(5)] = \frac{4}{3}\pi[r(5)]^3$$
$$= \frac{4}{3}\pi(15)^3$$
$$= 4500\pi$$

The volume is $4500\pi \approx 14137.17$ cu. in

(b) $$V[r(t)] = \frac{4}{3}\pi[r(t)]^3$$
$$= \frac{4}{3}\pi(3t)^3$$
$$= 36\pi t^3$$

The volume is $36\pi t^3$ cu. in.

57. $C = 8P$

59. $A = 1000$
$s^2 = 1000$
$s = \sqrt{1000} = 10\sqrt{10}$

$P = 4s$
$= 4(10\sqrt{10})$
$= 40\sqrt{10}$

$C = 40x\sqrt{10}$

63. $$\left(\frac{16x^2y^8}{x^{-1/2}}\right)^{1/4}$$
$= (16x^{5/2}y^8)^{1/4}$
$= 2x^{5/8}y^2$

65. $\sqrt{32x^8y^5}$
$= 4x^4y^2\sqrt{2y}$

Exercises 9.3

1. Linear function
3. Quadratic function
5. Polynomial function
7. Quadratic function
9. Square root function
11. Absolute value function
13. Linear function
15. Quadratic function

17. $f(x) = |4x - 1|$
$f(-2) = |4(-2) - 1| = 9$
$f(-1) = |4(-1) - 1| = 5$
$f(0) = |4(0) - 1| = 1$
$f(1) = |4(1) - 1| = 3$
$f(2) = |4(2) - 1| = 7$

19. $f(x) = |4x| - 1$
$f(-2) = |4(-2)| - 1 = 7$
$f(-1) = |4(-1)| - 1 = 3$
$f(0) = |4(0)| - 1 = -1$
$f(1) = |4(1)| - 1 = 3$
$f(2) = |4(2)| - 1 = 7$

21. $f(x) = |4x| + 1$
$f(-2) = |4(-2)| + 1 = 9$
$f(-1) = |4(-1)| + 1 = 5$
$f(0) = |4(0)| + 1 = 1$
$f(1) = |4(1)| + 1 = 5$
$f(2) = |4(2)| + 1 = 9$

23. $f(x) = |4x + 1|$
$f(-2) = |4(-2) + 1| = 7$
$f(-1) = |4(-1) + 1| = 3$
$f(0) = |4(0) + 1| = 1$
$f(1) = |4(1) + 1| = 5$
$f(2) = |4(2) + 1| = 9$

25. $f(x) = 4 - 3x$
Linear function
$(0, 4)$; $m = -3$

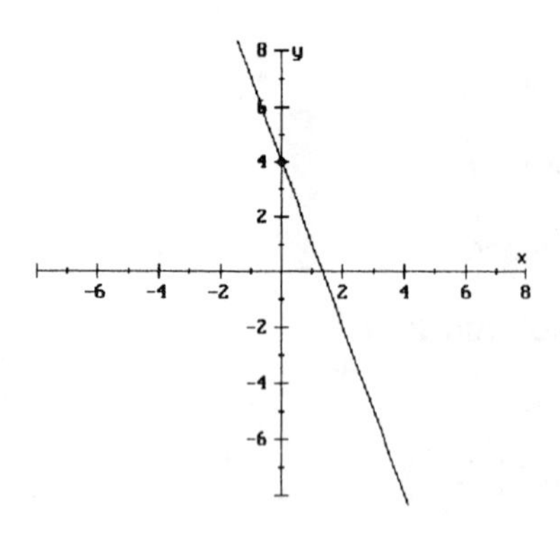

27. $f(x) = x^2 - 9$
Quadratic function

$$\frac{-b}{2a} = -\frac{0}{2(1)} = 0$$

$f(0) = 0^2 - 9 = -9$
Vertex: $(0, -9)$
x-intercepts: Let $y = 0$
$0 = x^2 - 9$
$9 = x^2$
$\pm 3 = x$
y-intercept: Let $x = 0$
$y = 0^2 - 9 = -9$

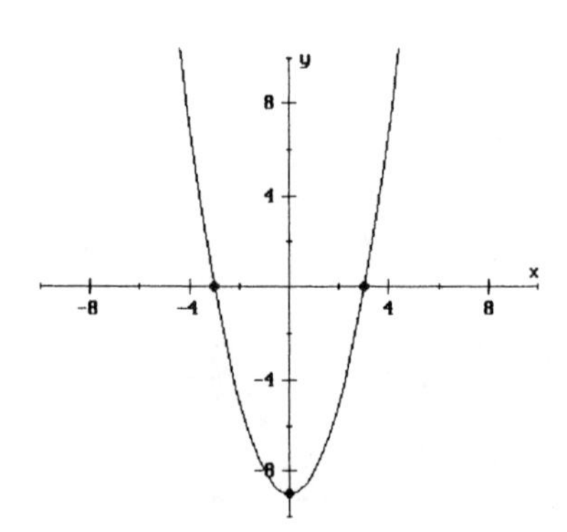

29. $f(x) = x^3$
Polynomial function

x	y
-2	-8
-1	-1
0	0
1	-1
2	-8

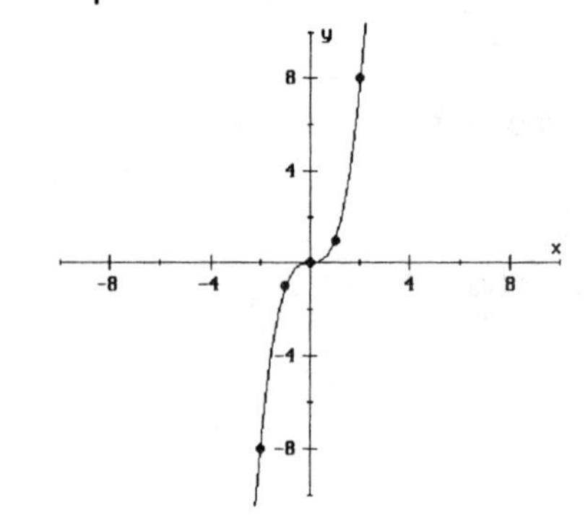

31. $f(x) = 2x^3$
Polynomial function

x	y
-2	-16
-1	-2
0	0
1	2
2	16

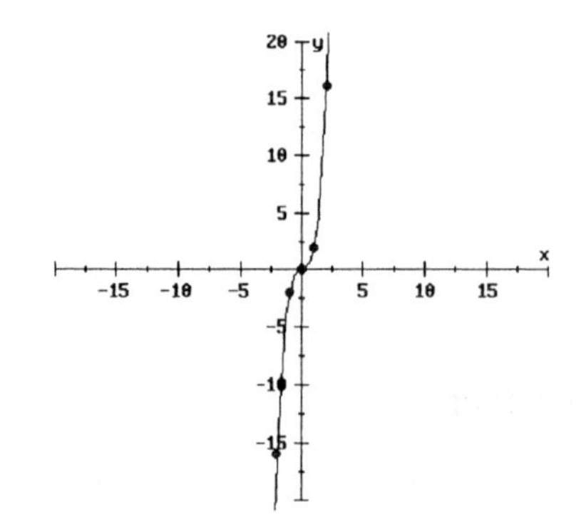

33. $f(x) = \sqrt{x - 5}$
Square root function

x	y
5	0
6	1
9	2

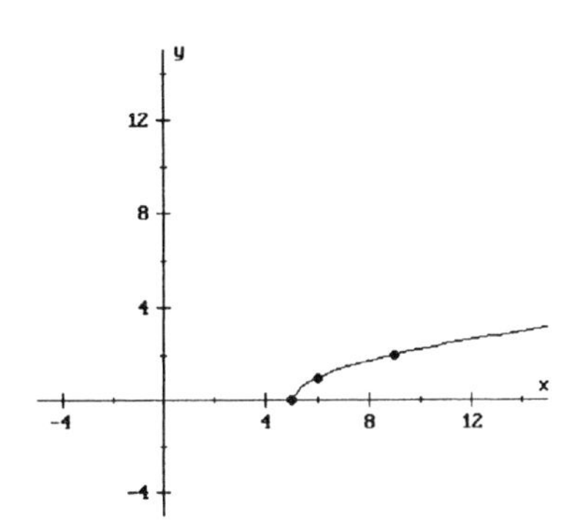

35. $f(x) = \sqrt{x + 5}$
Square root function

x	y
-5	0
-4	1
-1	2

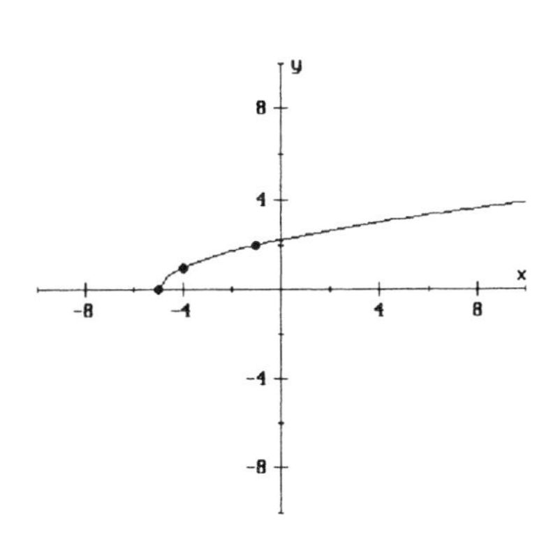

37. $f(x) = \sqrt{x} + 5$
Square root function

x	y
0	5
1	6
4	7

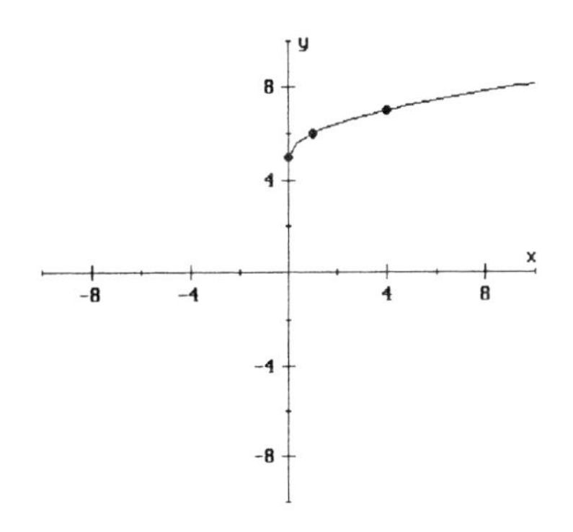

39. $f(x) = \sqrt{x} - 5$
Square root function

x	y
0	-5
1	-4
4	-3

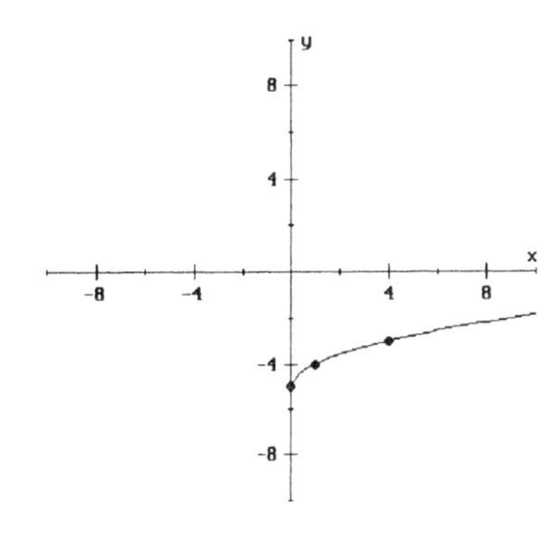

41. $f(x) = x^2 - 4x + 1$
Quadratic function
$\frac{-b}{2a} = \frac{-(-4)}{2(1)} = 2$
$f(2) = 2^2 - 4(2) + 1 = -3$
Vertex: $(2, -3)$
x-intercepts: Let $y = 0$
$0 = x^2 - 4x + 1$

$$x = \frac{-(-4) \pm \sqrt{(-4)^2 - 4(1)(1)}}{2(1)}$$

$$= \frac{4 \pm \sqrt{12}}{2}$$

$$= \frac{4 \pm 2\sqrt{3}}{2}$$

$$= 2 \pm \sqrt{3}$$

y - intercept: Let $x = 0$
$y = 0^2 - 4(0) + 1 = 1$

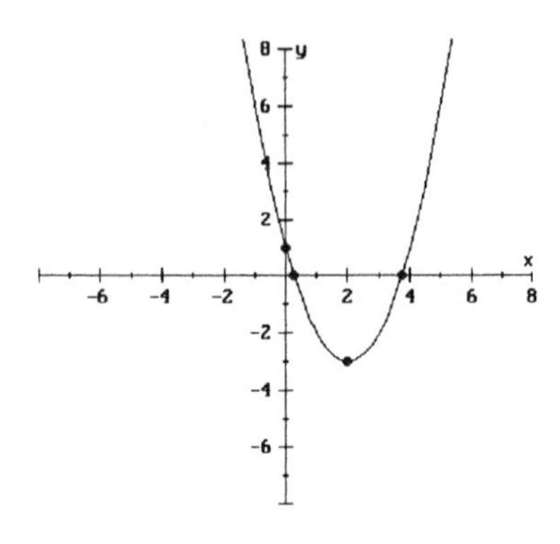

43. $f(x) = \sqrt{3x - 2}$
Square root function

x	y
2/3	0
1	1
2	2

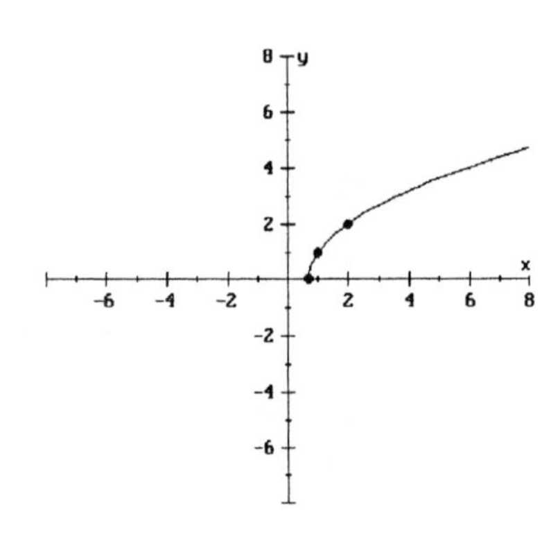

45. $f(x) = 8 - 2x - x^2$
Quadratic function

$$-\frac{b}{2a} = \frac{-(-2)}{2(-1)} = -1$$

$f(-1) = 8 - 2(-1) - (-1)^2 = 9$
Vertex: $(-1, 9)$

x-intercepts: Let $y = 0$
$0 = 8 - 2x - x^2$
$0 = x^2 + 2x - 8$
$0 = (x + 4)(x - 2)$
$x + 4 = 0, \quad x - 2 = 0$
$x = -4, \quad x = 2$

y-intercept: Let $x = 0$
$y = 8 - 2(0) - 0^2 = 8$

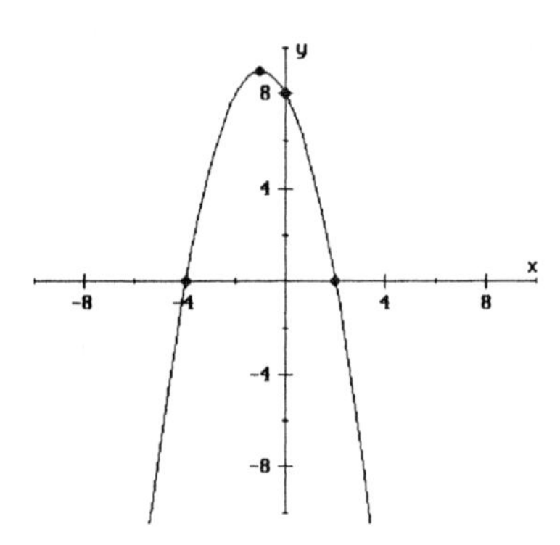

47. $f(x) = \sqrt{8 - 2x}$
Square root function

x	y
4	0
7/2	1
2	2

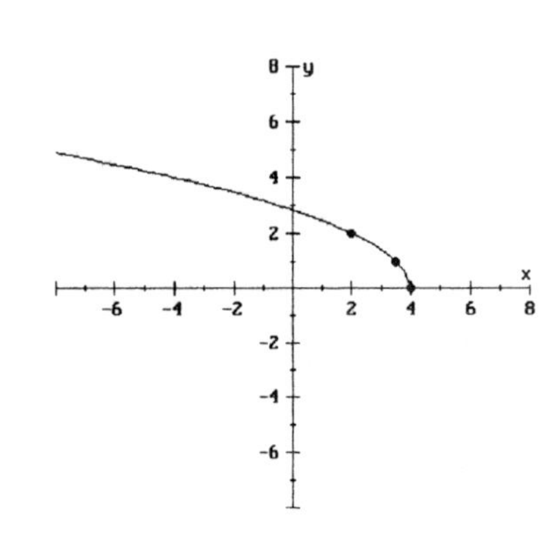

49. $f(x) = \sqrt{6 - 4x}$
Square root function

x	y
3/2	0
5/4	1
1/2	2

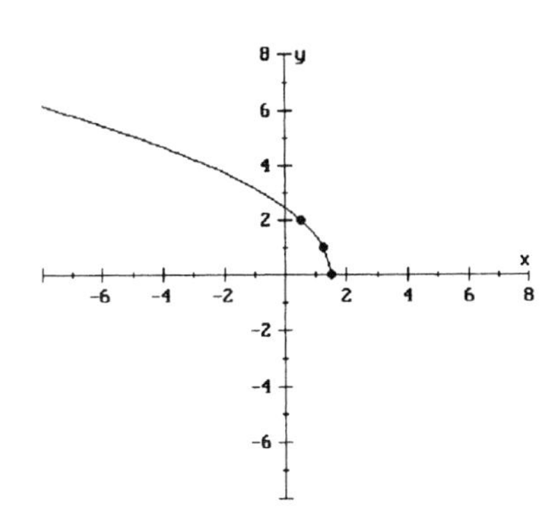

51. $f(x) = |x + 5|$
Absolute value function

x	y
-7	2
-5	0
-3	2

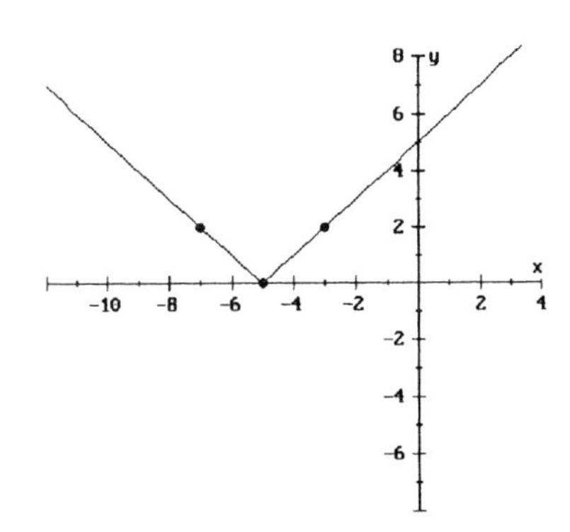

53. $f(x) = |x| + 5$
Absolute value function

x	y
-1	6
0	5
1	6

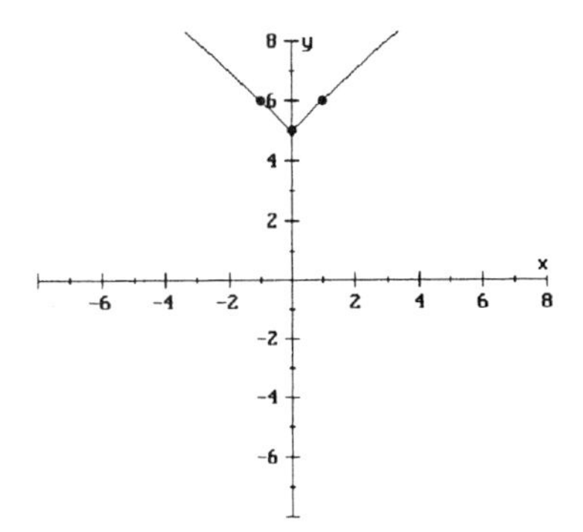

55. $f(x) = x + 5$
Linear function
$(0, 5)$; $m = 1$

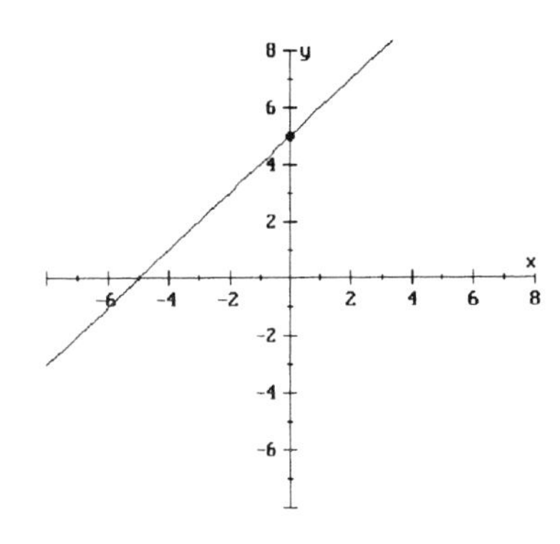

57. $f(x) = |5 - x|$
Absolute value function

x	y
4	1
5	0
6	1

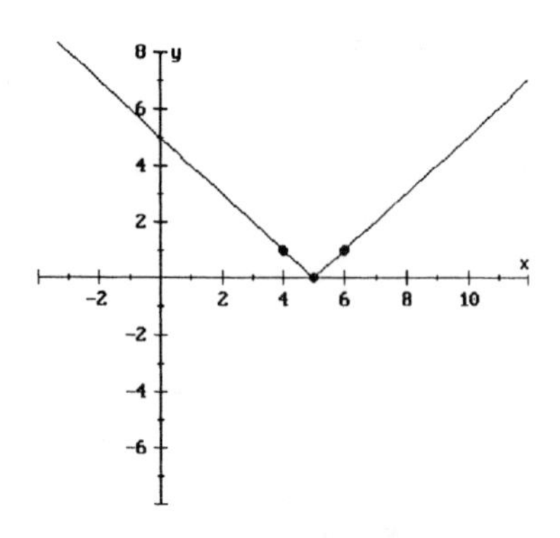

65. Let x = rate of medium speed sorter,
then + 150 = rate of high speed sorter

$$\frac{500}{x} = \frac{500}{x + 150} + \frac{1}{6}$$

$$6x(x + 150)\left(\frac{500}{x}\right) = 6x(x + 150)\left(\frac{500}{x + 150} + \frac{1}{6}\right)$$

$$\begin{aligned} 3000(x + 150) &= 3000x + x(x + 150) \\ 3000x + 450000 &= 3000x + x^2 + 150x \\ 0 &= x^2 + 150x - 450000 \\ 0 &= (x + 750)(x - 600) \end{aligned}$$

$x + 750 = 0$ or $x - 600 = 0$
$x = -750$ or $x = 600$

Since $x > 0$, $x = 600$
$x + 150 = 750$

$$\frac{500}{x} = \frac{500}{600} = \frac{5}{6}$$

$$\frac{500}{x + 150} = \frac{500}{750} = \frac{2}{3}$$

The medium speed sorter has a rate of 600 letters/min and takes $\frac{5}{6}$ hr = 50 min.

The high speed sorter has a rate of 750 letters/min and takes $\frac{2}{3}$ hr = 40 min.

67. $f(x) = 3x^2 - 2x + 5$

(a) $$\begin{aligned} f(x - 2) &= 3(x - 2)^2 - 2(x - 2) + 5 \\ &= 3(x^2 - 4x + 4) - 2x + 4 + 5 \\ &= 3x^2 - 12x + 12 - 2x + 9 \\ &= 3x^2 - 14x + 21 \end{aligned}$$

(b) $$\begin{aligned} f(x) - f(2) &= (3x^2 - 2x + 5) - [3(2)^2 - 2(2) + 5] \\ &= 3x^2 - 2x + 5 - 13 \\ &= 3x^2 - 2x - 8 \end{aligned}$$

Exercises 9.4

1. $\{(3, 1), (5, 2)\}$

3. $\{(2, 3), (-1, -3), (3, -1)\}$

5. function has no inverse

7. function (not one-to-one since $y = 1$ is associated with 2 values of x).

9. one-to-one function

11. neither ($x = 3$ is assigned 2 values of y)

13. function (not one-to-one since $y = -2$ is associated with 2 values of x.)

15. one-to-one function

17. one-to-one function (passes horizontal line test)

19. function (not one-to-one since it fails horizontal line test)

21. neither (fails vertical line test)

23. one-to-one (passes horizontal line test)

25. domain: $\{3, 2\}$
inverse: $\{(-2, 3), (-3, 2)\}$
domain of inverse: $\{-2, -3\}$

27. domain: $\{6, 2 -3\}$
inverse: $\{(-3, 6), (-4, 2), (6, -3\}$
domain of inverse: $\{-3, -4, 6\}$

29. domain: $\{2, 3 4\}$
inverse not a function

31. domain: all real numbers

$$f(x) = 3x + 4$$
$$y = 3x + 4$$
$$x = 3y + 4$$
$$x - 4 = 3y$$
$$\frac{x - 4}{3} = y$$
$$f^{-1}(x) = \frac{x - 4}{3}$$

domain of inverse: all real numbers

33. domain: all real numbers

$$g(x) = 2x - 3$$
$$y = 2x - 3$$
$$x = 2y - 3$$
$$x + 3 = 2y$$
$$\frac{x + 3}{2} = y$$
$$g^{-1}(x) = \frac{x + 3}{2}$$

domain of inverse: all real numbers

35. domain: all real numbers

$$h(x) = 4 - 5x$$
$$y = 4 - 5x$$
$$x = 4 - 5y$$
$$x - 4 = -5y$$
$$\frac{x - 4}{-5} = y$$
$$\frac{4 - x}{5} = y$$
$$h^{-1}(x) = \frac{4 - x}{5}$$

domain of inverse: all real numbers

37. domain: all real numbers
Since $f(x)$ is not one-to-one, the inverse is not a function.

39. domain: all real numbers

$$f(x) = x^3 + 4$$
$$y = x^3 + 4$$
$$x = y^3 + 4$$
$$x - 4 = y^3$$
$$\sqrt[3]{x - 4} = y$$
$$f^{-1}(x) = \sqrt[3]{x - 4}$$

domain: all real numbers

41. domain: $\{x \mid x \neq 0\}$

$$g(x) = \frac{1}{x}$$
$$y = \frac{1}{x}$$

$$x = \frac{1}{y}$$

$$xy = 1$$

$$y = \frac{1}{x}$$

$$g^{-1}(x) = \frac{1}{x}$$

domain of inverse: $\{x \mid x \neq 0\}$

43. domain: $x + 3 \neq 0$

$$x \neq -3$$

$$\{x \mid x \neq -3\}$$

$$g(x) = \frac{2}{x + 3}$$

$$y = \frac{2}{x + 3}$$

$$x = \frac{2}{y + 3}$$

$$x(y + 3) = 2$$

$$y + 3 = \frac{2}{x}$$

$$y = \frac{2}{x} - 3$$

$$g^{-1}(x) = \frac{2}{x} - 3 = \frac{2 - 3x}{x}$$

domain of inverse: $\{x \mid x \neq 0\}$

45. domain: $\{x \mid x \neq 0\}$

$$g(x) = \frac{x - 1}{x}$$

$$y = \frac{x - 1}{x}$$

$$x = \frac{y - 1}{y}$$

$$xy = y - 1$$

$$1 = y - xy$$

$$1 = y(1 - x)$$

$$\frac{1}{1 - x} = y$$

$$g^{-1}(x) = \frac{1}{1 - x}$$

domain of inverse: $1 - x \neq 0$

$$1 \neq x$$

$$\{x \mid x \neq 1\}$$

47. domain: $x - 1 \neq 0$

$$x \neq 1$$

$$\{x \mid x \neq 1\}$$

$$h(x) = \frac{x + 2}{x - 1}$$

$$y = \frac{x + 2}{x - 1}$$

$$x = \frac{y + 2}{y - 1}$$

$$x(y - 1) = y + 2$$

$$xy - x = y + 2$$

$$xy - y = x + 2$$

$$y(x - 1) = x + 2$$

$$y = \frac{x + 2}{x - 1}$$

$$h^{-1}(x) = \frac{x + 2}{x - 1}$$

domain of inverse: $\{x \mid x \neq 1\}$

49.

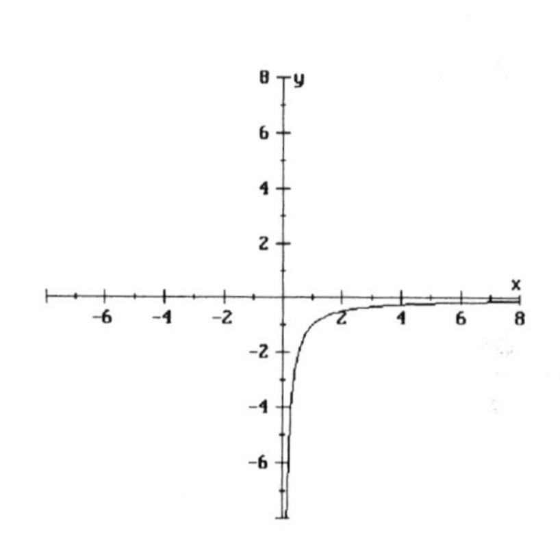

51.

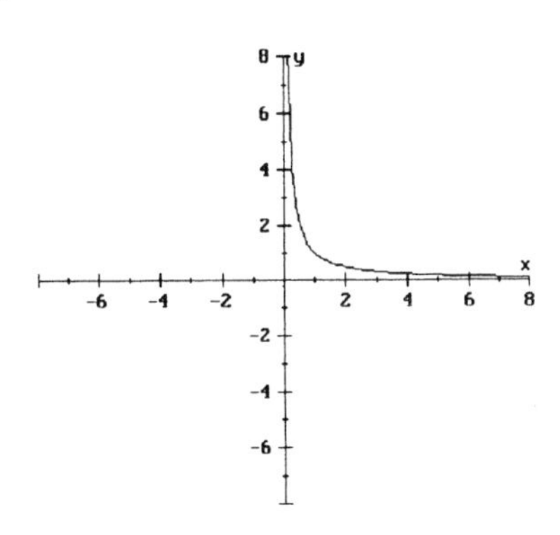

55. $(\sqrt{x} - 5)^2$
$= (\sqrt{x})^2 - 2(\sqrt{x})(5) + 5^2$
$= x - 10\sqrt{x} + 25$

57. Let x = other rate

$$0.062(2800) + x(1800) = 300$$
$$173.6 + 1800x = 300$$
$$1800x = 126.4$$
$$x = 0.070$$

The rate is 7%.

Exercises 9.5

1. $y = kx$
$8 = k(4)$
$2 = k$

$y = 2x$
$y = 2(3) = 6$

3. $y = kx$
$25 = k(15)$

$\frac{5}{3} = k$

$y = \frac{5}{3}x$

$y = \frac{5}{3}(8) = \frac{40}{3}$

5. $y = kx$
$22 = k(3)$

$\frac{22}{3} = k$

$y = \frac{22}{3}x$

$5 = \frac{22}{3}x$

$\frac{15}{22} = x$

7. $a = kb^2$
$4 = k(3)^2$
$4 = 9k$
$\frac{4}{9} = k$

$a = \frac{4}{9}b^2$

$a = \frac{4}{9}(9)^2 = 36$

9. $r = ks^4$
$12 = k(2)^4$
$12 = 16k$
$\frac{3}{4} = k$

$r = \frac{3}{4}s^4$

$r = \frac{3}{4}(3)^4 = \frac{243}{4}$

11. $y = \frac{k}{x}$

$20 = \frac{k}{4}$

$80 = k$

$y = \frac{80}{x}$

$y = \frac{80}{8} = 10$

13. $y = \frac{k}{x}$

$21 = \frac{k}{12}$

$252 = k$

$y = \frac{252}{x}$

$9x = 252$
$x = 28$

15. $y = \frac{k}{x}$

$25 = \frac{k}{10}$

$250 = k$

$y = \frac{250}{x}$

$12 = \frac{250}{x}$

$12x = 250$

$x = \frac{125}{6}$

17. $a = \frac{k}{b^3}$

$6 = \frac{k}{2^3}$

$6 = \frac{k}{8}$

$48 = k$

$a = \frac{48}{b^3}$

$a = \frac{48}{16^3} = \frac{3}{256}$

19. $a = \frac{k}{\sqrt{b}}$

$16 = \frac{k}{\sqrt{4}}$

$16 = \frac{k}{2}$

$32 = k$

$a = \frac{32}{\sqrt{b}}$

$a = \frac{32}{\sqrt{9}} = \frac{32}{3}$

21. $z = kxy$
$12 = k(2)(4)$
$12 = 8k$
$\frac{3}{2} = k$

$z = \frac{3}{2}xy$

$z = \frac{3}{2}(5)(2) = 15$

23. $z = kxy$
$20 = k(3)(4)$
$20 = 12k$
$\frac{5}{3} = k$

$z = \frac{5}{3}xy$

$z = \frac{5}{3}(2)(5)$

$z = \frac{50}{3}$

25. $a = kcd$
$20 = k(2)(4)$
$20 = 8k$
$\frac{5}{2} = k$

$a = \frac{5}{2}cd$

$25 = \frac{5}{2}(8)d$

$25 = 20d$

$\frac{5}{4} = d$

27. $z = kxy^2$
$20 = k(4)(2)^2$
$20 = 16k$
$\frac{5}{4} = k$

$z = \frac{5}{4}xy^2$

$z = \frac{5}{4}(2)(4)^2 = 40$

29. $z = \frac{kx}{y}$

$16 = \frac{k(3)}{2}$

$32 = 3k$

$\frac{32}{3} = k$

$z = \frac{\frac{32}{3}x}{y}$

$z = \frac{32x}{3y}$

$z = \frac{32(5)}{3(3)} = \frac{160}{9}$

31. $z = \frac{kx^2}{y}$

$20 = \frac{k(2)^2}{4}$

$80 = 4k$

$$20 = k$$

$$z = \frac{20x^2}{y}$$

$$z = \frac{20(4)^2}{2} = 160$$

33. $$z = \frac{kx}{y^2}$$

$$32 = \frac{k(4)}{2^2}$$

$$32 = k$$

$$z = \frac{32x}{y^2}$$

$$z = \frac{32(3)}{3^2} = \frac{32}{3}$$

35. $$V = kt$$
$$250 = k(30)$$

$$\frac{25}{3} = k$$

$$V = \frac{25}{3}t$$

$$V = \frac{25}{3}(40) = \frac{1000}{3}$$

The volume is $\frac{1000}{3}$ m^3.

37. $$V = kr^3$$
$$36\pi = k(3)^3$$

$$36\pi = 27k$$

$$\frac{4}{3}\pi = k$$

$$V = \frac{4}{3}\pi r^3$$

$$V = \frac{4}{3}\pi(4)^3 = \frac{256}{3}\pi$$

The volume is $\frac{256\pi}{3}$ cm^3.

39. $$d = kt^2$$
$$256 = k(4)^2$$
$$256 = 16k$$
$$16 = k$$

$$d = 16t^2$$
$$800 = 16t^2$$
$$50 = t^2$$
$$5\sqrt{2} = t$$

It takes $5\sqrt{2} \approx 7.07$ sec.

41. $$E = \frac{k}{d^2}$$

$$25 = \frac{k}{4^2}$$

$$400 = k$$

$$E = \frac{400}{d^2}$$

$$E = \frac{400}{8^2} = 6.25$$

The illumination is 6.25 footcandles.

43. $$V = khr^2$$
$$4\pi = k(3)(2)^2$$
$$4\pi = 12k$$
$$\frac{\pi}{3} = k$$

$$V = \frac{\pi}{3}hr^2$$

$$V = \frac{\pi}{3}(2)(3)^2 = 6\pi$$

The volume is 6π m^3.

45. $$R = \frac{kl}{d^2}$$

$$12 = \frac{k(80)}{(0.01)^2}$$

$$0.0012 = 80k$$
$$0.000015 = k$$

$$R = \frac{0.000015l}{d^2}$$

$$R = \frac{0.000015(100)}{(0.02)^2}$$

$$= 3.75$$

The resistance is 3.75 ohms.

47. $C = 2\pi r$

radius doubled: $C_1 = 2\pi(2r)$
$= 2(2\pi r)$
$= 2c$

The circumference is doubled.

radius tripled: $C_2 = 2\pi(3r)$
$= 3(2\pi r)$
$= 3C$

The circumference is tripled.

radius halved: $C_3 = 2\pi\left(\frac{1}{2}r\right)$

$= \frac{1}{2}(2\pi r)$

$= \frac{1}{2}C$

The circumference is halved.

49. $y = \frac{k}{x}$

$y_1 = \frac{c}{4x}$

$= \frac{1}{4}\left(\frac{k}{x}\right)$

$= \frac{1}{4}y$

y is divided by 4.

51. $z = kxy$

$z_1 = k(4x)(5y)$
$= 20kxy$
$= 20z$

z is multiplied by 20.

53. $s = \frac{kr}{t}$

$s_1 = \frac{k(2r)}{\frac{1}{2}t}$

$= \frac{4kr}{t}$

$= 4s$

s is multiplied by 4.

55. $\sqrt{5x} + 5\sqrt[3]{x}$
$= (5x)^{1/2} + 5x^{1/3}$

57. Let x = amount invested at 6.5%,
then $x + 5000$ = amount invested at 7%

$$\begin{aligned} 0.065x + 0.07(x + 5000) &= 1430 \\ 0.065x + 0.07x + 350 &= 1430 \\ 0.135x + 350 &= 1430 \\ 0.135x &= 1080 \\ x &= 8000 \\ x + 5000 &= 13000 \end{aligned}$$

She invests \$8000 at 6.5% and \$13,000 at 7%.

CHAPTER 9 REVIEW EXERCISES

1. $f(x) = 5x + 4$
$f(-3) = 5(-3) + 4 = -11$
$f(1) = 5(1) + 4 = 9$
$f(3) = 5(3) + 4 = 19$

3. $f(x) = \sqrt{x - 1}$
-3 is not in the domain of $f(x)$.

$f(1) = \sqrt{1 - 1}$
$= \sqrt{0}$
$= 0$

$f(3) = \sqrt{3 - 1} = \sqrt{2}$

5. $f(x) = \dfrac{3}{\sqrt{1 - x}}$

1 and 3 are not in the domain of $f(x)$.

$$f(-3) = \frac{3}{\sqrt{1 - (-3)}} = \frac{3}{\sqrt{4}} = \frac{3}{2}$$

7. all real numbers

9. all real numbers

11.
$$\begin{aligned} 3 - 5x &\geq 0 \\ -5x &\geq -3 \\ x &\leq \frac{3}{5} \end{aligned}$$
$\left\{x \mid x \leq \frac{3}{5}\right\}$

13.
$$\begin{aligned} 2x - 1 &> 0 \\ 2x &> 1 \\ x &> \frac{1}{2} \end{aligned}$$
$\left\{x \mid x > \frac{1}{2}\right\}$

15.
$$\begin{aligned} 2x - 3 &> 0 \\ 2x &> 3 \\ x &> \frac{3}{2} \end{aligned}$$
$\left\{x \mid x > \frac{3}{2}\right\}$

17.
$$\begin{aligned} f(x + 3) &= 5(x + 3) + 2 \\ &= 5x + 15 + 2 \\ &= 5x + 17 \end{aligned}$$

19.
$$\begin{aligned} f(x) + 3 &= (5x + 2) + 3 \\ &= 5x + 5 \end{aligned}$$

21.
$$\begin{aligned} f(x) + f(3) &= (5x + 2) + [5(3) + 2] \\ &= 5x + 2 + 17 \\ &= 5x + 19 \end{aligned}$$

23.
$$\begin{aligned} g(x + 3) &= \sqrt{3 - 2(x + 3)} \\ &= \sqrt{3 - 2x - 6} \\ &= \sqrt{-2x - 3} \end{aligned}$$

25.
$$\begin{aligned} g(3x) &= \sqrt{3 - 2(3x)} \\ &= \sqrt{3 - 6x} \end{aligned}$$

27. $3g(x) = 3\sqrt{3 - 2x}$

29.
$$\begin{aligned} g(x + h) &= \sqrt{3 - 2(x + h)} \\ &= \sqrt{3 - 2x - 2h} \end{aligned}$$

$$g(x + h) - g(x) = \sqrt{3 - 2x - 2h} - \sqrt{3 - 2x}$$

31.
$$\begin{aligned} g[f(x)] &= \sqrt{3 - 2[f(x)]} \\ &= \sqrt{3 - 2(5x + 2)} \\ &= \sqrt{3 - 10x - 4} \\ &= \sqrt{-10x - 1} \end{aligned}$$

33.
$$\begin{aligned} (f + g)(3) &= f(3) + g(3) \\ &= (3^2 - 3) + (3 - 8) \\ &= 6 - 5 \\ &= 1 \end{aligned}$$

35.
$$\left(\frac{f}{g}\right)(3) = \frac{f(3)}{g(3)} = \frac{3^2 - 3}{3 - 8} = -\frac{6}{5}$$

37.
$$\left(\frac{f}{g}\right)(8) = \frac{f(8)}{g(8)} = \frac{8^2 - 3}{8 - 8} = \frac{61}{0}$$

undefined

39. $g(3) = 3 + 1 = 4$

$f[g(3)] = [g(3)]^2 - [g(3)] - 5$
$= 4^2 - 4 - 5$
$= 7$

41. $g(-1) = -1 + 1 = 0$

$f[g(-1)] = [g(-1)]^2 - [g(-1)] - 5$
$= 0^2 - 0 - 5$
$= -5$

43. $$h\left(\frac{1}{2}\right) = \frac{\frac{1}{2} + 1}{2\left(\frac{1}{2}\right)} = \frac{3}{2}$$

$$f\left[h\left(\frac{1}{2}\right)\right] = \left[h\left(\frac{1}{2}\right)\right]^2 - \left[h\left(\frac{1}{2}\right)\right] - 5$$

$$= \left(\frac{3}{2}\right)^2 - \frac{3}{2} - 5$$

$$= -\frac{17}{4}$$

45. $$h(-1) = \frac{-1 + 1}{2(-1)} = 0$$

$g[h(-1)] = [h(-1)] + 1$
$= 0 + 1$
$= 1$

47. $g[f(x)] = [f(x)] + 1$
$= x^2 - x - 5 + 1$
$= x^2 - x - 4$

49. $g[h(x)] = [h(x)] + 1$

$$= \frac{x + 1}{2x} + 1$$

$$= \frac{x + 1 + 2x}{2x}$$

$$= \frac{3x + 1}{2x}$$

51. $A = 8t$

(a) $A = 8(4) = 32$

$\pi r^2 = 32$

$$r^2 = \frac{32}{\pi}$$

$$r = \sqrt{\frac{32}{\pi}}$$

$$r = \frac{\sqrt{32\pi}}{\pi} = \frac{4\sqrt{2\pi}}{\pi}$$

The radius is $\frac{4\sqrt{2\pi}}{\pi} \approx 3.2$ in.

(b) $A = 8t$

$\pi r^2 = 8t$

$$r^2 = \frac{8t}{\pi}$$

$$r = \sqrt{\frac{8t}{\pi}}$$

$$r = \frac{2\sqrt{2t\pi}}{\pi}$$

53. $A = \pi r^2$

$200\pi = \pi r^2$
$200 = r^2$
$\sqrt{200} = r^2$
$r = 10\sqrt{2}$

$C = 2\pi r$
$= 2\pi(10\sqrt{2})$
$= 20\pi\sqrt{2}$

Cost $= x(20\pi\sqrt{2})$ dollars

55. $f(x) = 2x - 3$

Linear function

$(0, -3)$; $m = 2$

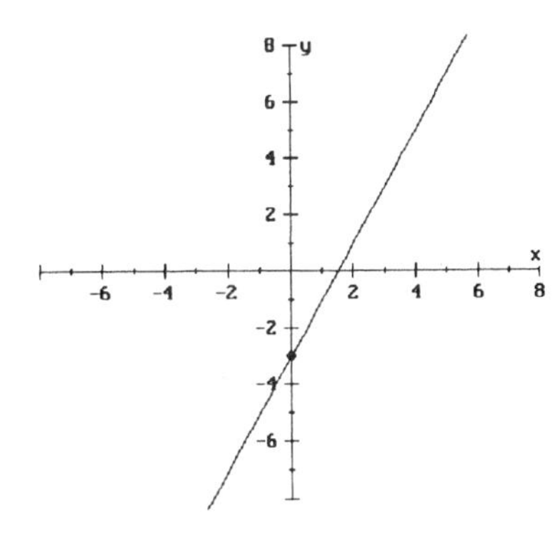

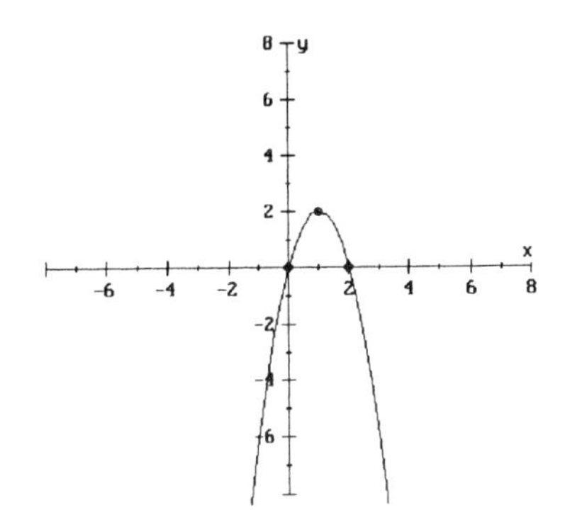

57. $f(x) = \sqrt{2x - 3}$
Square root function

x	y
3/2	0
2	1
7/2	2

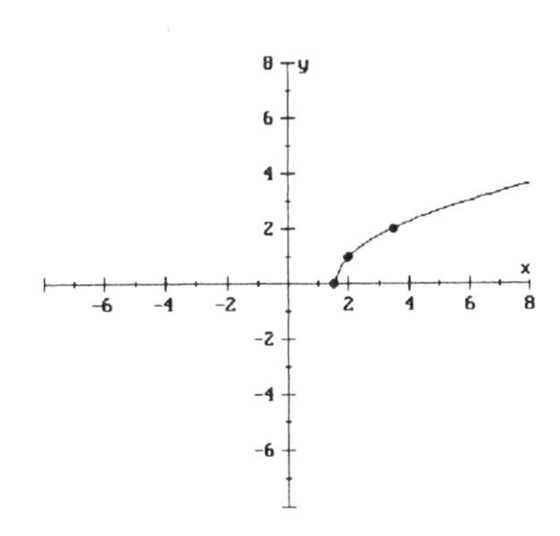

59. $f(x) = 4x - 2x^2$
Quadratic function

$$-\frac{b}{2a} = -\frac{4}{2(-2)} = 1$$

$$f(1) = 4(1) - 2(1)^2 = 2$$

Vertex: (1, 2)
x-intercepts: Let $y = 0$
$0 = 4x - 2x^2$
$0 = 2x(2 - x)$
$2x = 0 \qquad 2 - x = 0$
$x = 0 \qquad x = 2$

y-intercept: Let $x = 0$
$y = 4(0) - 2(0)^2 = 0$

61. $f(x) = |2x| - 3$
Absolute value function

x	y
-1	-1
0	-3
1	-1

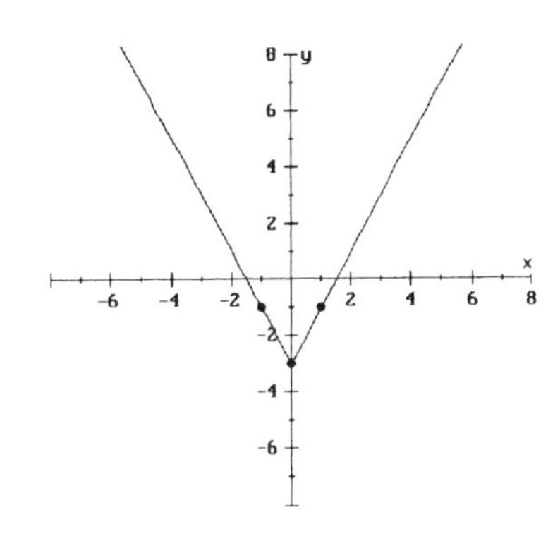

63. function

65. one-to-one function

67. one-to-one function (passes horizontal line test)

69. neither (fails vertical line test)

71. {(3, 2), (4, 3)}

73.
$$y = 3x + 8$$
$$x = 3y + 8$$
$$x - 8 = 3y$$
$$\frac{x - 8}{3} = y$$

75. $y = x^3$
$x = y^3$
$x^{1/3} = y$

77. $y = \dfrac{3}{x + 1}$

$x = \dfrac{3}{y + 1}$

$x(y + 1) = 3$
$xy + x = 3$
$xy = 3 - x$
$y = \dfrac{3 - x}{x}$

79.

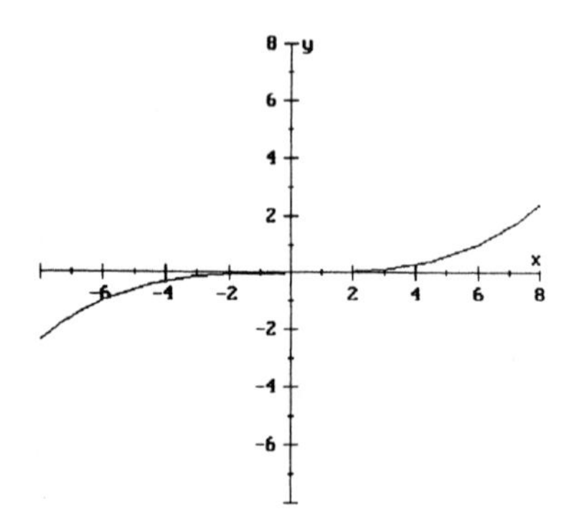

81. $x = ky$
$8 = k(3)$
$\dfrac{8}{3} = k$

$x = \dfrac{8}{3}y$

$x = \dfrac{8}{3}(2) = \dfrac{16}{3}$

83. $r = \dfrac{k}{s^2}$

$8 = \dfrac{k}{2^2}$

$32 = k$

$r = \dfrac{32}{s^2}$

$4 = \dfrac{32}{s^2}$

$s^2 = 8$
$s = \pm\sqrt{8} = \pm 2\sqrt{2}$

85. $z = kxy$
$16 = k(7)(2)$
$\dfrac{8}{7} = k$

$z = \dfrac{8}{7}xy$

$z = \dfrac{8}{7}(3)(4) = \dfrac{96}{7}$

87. $V = kr^3$

$\dfrac{500\pi}{3} = k(5)^3$

$\dfrac{4\pi}{3} = k$

$V = \dfrac{4\pi}{3}r^3$

$V = \dfrac{4\pi}{3}(6)^3 = 288\pi$

The volume is $288\pi \approx 904.78$ in.^{3}.

89. $V = \dfrac{kT}{P}$

$80 = \dfrac{k(20)}{30}$

$120 = k$

$V = \dfrac{120T}{P}$

$V = \dfrac{120(10)}{20} = 60$

The volume is 60 m^3.

CHAPTER 9 PRACTICE TEST

1. (a) all real numbers

(b) $2x - 3 \neq 0$

$2x \neq 3$

$x \neq \frac{3}{2}$

$\left\{x \mid x \neq \frac{3}{2}\right\}$

(c) $8 - 3x \geq 0$

$-3x \geq -8$

$x \leq \frac{8}{3}$

$\left\{x \mid x \leq \frac{8}{3}\right\}$

3. (a) $\left(\frac{f}{h}\right)(3) = \frac{f(3)}{h(3)}$

$= \frac{3^2 - 3}{\sqrt{3 - 2}}$

$= \frac{6}{1}$

$= 6$

(b) $h(6) = \sqrt{6 - 2} = 2$

$f[h(6)] = [h(6)]^2 - 3$

$= 2^2 - 3$

$= 1$

(c) $f(6) = 6^2 - 3 = 33$

$h[f(6)] = \sqrt{[f(6)] - 2}$

$= \sqrt{33 - 2}$

$= \sqrt{31}$

(d) $f[h(x)] = [h(x)]^2 - 3$

$= \left(\sqrt{x - 2}\right)^2 - 3$

$= x - 2 - 3$

$= x - 5$

(e) $h[f(x)] = \sqrt{[f(x)] - 2}$

$= \sqrt{x^2 - 3 - 2}$

$= \sqrt{x^2 - 5}$

5. $f(x) = |2x - 1|$

x	y
0	1
1/2	0
1	1

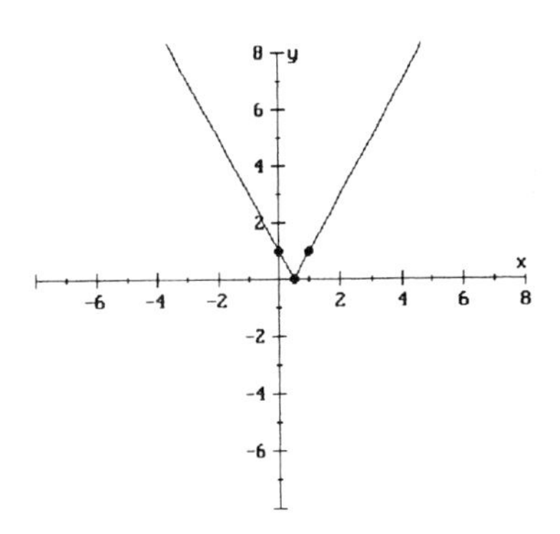

7. No, it fails the horizontal line test.

9. $y = \frac{k}{x^2}$

$24 = \frac{k}{2^2}$

$96 = k$

$y = \frac{96}{x^2}$

$y = \frac{96}{6^2} = \frac{8}{3}$

CHAPTERS 7 - 9 CUMULATIVE REVIEW

1. $(-2x^2y)(-3xy^2)^3$
$= (-2x^2y)(-27x^3y^6)$
$= 54x^5y^7$

3. $\left(\dfrac{2xy^2}{4x^2y^3}\right)^2$
$= \left(\dfrac{1}{2xy}\right)^2$
$= \dfrac{1}{4x^2y^2}$

5. $(x^{-5}y^{-2})(x^{-7}y)$
$= x^{-12}y^{-1}$
$= \dfrac{1}{x^{12}y}$

7. $\dfrac{r^{-3}s^{-2}}{r^{-2}s^0} = r^{-1}s^{-2}$
$= \dfrac{1}{rs^2}$

9. $(-2r^{-3}s)^{-1}(-4r^{-2}s^{-3})^2$
$= \dfrac{1}{-2r^{-3}s} \cdot 16r^{-4}s^{-6}$
$= -8r^{-1}s^{-7}$
$= -\dfrac{8}{rs^7}$

11. $(x^{-1} - y^{-1})(x + y)$
$= x^0 - xy^{-1} + x^{-1}y - y^0$
$= 1 - \dfrac{x}{y} + \dfrac{y}{x} - 1$
$= \dfrac{y}{x} - \dfrac{x}{y}$
$= \dfrac{y^2 - x^2}{xy}$

13. $56{,}429.32 = 5.642932 \times 10^4$

15. 1 day $= 24$ hr
$= 24(60 \text{ min})$
$= 1440$ min
$= 1440(60 \text{ sec})$
$= 86400$ sec

$d = (86400 \text{ sec})(186000 \text{ miles/sec})$
$= 1.60704 \times 10^{10}$ miles

17. $(-1000)^{1/3} = -10$

19. $a^{2/3}a^{-1/2} = a^{2/3 - 1/2}$
$= a^{1/6}$

21. $\left[(81)^{-1/3}3^2\right]^{-3}$
$= \left[(3^4)^{-1/3}3^2\right]^{-3}$
$= (3^{-4/3} \cdot 3^2)^{-3}$
$= (3^{2/3})^{-3}$
$= 3^{-2}$
$= \dfrac{1}{9}$

23. $x^{3/4} = \sqrt[4]{x^3}$

25. $\sqrt{16a^4b^8} = 4a^2b^4$

27. $\left(3x\sqrt{2x^2y}\right)\left(2x\sqrt{8xy^3}\right)$
$= \left(3x^2\sqrt{2y}\right)\left(4xy\sqrt{2xy}\right)$
$= 12x^3y\sqrt{4xy^2}$
$= 24x^3y^2\sqrt{x}$

29. $\dfrac{\sqrt{48}}{\sqrt{3}} = \sqrt{\dfrac{48}{3}}$
$= \sqrt{16}$
$= 4$

31. $\sqrt[8]{x^4} = \sqrt{x}$

33. $2\sqrt{5} - 3\sqrt{5} + 8\sqrt{5}$
$= (2 - 3 + 8)\sqrt{5}$
$= 7\sqrt{5}$

35. $3a\sqrt[3]{a^4} - a^2\sqrt[3]{a}$
$= 3a^2\sqrt[3]{a} - a^2\sqrt[3]{a}$
$= (3a^2 - a^2)\sqrt[3]{a}$
$= 2a^2\sqrt[3]{a}$

37. $\sqrt{2}(\sqrt{2} - 1) + 2\sqrt{2}$
$= 2 - \sqrt{2} + 2\sqrt{2}$
$= 2 + \sqrt{2}$

39. $(\sqrt{5} - \sqrt{3})(\sqrt{5} + \sqrt{3})$
$= (\sqrt{5})^2 - (\sqrt{3})^2$
$= 5 - 3$
$= 2$

41. $$\frac{5}{\sqrt{3} + \sqrt{2}} - \frac{3}{\sqrt{3}}$$
$$= \frac{5(\sqrt{3} - \sqrt{2})}{(\sqrt{3} + \sqrt{2})(\sqrt{3} - \sqrt{2})} - \frac{3 \cdot \sqrt{3}}{\sqrt{3} \cdot \sqrt{3}}$$
$$= \frac{5\sqrt{3} - 5\sqrt{2}}{3 - 2} - \frac{3\sqrt{3}}{3}$$
$= 5\sqrt{3} - 5\sqrt{2} - \sqrt{3}$
$= 4\sqrt{3} - 5\sqrt{2}$

43. $\sqrt{3x + 2} = 7$
$(\sqrt{3x + 2})^2 = 7^2$
$3x + 2 = 49$
$3x = 47$
$x = \frac{47}{3}$

45. $\sqrt[3]{x - 1} = -2$
$(\sqrt[3]{x - 1})^3 = (-2)^3$
$x - 1 = -8$
$x = -7$

47. $i^{35} = (i^4)^8 \cdot i^3$
$= 1^8 \cdot (-i)$
$= -i$

49. $(3 - 2i)(5 + i)$
$= 15 + 3i - 10i - 2i^2$
$= 15 + 3i - 10i + 2$
$= 17 - 7i$

51. $a^2 - 2a - 15 = 0$
$(a - 5)(a + 3) = 0$
$a - 5 = 0$ or $a + 3 = 0$
$a = 5$ or $a = -3$

53. $2x^2 - 3x - 4 = 9 - 3(x - 2)$
$2x^2 - 3x - 4 = 9 - 3x + 6$
$2x^2 = 19$
$x^2 = \frac{19}{2}$
$$x = \pm\sqrt{\frac{19}{2}}$$
$$x = \pm\frac{\sqrt{38}}{2}$$

55. $y^2 + 6y - 1 = 0$
$y^2 + 6y = 1$
$y^2 + 6y + 9 = 1 + 9$
$(y + 3)^2 = 10$
$y + 3 = \pm\sqrt{10}$
$y = -3 \pm \sqrt{10}$

57. $3a^2 - 2a - 2 = 0$
$$a = \frac{-(-2) \pm \sqrt{(-2)^2 - 4(3)(-2)}}{2(3)}$$
$$= \frac{2 \pm \sqrt{4 + 24}}{6}$$
$$= \frac{2 \pm \sqrt{28}}{6}$$
$$= \frac{2 \pm 2\sqrt{7}}{6}$$
$$= \frac{1 \pm \sqrt{7}}{3}$$

59. $$\frac{1}{x+2} = x + 2$$

$$(x+2)\left(\frac{1}{x+2}\right) = (x+2)(x+2)$$

$$1 = (x+2)^2$$
$$\pm\sqrt{1} = x + 2$$
$$\pm 1 = x + 2$$
$$x + 2 = 1 \quad \text{or} \quad x + 2 = -1$$
$$x = -1 \quad \text{or} \quad x = -3$$

61. $$\frac{2}{x+3} - \frac{3}{x} = -\frac{2}{3}$$

$$3x(x+3)\left(\frac{2}{x+3} - \frac{3}{x}\right) = 3x(x+3)\left(-\frac{2}{3}\right)$$

$$6x - 9(x+3) = -2x(x+3)$$
$$6x - 9x - 27 = -2x^2 - 6x$$
$$2x^2 + 3x - 27 = 0$$
$$(2x+9)(x-3) = 0$$
$$2x + 9 = 0 \quad \text{or} \quad x - 3 = 0$$
$$2x = -9 \qquad x = 3$$
$$x = -\frac{9}{2} \quad \text{or} \quad x = 3$$

63. $$\frac{3}{x-3} + \frac{2x}{x+3} = \frac{5}{x-3}$$

$$(x-3)(x+3)\left(\frac{3}{x-3} + \frac{2x}{x+3}\right) = (x-3)(x+3)\left(\frac{5}{x-3}\right)$$

$$3(x+3) + 2x(x-3) = 5(x+3)$$
$$3x + 9 + 2x^2 - 6x = 5x + 15$$
$$2x^2 - 8x - 6 = 0$$
$$x^2 - 4x - 3 = 0$$

$$x^2 - 4x - 3 = 0$$

$$x = \frac{-(-4) \pm \sqrt{(-4)^2 - 4(1)(-3)}}{2(1)}$$

$$= \frac{4 \pm \sqrt{16 + 12}}{2}$$

$$= \frac{4 \pm \sqrt{28}}{2}$$

$$= \frac{4 \pm 2\sqrt{7}}{2}$$

$$= 2 \pm \sqrt{7}$$

65. $y = 2x^2$
$y = 2(x - 0)^2$

Vertex: $(0, 0)$
Axis of Symmetry: $x = 0$

x-intercepts: Let $y = 0$
$0 = 2x^2$
$0 = x^2$
$0 = x$

y-intercept: Let $x = 0$
$y = 2(0)^2$
$y = 0$

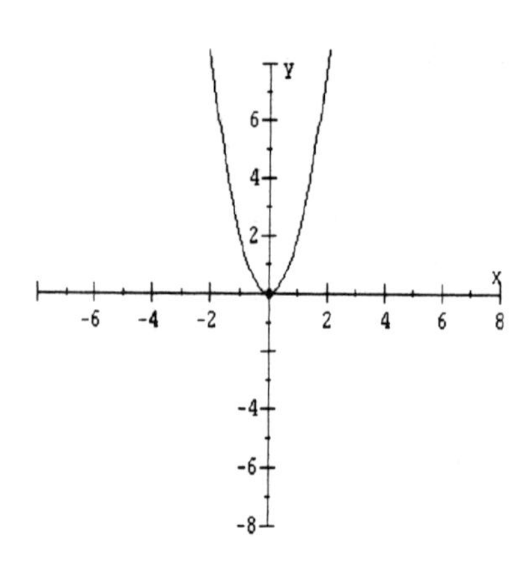

67. $y = 2x$

x	y
-1	-2
0	0
1	2

x-intercept, y-intercept: $(0, 0)$

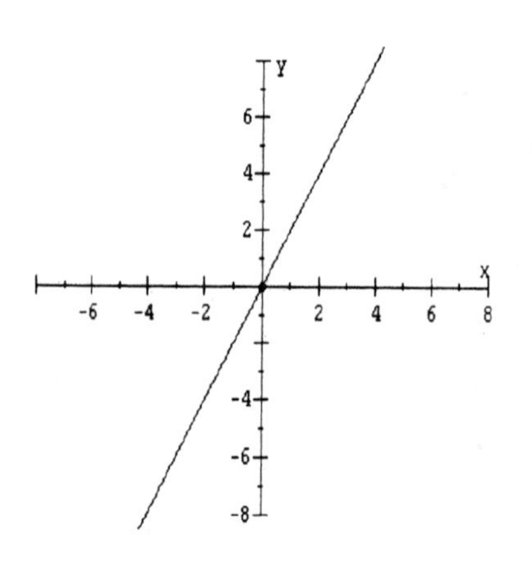

69. $y = -\frac{1}{3}x^2$

$y = -\frac{1}{3}(x - 0)^2$

Vertex: (0, 0)
Axis of symmetry: $x = 0$

x-intercepts: Let $y = 0$

$0 = -\frac{1}{3}x^2$

$0 = x^2$

$0 = x$

y-intercept: Let $x = 0$

$y = -\frac{1}{3}(0)^2$

$y = 0$

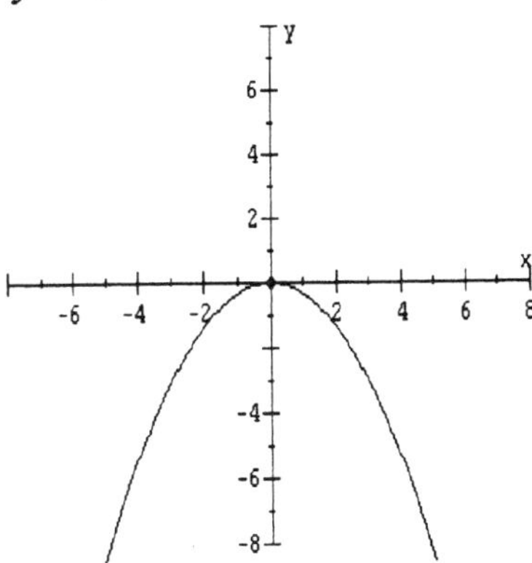

71. $y = (x - 2)^2 + 4$

Vertex: (2, 4)
Axis of symmetry: $x = 2$

x-intercepts: Let $y = 0$

$0 = (x - 2)^2 + 4$

$-4 = (x - 2)^2$

none

y-intercepts: Let $x = 0$

$y = (0 - 2)^2 + 4$

$y = 8$

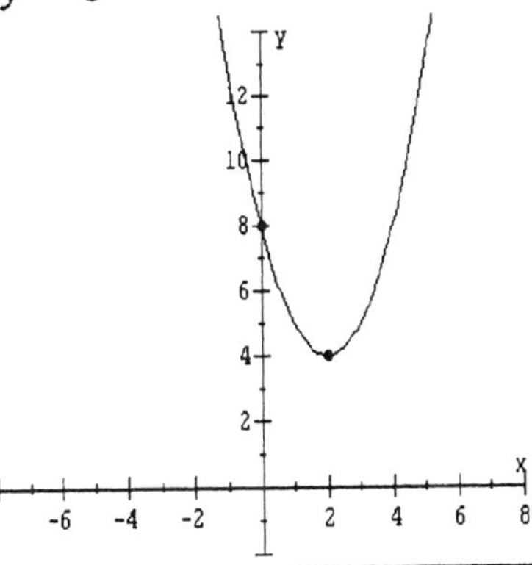

73. $y = -(x - 3)^2 + 5$

Vertex: (3, 5)
Axis of symmetry: $x = 3$

x-intercepts: Let $y = 0$

$0 = -(x - 3)^2 + 5$

$(x - 3)^2 = 5$

$x - 3 = \pm\sqrt{5}$

$x = 3 \pm \sqrt{5}$

y-intercept: Let $x = 0$

$y = -(0 - 3)^2 + 5$

$y = -4$

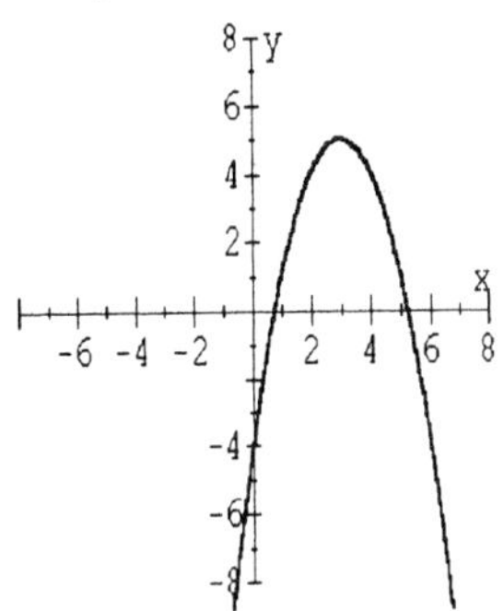

75. $y = 2x^2 - 4x + 5$

Vertex: $x = \frac{-b}{2a}$

$x = \frac{-(-4)}{2(2)} = 1$

$y = 2(1)^2 - 4(1) + 5 = 3$

(1, 3)

Axis of symmetry: $x = 1$

x-intercepts: Let $y = 0$

$0 = 2x^2 - 4x + 5$

$x = \frac{-(-4) \pm \sqrt{(-4)^2 - 4(2)(5)}}{2(2)}$

$= \frac{4 \pm \sqrt{-24}}{4}$

none

y-intercept: Let $x = 0$

$y = 2(0)^2 - 4(0) + 5$

$y = 5$

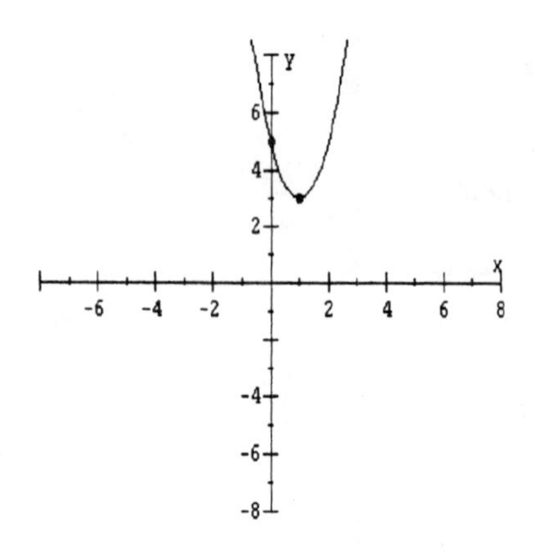

77. $(x - 7)(2x + 1) \le 0$

$x - 7 = 0 \qquad 2x + 1 = 0$
$x = 7 \qquad 2x = -1$
$x = -\frac{1}{2}$

The intervals are $x < -\frac{1}{2}$, $-\frac{1}{2} < x < 7$, $x > 7$

Test points:

	$(x - 7)(2x + 1)$	
-1	$(-1 - 7)[2(-1) + 1]$	pos
0	$(0 - 7)[2(0) + 1]$	neg
8	$(8 - 7)[2(8) + 1]$	pos

We want $(x - 7)(2x + 1) \le 0$; neg or 0

Solution: $-\frac{1}{2} \le x \le 7$

79. $2x^2 - x > 6$
$2x^2 - x - 6 > 0$
$(2x + 3)(x - 2) > 0$
$2x + 3 = 0 \qquad x - 2 = 0$
$2x = -3 \qquad x = 2$
$x = -\frac{3}{2}$

The intervals are $x < -\frac{3}{2}$, $-\frac{3}{2} < x < 2$, $x > 2$

Test points:

	$(2x + 3)(x - 2)$	
-2	$[2(-2) + 3](-2 - 2)$	pos
0	$[2(0) + 3](0 - 2)$	neg
3	$[2(3) + 3](3 - 2)$	pos

We want $(2x + 3)(x - 2) > 0$: pos

Solution: $x < -\frac{3}{2}$ or $x > 2$

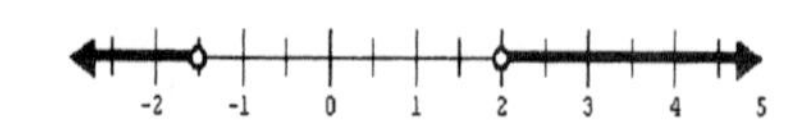

81. $\frac{x - 1}{x + 3} \le 2$

$\frac{x - 1}{x + 3} - 2 \le 0$

$\frac{x - 1 - 2(x + 3)}{x + 3} \le 0$

$\frac{x - 1 - 2x - 6}{x + 3} \le 0$

$\frac{-x - 7}{x + 3} \le 0$

$-x - 7 = 0 \qquad x + 3 = 0$
$-7 = x \qquad x = -3$

The intervals are $x < -7$, $-7 < x < -3$, $x > -3$

Test points:

	$\frac{-x - 7}{x + 3}$	
-8	$\frac{-(-8) - 7}{-8 + 3}$	neg
-5	$\frac{-(-5) - 7}{-5 + 3}$	pos
0	$\frac{-0 - 7}{0 + 3}$	neg

We want $\frac{-x - 7}{x + 3} \le 0$; neg or 0

Solution: $x \le -7$ or $x > -3$

Do not include -3 since it would cause the denominator to equal 0.

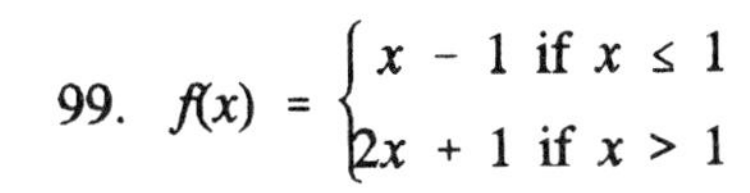

83. $P(x) = -200x^2 + 500x$

$$-\frac{b}{2a} = -\frac{500}{2(-200)} = \frac{5}{4} = 1.25$$

$$P\left(\frac{5}{4}\right) = -200\left(\frac{5}{4}\right)^2 + 500\left(\frac{5}{4}\right)$$

$$= 312.50$$

Vertex: (1.25, 312.50); maximum

The ticket price \$1.25 would produce a maximum profit of \$312.50.

85. $r = \sqrt{(1 - 0)^2 + (-3 - 2)^2}$
$= \sqrt{1 + 25}$
$= \sqrt{26}$

$(x - h)^2 + (y - k)^2 = r^2$
$(x - 0)^2 + (y - 2)^2 = (\sqrt{26})^2$
$x^2 + (y - 2)^2 = 26$

87. Center = midpoint

$$= \left(\frac{-2 + 4}{2}, \frac{1 + 5}{2}\right)$$

$$= (1, 3)$$

$r = \sqrt{(4 - 1)^2 + (5 - 3)^2} = \sqrt{13}$
$r^2 = 13$

$(x - 1)^2 + (y - 3)^2 = 13$

89. $f(x) + g(x)$
$= (2x^2 - 4x) + (3x - 6)$
$= 2x^2 - x - 6$

91. $f[g(x)] = 2[g(x)]^2 - 4[g(x)]$
$= 2(3x - 6)^2 - 4(3x - 6)$
$= 2(9x^2 - 36x + 36) - 12x + 24$
$= 18x^2 - 72x + 72 - 12x + 24$
$= 18x^2 - 84x + 96$

93. $h(x) \cdot r(x)$

$$= \left(\frac{2x}{x + 1}\right)\left(\frac{1}{x}\right)$$

$$= \frac{2}{x + 1}$$

95. $$\frac{f(x)}{g(x)} = \frac{2x^2 - 4x}{3x - 6}$$

$$= \frac{2x(x - 2)}{3(x - 2)}$$

$$= \frac{2x}{3}$$

97. $$h[r(x)] = \frac{2[r(x)]}{r(x) + 1}$$

$$= \frac{2\left(\frac{1}{x}\right)}{\frac{1}{x} + 1}$$

$$= \frac{\left(\frac{2}{x}\right)x}{\left(\frac{1}{x} + 1\right)x}$$

$$= \frac{2}{1 + x}$$

99. $$f(x) = \begin{cases} x - 1 \text{ if } x \le 1 \\ 2x + 1 \text{ if } x > 1 \end{cases}$$

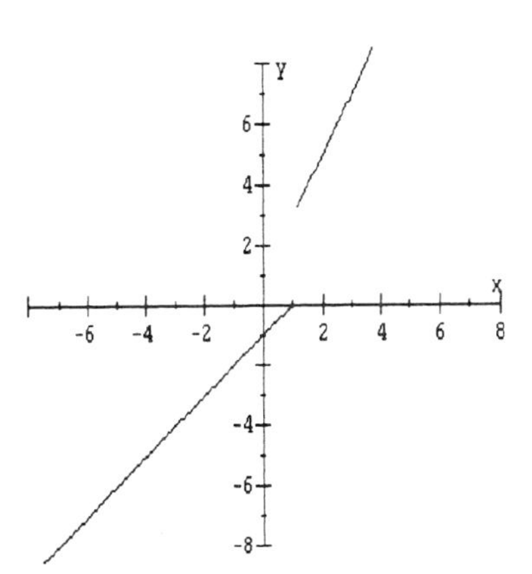

$$f(-2) = -2 - 1 = -3$$
$$f(1) = 1 - 1 = 0$$
$$f(3) = 2(3) + 1 = 7$$

101.
$$y = x^3 - 1$$
$$x = y^3 - 1$$
$$x + 1 = y^3$$
$$\sqrt[3]{x + 1} = y$$

103.
$$y = \frac{x + 1}{x}$$
$$x = \frac{y + 1}{y}$$
$$xy = y + 1$$
$$xy - y = 1$$
$$y(x - 1) = 1$$
$$y = \frac{1}{x - 1}$$

105. $y = \sqrt{x + 4}$

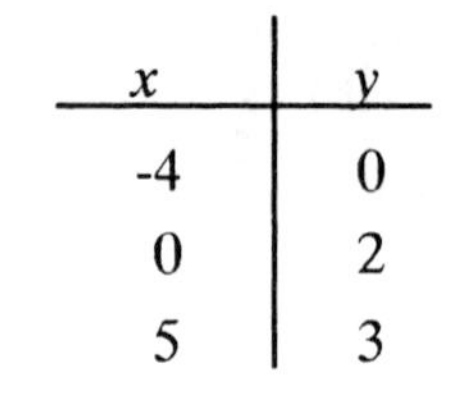

x	y
-4	0
0	2
5	3

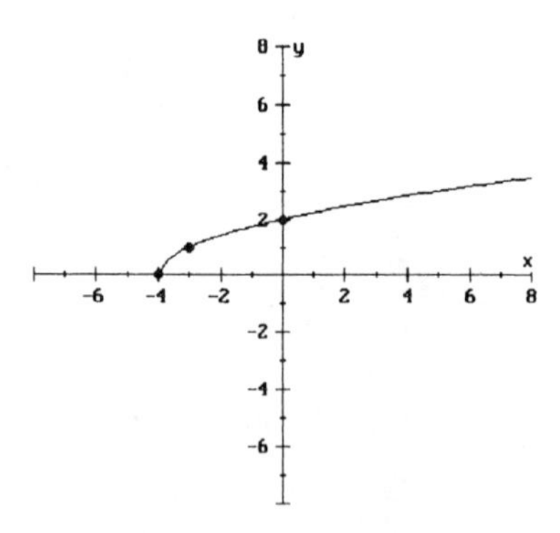

107.
$$y = x^2 - 4x - 5$$
$$-\frac{b}{2a} = \frac{-(-4)}{2(1)} = 2$$
$$y = 2^2 - 4(2) - 5 = -9$$

Vertex: $(2, -9)$

x-intercepts:
$$0 = x^2 - 4x - 5$$
$$0 = (x - 5)(x + 1)$$
$$x - 5 = 0 \quad \text{or} \quad x + 1 = 0$$
$$x = 5 \quad \text{or} \quad x = -1$$

y-intercept:
$$y = 0^2 - 4(0) - 5$$
$$y = -5$$

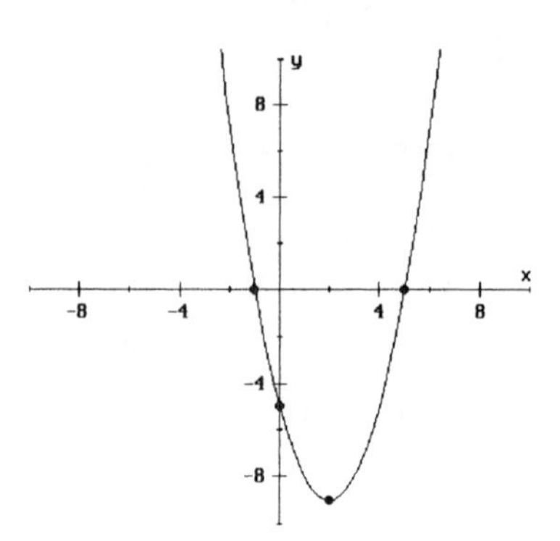

109. $(x - 2)^2 + (y + 3)^2 = 4$

$C(2, -3)$

$r = \sqrt{4} = 2$

x - intercepts:
$$(x - 2)^2 + (0 + 3)^2 = 4$$
$$(x - 2)^2 = -5$$
None

y-intercepts:
$$(0 - 2)^2 + (y + 3)^2 = 4$$
$$(y + 3)^2 = 0$$
$$y + 3 = 0$$
$$y = -3$$

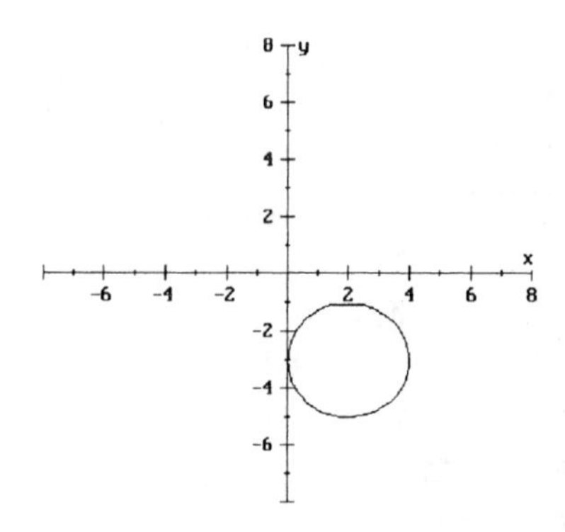

111. $y = kx$
$8 = k(6)$
$\frac{4}{3} = k$

$y = \frac{4}{3}x$

$y = \frac{4}{3}(20) = \frac{80}{3}$

113. $x = kyz$
$10 = k(4)(15)$
$\frac{1}{6} = k$

$x = \frac{1}{6}yz$

$x = \frac{1}{6}(6)(20) = 20$

115. $A = kbh$

$A_1 = k(2b)(3h)$
$A_1 = 6kbh$
$A_1 = 6A$

The area is multiplied by 6.

CHAPTERS 7 - 9 CUMULATIVE PRACTICE TEST

1. (a) $(3x^2y)^2(-2xy^3)^3$
$= (9x^4y^2)(-8x^3y^9)$
$= -72x^7y^{11}$

(b) $(-2x^{-1}y^3)(-3x^2y^{-3})^2$
$= (-2x^{-1}y^3)(9x^4y^{-6})$
$= -18x^3y^{-3}$
$= \frac{-18x^3}{y^3}$

(c) $\left(\frac{3x^{-2}y^{-1}}{9xy^{-3}}\right)^{-2}$
$= \left(\frac{y^2}{3x^3}\right)^{-2}$
$= \left(\frac{3x^3}{y^2}\right)^2$
$= \frac{9x^6}{y^4}$

(d) $\frac{a^{-1}}{a^{-1} + b^{-1}}$
$= \frac{\frac{1}{a}}{\frac{1}{a} + \frac{1}{b}}$
$= \frac{\left(\frac{1}{a}\right)ab}{\left(\frac{1}{a} + \frac{1}{b}\right)ab}$
$= \frac{b}{b + a}$

3. $\frac{(150{,}000)(0.00028)}{(0.07)(0.0002)}$

$= \frac{(1.5 \times 10^5)(2.8 \times 10^{-4})}{(7.0 \times 10^{-2})(2.0 \times 10^{-4})}$

$= \frac{(1.5)(2.8)}{(7)(2)} \times \left(\frac{10^5 \cdot 10^{-4}}{10^{-2} \cdot 10^{-4}}\right)$
$= 0.3 \times 10^7$
$= 3{,}000{,}000$

5. (a) $(x^{1/3}x^{-1/5})^5$
$= (x^{2/15})^5$
$= x^{2/3}$

(b) $\frac{x^{-2/3}y^{-3/4}}{x^{1/2}}$
$= \frac{1}{x^{7/6}y^{3/4}}$

7. (a) $5\sqrt{8} - 5\sqrt{2} - \sqrt{50}$
$= 10\sqrt{2} - 5\sqrt{2} - 5\sqrt{2}$
$= 0$

(b) $\sqrt{\frac{x}{y}} - \sqrt{\frac{x}{2}}$

$= \frac{\sqrt{x}}{\sqrt{y}} - \frac{\sqrt{x}}{\sqrt{2}}$

$= \frac{\sqrt{xy}}{y} - \frac{\sqrt{2x}}{2}$
$= \frac{2\sqrt{xy} - y\sqrt{2x}}{2y}$

(c) $(2 - \sqrt{3})(\sqrt{3} - 2)$
$= 2\sqrt{3} - 4 - 3 + 2\sqrt{3}$
$= -7 + 4\sqrt{3}$

(d) $\dfrac{5\sqrt{2}}{\sqrt{5} - \sqrt{2}}$

$= \dfrac{5\sqrt{2}(\sqrt{5} + \sqrt{2})}{(\sqrt{5} - \sqrt{2})(\sqrt{5} + \sqrt{2})}$

$= \dfrac{5\sqrt{2}(\sqrt{5} + \sqrt{2})}{5 - 2}$

$= \dfrac{5\sqrt{10} + 10}{3}$

9. $\dfrac{2 - i}{3 - i}$

$= \dfrac{(2 - i)(3 + i)}{(3 - i)(3 + i)}$

$= \dfrac{6 - i - i^2}{9 - i^2}$

$= \dfrac{7 - i}{10}$

$= \dfrac{7}{10} - \dfrac{1}{10}i$

11. (a) $y = 5x^2$

Vertex: (0, 0)
Axis of symmetry: $x = 0$

x-intercepts: Let $y = 0$
$0 = 5x^2$
$0 = x^2$
$0 = x$

y-intercept: Let $x = 0$

$y = 5(0)^2 = 0$

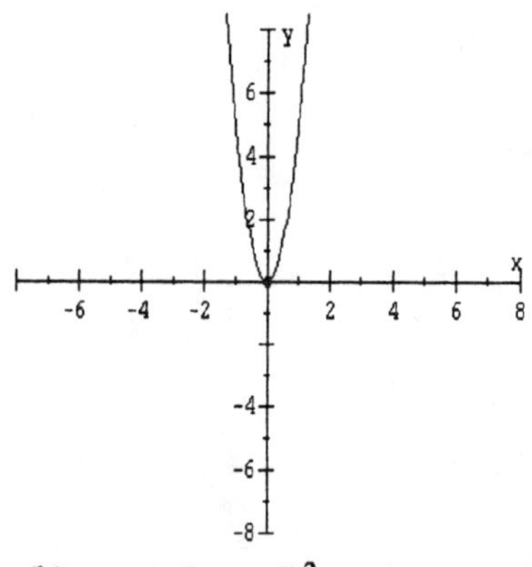

(b) $y = (x - 4)^2 + 1$

Vertex: (4, 1)
Axis of symmetry: $x = 4$

x-intercepts: Let $y = 0$

$0 = (x - 4)^2 + 1$
$-1 = (x - 4)^2$
none

y-intercept: Let $x = 0$

$y = (0 - 4)^2 + 1 = 17$

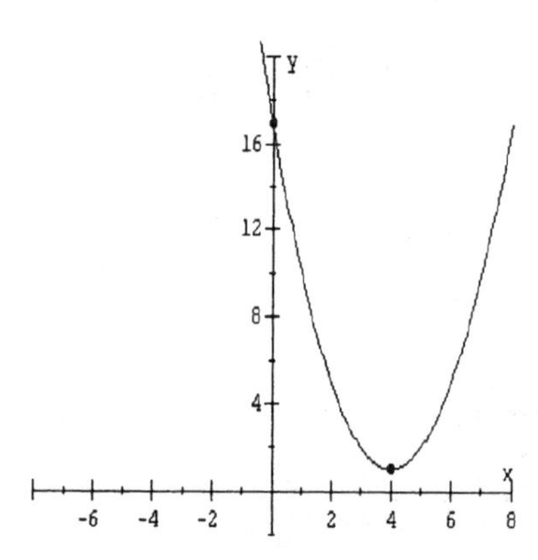

(c) $y = -(x - 1)^2$

Vertex: (1, 0)
Axis of symmetry: $x = 1$

x-intercepts: Let $y = 0$

$0 = -(x - 1)^2$
$0 = (x - 1)^2$
$0 = x - 1$
$1 = x$

y-intercept: Let $x = 0$

$y = -(0 - 1)^2 = -1$

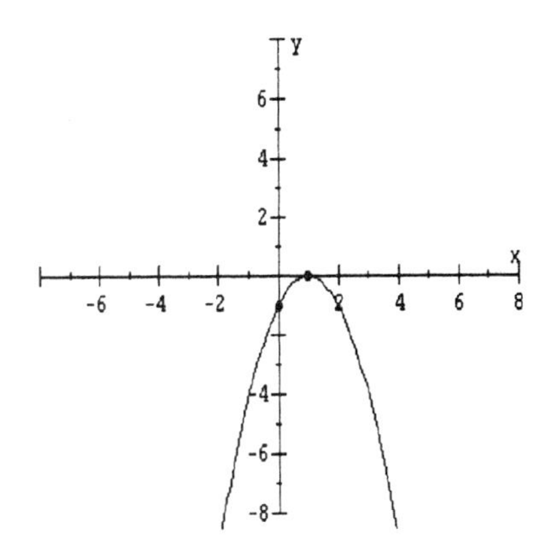

(d) $y = 3x^2 - 6x + 4$

$$x = \frac{-b}{2a} = \frac{-(-6)}{2(3)} = 1$$

$$y = 3(1)^2 - 6(1) + 4 = 1$$

Vertex: $(1, 1)$
Axis of symmetry: $x = 1$

x-intercepts: Let $y = 0$

$$0 = 3x^2 - 6x + 4$$

$$x = \frac{-(-6) \pm \sqrt{(-6)^2 - 4(3)(4)}}{2(3)}$$

$$= \frac{6 \pm \sqrt{-12}}{6}$$

none

y-intercept: Let $x = 0$

$$y = 3(0)^2 - 6(0) + 4 = 4$$

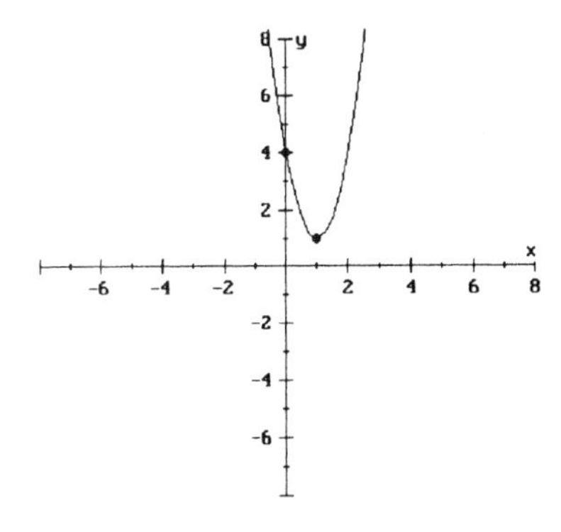

13. $r = \sqrt{(-2 - 5)^2 + [0 - (-1)]^2}$
$r = \sqrt{50}$
$r^2 = 50$
$(x + 2)^2 + y^2 = 50$

15. $f(x) = \begin{cases} x + 1 \text{ if } x \le 2 \\ x^2 - 2 \text{ if } x > 2 \end{cases}$

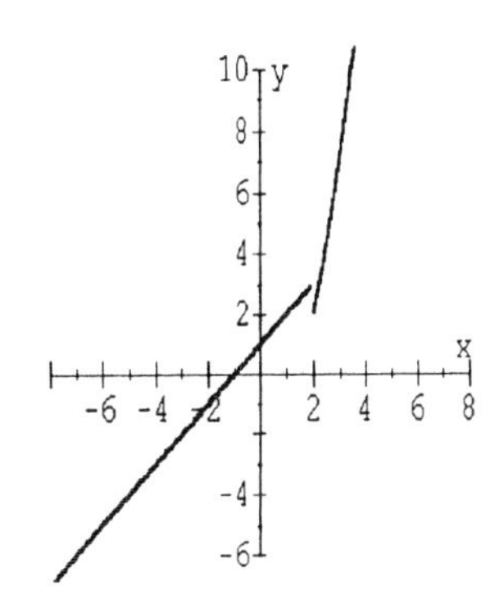

$f(0) = 0 + 1 = 1$
$f(2) = 2 + 1 = 3$
$f(3) = 3^2 - 2 = 7$

17. (a) $y = \sqrt{x} - 2$

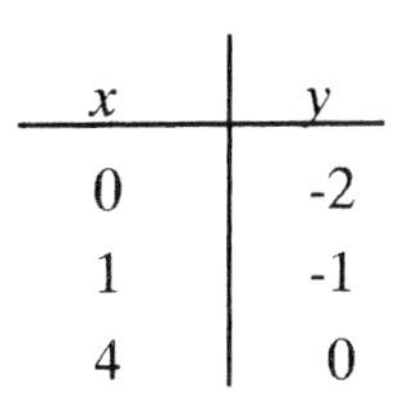

x	y
0	-2
1	-1
4	0

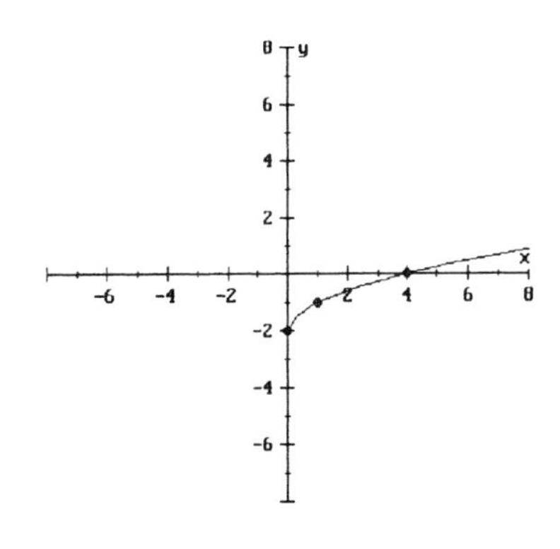

(b) $y = x^2 - 6x$

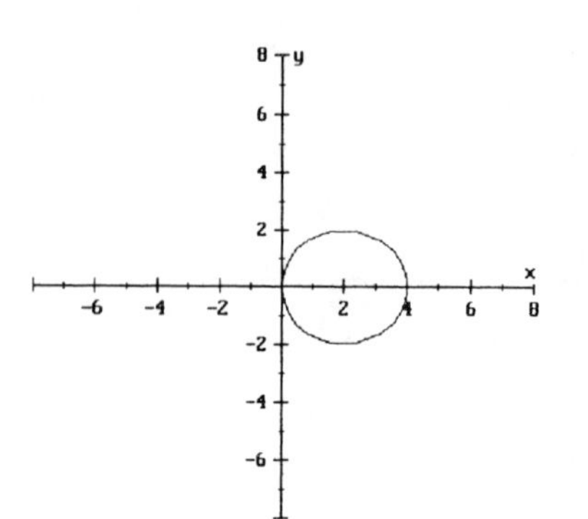

$$\frac{-b}{2a} = \frac{-(-6)}{2(1)} = 3$$

$y = 3^2 - 6(3) = -9$
Vertex: $(3, -9)$
x-intercepts: Let $y = 0$
$0 = x^2 - 6x$
$0 = x(x - 6)$
$x = 0$ or $x - 6 = 0$
$x = 6$

y-intercept: Let $x = 0$
$y = 0^2 - 6(0) = 0$

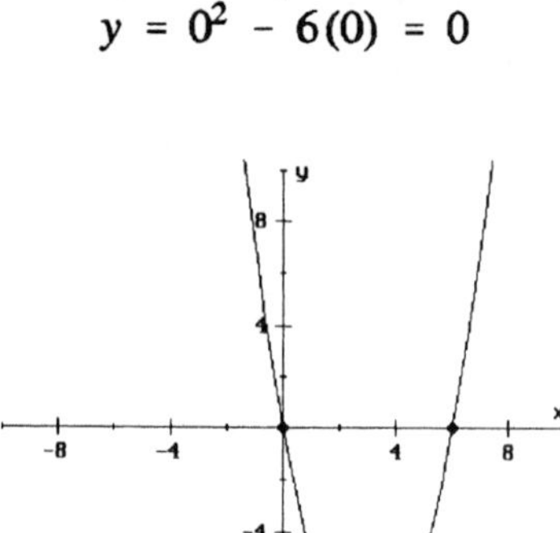

(c) $x^2 - 4x + y^2 = 0$
$x^2 - 4x + 4 + y^2 = 4$
$(x - 2)^2 + y^2 = 4$
Center: $(2, 0)$
$r = \sqrt{4} = 2$

x-intercepts: Let $y = 0$
$x^2 - 4x + 0^2 = 0$
$x^2 - 4x = 0$
$x(x - 4) = 0$
$x = 0$ or $x - 4 = 0$
$x = 4$

y-intercept: Let $x = 0$
$0^2 - 4(0) + y^2 = 0$
$y^2 = 0$
$y = 0$

CHAPTER 10

Exercises 10.1

1. $f(x) = 4^x$

x	y
-2	1/16
-1	1/4
0	1
1	4
2	16

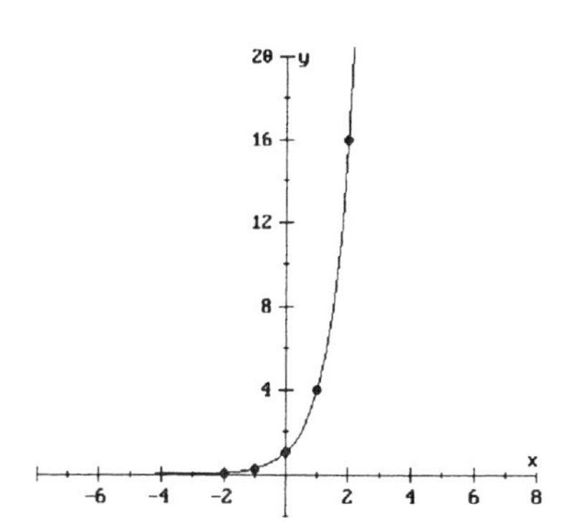

3. $f(x) = \left(\frac{1}{5}\right)^x$

x	y
-2	25
-1	5
0	1
1	1/5
2	1/25

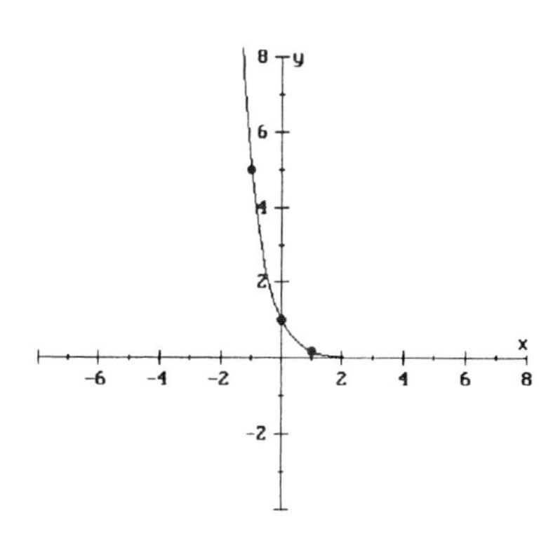

5. $f(x) = 3^{-x} = \left(\frac{1}{3}\right)^x$

x	y
-2	9
-1	3
0	1
1	1/3
2	1/9

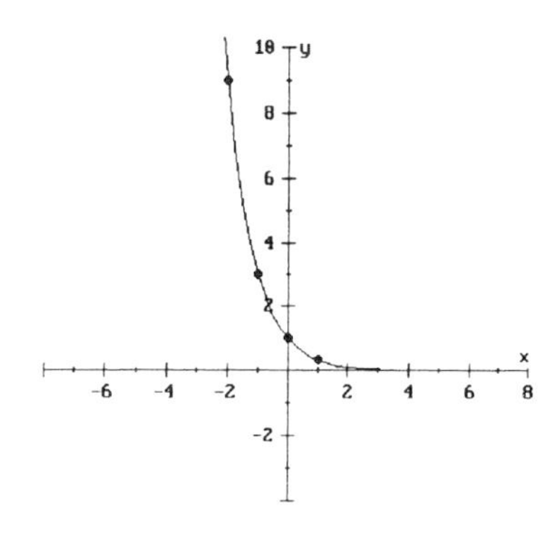

7. $y = 2^{x+1}$

x	y
-3	1/4
-2	1/2
-1	1
0	2
1	4

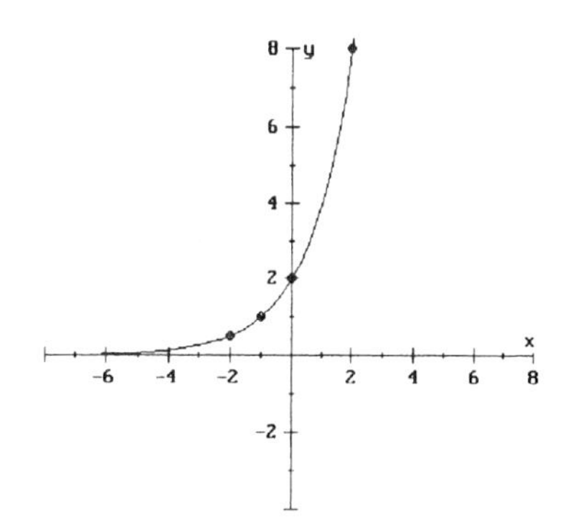

9. $f(x) = 2^x + 1$

x	y
-2	5/4
-1	3/2
0	2
1	3
2	5

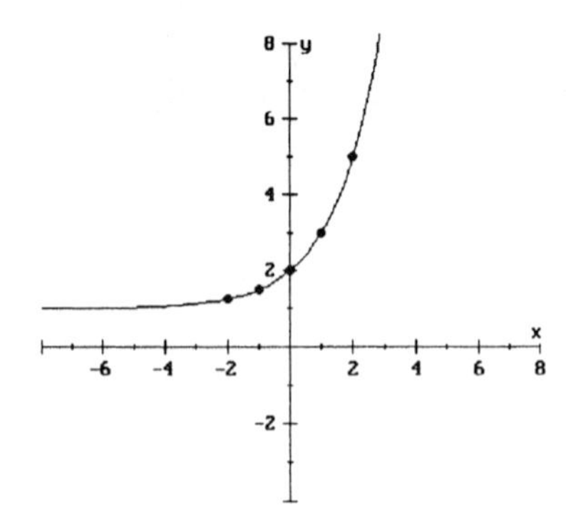

11. (a) $P(t) = 4.28(2.0263)^t$

$$P(0) = 4.28(2.0263)^0 = 4.28$$
$$P(2) = 4.28(2.0263)^2 = 17.6$$
$$P(4) = 4.28(2.0263)^4 = 72.2$$
$$P(6) = 4.28(2.0263)^6 = 296.3$$
$$P(8) = 4.28(2.0263)^8 = 1216.4$$
$$P(10) = 4.28(2.0263)^{10} = 4994.4$$

They are very close in value.

(b)

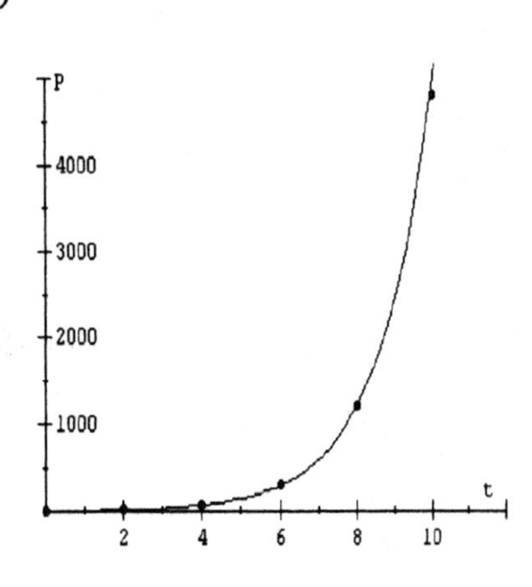

13.
$$2^x = 2^{3x-2}$$
$$x = 3x - 2$$
$$-2x = -2$$
$$x = 1$$

15.
$$5^x = 25^{x-1}$$
$$5x = (5^2)^{x-1}$$
$$5^x = 5^{2x-2}$$
$$x = 2x - 2$$
$$-x = -2$$
$$x = 2$$

17.
$$4^{1-x} = 16$$
$$4^{1-x} = 4^2$$
$$1 - x = 2$$
$$-x = 1$$
$$x = -1$$

19.
$$8^x = 4^{x+1}$$
$$(2^3)^x = (2^2)^{x+1}$$
$$2^{3x} = 2^{2x+2}$$
$$3x = 2x + 2$$
$$x = 2$$

21.
$$\frac{9^{x^2}}{9^x} = 81$$

$$9^{x^2-x} = 9^2$$
$$x^2 - x = 2$$
$$x^2 - x - 2 = 0$$
$$(x - 2)(x + 1) = 0$$
$$x - 2 = 0 \quad \text{or} \quad x + 1 = 0$$
$$x = 2 \quad \text{or} \quad x = -1$$

23.
$$4^{\sqrt{x}} = 4^{2x-3}$$
$$\sqrt{x} = 2x - 3$$
$$(\sqrt{x})^2 = (2x - 3)^2$$
$$x = 4x^2 - 12x + 9$$
$$0 = 4x^2 - 13x + 9$$
$$0 = (4x - 9)(x - 1)$$
$$4x - 9 = 0 \quad \text{or} \quad x - 1 = 0$$
$$4x = 9 \qquad x = 1$$
$$x = \frac{9}{4}$$

When you check $x = 1$, you find it to be extraneous. The only solution is $x = \frac{9}{4}$.

25.
$$\begin{aligned}
16^{x^2-1} &= 8^{x-1}\\
(2^4)^{x^2-1} &= (2^3)^{x-1}\\
2^{4x^2-4} &= 2^{3x-3}\\
4x^2 - 4 &= 3x - 3\\
4x^2 - 3x - 1 &= 0\\
(4x+1)(x-1) &= 0
\end{aligned}$$

$4x + 1 = 0$ or $x - 1 = 0$

$4x = -1 \qquad x = 1$

$x = -\dfrac{1}{4}$ or $x = 1$

27. After 10 hours: $2(2500) = 5000$ bacteria
After 20 hours: $2[2(2500)] = 2^2(2500)$
$= 10{,}000$ bacteria
After 50 hours: $2^5(2500) = 80{,}000$ bacteria
After t hours: $2^{t/10}(2500)$

29. 2010 is 14 years after 1996:
$2(20{,}000) = 40{,}000$
2024 is 28 years after 1996:
$2[2(20{,}000)] = 2^2(20{,}000)$
$= 80{,}000$

In year Y: $2^{(Y-1996)/14}(20{,}000)$

31. After 2 hours: $\frac{1}{2}\left(\frac{1}{2}\right)(10{,}000) = \left(\frac{1}{2}\right)^2(10{,}000)$

$= 2500$ bacteria

After 3 hours: $\left(\frac{1}{2}\right)^3(10{,}000) = 1250$ bacteria

After t hours: $\left(\frac{1}{2}\right)^t(10{,}000)$

33. After each year it is worth 5/6 of its previous year's value.

$$V = \left(\frac{5}{6}\right)^t(16000)$$

After 4 years: $V = \left(\frac{5}{6}\right)^4(16000)$

$= \$7716.05$

35. Each year it is 4/5 of its previous year's value.

$$V = \left(\frac{4}{5}\right)^t(24{,}000)$$

In 3 years:

$$V = \left(\frac{4}{5}\right)^3(24{,}000)$$

$= \$12{,}288$

37. Each year the population is 105% of the previous year's population.

$105\% = 1.05$
$P = (1.05)^t(100{,}000)$

4 years from now:

$P = (1.05)^4(100{,}000)$
$= 121{,}551$

39. Each year the money is 108% of the previous year's value. This is interest compounded annually (not simple yearly interest.)

$108\% = 1.08$

After 5 years:
$V = (1.08)^5(10{,}000) = \$14{,}693.28$

41. Each year the money is 110% of the previous year's value. This is interest compounded annually (not simple yearly interest.)

$110\% = 1.10$

After 5 years:
$V = (1.10)^5(5{,}000) = \$8{,}052.55$

43. $A(t) = (10{,}000)1.005^{12t}$

(a) $A(10) = (10{,}000)1.005^{12(10)} = \$18{,}193.97$

(b) It appears that $A = \$30{,}000$ when $t = 18$ years.

45. $(2^{-2} + 3^{-2})^{-1}$

$= \left(\frac{1}{8} + \frac{1}{9}\right)^{-1}$

$= \left(\frac{17}{72}\right)^{-1}$

$= \frac{72}{17}$

47. $\frac{x}{2} - \frac{y}{3} = 4$

$3x - 2y = 24$

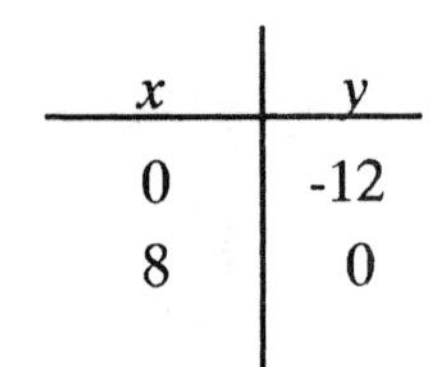

x	y
0	-12
8	0

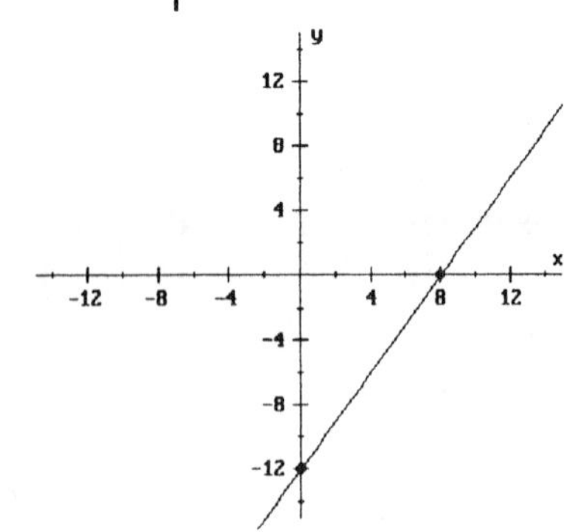

Exercises 10.2

1. $7^2 = 49$

3. $\log_3 81 = 4$

5. $10^4 = 10{,}000$

7. $\log_{10} 1000 = 3$

9. $9^2 = 81$

11. $81^{1/2} = 9$

13. $\log_6 \frac{1}{36} = -2$

15. $3^{-1} = \frac{1}{3}$

17. $\log_{25} 5 = \frac{1}{2}$

19. $8^1 = 8$

21. $\log_8 1 = 0$

23. $16^{3/4} = 8$

25. $\log_{27} \frac{1}{9} = -\frac{2}{3}$

27. $8^{-1/3} = \frac{1}{2}$

29. $\left(\frac{1}{2}\right)^{-2} = 4$

31. $\log_3 1 = 0$

33. $7^0 = 1$

35. $6^{1/2} = \sqrt{6}$

37. $\log_6 \sqrt{6} = \frac{1}{2}$

39. $\log_2 8 = \log_2 2^3$
$= 3$

41. $\log_9 81 = \log_9 9^2$
$= 2$

43. $\log_4 \frac{1}{4} = \log_4 4^{-1}$
$= -1$

45. $\log_5 \frac{1}{125} = \log_5 5^{-3}$
$= -3$

47. $\log_4 \frac{1}{2} = t$

$4^t = \frac{1}{2}$

$(2^2)^t = 2^{-1}$
$2^{2t} = 2^{-1}$
$2t = -1$

$t = -\frac{1}{2}$

49. $\log_8 4 = t$
$8^t = 4$
$(2^3)^t = 2^2$
$2^{3t} = 2^2$
$3t = 2$
$t = \frac{2}{3}$

51. $\log_9 (-27) = t$
$9^t = -27$

Not defined

53. $\log_4 \frac{1}{8} = t$

$4^t = \frac{1}{8}$

$(2^2)^t = 2^{-3}$
$2^{2t} = 2^{-3}$
$2t = -3$
$t = -\frac{3}{2}$

55. $\log_6 \sqrt{6} = \log_6 6^{1/2}$
$= \frac{1}{2}$

57. $\log_5 \sqrt[3]{25} = t$
$5t = \sqrt[3]{25}$
$5^t = 5^{2/3}$
$t = \frac{2}{3}$

59. $\log_5 (\log_3 243)$
$= \log_5 (\log_3 3^5)$
$= \log_5 5$
$= 1$

61. $\log_8 (\log_7 7)$
$= \log_8 1$
$= 0$

63. $5^{\log_5 7} = 7$

65. $\log_5 x = 3$
$$5^3 = x$$
$$125 = x$$

67. $y = \log_{10} 1000$
$$10^y = 1000$$
$$10^y = 10^3$$
$$y = 3$$

69. $\log_b 64 = 3$
$$b^3 = 64$$
$$b^3 = 4^3$$
$$b = 4$$

71. $\log_6 x = -2$
$$6^{-2} = x$$
$$\frac{1}{36} = x$$

73. $\log_4 x = \frac{3}{2}$
$$4^{3/2} = x$$
$$8 = x$$

75. $y = \log_8 32$
$$8^y = 32$$
$$(2^3)^y = 2^5$$
$$2^{3y} = 2^5$$
$$3y = 5$$
$$y = \frac{5}{3}$$

77. $\log_b \frac{1}{8} = -3$
$$b^{-3} = \frac{1}{8}$$
$$b^{-3} = 2^{-3}$$
$$b = 2$$

79. $\log_5 x = 0$
$$5^0 = x$$
$$1 = x$$

81. $\log_{10} 23596 = 4.3728$

83. $\log_{10} 0.000925 = -3.0339$

85. $f(x) = \log_2 x$ and $g(x) = \log_5 x$

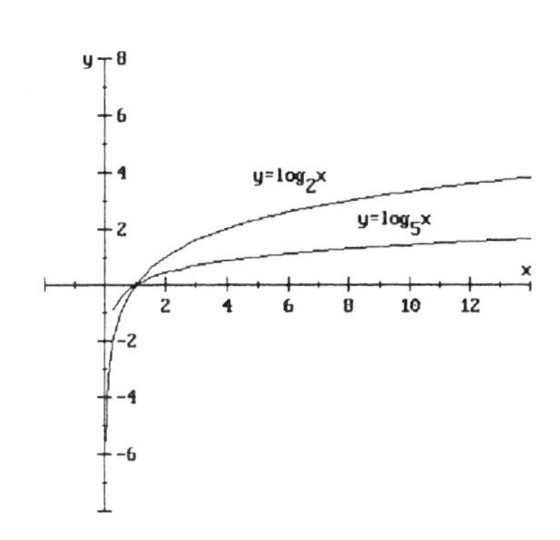

91. $f(x) = \frac{\sqrt{x+2}}{\sqrt{x}+2}$

$$f(7) = \frac{\sqrt{7+2}}{\sqrt{7}+2}$$
$$= \frac{\sqrt{9}}{\sqrt{7}+2}$$
$$= \frac{3}{\sqrt{7}+2}$$
$$= \frac{3(\sqrt{7}-2)}{(\sqrt{7}+2)(\sqrt{7}-2)}$$
$$= \frac{3(\sqrt{7}-2)}{7-4}$$
$$= \frac{3(\sqrt{7}-2)}{3}$$
$$= \sqrt{7}-2$$

93. $x + 3 \overline{\smash{)}x^3 + 0x^2 + 0x + 27}$ with quotient $x^2 - 3x + 9$

$$\begin{array}{l} \underline{-(x^3 + 3x^2)} \\ \quad -3x^2 + 0x \\ \quad \underline{-(-3x^2 - 9x)} \\ \qquad 9x + 27 \\ \qquad \underline{-(9x + 27)} \\ \qquad\qquad 0 \end{array}$$

Exercises 10.3

1. $\log_5 (xyz) = \log_5 x + \log_5 y + \log_5 z$

3. $\log_7 \dfrac{2}{3} = \log_7 2 - \log_7 3$

5. $\log_3 x^3 = 3 \log_3 x$

7. $\log_b a^{2/3} = \dfrac{2}{3} \log_b a$

9. $\log_b b^8 = 8$

11. $\log_s s^{-1/4} = -\dfrac{1}{4}$

13. $\log_b (x^2y^3) = \log_b x^2 + \log_b y^3$
$= 2 \log_b x + 3 \log_b y$

15. $\log_b \dfrac{m^4}{n^2} = \log_b m^4 - \log_b n^2$
$= 4 \log_b m - 2 \log_b n$

17. $\log_b \sqrt{xy} = \log_b (xy)^{1/2}$

$= \dfrac{1}{2} \log_b (xy)$

$= \dfrac{1}{2}(\log_b x + \log_b y)$

19. $\log_2 \sqrt[5]{\dfrac{x^2y}{z^3}}$

$= \log_2 \left(\dfrac{x^2y}{z^3}\right)^{1/5}$

$= \dfrac{1}{5} \log_2 \dfrac{x^2y}{z^3}$

$= \dfrac{1}{5}\left(\log_2 x^2y - \log_2 z^3\right)$

$= \dfrac{1}{5}\left(\log_2 x^2 + \log_2 y - \log_2 z^3\right)$

$= \dfrac{1}{5}(2 \log_2 x + \log_2 y - 3 \log_2 z)$

$= \dfrac{2}{5} \log_2 x + \dfrac{1}{5} \log_2 y - \dfrac{3}{5} \log_2 z$

21. $\log_b (xy + z^2)$

23. $\log_b \dfrac{x^2}{yz} = \log_b x^2 - \log_b yz$
$= 2 \log_b x - (\log_b y + \log_b z)$
$= 2 \log_b x - \log_b y - \log_b z$

25. $\log_6 \sqrt{\dfrac{6m^2n}{p^5q}}$

$= \log_6 \left(\dfrac{6m^2n}{p^5q}\right)^{1/2}$

$= \dfrac{1}{2} \log_6 \dfrac{6m^2n}{p^5q}$

$= \dfrac{1}{2}(\log_6 6m^2n - \log_6 p^5q)$

$= \dfrac{1}{2}\left[\log_6 6 + \log_6 m^2 + \log_6 n - (\log_6 p^5 + \log_6 q)\right]$

$= \dfrac{1}{2}(1 + 2 \log_6 m + \log_6 n - \log_6 p^5 - \log_6 q)$

$= \dfrac{1}{2} + \log_6 m + \dfrac{1}{2} \log_6 n - \dfrac{5}{2} \log_6 p - \dfrac{1}{2} \log_6 q$

27. $\log_b x + \log_b y = \log_b xy$

29. $2 \log_b m - 3 \log_b n$
$= \log_b m^2 - \log_b n^3$
$= \log_b \dfrac{m^2}{n^3}$

31. $4 \log_b 2 + \log_b 5$
$= \log_b 2^4 + \log_b 5$
$= \log_b 2^4 \cdot 5$
$= \log_b 80$

33. $\dfrac{1}{3} \log_b x + \dfrac{1}{4} \log_b y - \dfrac{1}{5} \log_b z$

$= \log_b x^{1/3} + \log_b y^{1/4} - \log_b z^{1/5}$
$= \log_b (x^{1/3}y^{1/4}) - \log_b z^{1/5}$

$= \log_b \left(\dfrac{x^{1/3}y^{1/4}}{z^{1/5}}\right)$

35. $\frac{1}{2}(\log_b x + \log_b y) - 2 \log_b z$

$= \frac{1}{2} \log_b (xy) - 2 \log_b z$

$= \log_b (xy)^{1/2} - \log_b z^2$

$= \log_b \frac{(xy)^{1/2}}{z^2}$

$= \log_b \frac{\sqrt{xy}}{z^2}$

37. $2 \log_b x - (\log_b y + 3 \log_b z)$
$= \log_b x^2 - (\log_b y + \log_b z^3)$
$= \log_b x^2 - \log_b (yz^3)$
$= \log_b \left(\frac{x^2}{yz^3}\right)$

39. $\frac{2}{3} \log_p x + \frac{4}{3} \log_p y - \frac{3}{7} \log_p z$

$= \log_p x^{2/3} + \log_p y^{4/3} - \log_p z^{3/7}$

$= \log_p (x^{2/3}y^{4/3}) - \log_p z^{3/7}$

$= \log_p \left(\frac{x^{2/3}y^{4/3}}{z^{3/7}}\right)$

41. $\log_b 10$
$= \log_b (2 \cdot 5)$
$= \log_b 2 + \log_b 5$
$= 1.2 + 2.1$
$= 3.3$

43. $\log_b \frac{2}{5}$
$= \log_b 2 - \log_b 5$
$= 1.2 - 2.1$
$= -0.9$

45. $\log_b \frac{1}{3}$
$= \log_b 1 - \log_b 3$
$= 0 - 1.42$
$= -1.42$

47. $\log_b 32$
$= \log_b 2^5$
$= 5 \log_b 2$
$= 5(1.2)$
$= 6$

49. $\log_b 100$
$= \log_b 10^2$
$= 2 \log_b 10$
$= 2 \log_b (2 \cdot 5)$
$= 2(\log_b 2 + \log_b 5)$
$= 2(1.2 + 2.1)$
$= 6.6$

51. $\log_b \sqrt{20}$
$= \log_b 20^{1/2}$

$= \frac{1}{2} \log_b 20$

$= \frac{1}{2} \log_b (2^2 \cdot 5)$

$= \frac{1}{2}(\log_b 2^2 + \log_b 5)$

$= \frac{1}{2}(2 \log_b 2 + \log_b 5)$

$= \log_b 2 + \frac{\log_b 5}{2}$

$= 1.2 + \frac{2.1}{2}$

$= 2.25$

53. $\log_b \sqrt[3]{x^2}$
$= \log_b x^{2/3}$

$= \frac{2}{3} \log_b x$

$= \frac{2}{3} A$

55. $\log_b \frac{x^3y^2}{z}$
$= \log_b x^3 + \log_b y^2 - \log_b z$

$= 3 \log_b x + 2 \log_b y - \log_b z$
$= 3A + 2B - C$

57. $\log_b \sqrt{\dfrac{x^5 y}{z^3}}$

$= \log_b \left(\dfrac{x^5 y}{z^3}\right)^{1/2}$

$= \dfrac{1}{2} \log_b \left(\dfrac{x^5 y}{z^3}\right)$

$= \dfrac{1}{2}\left[\log_b (x^5 y) - \log_b z^3\right]$

$= \dfrac{1}{2}(\log_b x^5 + \log_b y - \log_b z^3)$

$= \dfrac{1}{2}(5 \log_b x + \log_b y - 3 \log_b z)$

$= \dfrac{5}{2} \log_b x + \dfrac{1}{2} \log_b y - \dfrac{3}{2} \log_b z$

$= \dfrac{5}{2}A + \dfrac{1}{2}B - \dfrac{3}{2}C$

63. $(2\sqrt{x} - 5)(3\sqrt{x} + 4)$
$= 6x + 8\sqrt{x} - 15\sqrt{x} - 20$
$= 6x - 7\sqrt{x} - 20$

65. (s, N); $(15, 80)$ and $(18, 85)$

(a) $m = \dfrac{85 - 80}{18 - 15} = \dfrac{5}{3}$

$N - 80 = \dfrac{5}{3}(s - 15)$

$N - 80 = \dfrac{5}{3}s - 25$

$N = \dfrac{5}{3}s + 55$

(b) When $s = 30$:
$N = \dfrac{5}{3}(30) + 55$
$= 105$
The heat rate should be 105 beats per minute.

Exercises 10.4

1. $\log 584 = 2.7664$

3. $\log 0.00371 = -2.4306$

5. $\log 280{,}000 = 5.4472$

7. $\log 0.0000553 = -4.2573$

9. $\log 8 = 0.9031$

11. $\ln 0.941 = -0.0608$

13. $e^{0.941} = 2.5625$

15. $e^{1.24} = 3.4556$

17. $\ln 4.5 = 1.5041$

19. $\ln(-23)$ undefined

21. $\log x = 0.941$
$10^{0.941} = x$
$x \approx 8.7297$

23. $\ln x = 4$
$e^4 = x$
$x \approx 54.5982$

25. $\log_5 x = \dfrac{\log x}{\log 5}$

27. $\log_7 8 = \dfrac{\log 8}{\log 7}$

29. $\log_5 8 = x$

$5^x = 8$
$(5^x)^2 = 8^2$
$5^{2x} = 8^2$
$(5^2)^x = 8^2$
$25^x = 8^2$

$x = \log_{25} 8^2 = 2 \log_{25} 8$

31. $\log_5 87 = \dfrac{\log 87}{\log 5}$

≈ 2.7748

33. $\log_4 265 = \dfrac{\log 265}{\log 4}$

≈ 4.0249

35. $\log_3 821 = \dfrac{\log 821}{\log 3}$

≈ 6.1082

37. $\log_7 52 = \dfrac{\log 52}{\log 7}$

≈ 2.0305

39. (a) $y = \log_a x$
(b) $a^y = x$
(c) $\log_b a^y = \log_b x$
(d) $y \log_b a = \log_b x$
(e) $y = \dfrac{\log_b x}{\log_b a}$

41. $y = \log_8 x = \dfrac{\log x}{\log 8}$

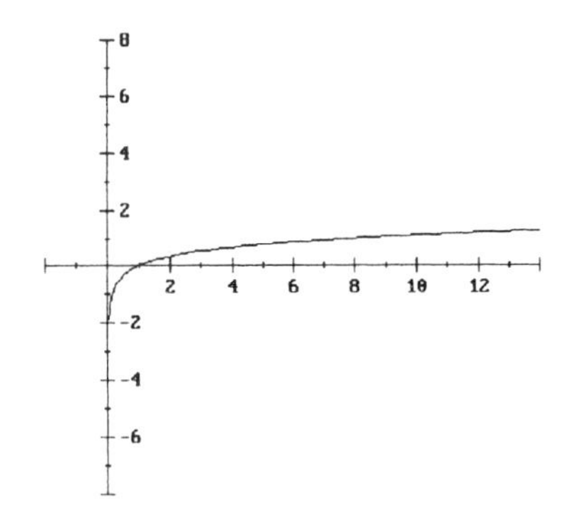

43. $y = -4 + \log_3 x$

$y = -4 + \dfrac{\log x}{\log 3}$

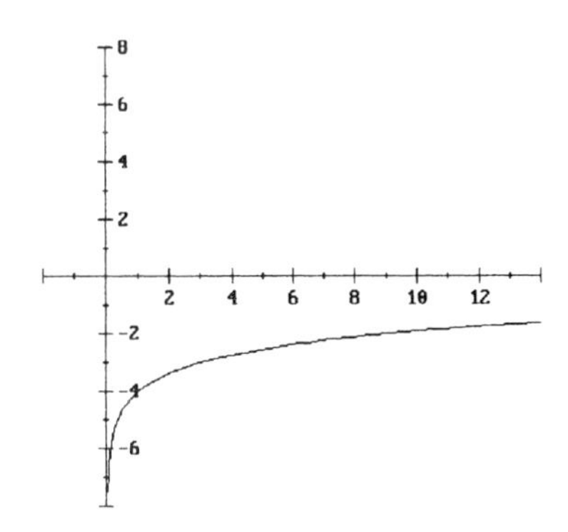

45. $y = \log_3 (x - 4)$

$y = \dfrac{\log (x - 4)}{\log 3}$

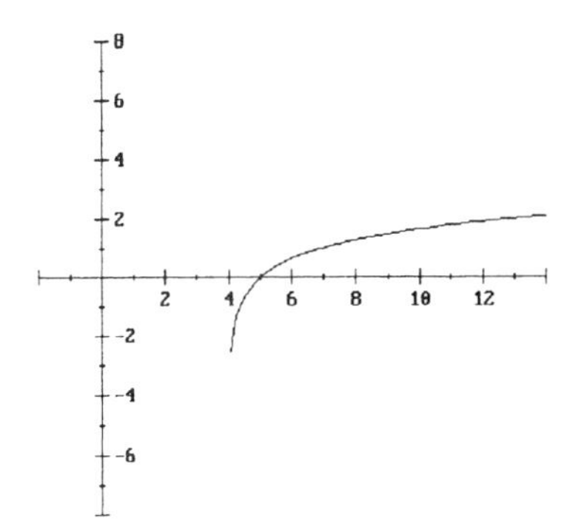

47. $\text{midpoint} = \left(\dfrac{2 + 5}{2}, \dfrac{4 - 6}{2}\right)$

$= \left(\dfrac{7}{2}, -1\right)$

49.
$$\begin{aligned}
\sqrt{7x + 4} - 2 &= x \\
\sqrt{7x + 4} &= x + 2 \\
\left(\sqrt{7x + 4}\right)^2 &= (x + 2)^2 \\
7x + 4 &= x^2 + 4x + 4 \\
0 &= x^2 - 3x \\
0 &= x(x - 3)
\end{aligned}$$

$x = 0$ or $x - 3 = 0$

$x = 0$ or $x = 3$

Exercises 10.5

1.
$$\begin{aligned}
\log_3 5 + \log_3 x &= 2 \\
\log_3 5x &= 2 \\
3^2 &= 5x \\
9 &= 5x \\
\frac{9}{5} &= x
\end{aligned}$$

3.
$$\begin{aligned}
\log_2 x &= 2 + \log_2 3 \\
\log_2 x - \log_2 3 &= 2 \\
\log_2 \frac{x}{3} &= 2 \\
2^2 &= \frac{x}{3} \\
12 &= x
\end{aligned}$$

5. $2\log_5 x = \log_5 36$
$$\log_5 x^2 = \log_5 36$$
$$x^2 = 36$$
$$x = 6$$

(Remember: domain of log function is $x > 0$.)

7. $\log_3 x + \log_3 (x - 8) = 2$
$$\log_3 [x(x - 8)] = 2$$
$$3^2 = x(x - 8)$$
$$9 = x^2 - 8x$$
$$0 = x^2 - 8x - 9$$
$$0 = (x - 9)(x + 1)$$
$$x - 9 = 0 \quad \text{or} \quad x + 1 = 0$$
$$x = 9 \quad \text{or} \quad x = -1$$

$x = -1$ is not in the domain of $\log_3 x$, hence the only solution is $x = 9$.

9. $\log_2 a + \log_2 (a + 2) = 3$
$$\log_2 [a(a + 2)] = 3$$
$$2^3 = a(a + 2)$$
$$8 = a^2 + 2a$$
$$0 = a^2 + 2a - 8$$
$$0 = (a + 4)(a - 2)$$
$$a + 4 = 0 \quad \text{or} \quad a - 2 = 0$$
$$a = -4 \quad \text{or} \quad a = 2$$

$a = -4$ is not in the domain of $\log_2 a$, hence the only solution is $a = 2$.

11. $\log_2 y - \log_2 (y - 2) = 3$
$$\log_2 \frac{y}{y - 2} = 3$$
$$2^3 = \frac{y}{y - 2}$$
$$8 = \frac{y}{y - 2}$$
$$8(y - 2) = y$$
$$8y - 16 = y$$
$$-16 = -7y$$
$$\frac{16}{7} = y$$

13. $\log_3 x - \log_3 (x + 3) = 5$
$$\log_3 \frac{x}{x + 3} = 5$$
$$3^5 = \frac{x}{x + 3}$$
$$243 = \frac{x}{x + 3}$$
$$243(x + 3) = x$$
$$243x + 729 = x$$
$$729 = -242x$$
$$-\frac{729}{242} = x$$

No solution, since $x = -\frac{729}{242}$ is not in the domain of $\log_3 x$.

15. $\log_b 5 + \log_b x = \log_b 10$
$$\log_b 5x = \log_b 10$$
$$5x = 10$$
$$x = 2$$

17. $\log_p x - \log_p 2 = \log_p 7$
$$\log_p \frac{x}{2} = \log_p 7$$
$$\frac{x}{2} = 7$$
$$x = 14$$

19. $\log_5 x + \log_5 (x + 1) = \log_5 2$
$$\log_5 [x(x + 1)] = \log_5 2$$
$$x(x + 1) = 2$$
$$x^2 + x = 2$$
$$x^2 + x - 2 = 0$$
$$(x + 2)(x - 1) = 0$$
$$x + 2 = 0 \quad \text{or} \quad x - 1 = 0$$
$$x = -2 \quad \text{or} \quad x = 1$$

$x = -2$ is not in the domain of $\log_5 x$, hence the only solution is $x = 1$.

21. $\log_3 x - \log_3(x-2) = \log_3 4$

$$\log_3 \frac{x}{x-2} = \log_3 4$$

$$\frac{x}{x-2} = 4$$

$$x = 4(x-2)$$
$$x = 4x - 8$$
$$-3x = -8$$
$$x = \frac{8}{3}$$

23. $\log_4 x - \log_4(x-4) = \log_4(x-6)$

$$\log_4\left(\frac{x}{x-4}\right) = \log_4(x-6)$$

$$\frac{x}{x-4} = x - 6$$
$$x = (x-4)(x-6)$$
$$x = x^2 - 10x + 24$$
$$0 = x^2 - 11x + 24$$
$$0 = (x-8)(x-3)$$
$$x - 8 = 0 \quad \text{or} \quad x - 3 = 0$$
$$x = 8 \quad \text{or} \quad x = 3$$

$x = 3$ is not in the domain of $\log_4(x-6)$, hence $x = 8$ is the only solution.

25.
$$2\log_2 x = \log_2(2x-1)$$
$$\log_2 x^2 = \log_2(2x-1)$$
$$x^2 = 2x - 1$$
$$x^2 - 2x + 1 = 0$$
$$(x-1)^2 = 0$$
$$x - 1 = 0$$
$$x = 1$$

27. $\frac{1}{2}\log_3 x = \log_3(x-6)$

$$\log_3 x^{1/2} = \log_3(x-6)$$
$$x^{1/2} = x - 6$$
$$(x^{1/2})^2 = (x-6)^2$$
$$x = x^2 - 12x + 36$$
$$0 = x^2 - 13x + 36$$
$$0 = (x-9)(x-4)$$
$$x - 9 = 0 \quad \text{or} \quad x - 4 = 0$$
$$x = 9 \quad \text{or} \quad x = 4$$

$x = 4$ is not in the domain of $\log_3(x-6)$, hence $x = 9$ is the only solution.

29.
$$2\log_b x = \log_b(6x-5)$$
$$\log_b x^2 = \log_b(6x-5)$$
$$x^2 = 6x - 5$$
$$x^2 - 6x + 5 = 0$$
$$(x-5)(x-1) = 0$$
$$x - 5 = 0 \quad \text{or} \quad x - 1 = 0$$
$$x = 5 \quad \text{or} \quad x = 1$$

31.
$$2^x = 5$$
$$\log 2^x = \log 5$$
$$x \log 2 = \log 5$$
$$x = \frac{\log 5}{\log 2}$$
$$x \approx 2.3219$$

33.
$$2^{x+1} = 6$$
$$\log 2^{x+1} = \log 6$$
$$(x+1)\log 2 = \log 6$$
$$x + 1 = \frac{\log 6}{\log 2}$$
$$x = \frac{\log 6}{\log 2} - 1$$
$$x \approx 1.5850$$

35.
$$4^{2x+3} = 5$$
$$\log 4^{2x+3} = \log 5$$
$$(2x+3)\log 4 = \log 5$$
$$2x + 3 = \frac{\log 5}{\log 4}$$
$$2x = \frac{\log 5}{\log 4} - 3$$
$$x = \frac{\log 5}{2\log 4} - \frac{3}{2}$$
$$x \approx -0.9195$$

37.
$$7^{y+1} = 3^y$$
$$\log 7^{y+1} = \log 3^y$$
$$(y+1)\log 7 = y\log 3$$
$$y\log 7 + \log 7 = y\log 3$$
$$\log 7 = y\log 3 - y\log 7$$
$$\log 7 = y(\log 3 - \log 7)$$

$$\log 7 = y\left(\log \frac{3}{7}\right)$$

$$\frac{\log 7}{\log \frac{3}{7}} = y$$

$$y \approx -2.2966$$

39.
$$6^{2x+1} = 5^{x+2}$$
$$\log 6^{2x+1} = \log 5^{x+2}$$
$$(2x + 1) \log 6 = (x + 2) \log 5$$
$$2x \log 6 + \log 6 = x \log 5 + 2 \log 5$$
$$2x \log 6 - x \log 5 = 2 \log 5 - \log 6$$
$$x(2 \log 6 - \log 5) = 2 \log 5 - \log 6$$
$$x = \frac{2 \log 5 - \log 6}{2 \log 6 - \log 5}$$
$$x \approx 0.7229$$

41.
$$8^{3x-2} = 9^{x+2}$$
$$\log 8^{3x-2} = \log 9^{x+2}$$
$$(3x - 2) \log 8 = (x + 2) \log 9$$
$$3x \log 8 - 2 \log 8 = x \log 9 + 2 \log 9$$
$$3x \log 8 - x \log 9 = 2 \log 9 + 2 \log 8$$
$$x(3 \log 8 - \log 9) = 2 \log 9 + 2 \log 8$$
$$x = \frac{2 \log 9 + 2 \log 8}{3 \log 8 - \log 9}$$
$$x \approx 2.1166$$

43.
$$3^x = 5 \cdot 2^x$$
$$\frac{3^x}{2^x} = 5$$
$$\left(\frac{3}{2}\right)^x = 5$$
$$\log\left(\frac{3}{2}\right)^x = \log 5$$
$$x \log \frac{3}{2} = \log 5$$
$$x = \frac{\log 5}{\log \frac{3}{2}}$$
$$x \approx 3.9694$$

45.
$$2^y 5^y = 3$$
$$(2 \cdot 5)^y = 3$$
$$10^y = 3$$
$$\log 10^y = \log 3$$
$$y = \log 3$$
$$y \approx 0.4771$$

47.
$$4^a 3^{a+1} = 2$$
$$4^a \cdot 3^a \cdot 3 = 2$$
$$4^a \cdot 3^a = \frac{2}{3}$$
$$(4 \cdot 3)^a = \frac{2}{3}$$
$$12^a = \frac{2}{3}$$
$$\log 12^a = \log \frac{2}{3}$$
$$a \log 12 = \log \frac{2}{3}$$
$$a = \frac{\log^{2/3}}{\log 12}$$
$$a \approx -0.1632$$

49.
$$pH = -\log\left[H_3O^+\right]$$
$$= -\log\left[3.98 \times 10^{-6}\right]$$
$$= 5.4$$

51.
$$pH = -\log\left[H_3O^+\right]$$
$$7 = -\log\left[H_3O^+\right]$$
$$-7 = \log\left[H_3O^+\right]$$
$$10^{-7} = H_3O^+$$

53.
$$N = 10 \log I + 160$$
$$200 = 10 \log I + 160$$
$$40 = 10 \log I$$
$$4 = \log I$$
$$10^4 = I$$
The intensity is 10^4 watts/cm^2.

55. (a) $N(m) = 10(2^m)$

(b) $N(2) = 10(2^{24})$
$= 167{,}772{,}160$ mice

(c)
$$10{,}000 = 10(2^m)$$
$$1000 = 2^m$$
$$\log 1000 = \log 2^m$$
$$3 = m \log 2$$
$$\frac{3}{\log 2} = m$$
$$m \approx 10$$
It takes approximately 10 months.

57.
$$(x-h)^2+(y-k)^2=r^2$$
$$(x-2)^2+[y-(-3)]^2=6^2$$
$$(x-2)^2+(y+3)^2=36$$

59.
$$2x^2-5x\le 3$$
$$2x^2-5x-3\le 0$$

$$2x^2-5x-3=0$$
$$(2x+1)(x-3)=0$$
$$2x+1=0 \qquad x-3=0$$
$$x=-\frac{1}{2} \qquad x=3$$

The intervals are $x<-\frac{1}{2}$, $-\frac{1}{2}<x<3$, $x>3$

Test points:

	$(2x+1)(x-3)$	
-1	$[2(-1)+1](-1-3)$	pos
0	$(2\cdot 0+1)(0-3)$	neg
4	$(2\cdot 4+1)(4-3)$	pos

We want $(2x+1)(x-3)\le 0$: neg or 0

Solution: $-\frac{1}{2}\le x\le 3$

Exercises 10.6

1. $A=P\left(1+\frac{r}{n}\right)^{nt}$
$$=8000\left(1+\frac{0.06}{2}\right)^{2(5)}$$
$$=\$10{,}751.33$$

3. $A=P\left(1+\frac{r}{n}\right)^{nt}$
$$=8000\left(1+\frac{0.06}{12}\right)^{12(5)}$$
$$=\$10{,}790.80$$

5.
$$A=P\left(1+\frac{r}{n}\right)^{nt}$$
$$2(9000)=9000\left(1+\frac{0.062}{12}\right)^{12t}$$
$$2=\left(1+\frac{0.062}{12}\right)^{12t}$$
$$\ln 2=\ln\left(1+\frac{0.062}{12}\right)^{12t}$$
$$\ln 2=12t\ln\left(1+\frac{0.062}{12}\right)$$
$$t=\frac{\ln 2}{12\ln\left(1+\frac{0.062}{12}\right)}\approx 11.21 \text{ years}$$

7. $2P=P\left(1+\frac{0.062}{12}\right)^{12t}$
$$2=\left(1+\frac{0.062}{12}\right)^{12t}$$
See Exercise #5.
This is the same equation, hence it would take approximately 11.21 years.

9.
$$2P=P\left(1+\frac{0.10}{12}\right)^{12t}$$
$$2=\left(1+\frac{0.10}{12}\right)^{12t}$$
$$\ln 2=\ln\left(1+\frac{0.10}{12}\right)^{12t}$$
$$\ln 2=12t\ln\left(1+\frac{0.10}{12}\right)$$
$$t=\frac{\ln 2}{12\ln\left(1+\frac{0.10}{12}\right)}\approx 6.96 \text{ years}$$

11.
$$A=Pe^{rt}$$
$$20000=5000e^{0.073t}$$
$$4=e^{0.073t}$$
$$\ln 4=\ln e^{0.073t}$$
$$\ln 4=0.073t$$
$$\frac{\ln 4}{0.073}=t$$
$$t\approx 18.99$$
It would take approximately 19 years.

13. $P = P_0e^{rt}$

$P = 2000e^{0.08(15)}$

$P \approx 6640$

There will be approximately 6640 inhabitants.

15. $A = 10000e^{0.0542t}$

$A(5) = 10000e^{0.0542(5)}$

≈ 13112.75

There are approximately 13113 bacteria present.

17. $A = 10000e^{0.0542t}$

$100000 = 10000e^{0.0542t}$

$10 = e^{0.0542t}$

$\ln 10 = \ln e^{0.0542t}$

$\ln 10 = 0.0542t$

$\dfrac{\ln 10}{0.0542} = t$

$t \approx 42.5$ hr

$A = 10000e^{0.122t}$

$100000 = 10000e^{0.122t}$

$10 = e^{0.122t}$

$\ln 10 = \ln e^{0.122t}$

$\ln 10 = 0.122t$

$\dfrac{\ln 10}{0.122} = t$

$t \approx 18.9$ hr

19. $P = P_0e^{rt}$

$10000 = 1000e^{0.08t}$

$10 = e^{0.08t}$

$\ln 10 = \ln e^{0.08t}$

$\ln 10 = 0.08t$

$\dfrac{\ln 10}{0.08} = t$

$t \approx 28.8$ hr

21. $P = P_0e^{rt}$

$3000 = 1000e^{r(5)}$

$3 = e^{5r}$

$\ln 3 = \ln e^{5r}$

$\ln 3 = 5r$

$\dfrac{\ln 3}{5} = r$

$r = 0.2197$

$P = 1000e^{0.2197t}$

$50000 = 1000e^{0.2197t}$

$50 = e^{0.2197t}$

$\ln 50 = \ln e^{0.2197t}$

$\ln 50 = 0.2197t$

$\dfrac{\ln 50}{0.2197} = t$

$t \approx 17.8$ hr

23. $A = A_0e^{-0.0004t}$

$\dfrac{1}{2}A_0 = A_0e^{-0.0004t}$

$\dfrac{1}{2} = e^{-0.0004t}$

$\ln \dfrac{1}{2} = \ln e^{-0.0004t}$

$\ln \dfrac{1}{2} = -0.0004t$

$\dfrac{\ln \frac{1}{2}}{-0.0004} = t$

$t \approx 1732.87$

25. $A = A_0e^{-rt}$

$\dfrac{1}{2}A_0 = A_0e^{-r(4)}$

$\dfrac{1}{2} = e^{-4r}$

$\ln \dfrac{1}{2} = \ln e^{-4r}$

$\ln \dfrac{1}{2} = -4r$

$\dfrac{\ln \frac{1}{2}}{-4} = r$

$r = 0.173 = 17.3\%$ per day

27. $R = \log I$

$3.6 = \log I_1$

$10^{3.6} = I_1$

$R = \log I$

$7.2 = \log I_2$

$10^{7.2} = I_2$

$10^{7.2} = (10^{3.6})^2$

Hence $I_2 = I_1^2$.

29. (a) $A = A_0 e^{rt}$

$$\frac{1}{2}A_0 = A_0 e^{r(5730)}$$

$$\frac{1}{2} = e^{5730r}$$

$$\ln \frac{1}{2} = \ln e^{5730r}$$

$$\ln \frac{1}{2} = 5730r$$

$$\frac{\ln \frac{1}{2}}{5730} = r$$

$$r = -0.000121$$

$$A = A_0 e^{-0.000121t}$$

(b)

$$0.35A_0 = A_0 e^{-0.000121t}$$
$$0.35 = e^{0.000121t}$$
$$\ln 0.35 = \ln e^{-0.000121t}$$
$$\ln 0.35 = -0.000121t$$

$$\frac{\ln 0.35}{-0.000121} = t$$

$$t \approx 8676.2 \text{ years}$$

31. $y = 6 - x$

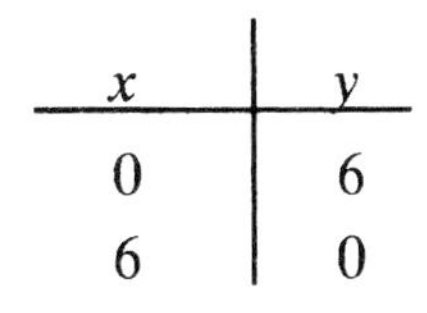

x	y
0	6
6	0

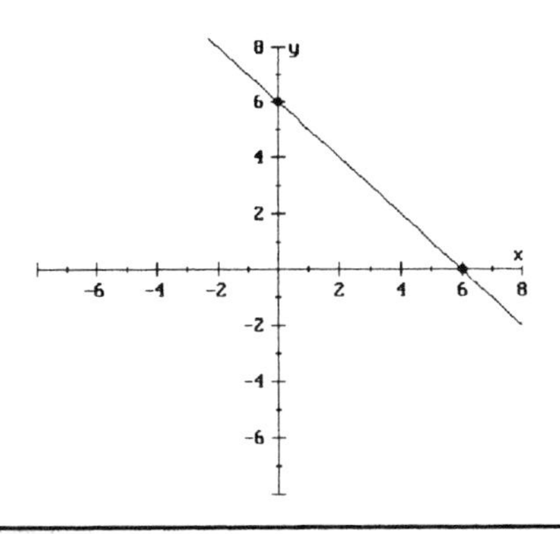

33. $(3x - 4)^2 = 5$

$$3x - 4 = \pm\sqrt{5}$$
$$3x = 4 \pm \sqrt{5}$$
$$x = \frac{4 \pm \sqrt{5}}{3}$$

CHAPTER 10 REVIEW EXERCISES

1. $y = 6^x$

x	y
-2	1/36
-1	1/6
0	1
1	6
2	36

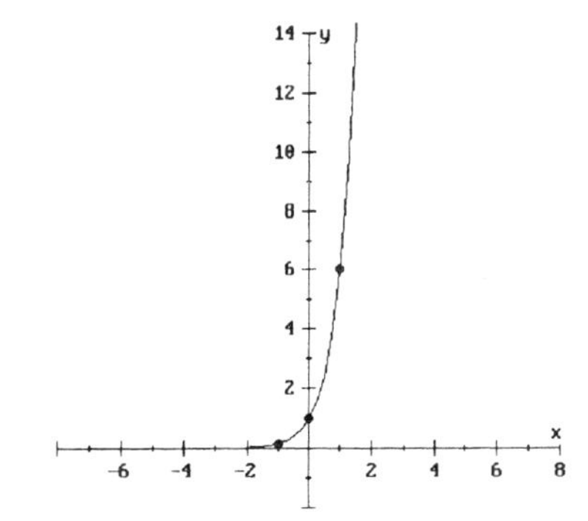

3. $y = \log_{10} x$

$10^y = x$

x	y
1/100	-2
1/10	-1
1	0
10	1
100	2

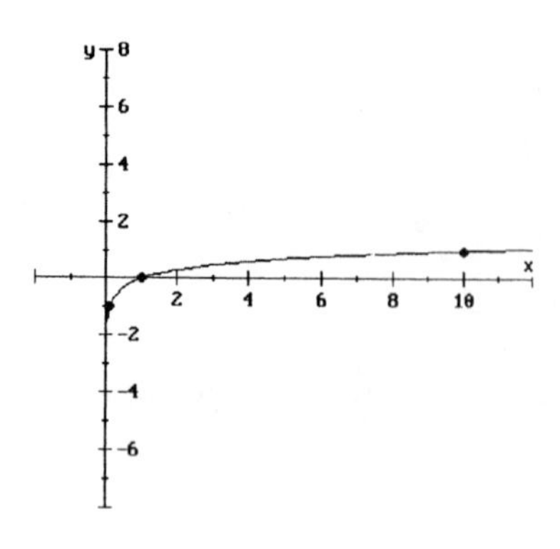

5. $y = 2^{-x}$

x	y
-2	4
-1	2
0	1
1	1/2
2	1/4

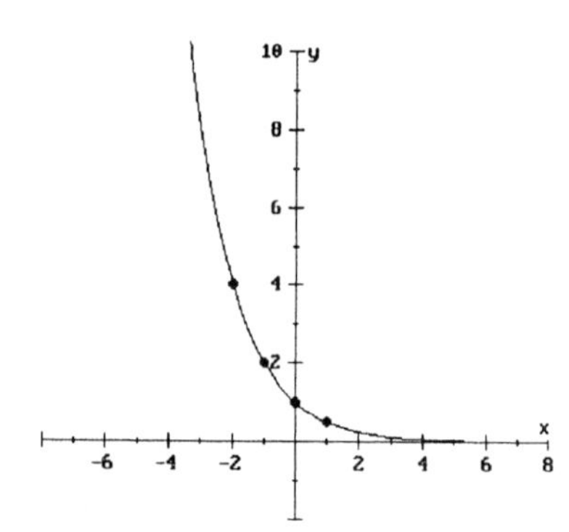

7. $\log_3 81 = 4$
$$3^4 = 81$$

9. $$4^{-3} = \frac{1}{64}$$
$$\log_4 \frac{1}{64} = -3$$

11. $\log_8 4 = \frac{2}{3}$
$$8^{2/3} = 4$$

13. $$25^{1/2} = 5$$
$$\log_{25} 5 = \frac{1}{2}$$

15. $\log_7 \sqrt{7} = \frac{1}{2}$
$$7^{1/2} = \sqrt{7}$$

17. $\log_6 1 = 0$
$$6^0 = 1$$

19. $\log_{10} 1000 = x$
$$10^x = 1000$$
$$10^x = 10^3$$
$$x = 3$$

21. $\log_3 \frac{1}{9} = x$
$$3^x = \frac{1}{9}$$
$$3^x = 3^{-2}$$
$$x = -2$$

23. $\log_2 \frac{1}{4} = x$
$$2^x = \frac{1}{4}$$
$$2^x = 2^{-2}$$
$$x = -2$$

25. $\log_{1/3} 9 = x$
$$\left(\frac{1}{3}\right)^x = 9$$
$$3^{-x} = 9$$
$$3^{-x} = 3^2$$
$$-x = 2$$
$$x = -2$$

27. $\log_3 \frac{1}{9} = x$
$$3^x = \frac{1}{9}$$
$$3^x = 3^{-2}$$
$$x = -2$$

29. $\log_b \sqrt{b} = x$

$$b^x = \sqrt{b}$$
$$b^x = b^{1/2}$$
$$x = \frac{1}{2}$$

31. $\log_{16} 32 = x$

$$16^x = 32$$
$$(2^4)^x = 2^5$$
$$2^{4x} = 2^5$$
$$4x = 5$$
$$x = \frac{5}{4}$$

33. $\log_b (x^3y^7) = \log_b x^3 + \log_b y^7$

$$= 3 \log_b x + 7 \log_b y$$

35. $\log_b \dfrac{u^2v^5}{w^3} = \log_b (u^2v^5) - \log_b w^3$

$$= \log_b u^2 + \log_b v^5 - \log_b w^3$$
$$= 2 \log_b u + 5 \log_b v - 3 \log_b w$$

37. $\log_b \sqrt[3]{xy} = \log_b (xy)^{1/3}$

$$= \frac{1}{3} \log_b (xy)$$
$$= \frac{1}{3}(\log_b x + \log_b y)$$
$$= \frac{1}{3} \log_b x + \frac{1}{3} \log_b y$$

39. $\log_b (x^3 + y^4)$

41. $\log_b \sqrt[4]{\dfrac{x^6y^2}{z^2}}$

$$= \log_b \left(\frac{x^6y^2}{z^2}\right)^{1/4}$$
$$= \frac{1}{4} \log_b \left(\frac{x^6y^2}{z^2}\right)$$
$$= \frac{1}{4}\left[\log_b (x^6y^2) - \log_b z^2\right]$$
$$= \frac{1}{4}(\log_b x^6 + \log_b y^2 - \log_b z^2)$$
$$= \frac{1}{4}(6 \log_b x + 2 \log_b y - 2 \log_b z)$$
$$= \frac{3}{2} \log_b x + \frac{1}{2} \log_b y - \frac{1}{2} \log_b z$$

43. $\log_b 28 = \log_b 2^2 \cdot 7$

$$= \log_b 2^2 + \log_b 7$$
$$= 2 \log_b 2 + \log_b 7$$
$$= 2(1.1) + 1.32$$
$$= 3.52$$

45. $9^x = \dfrac{1}{81}$

$$9^x = 9^{-2}$$
$$x = -2$$

47. $16^x = 32$

$$(2^4)^x = 2^5$$
$$2^{4x} = 2^5$$
$$4x = 5$$
$$x = \frac{5}{4}$$

49. $5^{x+1} = 3$

$$\log 5^{x+1} = \log 3$$
$$(x + 1) \log 5 = \log 3$$
$$x + 1 = \frac{\log 3}{\log 5}$$
$$x = \frac{\log 3}{\log 5} - 1$$
$$x \approx -0.3174$$

51. $\log (x + 10) - \log (x + 1) = 1$

$$\log \frac{x + 10}{x + 1} = 1$$
$$10^1 = \frac{x + 10}{x + 1}$$
$$10(x + 1) = x + 10$$
$$10x + 10 = x + 10$$
$$9x = 0$$
$$x = 0$$

53. $\log_2 (t + 1) + \log_2 (t - 1) = 3$

$$\log_2 [(t + 1)(t - 1)] = 3$$
$$2^3 = (t + 1)(t - 1)$$
$$8 = t^2 - 1$$
$$9 = t^2$$
$$t = \pm 3$$

$t = -3$ is not in the domains of the above log functions, hence $t = 3$ is the only solution.

55. $\log_b 3x + \log_b (x + 2) = \log_b 9$

$$\log_b [3x(x + 2)] = \log_b 9$$
$$3x(x + 2) = 9$$
$$3x^2 + 6x = 9$$
$$3x^2 + 6x - 9 = 0$$
$$x^2 + 2x - 3 = 0$$
$$(x + 3)(x - 1) = 0$$
$$x + 3 = 0 \quad \text{or} \quad x - 1 = 0$$
$$x = -3 \quad \text{or} \quad x = 1$$

$x = -3$ is not in the domains of the above log functions, hence the only solution is $x = 1$.

57. $\log 783 \approx 2.8938$

59. $\log 0.00499 \approx -2.3019$

61. $\ln 0.0063 \approx -5.0672$

63. $e^{7.8} \approx 2440.6020$

65. $\log x = -3$

$$10^{-3} = x$$
$$x = 0.001$$

67. $\log_5 73 = \dfrac{\log 73}{\log 5}$

$$\approx 2.6658$$

69. $\log_{12} 764 = \dfrac{\log 764}{\log 12}$

$$\approx 2.6716$$

71. $\log_{0.2} 190 = \dfrac{\log 190}{\log 0.2}$

$$\approx -3.2602$$

73. $A = P\left(1 + \dfrac{r}{n}\right)^{nt}$

$$= 6000\left(1 + \frac{0.082}{2}\right)^{2(8)}$$
$$= \$11{,}412.03$$

75. $A = A_0 e^{-0.045t}$

$$25 = 100e^{-0.045t}$$
$$0.25 = e^{-0.045t}$$
$$\ln 0.25 = \ln e^{-0.045t}$$
$$\ln (0.25) = -0.045t$$
$$\frac{\ln 0.25}{-0.045} = t$$
$$t \approx 30.8$$

77. $pH = -\log \left[H_3 0^+\right]$

$$= -\log \left[6.21 \times 10^{-9}\right]$$
$$= 8.2069$$

CHAPTER 10 PRACTICE TEST

1. $y = \left(\dfrac{1}{3}\right)^x$

x	y
-2	9
-1	3
0	1
1	1/3
2	1/9

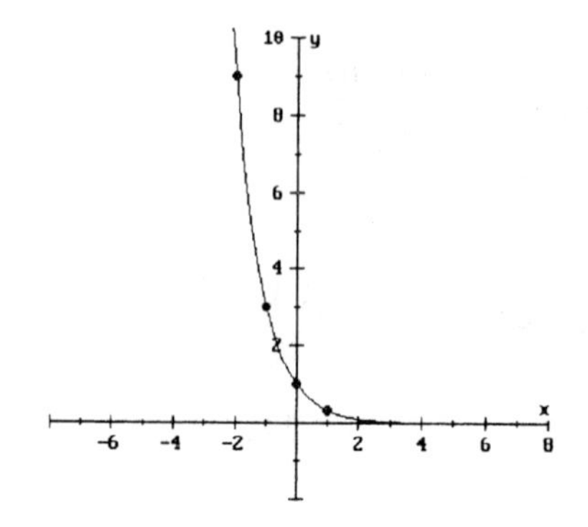

3. (a) $\log_2 16 = 4$

$2^4 = 16$

(b) $\log_9 \frac{1}{3} = -\frac{1}{2}$

$9^{-1/2} = \frac{1}{3}$

5. (a) $\log_3 \frac{1}{3} = x$

$3^x = \frac{1}{3}$

$3^x = 3^{-1}$

$x = -1$

(b) $\log_{81} 9 = x$

$81^x = 9$

$(9^2)^x = 9$

$9^{2x} = 9^1$

$2x = 1$

$x = \frac{1}{2}$

(c) $\log_8 32 = x$

$8^x = 32$

$(2^3)^x = 2^5$

$2^{3x} = 2^5$

$3x = 5$

$x = \frac{5}{3}$

7. (a) $\log 27{,}900 \approx 4.4456$
(b) $\ln 0.004 \approx -5.5215$
(c) $e^{-0.02} \approx 0.9802$

9. (a) $\log_5 67 = \dfrac{\log 67}{\log 5}$

≈ 2.613

(b) $\log_8 0.0034 = \dfrac{\log 0.0034}{\log 8}$

≈ -2.733

11. $A = P\left(1 + \dfrac{r}{n}\right)^{nt}$

$= 5000\left(1 + \dfrac{0.084}{4}\right)^{4(7)}$

$= \$8947.27$

CHAPTER 11

Exercises 11.1

1. $\begin{cases} x + y + z = 9 \\ 2x - y + z = 9 \\ x - y + z = 3 \end{cases}$

$$\begin{array}{l} x + y + z = 9 \\ \underline{2x - y + z = 9} \\ 3x + 2z = 18 \end{array}$$

$$\begin{array}{l} x + y + z = 9 \\ \underline{2x - y + z = 3} \\ 2x + 2z = 12 \end{array}$$

$$\begin{cases} 3x + 2z = 18 \\ 2x + 2z = 12 \end{cases}$$

$$\begin{array}{l} 3x + 2z = 18 \\ \underline{-2x - 2z = -12} \\ x = 6 \end{array}$$

$$\begin{aligned} 3(6) + 2z &= 18 \\ 18 + 2z &= 18 \\ 2z &= 0 \\ z &= 0 \end{aligned}$$

$$\begin{aligned} 6 + y + 0 &= 9 \\ 6 + y &= 9 \\ y &= 3 \end{aligned}$$

$x = 6, \quad y = 3, \quad z = 0$

3. $\begin{cases} -x + 2y + z = 0 \\ x - y + 2z = 1 \\ x + 3y + z = 5 \end{cases}$

$$\begin{array}{l} -x + 2y + z = 0 \\ \underline{x - y + 2z = 1} \\ y + 3z = 1 \end{array}$$

$$\begin{array}{l} -x + 2y + z = 0 \\ \underline{x + 3y + z = 5} \\ 5y + 2z = 5 \end{array}$$

$$\begin{cases} y + 3z = 1 \\ 5y + 2z = 5 \end{cases}$$

$$\begin{array}{l} -5y - 15z = -5 \\ \underline{5y + 2z = 5} \\ -13z = 0 \\ z = 0 \end{array}$$

$$\begin{aligned} y + 3(0) &= 1 \\ y &= 1 \end{aligned}$$

$$\begin{aligned} x - 1 + 2(0) &= 1 \\ x - 1 &= 1 \\ x &= 2 \end{aligned}$$

$x = 2, \quad y = 1, \quad z = 0$

5. $\begin{cases} x + y - z = 1 \\ 2x + 2y + 2z = 0 \\ x - y + z = 3 \end{cases}$

$$\begin{cases} x + y - z = 1 \\ x + y + z = 0 \\ x - y + z = 3 \end{cases}$$

$$\begin{array}{l} x + y - z = 1 \\ \underline{x + y + z = 0} \\ 2x + 2y = 1 \end{array}$$

$$\begin{array}{l} x + y - z = 1 \\ \underline{x - y + z = 3} \\ 2x = 4 \\ x = 2 \end{array}$$

$$\begin{aligned} 2(2) + 2y &= 1 \\ 4 + 2y &= 1 \\ 2y &= -3 \\ y &= -\frac{3}{2} \end{aligned}$$

$$2 - \left(-\frac{3}{2}\right) + z = 3$$

$$\frac{7}{2} + z = 3$$

$$z = -\frac{1}{2}$$

$x = 2, \quad y = -\frac{3}{2}, \quad z = -\frac{1}{2}$

7. $\begin{cases} x + 2y + 3z = 1 \\ 3x + 6y + 9z = 3 \\ 4x + 8y + 12z = 4 \end{cases}$

$$\begin{cases} x + 2y + 3z = 1 \\ x + 2y + 3z = 1 \\ x + 2y + 3z = 1 \end{cases}$$

$\{(x, y, z) \mid x + 2y + 3z = 1\}$

9. $\begin{cases} 3x - 2y + 5z = 2 \\ 4x - 7y - z = 19 \\ 5x - 6y + 4z = 13 \end{cases}$

$$\begin{array}{r} 3x - 2y + 5z = 2 \\ \underline{20x - 35y - 5z = 95} \\ 23x - 37y \qquad = 97 \end{array}$$

$$\begin{array}{r} 16x - 28y - 4z = 76 \\ \underline{5x - 6y + 4z = 13} \\ 21x - 34y \qquad = 89 \end{array}$$

$$\begin{cases} 23x - 37y = 97 \\ 21x - 34y = 89 \end{cases}$$

$$\begin{array}{r} -483x + 777y = -2037 \\ \underline{483x - 782y = 2047} \\ -5y = 10 \\ y = -2 \end{array}$$

$$\begin{array}{r} 21x - 34(-2) = 89 \\ 21x + 68 = 89 \\ 21x = 21 \\ x = 1 \end{array}$$

$$\begin{array}{r} 3(1) - 2(-2) + 5z = 2 \\ 7 + 5z = 2 \\ 5z = -5 \\ z = -1 \end{array}$$

$x = 1, \quad y = -2, \quad z = -1$

11. $\begin{cases} 2a + b - 3c = -6 \\ 4a - 4b + 2c = 10 \\ 6a - 7b + c = 12 \end{cases}$

$$\begin{array}{r} 8a + 4b - 12c = -24 \\ \underline{4a - 4b + 2c = 10} \\ 12a \qquad - 10c = -14 \\ 6a \qquad - 5c = -7 \end{array}$$

$$\begin{array}{r} 14a + 7b - 21c = -42 \\ \underline{6a - 7b + c = 12} \\ 20a \qquad - 20c = -30 \\ 2a \qquad - 2c = -3 \end{array}$$

$$\begin{cases} 6a - 5c = -7 \\ 2a - 2c = -3 \end{cases}$$

$$\begin{array}{r} 6a - 5c = -7 \\ \underline{-6a + 6c = 9} \\ c = 2 \end{array}$$

$$\begin{array}{r} 2a - 2(2) = -3 \\ 2a - 4 = -3 \\ 2a = 1 \\ a = \frac{1}{2} \end{array}$$

$$\begin{array}{r} 2\left(\frac{1}{2}\right) + b - 3(2) = -6 \\ 1 + b - 6 = -6 \\ b = -1 \end{array}$$

$a = \frac{1}{2}, \quad b = -1, \quad c = 2$

13. $\begin{cases} x + 3y + 2z = 3 \\ x \qquad + 3z = 4 \\ x - 4y - z = 0 \end{cases}$

$$\begin{array}{r} 4x + 12y + 8z = 12 \\ \underline{3x - 12y - 3z = 0} \\ 7x \qquad + 5z = 12 \end{array}$$

$$\begin{cases} x + 3z = 4 \\ 7x + 5z = 12 \end{cases}$$

$$\begin{array}{r} -7x - 21z = -28 \\ \underline{7x + 5z = 12} \\ -16z = -16 \\ z = 1 \end{array}$$

$$\begin{array}{r} x + 3(1) = 4 \\ x = 1 \end{array}$$

$$\begin{array}{r} 1 + 3y + 2(1) = 3 \\ 3y = 0 \\ y = 0 \end{array}$$

$x = 1, \quad y = 0, \quad z = 1$

15. $\begin{cases} \frac{1}{2}s + \frac{1}{3}t + u = 3 \\ \frac{1}{3}s - \frac{1}{2}t - 2u = 1 \\ \frac{2}{3}s - \frac{1}{6}t + \frac{1}{2}u = 6 \end{cases}$

$$\begin{cases} 3s + 2t + 6u = 18 \\ 2s - 3t - 12u = 6 \\ 4s - t + 3u = 36 \end{cases}$$

$$\begin{array}{r} 3s + 2t + 6u = 18 \\ \underline{8s - 2t + 6u = 72} \\ 11s \qquad + 12u = 90 \end{array}$$

$$\begin{array}{r} 2s - 3t - 12u = 6 \\ \underline{-12s + 3t - 9u = -108} \\ -10s \qquad - 21u = -102 \end{array}$$

$$\begin{cases} 11s + 12u = 90 \\ -10s - 21u = -102 \end{cases}$$

$$\begin{array}{r} 77s + 84u = 630 \\ \underline{-40s - 84u = -408} \\ 37s \qquad = 222 \\ s = 6 \end{array}$$

$$\begin{aligned} 11(6) + 12u &= 90 \\ 12u &= 24 \\ u &= 2 \end{aligned}$$

$$\begin{aligned} 4(6) - t + 3(2) &= 36 \\ 30 - t &= 36 \\ -t &= 6 \\ t &= -6 \end{aligned}$$

$s = 6,\ t = -6,\ u = 2$

17. $\begin{cases} p + q + r = 6 \\ 2q + r - p = 6 \\ r - p + q = 4 \end{cases}$

$$\begin{cases} p + q + r = 6 \\ -p + 2q + r = 6 \\ -p + q + r = 4 \end{cases}$$

$$\begin{array}{r} p + q + r = 6 \\ \underline{-p + 2q + r = 6} \\ 3q + 2r = 12 \end{array}$$

$$\begin{array}{r} p + q + r = 6 \\ \underline{-p + q + r = 4} \\ 2q + 2r = 10 \end{array}$$

$$\begin{cases} 3q + 2r = 12 \\ 2q + 2r = 10 \end{cases}$$

$$\begin{array}{r} 3q + 2r = 12 \\ \underline{-2q - 2r = -10} \\ q \qquad = 2 \end{array}$$

$$\begin{aligned} 2(2) + 2r &= 10 \\ 2r &= 6 \\ r &= 3 \end{aligned}$$

$$\begin{aligned} p + 2 + 3 &= 6 \\ p &= 1 \end{aligned}$$

$p = 1,\ q = 2,\ r = 3$

19. $\begin{cases} x + y \qquad = 0 \\ \quad y + z = 0 \\ x \qquad + z = 2 \end{cases}$

$$\begin{array}{r} -y - z = 0 \\ \underline{x \qquad + z = 2} \\ x - y \qquad = 2 \end{array}$$

$$\begin{cases} x + y = 0 \\ x - y = 2 \end{cases}$$

$$\begin{array}{r} x + y = 0 \\ \underline{x - y = 2} \\ 2x \qquad = 2 \\ x = 1 \end{array}$$

$$\begin{aligned} 1 + y &= 0 \\ y &= -1 \end{aligned}$$

$$\begin{aligned} -1 + z &= 0 \\ z &= 1 \end{aligned}$$

$x = 1,\ y = -1,\ z = 1$

21. $\begin{cases} a + b = 2b + c \\ a - 2b = c + 3 \\ 2a - b = 3c - 9 \end{cases}$

$$\begin{cases} a - b - c = 0 \\ a - 2b - c = 3 \\ 2a - b - 3c = -9 \end{cases}$$

$$\begin{array}{r} a - b - c = 0 \\ \underline{-a + 2b + c = -3} \\ b \qquad = -3 \end{array}$$

$$\begin{array}{r} -2a + 2b + 2c = 0 \\ \underline{2a - b - 3c = -9} \\ b - c = -9 \\ -3 - c = -9 \\ -c = -6 \\ c = 6 \end{array}$$

$$\begin{aligned} a - (-3) - 6 &= 0 \\ a &= 3 \end{aligned}$$

$a = 3,\ b = -3,\ c = 6$

23. $\begin{cases} 12a + 5b + 3c = 24000 \\ 10a + 6b + 4c = 13300 \\ 8a + 7b + 5c = 8700 \end{cases}$

$$\begin{array}{r} 60a + 25b + 15c = 120000 \\ \underline{-60a - 36b - 24c = -79800} \\ -11b - 9c = 40200 \end{array}$$

$$\begin{array}{r} 24a + 10b + 6c = 48000 \\ \underline{-24a - 21b - 15c = -26100} \\ -11b - 9c = 21900 \end{array}$$

$$\begin{cases} -11b - 9c = 40200 \\ -11b - 9c = 21900 \end{cases}$$

$$\begin{array}{r} -11b - 9c = 40200 \\ \underline{11b + 9c = -21900} \\ 0 = 18300 \end{array}$$

No solution

25. $$\begin{cases} 0.06x + 0.07y + 0.08z = 440 \\ 0.05x + 0.06y + 0.08z = 410 \\ 0.04x + 0.05y + 0.06z = 320 \end{cases}$$

$$\begin{cases} 6x + 7y + 8z = 44000 \\ 5x + 6y + 8z = 41000 \\ 4x + 5y + 6z = 32000 \end{cases}$$

$$\begin{array}{r} 6x + 7y + 8z = 44000 \\ \underline{-5x - 6y - 8z = -41000} \\ x + y = 3000 \end{array}$$

$$\begin{array}{r} 18x + 21y + 24z = 132000 \\ \underline{-16x - 20y - 24z = -128000} \\ 2x + y = 4000 \end{array}$$

$$\begin{cases} x + y = 3000 \\ 2x + y = 4000 \end{cases}$$

$$\begin{array}{r} x + y = 3000 \\ \underline{-2x - y = -4000} \\ -x = -1000 \\ x = 1000 \end{array}$$

$$\begin{array}{r} 1000 + y = 3000 \\ y = 2000 \end{array}$$

$$\begin{array}{r} 4(1000) + 5(2000) + 6z = 32000 \\ 14000 + 6z = 32000 \\ 6z = 18000 \\ z = 3000 \end{array}$$

$x = 1000, \quad y = 2000, \quad z = 3000$

27. d = number of dimes
q = number of quarters
h = number of half-dollars

$$\begin{cases} d + q + h = 48 \\ 10d + 25q + 50h = 1055 \\ d = q + h - 2 \end{cases}$$

$$\begin{cases} d + q + h = 48 \\ 10d + 25q + 50h = 1055 \\ d - q - h = -2 \end{cases}$$

$$\begin{array}{r} d + q + h = 48 \\ \underline{d - q - h = -2} \\ 2d = 46 \\ d = 23 \end{array}$$

$$\begin{array}{r} -50d - 50q - 50h = -2400 \\ \underline{10d + 25q + 50h = 1055} \\ -40d - 25q = -1345 \\ -40(23) - 25q = -1345 \\ -25q = -425 \\ q = 17 \\ 23 + 17 + h = 48 \\ h = 8 \end{array}$$

There are 23 dimes, 17 quarters and 8 half-dollars.

29. x = amount at 8.7%
y = amount at 9.3%
z = amount at 12.66%

$$\begin{cases} x + y + z = 12000 \\ 0.087x + 0.093y + 0.1266z = 1266 \\ 0.1266z = 0.087x + 0.093y \end{cases}$$

$$\begin{cases} x + y + z = 12000 \\ 870x + 930y + 1266z = 12660000 \\ -870x - 930y + 1266z = 0 \end{cases}$$

$$\begin{array}{r} 870x + 930y + 1266z = 12660000 \\ \underline{-870x - 930y + 1266z = 0} \\ 2532z = 12660000 \\ z = 5000 \\ -870x - 870y - 870z = -10440000 \\ \underline{870x + 930y + 1266z = 12660000} \\ 60y + 396z = 2220000 \end{array}$$

$$\begin{array}{r} 60y + 396(5000) = 2220000 \\ 60y = 240000 \\ y = 4000 \\ x + 4000 + 5000 = 12000 \\ x = 3000 \end{array}$$

She has \$3000 at 8.7%, \$4000 at 9.3% and \$5000 at 12.66%.

31. x = number of orchestra seats
y = number of mezzanine seats
z = number of balcony seats

$$\begin{cases} x + y + z = 750 \\ 12x + 8y + 6z = 7290 \\ x = y + z + 100 \end{cases}$$

$$\begin{cases} x + y + z = 750 \\ 12x + 8y + 6z = 7290 \\ x - y - z = 100 \end{cases}$$

$$\begin{aligned} x + y + z &= 750 \\ x - y - z &= 100 \\ \hline 2x &= 850 \\ x &= 425 \end{aligned}$$

$$\begin{aligned} 12x + 8y + 6z &= 7290 \\ 6x - 6y - 6z &= 600 \\ \hline 18x + 2y &= 7890 \\ 18(425) + 2y &= 7890 \\ 2y &= 240 \\ y &= 120 \end{aligned}$$

$$\begin{aligned} 425 + 120 + z &= 750 \\ z &= 205 \end{aligned}$$

There are 425 orchestra seats, 120 mezzanine seats and 205 balcony seats.

33. $$\begin{cases} 2.1A + 2.8B + 3.2C = 721 \\ 3.2A + 3.6B + 4C = 974 \\ 0.5A + 0.6B + 0.8C = 168 \end{cases}$$

$$\begin{cases} 21A + 28B + 32C = 7210 \\ 32A + 36B + 40C = 9740 \\ 5A + 6B + 8C = 1680 \end{cases}$$

$$\begin{cases} 21A + 28B + 32C = 7210 \\ 8A + 9B + 10C = 2435 \\ 5A + 6B + 8C = 1680 \end{cases}$$

$$\begin{aligned} 105A + 140B + 160C &= 36050 \\ -128A - 144B - 160C &= -38960 \\ \hline -23A - 4B &= -2910 \end{aligned}$$

$$\begin{aligned} 32A + 36B + 40C &= 9740 \\ -25A - 30B - 40C &= -8400 \\ \hline 7A + 6B &= 1340 \end{aligned}$$

$$\begin{cases} -23A - 4B = -2910 \\ 7A + 6B = 1340 \end{cases}$$

$$\begin{aligned} -69A - 12B &= -8730 \\ 14A + 12B &= 2680 \\ \hline -55A &= -6050 \\ A &= 110 \end{aligned}$$

$$\begin{aligned} 7(110) + 6B &= 1340 \\ 6B &= 570 \\ B &= 95 \end{aligned}$$

$$\begin{aligned} 5(110) + 6(95) + 8C &= 1680 \\ 1120 + 8C &= 1680 \\ 8C &= 560 \\ C &= 70 \end{aligned}$$

110 model A's, 95 model B's and 70 model C's

35. x = number of grams of Food A
y = number of grams of Food B
z = number of grams of Food C

$$\begin{cases} 0.05x + 0.06y + 0.04z = 9 \\ 0.20x + 0.15y + 0.10z = 28.5 \\ 0.40x + 0.60y + 0.70z = 97 \end{cases}$$

$$\begin{cases} 5x + 6y + 4z = 900 \\ 20x + 15y + 10z = 2850 \\ 40x + 60y + 70z = 9700 \end{cases}$$

$$\begin{cases} 5x + 6y + 4z = 900 \\ 4x + 3y + 2z = 570 \\ 4x + 6y + 7z = 970 \end{cases}$$

$$\begin{aligned} 5x + 6y + 4z &= 900 \\ -8x - 6y - 4z &= -1140 \\ \hline -3x &= -240 \\ x &= 80 \end{aligned}$$

$$\begin{aligned} -5x - 6y - 4z &= -900 \\ 4x + 6y + 7z &= 970 \\ \hline -x + 3z &= 70 \end{aligned}$$

$$\begin{aligned} -80 + 3z &= 70 \\ 3z &= 150 \\ z &= 50 \end{aligned}$$

$$\begin{aligned} 5(80) + 6y + 4(50) &= 900 \\ 6y + 600 &= 900 \\ 6y &= 300 \\ y &= 50 \end{aligned}$$

80 gm of Food A, 50 gm of Food B, 50 gm of Food C

37. x = number of acres of A
y = number of acres of B
z = number of acres of C

$$\begin{cases} 90x + 110y + 75z = 94000 \\ 6x + 10y + 5z = 7200 \\ 400x + 500y + 600z = 555000 \end{cases}$$

$$\begin{cases} 18x + 22y + 15z = 18800 \\ 6x + 10y + 5z = 7200 \\ 4x + 5y + 6z = 5550 \end{cases}$$

$$\begin{aligned} 18x + 22y + 15z &= 18800 \\ -18x - 30y - 15z &= -21600 \\ \hline -8y &= -2800 \\ y &= 350 \end{aligned}$$

$$\begin{aligned} 36x + 44y + 30z &= 37600 \\ -20x - 25y - 30z &= -27750 \\ \hline 16x + 19y &= 9850 \\ 16x + 19(350) &= 9850 \\ 16x &= 3200 \\ x &= 200 \end{aligned}$$

$$\begin{aligned} 6(200) + 10(350) + 5z &= 7200 \\ 4700 + 5z &= 7200 \\ 5z &= 2500 \\ z &= 500 \end{aligned}$$

200 acres of crop A, 350 acres of crop B, 500 acres of crop C

39. $2x^2 - x - 3 \neq 0$

$(2x - 3)(x + 1) \neq 0$

$2x - 3 \neq 0 \quad x + 1 \neq 0$

$x \neq \frac{3}{2} \qquad x \neq -1$

$\left\{x \mid x \neq \frac{3}{2}, -1\right\}$

41. $y = 8 - 3x - x^2$

$-\frac{b}{2a} = \frac{-(-3)}{2(-1)} = -\frac{3}{2}$

$y = 8 - 3\left(-\frac{3}{2}\right) - \left(-\frac{3}{2}\right)^2 = \frac{41}{4}$

Vertex: $\left(-\frac{3}{2}, \frac{41}{4}\right)$

x-intercepts: Let $y = 0$

$0 = 8 - 3x - x^2$

$x = \frac{-\left(-3 \pm \sqrt{(-3)^2 - 4(-1)(8)}\right)}{2(-1)}$

$= \frac{3 \pm \sqrt{41}}{-2}$

$= \frac{-3 \pm \sqrt{41}}{2}$

y-intercept: Let $x = 0$

$y = 8 - 3(0) - 0^2 = 8$

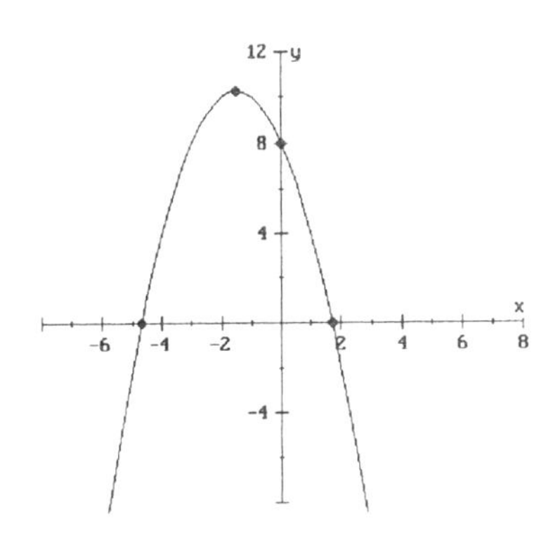

Exercises 11.2

1. $\left[\begin{array}{cc|c} 3 & -2 & 5 \\ 1 & -1 & 8 \end{array}\right]$

3. $\left[\begin{array}{ccc|c} 1 & -2 & 3 & 4 \\ 0 & 1 & -1 & -3 \\ 2 & 3 & 0 & 8 \end{array}\right]$

5. $\left[\begin{array}{cc|c} 1 & -2 & 7 \\ 2 & -3 & 12 \end{array}\right]$

$-2R_1 + R_2 \to R_2$

$\left[\begin{array}{cc|c} 1 & -2 & 7 \\ 0 & 1 & -2 \end{array}\right]$

$\begin{cases} x - 2y = 7 \\ y = -2 \end{cases}$

$x - 2(-2) = 7$

$x = 3$

$(3, -2)$

7. $\left[\begin{array}{cc|c} 1 & -3 & 6 \\ 3 & 5 & -10 \end{array}\right]$

$-3R_1 + R_2 \to R_2$

$\left[\begin{array}{cc|c} 1 & -3 & 6 \\ 0 & 14 & -28 \end{array}\right]$

$\begin{cases} x - 3y = 6 \\ 14y = -28 \end{cases}$

$$14y = -28$$
$$y = -2$$

$$x - 3(-2) = 6$$
$$x = 0$$

$(0, -2)$

9. $\begin{bmatrix} 2 & -5 & | & -8 \\ 7 & 3 & | & -28 \end{bmatrix}$

$-7R_1 + 2R_2 \to R_2$

$\begin{bmatrix} 2 & -5 & | & -8 \\ 0 & 41 & | & 0 \end{bmatrix}$

$$\begin{cases} 2x - 5y = -8 \\ \quad 41y = 0 \end{cases}$$

$$41y = 0$$
$$y = 0$$

$$2x - 5(0) = -8$$
$$2x = -8$$
$$x = -4$$

$(-4, 0)$

11. $\begin{bmatrix} 6 & 2 & | & 9 \\ 4 & -1 & | & -1 \end{bmatrix}$

$-2R_1 + 3R_2 \to R_2$

$\begin{bmatrix} 6 & 2 & | & 9 \\ 0 & -7 & | & -21 \end{bmatrix}$

$$\begin{cases} 6x + 2y = 9 \\ \quad -7y = -21 \end{cases}$$

$$-7y = -21$$
$$y = 3$$

$$6x + 2(3) = 9$$
$$6x = 3$$
$$x = \frac{1}{2}$$

$\left(\frac{1}{2}, 3\right)$

13. $\begin{bmatrix} 6 & 2 & | & 5 \\ 3 & -4 & | & 0 \end{bmatrix}$

$-2R_2 + R_1 \to R_2$

$\begin{bmatrix} 6 & 2 & | & 5 \\ 0 & 10 & | & 5 \end{bmatrix}$

$$\begin{cases} 6x + 2y = 5 \\ \quad 10y = 5 \end{cases}$$

$$10y = 5$$
$$y = \frac{1}{2}$$

$$6x + 2\left(\frac{1}{2}\right) = 5$$
$$6x = 4$$
$$x = \frac{2}{3}$$

$\left(\frac{2}{3}, \frac{1}{2}\right)$

15. $\begin{bmatrix} 1 & 3 & 1 & | & 8 \\ 1 & 2 & 1 & | & 7 \\ 1 & -2 & 2 & | & 6 \end{bmatrix}$

$R_1 - R_2 \to R_2$
$R_1 - R_3 \to R_3$

$\begin{bmatrix} 1 & 3 & 1 & | & 8 \\ 0 & 1 & 0 & | & 1 \\ 0 & 5 & -1 & | & 2 \end{bmatrix}$

$5R_2 + R_3 \to R_3$

$\begin{bmatrix} 1 & 3 & 1 & | & 8 \\ 0 & 1 & 0 & | & 1 \\ 0 & 0 & -1 & | & -3 \end{bmatrix}$

$$\begin{cases} x + 3y + z = 8 \\ \quad y = 1 \\ \quad -z = -3 \end{cases}$$

$$-z = -3$$
$$z = 3$$

$$x + 3(1) + 3 = 8$$
$$x = 2$$

(2, 1, 3)

17. $\begin{bmatrix} 1 & 1 & -1 & | & 2 \\ 3 & 1 & -1 & | & -2 \\ 4 & -2 & 1 & | & -13 \end{bmatrix}$

$-3R_1 + R_2 \to R_2$
$-4R_1 + R_3 \to R_3$

$$\begin{bmatrix} 1 & 1 & -1 & | & 2 \\ 0 & -2 & 2 & | & -8 \\ 0 & -6 & 5 & | & -21 \end{bmatrix}$$

$-3R_2 + R_3 \to R_3$

$$\begin{bmatrix} 1 & 1 & -1 & | & 2 \\ 0 & -2 & 2 & | & -8 \\ 0 & 0 & -1 & | & 3 \end{bmatrix}$$

$$\begin{cases} x + y - z = 2 \\ -2y + z = -8 \\ -z = 3 \end{cases}$$

$$-z = 3$$
$$z = -3$$

$$-2y + 2(-3) = -8$$
$$-2y = -2$$
$$y = 1$$

$$x + 1 - (-3) = 2$$
$$x = -2$$

(−2, 1, −3)

19. $\begin{bmatrix} 1 & -2 & 3 & | & 7 \\ 2 & 3 & -1 & | & 0 \\ 1 & 1 & 1 & | & 1 \end{bmatrix}$

$-2R_1 + R_2 \to R_2$
$R_1 - R_3 \to R_3$

$$\begin{bmatrix} 1 & -2 & 3 & | & 7 \\ 0 & 7 & -7 & | & -14 \\ 0 & -3 & 2 & | & 6 \end{bmatrix}$$

$\frac{1}{7} R_2 \to R_2$

$$\begin{bmatrix} 1 & -2 & 3 & | & 7 \\ 0 & 1 & -1 & | & -2 \\ 0 & -3 & 2 & | & 6 \end{bmatrix}$$

$3R_2 + R_3 \to R_3$

$$\begin{bmatrix} 1 & -2 & 3 & | & 7 \\ 0 & 1 & -1 & | & -2 \\ 0 & 0 & -1 & | & 0 \end{bmatrix}$$

$$\begin{cases} x - 2y + 3z = 7 \\ y - z = -2 \\ -z = 0 \end{cases}$$

$$-z = 0$$
$$z = 0$$

$$y - 0 = -2$$
$$y = -2$$

$$x - 2(-2) + 3(0) = 7$$
$$x + 4 = 7$$
$$x = 3$$

(3, −2, 0)

21. $\begin{bmatrix} 4 & -1 & 2 & | & 6 \\ 2 & 3 & -1 & | & 4 \\ 2 & -2 & 1 & | & 0 \end{bmatrix}$

$-2R_2 + R_1 \to R_2$
$R_2 - R_3 \to R_3$

$$\begin{bmatrix} 4 & -1 & 2 & | & 6 \\ 0 & -7 & 4 & | & -2 \\ 0 & 5 & -2 & | & 4 \end{bmatrix}$$

$5R_2 + 7R_3 \to R_3$

$$\begin{bmatrix} 4 & -1 & 2 & | & 6 \\ 0 & -7 & 4 & | & -2 \\ 0 & 0 & 6 & | & 18 \end{bmatrix}$$

$$\begin{cases} 4x - y + 2z = 6 \\ -7y + 4z = -2 \\ 6z = 18 \end{cases}$$

$$6z = 18$$
$$z = 3$$

$$-7y + 4(3) = -2$$
$$-7y = -14$$
$$y = 2$$

$$4x - 2 + 2(3) = 6$$
$$4x = 2$$
$$x = \frac{1}{2}$$

$$\left(\frac{1}{2}, 2, 3\right)$$

23. $\begin{bmatrix} 2 & 3 & 2 & | & 4 \\ 4 & 6 & 4 & | & 8 \\ 2 & 3 & 5 & | & 13 \end{bmatrix}$

$-2R_1 + R_2 \to R_2$
$R_1 - R_3 \to R_3$

$$\begin{bmatrix} 2 & 3 & 2 & | & 4 \\ 0 & 0 & 0 & | & 0 \\ 0 & 0 & -3 & | & -9 \end{bmatrix}$$

$$\begin{cases} -3z = -9 \\ 2x + 3y + 2z = 4 \end{cases}$$

$$-3z = -9$$
$$z = 3$$

$$2x + 3y + 2(3) = 4$$
$$2x + 3y = -2$$
$$\{(x, y, 3) \mid 2x + 3y = -2\}$$

25. $\begin{bmatrix} 1 & -2 & 1 & -1 & | & 2 \\ 1 & 2 & 2 & 1 & | & 0 \\ 2 & -2 & 1 & -1 & | & 3 \\ 2 & 0 & -2 & 1 & | & 5 \end{bmatrix}$

$R_1 - R_2 \to R_2$
$-2R_1 + R_3 \to R_3$
$R_3 - R_4 \to R_4$

$$\begin{bmatrix} 1 & -2 & 1 & -1 & | & 2 \\ 0 & -4 & -1 & -2 & | & 2 \\ 0 & 2 & -1 & 1 & | & -1 \\ 0 & -2 & 3 & -2 & | & -2 \end{bmatrix}$$

$R_2 + 2R_3 \to R_3$
$R_3 + R_4 \to R_4$

$$\begin{bmatrix} 1 & -2 & 1 & -1 & | & 2 \\ 0 & -4 & -1 & -2 & | & 2 \\ 0 & 0 & -3 & 0 & | & 0 \\ 0 & 0 & 2 & -1 & | & -3 \end{bmatrix}$$

$2R_3 + 3R_4 \to R_4$

$$\begin{bmatrix} 1 & -2 & 1 & -1 & | & 2 \\ 0 & -4 & -1 & -2 & | & 2 \\ 0 & 0 & -3 & 0 & | & 0 \\ 0 & 0 & 0 & -3 & | & -9 \end{bmatrix}$$

$$\begin{cases} w - 2x + y - z = 2 \\ -4x - y - 2z = 2 \\ -3y = 0 \\ -3z = -9 \end{cases}$$

$$-3z = -9$$
$$z = 3$$

$$-3y = 0$$
$$y = 0$$

$$-4x - 0 - 2(3) = 2$$
$$-4x = 8$$
$$x = -2$$

$$w - 2(-2) + 0 - 3 = 2$$
$$w = 1$$
$$(1, -2, 0, 3)$$

27. $\begin{bmatrix} 2 & 3 & | & 5 \\ 1 & 4 & | & 0 \end{bmatrix}$

$-2R_2 + R_1 \to R_2$
$1/2R_1 \to R_1$

$$\begin{bmatrix} 1 & 3/2 & | & 5/2 \\ 0 & -5 & | & 5 \end{bmatrix}$$

$(-1/5)R_2 \to R_2$

$$\begin{bmatrix} 1 & 3/2 & | & 5/2 \\ 0 & 1 & | & -1 \end{bmatrix}$$

$(-3/2)R_2 + R_1 \to R_1$

$$\begin{bmatrix} 1 & 0 & | & 4 \\ 0 & 1 & | & -1 \end{bmatrix}$$

29. $$\begin{bmatrix} 1 & 1 & 2 & | & 1 \\ 2 & 4 & 2 & | & 6 \\ 3 & 1 & 2 & | & 5 \end{bmatrix}$$

$-2R_1 + R_2 \to R_2$
$-3R_1 + R_3 \to R_3$

$$\begin{bmatrix} 1 & 1 & 2 & | & 1 \\ 0 & 2 & -2 & | & 4 \\ 0 & -2 & -4 & | & 2 \end{bmatrix}$$

$(1/2)R_2 \to R_2$
$R_2 + R_3 \to R_3$

$$\begin{bmatrix} 1 & 1 & 2 & | & 1 \\ 0 & 1 & -1 & | & 2 \\ 0 & 0 & -6 & | & 6 \end{bmatrix}$$

$(-1/6)R_3 \to R_3$

$$\begin{bmatrix} 1 & 1 & 2 & | & 1 \\ 0 & 1 & -1 & | & 2 \\ 0 & 0 & 1 & | & -1 \end{bmatrix}$$

$R_2 + R_3 \to R_2$
$R_1 - R_2 \to R_1$

$$\begin{bmatrix} 1 & 0 & 3 & | & -1 \\ 0 & 1 & 0 & | & 1 \\ 0 & 0 & 1 & | & -1 \end{bmatrix}$$

$-3R_3 + R_1 \to R_1$

$$\begin{bmatrix} 1 & 0 & 0 & | & 2 \\ 0 & 1 & 0 & | & 1 \\ 0 & 0 & 1 & | & -1 \end{bmatrix}$$

31. $$\begin{bmatrix} 1 & -3 & | & 5 \\ 3 & 5 & | & 1 \end{bmatrix}$$

$-3R_1 + R_2 \to R_2$

$$\begin{bmatrix} 1 & -3 & | & 5 \\ 0 & 14 & | & -14 \end{bmatrix}$$

$(-1/14)R_2 \to R_2$

$$\begin{bmatrix} 1 & -3 & | & 5 \\ 0 & 1 & | & -1 \end{bmatrix}$$

$3R_2 + R_1 \to R_1$

$$\begin{bmatrix} 1 & 0 & | & 2 \\ 0 & 1 & | & -1 \end{bmatrix}$$

(2, −1)

33. $$\begin{bmatrix} 1 & 2 & 1 & | & 8 \\ 1 & 4 & -1 & | & 12 \\ 1 & -2 & 1 & | & -4 \end{bmatrix}$$

$R_1 - R_2 \to R_2$
$R_2 - R_3 \to R_3$

$$\begin{bmatrix} 1 & 2 & 1 & | & 8 \\ 0 & -2 & 2 & | & -4 \\ 0 & 6 & -2 & | & 16 \end{bmatrix}$$

$-(1/2)R_2 \to R_2$
$3R_2 + R_3 \to R_3$

$$\begin{bmatrix} 1 & 2 & 1 & | & 8 \\ 0 & 1 & -1 & | & 2 \\ 0 & 0 & 4 & | & 4 \end{bmatrix}$$

$-(1/4)R_3 \to R_3$

$$\begin{bmatrix} 1 & 2 & 1 & | & 8 \\ 0 & 1 & -1 & | & 2 \\ 0 & 0 & 1 & | & 1 \end{bmatrix}$$

$R_2 + R_3 \to R_2$
$-2R_2 + R_1 \to R_1$

$$\begin{bmatrix} 1 & 0 & 3 & | & 4 \\ 0 & 1 & 0 & | & 3 \\ 0 & 0 & 1 & | & 1 \end{bmatrix}$$

$-3R_3 + R_1 \rightarrow R_1$

$$\begin{bmatrix} 1 & 0 & 0 & | & 1 \\ 0 & 1 & 0 & | & 3 \\ 0 & 0 & 1 & | & 1 \end{bmatrix}$$

(1, 3, 1)

35. $$\begin{bmatrix} 1 & -2 & 0 & | & -4 \\ 0 & 2 & 2 & | & 4 \\ 1 & 0 & 1 & | & 1 \end{bmatrix}$$

$R_1 + R_2 \rightarrow R_1$
$R_1 - R_3 \rightarrow R_3$
$(1/2)R_2 \rightarrow R_2$

$$\begin{bmatrix} 1 & 0 & 2 & | & 0 \\ 0 & 1 & 1 & | & 2 \\ 0 & -2 & -1 & | & -5 \end{bmatrix}$$

$2R_2 + R_3 \rightarrow R_3$

$$\begin{bmatrix} 1 & 0 & 2 & | & 0 \\ 0 & 1 & 1 & | & 2 \\ 0 & 0 & 1 & | & -1 \end{bmatrix}$$

$R_2 - R_3 \rightarrow R_2$
$-2R_3 + R_1 \rightarrow R_1$

$$\begin{bmatrix} 1 & 0 & 0 & | & 2 \\ 0 & 1 & 0 & | & 3 \\ 0 & 0 & 1 & | & -1 \end{bmatrix}$$

(2, 3, −1)

37. $$\begin{bmatrix} 1 & 1 & 1 & 1 & | & 7 \\ 2 & 1 & -1 & 2 & | & 6 \\ 3 & 2 & 1 & -1 & | & 3 \\ 1 & -1 & -1 & -2 & | & -10 \end{bmatrix}$$

$-2R_1 + R_2 \rightarrow R_2$
$-3R_1 + R_3 \rightarrow R_3$
$R_1 - R_4 \rightarrow R_4$

$$\begin{bmatrix} 1 & 1 & 1 & 1 & | & 7 \\ 0 & -1 & -3 & 0 & | & -8 \\ 0 & -1 & -2 & -4 & | & -18 \\ 0 & 2 & 2 & 3 & | & 17 \end{bmatrix}$$

$R_1 + R_2 \rightarrow R_1$
$-R_2 \rightarrow R_2$
$R_3 - R_2 \rightarrow R_3$
$2R_3 + R_4 \rightarrow R_4$

$$\begin{bmatrix} 1 & 0 & -2 & 1 & | & -1 \\ 0 & 1 & 3 & 0 & | & 8 \\ 0 & 0 & 1 & -4 & | & -10 \\ 0 & 0 & -2 & -5 & | & -19 \end{bmatrix}$$

$2R_3 + R_1 \rightarrow R_1$
$-3R_3 + R_2 \rightarrow R_2$
$2R_3 + R_4 \rightarrow R_4$

$$\begin{bmatrix} 1 & 0 & 0 & -7 & | & -21 \\ 0 & 1 & 0 & 12 & | & 38 \\ 0 & 0 & 1 & -4 & | & -10 \\ 0 & 0 & 0 & -13 & | & -39 \end{bmatrix}$$

$-(1/13)R_4 \rightarrow R_4$

$$\begin{bmatrix} 1 & 0 & 0 & -7 & | & -21 \\ 0 & 1 & 0 & 12 & | & 38 \\ 0 & 0 & 1 & -4 & | & -10 \\ 0 & 0 & 0 & 1 & | & 3 \end{bmatrix}$$

$7R_4 + R_1 \rightarrow R_1$
$-12R_4 + R_2 \rightarrow R_2$
$4R_4 + R_3 \rightarrow R_3$

$$\begin{bmatrix} 1 & 0 & 0 & 0 & | & 0 \\ 0 & 1 & 0 & 0 & | & 2 \\ 0 & 0 & 1 & 0 & | & 2 \\ 0 & 0 & 0 & 1 & | & 3 \end{bmatrix}$$
(0, 2, 2, 3)

39. $f[g(x)] = \dfrac{3}{g(x) + 1}$

$$= \frac{3}{\frac{x - 1}{x} + 1}$$

$$= \frac{3 \cdot x}{\left(\frac{x - 1}{x} + 1\right) \cdot x}$$

$$= \frac{3x}{x - 1 + x}$$

$$= \frac{3x}{2x - 1}$$

$g[f(x)] = \dfrac{f(x) - 1}{f(x)}$

$$= \frac{\frac{3}{x + 1} - 1}{\frac{3}{x + 1}}$$

$$= \frac{\left(\frac{3}{x + 1} - 1\right)(x + 1)}{\frac{3}{x + 1} \cdot (x + 1)}$$

$$= \frac{3 - (x + 1)}{3}$$

$$= \frac{3 - x - 1}{3}$$

$$= \frac{2 - x}{3}$$

41. $5x + 3y = 7$

$3y = -5x + 7$

$y = -\frac{5}{3}x + \frac{7}{3}$

$m = -\frac{5}{3}$

$m_{\perp} = \frac{3}{5}$

$(-2, 3)$

$y - 3 = \frac{3}{5}[x - (-2)]$

$y - 3 = \frac{3}{5}(x + 2)$

$y - 3 = \frac{3}{5}x + \frac{6}{5}$

$y = \frac{3}{5}x + \frac{21}{5}$

Exercises 11.3

1. $\begin{bmatrix} 1 & -3 \\ 5 & 2 \end{bmatrix} + \begin{bmatrix} 2 & 5 \\ 3 & 0 \end{bmatrix}$

$$= \begin{bmatrix} 1 + 2 & -3 + 5 \\ 5 + 3 & 2 + 0 \end{bmatrix}$$

$$= \begin{bmatrix} 3 & 2 \\ 8 & 2 \end{bmatrix}$$

3. $\begin{bmatrix} 1 & 3 \\ -5 & 0 \\ 2 & -7 \end{bmatrix} + \begin{bmatrix} -1 & -3 \\ 5 & 0 \\ -2 & 7 \end{bmatrix}$

$$= \begin{bmatrix} 1 + (-1) & 3 + (-3) \\ -5 + 5 & 0 + 0 \\ 2 + (-2) & -7 + 7 \end{bmatrix}$$

$$= \begin{bmatrix} 0 & 0 \\ 0 & 0 \\ 0 & 0 \end{bmatrix}$$

5. Not defined

7. $3\begin{bmatrix} 0 & 3 \\ 17 & 0 \end{bmatrix}$

$$= \begin{bmatrix} 3(0) & 3(3) \\ 3(17) & 3(0) \end{bmatrix}$$

$$= \begin{bmatrix} 0 & 9 \\ 51 & 0 \end{bmatrix}$$

9. $-A = -\begin{bmatrix} 2 & -3 \\ -1 & 0 \end{bmatrix}$

$$= \begin{bmatrix} -2 & -(-3) \\ -(-1) & -0 \end{bmatrix}$$

$$= \begin{bmatrix} -2 & 3 \\ 1 & 0 \end{bmatrix}$$

11. $\frac{2}{3}B = \frac{2}{3}\begin{bmatrix} 5 & 1 \\ -2 & 3 \end{bmatrix}$

$= \begin{bmatrix} \frac{10}{3} & \frac{2}{3} \\ -\frac{4}{3} & 2 \end{bmatrix}$

13. $A - B$

$= \begin{bmatrix} 2 & -3 \\ -1 & 0 \end{bmatrix} - \begin{bmatrix} 5 & 1 \\ -2 & 3 \end{bmatrix}$

$= \begin{bmatrix} 2 - 5 & -3 - 1 \\ -1 - (-2) & 0 - 3 \end{bmatrix}$

$= \begin{bmatrix} -3 & -4 \\ 1 & -3 \end{bmatrix}$

15. $2A - B$

$= 2\begin{bmatrix} 2 & -3 \\ -1 & 0 \end{bmatrix} - \begin{bmatrix} 5 & 1 \\ -2 & 3 \end{bmatrix}$

$= \begin{bmatrix} 4 & -6 \\ -2 & 0 \end{bmatrix} - \begin{bmatrix} 5 & 1 \\ -2 & 3 \end{bmatrix}$

$= \begin{bmatrix} -1 & -7 \\ 0 & -3 \end{bmatrix}$

17. $[3 \ -3 \ 1]\begin{bmatrix} 5 \\ 1 \\ -2 \end{bmatrix}$

$= [3(5) - 3(1) + 1(-2)]$

$= [10]$

19. $[0 \ -1 \ 1 \ 2]\begin{bmatrix} 1 \\ -1 \\ -2 \\ 2 \end{bmatrix}$

$= [0(1) - 1(-1) + 1(-2) + 2(2)]$

$= [3]$

21. $\begin{bmatrix} 0 & -1 \\ 3 & 2 \end{bmatrix}\begin{bmatrix} -1 & 3 \\ 0 & -2 \end{bmatrix}$

$= \begin{bmatrix} 0(-1) + (-1)(0) & 0(3) + (-1)(-2) \\ 3(-1) + 2(0) & 3(3) + 2(-2) \end{bmatrix}$

$= \begin{bmatrix} 0 & 2 \\ -3 & 5 \end{bmatrix}$

23. Not defined

25. $\begin{bmatrix} 0 & 1 & -5 \\ -3 & 1 & 6 \end{bmatrix}\begin{bmatrix} -1 & 3 \\ 0 & -2 \\ 6 & 0 \end{bmatrix}$

$= \begin{bmatrix} 0(-1) + 1(0) + (-5)(6) & 0(3) + 1(-2) + (-5)(0) \\ -3(-1) + 1(0) + 6(6) & -3(3) + 1(-2) + 6(0) \end{bmatrix}$

$= \begin{bmatrix} -30 & -2 \\ 39 & -11 \end{bmatrix}$

27. $\begin{bmatrix} 1 & 0 & -2 \\ 3 & 1 & -1 \\ 2 & 0 & 1 \end{bmatrix}\begin{bmatrix} 1 & -3 & 0 \\ 0 & 2 & -2 \\ -6 & 0 & 5 \end{bmatrix}$

$$= \begin{bmatrix} 1(1) + 0(0) + (-2)(-6) & 1(-3) + 0(2) + (-2)(0) & 1(0) + 0(-2) + (-2)(5) \\ 3(1) + 1(0) + (-1)(-6) & 3(-3) + 1(2) + (-1)(0) & 3(0) + 1(-2) + (-1)(5) \\ 2(1) + 0(0) + (1)(-6) & 2(-3) + 0(2) + 1(0) & 2(0) + 0(-2) + 1(5) \end{bmatrix}$$

$$= \begin{bmatrix} 13 & -3 & -10 \\ 9 & -7 & -7 \\ -4 & -6 & 5 \end{bmatrix}$$

29. $\begin{bmatrix} 2 & 0 & -2 & 1 \\ -2 & 3 & -1 & 1 \end{bmatrix}\begin{bmatrix} 1 & 3 \\ 1 & 0 \\ -1 & 5 \\ 1 & -3 \end{bmatrix}$

$$= \begin{bmatrix} 2(1) + 0(1) + (-2)(-1) + 1(1) & 2(3) + 0(0) + (-2)(5) + 1(-3) \\ -2(1) + 3(1) + (-1)(-1) + 1(1) & -2(3) + 3(0) + (-1)(5) + 1(-3) \end{bmatrix}$$

$$= \begin{bmatrix} 5 & -7 \\ 3 & -14 \end{bmatrix}$$

31. $\begin{bmatrix} 1 & 0 & 0 \\ 0 & 1 & 0 \\ 0 & 0 & 1 \end{bmatrix}\begin{bmatrix} 3 & 1 & 0 \\ -2 & 4 & 6 \\ 3 & 8 & -10 \end{bmatrix}$

$$= \begin{bmatrix} 1(3) + 0(-2) + 0(3) & 1(1) + 0(4) + 0(8) & 1(0) + 0(6) + 0(-10) \\ 0(3) + 1(-2) + 0(3) & 0(1) + 1(4) + 0(8) & 0(0) + 1(6) + 0(-10) \\ 0(3) + 0(-2) + 1(3) & 0(1) + 0(4) + 1(8) & 0(0) + 0(6) + 1(-10) \end{bmatrix}$$

$$= \begin{bmatrix} 3 & 1 & 0 \\ -2 & 4 & 6 \\ 3 & 8 & -10 \end{bmatrix}$$

33. $CB = \begin{bmatrix} 2 & -2 \\ 4 & 1 \end{bmatrix}\begin{bmatrix} 3 & 2 \\ -5 & 0 \end{bmatrix}$

$$= \begin{bmatrix} 2(3) + (-2)(-5) & 2(2) + (-2)(0) \\ 4(3) + 1(-5) & 4(2) + 1(0) \end{bmatrix}$$

$$= \begin{bmatrix} 16 & 4 \\ 7 & 8 \end{bmatrix}$$

$$BC = \begin{bmatrix} 3 & 2 \\ -5 & 0 \end{bmatrix}\begin{bmatrix} 2 & -2 \\ 4 & 1 \end{bmatrix}$$

$$= \begin{bmatrix} 3(2) + 2(4) & 3(-2) + 2(1) \\ -5(2) + 0(4) & -5(-2) + 0(1) \end{bmatrix}$$

$$= \begin{bmatrix} 14 & -4 \\ -10 & 10 \end{bmatrix}$$

$CB \neq BC$
False

35. $$B + C = \begin{bmatrix} 3 & 2 \\ -5 & 0 \end{bmatrix} + \begin{bmatrix} 2 & -2 \\ 4 & 1 \end{bmatrix}$$

$$= \begin{bmatrix} 5 & 0 \\ -1 & 1 \end{bmatrix}$$

$$(B + C)A = \begin{bmatrix} 5 & 0 \\ -1 & 1 \end{bmatrix}\begin{bmatrix} 2 & 1 & 0 \\ 3 & 1 & -2 \end{bmatrix}$$

$$= \begin{bmatrix} 10 & 5 & 0 \\ 1 & 0 & -2 \end{bmatrix}$$

$$BA = \begin{bmatrix} 3 & 2 \\ -5 & 0 \end{bmatrix}\begin{bmatrix} 2 & 1 & 0 \\ 3 & 1 & -2 \end{bmatrix}$$

$$= \begin{bmatrix} 12 & 5 & -4 \\ -10 & -5 & 0 \end{bmatrix}$$

$$CA = \begin{bmatrix} 2 & -2 \\ 4 & 1 \end{bmatrix}\begin{bmatrix} 2 & 1 & 0 \\ 3 & 1 & -2 \end{bmatrix}$$

$$= \begin{bmatrix} -2 & 0 & 4 \\ 11 & 5 & -2 \end{bmatrix}$$

$BA + CA$

$$= \begin{bmatrix} 12 & 5 & -4 \\ -10 & -5 & 0 \end{bmatrix} + \begin{bmatrix} -2 & 0 & 4 \\ 11 & 5 & -2 \end{bmatrix}$$

$$= \begin{bmatrix} 10 & 5 & 0 \\ 1 & 0 & -2 \end{bmatrix}$$

$(B + C)A -= BA + CA$
True

37. $$\left(\frac{1}{2}\begin{bmatrix} 3 & 5 \\ 2 & 4 \end{bmatrix}\right)\begin{bmatrix} 0 & 2 \\ 1 & -1 \end{bmatrix}$$

$$= \begin{bmatrix} \frac{3}{2} & \frac{5}{2} \\ 1 & 2 \end{bmatrix}\begin{bmatrix} 0 & 2 \\ 1 & -1 \end{bmatrix}$$

$$= \begin{bmatrix} \frac{5}{2} & \frac{1}{2} \\ 2 & 0 \end{bmatrix}$$

$$= \frac{1}{2}\left(\begin{bmatrix} 3 & 5 \\ 2 & 4 \end{bmatrix}\begin{bmatrix} 0 & 2 \\ 1 & -1 \end{bmatrix}\right)$$

$$= \frac{1}{2}\begin{bmatrix} 5 & 1 \\ 4 & 0 \end{bmatrix}$$

$$= \begin{bmatrix} \frac{5}{2} & \frac{1}{2} \\ 2 & 0 \end{bmatrix}$$

$$\begin{bmatrix} \frac{5}{2} & \frac{1}{2} \\ 2 & 0 \end{bmatrix} = \begin{bmatrix} \frac{5}{2} & \frac{1}{2} \\ 2 & 0 \end{bmatrix}$$

39. *AB*
$5 \times 4 \quad 3 \times 5$
$\neq$
Not defined

BA

$3 \times 5 \quad 5 \times 4$
3×4

41. Tom's Office Supply:

$$\begin{bmatrix} 7 & 6 & 2 \\ 4 & 10 & 9 \end{bmatrix}\begin{bmatrix} 230 \\ 65 \\ 18 \end{bmatrix}$$

$$= \begin{bmatrix} 2036 \\ 1732 \end{bmatrix}$$

Kuma's Office Supply:

$$\begin{bmatrix} 7 & 6 & 2 \\ 4 & 10 & 9 \end{bmatrix}\begin{bmatrix} 250 \\ 56 \\ 16 \end{bmatrix} = \begin{bmatrix} 2118 \\ 1704 \end{bmatrix}$$

Tom's Office Supply is less expensive for the Sociology Department and Kuma's Office Supply is less expensive for the Political Science Department.

45. $(2x - 1)(x + 3) = (3x - 2)(x + 4)$

$$2x^2 + 5x - 3 = 3x^2 + 10x - 8$$
$$0 = x^2 + 5x - 5$$
$$x = \frac{-5 \pm \sqrt{5^2 - 4(1)(-5)}}{2(1)}$$
$$= \frac{-5 \pm \sqrt{45}}{2}$$
$$= \frac{-5 \pm 3\sqrt{5}}{2}$$

47. Center = midpoint

$$= \left(\frac{-3 + 5}{2}, \frac{4 - 6}{2}\right)$$
$$= (1, -1)$$

$$r = \sqrt{(-3 - 1)^2 + [4 - (-1)]^2}$$
$$= \sqrt{16 + 25}$$
$$= \sqrt{41}$$

$$(x - h)^2 + (y - k)^2 = r^2$$
$$(x - 1)^2 + [y - (-1)]^2 = (\sqrt{41})^2$$
$$(x - 1)^2 + (y + 1)^2 = 41$$

Exercises 11.4

1. $\left[\begin{array}{cc|cc} 1 & 2 & 1 & 0 \\ 2 & 0 & 0 & 1 \end{array}\right]$

$-2R_1 + R_2 \to R_2$

$$\left[\begin{array}{cc|cc} 1 & 2 & 1 & 0 \\ 0 & -4 & -2 & 1 \end{array}\right]$$

$-\frac{1}{4}R_2 \to R_2$

$$\left[\begin{array}{cc|cc} 1 & 2 & 1 & 0 \\ 0 & 1 & \frac{1}{2} & -\frac{1}{4} \end{array}\right]$$

$-2R_2 + R_1 \to R_1$

$$\left[\begin{array}{cc|cc} 1 & 0 & 0 & \frac{1}{2} \\ 0 & 1 & \frac{1}{2} & -\frac{1}{4} \end{array}\right]$$

$$A^{-1} = \begin{bmatrix} 0 & \frac{1}{2} \\ \frac{1}{2} & -\frac{1}{4} \end{bmatrix}$$

3. $\left[\begin{array}{cc|cc} 2 & -1 & 1 & 0 \\ 0 & 1 & 0 & 1 \end{array}\right]$

$\frac{1}{2}R_1 \to R_1$

$$\left[\begin{array}{cc|cc} 1 & -\frac{1}{2} & \frac{1}{2} & 0 \\ 0 & 1 & 0 & 1 \end{array}\right]$$

$\frac{1}{2}R_2 + R_1 \to R_1$

$$\left[\begin{array}{cc|cc} 1 & 0 & \frac{1}{2} & \frac{1}{2} \\ 0 & 1 & 0 & 1 \end{array}\right]$$

$$A^{-1} = \begin{bmatrix} \frac{1}{2} & \frac{1}{2} \\ 0 & 1 \end{bmatrix}$$

5. $\left[\begin{array}{ccc|ccc} 1 & 1 & 2 & 1 & 0 & 0 \\ 2 & 0 & 1 & 0 & 1 & 0 \\ 3 & -1 & 1 & 0 & 0 & 1 \end{array}\right]$

$-2R_1 + R_2 \to R_2$
$-3R_1 + R_3 \to R_3$

$$\left[\begin{array}{ccc|ccc}1 & 1 & 2 & 1 & 0 & 0\\0 & -2 & -3 & -2 & 1 & 0\\0 & -4 & -5 & -3 & 0 & 1\end{array}\right]$$

$-2R_2 + R_3 \to R_3$

$-\frac{1}{2}R_2 \to R_2$

$$\left[\begin{array}{ccc|ccc}1 & 1 & 2 & 1 & 0 & 0\\0 & 1 & \frac{3}{2} & 1 & -\frac{1}{2} & 0\\0 & 0 & 1 & 1 & -2 & 1\end{array}\right]$$

$R_1 - R_2 \to R_1$

$$\left[\begin{array}{ccc|ccc}1 & 0 & \frac{1}{2} & 0 & \frac{1}{2} & 0\\0 & 1 & \frac{3}{2} & 1 & -\frac{1}{2} & 0\\0 & 0 & 1 & 1 & -2 & 1\end{array}\right]$$

$-\frac{1}{2}R_3 + R_1 \to R_1$

$-\frac{3}{2}R_3 + R_2 \to R_2$

$$\left[\begin{array}{ccc|ccc}1 & 0 & 0 & -\frac{1}{2} & \frac{3}{2} & -\frac{1}{2}\\0 & 1 & 0 & -\frac{1}{2} & \frac{5}{2} & -\frac{3}{2}\\0 & 0 & 1 & 1 & -2 & 1\end{array}\right]$$

$$A^{-1} = \begin{bmatrix}-\frac{1}{2} & \frac{3}{2} & -\frac{1}{2}\\-\frac{1}{2} & \frac{5}{2} & -\frac{3}{2}\\1 & -2 & 1\end{bmatrix}$$

7. $A = \begin{bmatrix}1 & 2\\1 & -1\end{bmatrix}$, $X = \begin{bmatrix}x\\y\end{bmatrix}$, $K = \begin{bmatrix}3\\6\end{bmatrix}$

$$\left[\begin{array}{cc|cc}1 & 2 & 1 & 0\\1 & -1 & 0 & 1\end{array}\right]$$

$R_1 - R_2 \to R_2$

$$\left[\begin{array}{cc|cc}1 & 2 & 1 & 0\\0 & 3 & 1 & -1\end{array}\right]$$

$\frac{1}{3}R_2 \to R_2$

$$\left[\begin{array}{cc|cc}1 & 2 & 1 & 0\\0 & 1 & \frac{1}{3} & -\frac{1}{3}\end{array}\right]$$

$-2R_2 + R_1 \to R_1$

$$\left[\begin{array}{cc|cc}1 & 0 & \frac{1}{3} & \frac{2}{3}\\0 & 1 & \frac{1}{3} & -\frac{1}{3}\end{array}\right]$$

$X = A^{-1}K$

$$= \begin{bmatrix}\frac{1}{3} & \frac{2}{3}\\\frac{1}{3} & -\frac{1}{3}\end{bmatrix}\begin{bmatrix}3\\6\end{bmatrix}$$

$$= \begin{bmatrix}\frac{1}{3}(3) + \frac{2}{3}(6)\\\frac{1}{3}(3) + \left(-\frac{1}{3}\right)(6)\end{bmatrix}$$

$$= \begin{bmatrix}5\\-1\end{bmatrix}$$

(5, −1)

9. $A = \begin{bmatrix}1 & -2\\2 & 1\end{bmatrix}$, $X = \begin{bmatrix}x\\y\end{bmatrix}$, $K = \begin{bmatrix}4\\13\end{bmatrix}$

$$\left[\begin{array}{cc|cc}1 & -2 & 1 & 0\\2 & 1 & 0 & 1\end{array}\right]$$

$-2R_1 + R_2 \to R_2$

$$\left[\begin{array}{cc|cc}1 & -2 & 1 & 0\\0 & 5 & -2 & 1\end{array}\right]$$

$\frac{1}{5}R_2 \to R_2$

$$\left[\begin{array}{cc|cc} 1 & -2 & 1 & 0 \\ 0 & 1 & -\frac{2}{5} & \frac{1}{5} \end{array}\right]$$

$2R_2 + R_1 \to R_1$

$$\left[\begin{array}{cc|cc} 1 & 0 & \frac{1}{5} & \frac{2}{5} \\ 0 & 1 & -\frac{2}{5} & \frac{1}{5} \end{array}\right]$$

$X = A^{-1}K$

$$= \begin{bmatrix} \frac{1}{5} & \frac{2}{5} \\ -\frac{2}{5} & \frac{1}{5} \end{bmatrix} \begin{bmatrix} 4 \\ 13 \end{bmatrix}$$

$$= \begin{bmatrix} \frac{1}{5}(4) + \frac{2}{5}(13) \\ -\frac{2}{5}(4) + \frac{1}{5}(13) \end{bmatrix}$$

$$= \begin{bmatrix} 6 \\ 1 \end{bmatrix}$$

(6, 1)

11. $A = \begin{bmatrix} 5 & -2 \\ 1 & -5 \end{bmatrix}, X = \begin{bmatrix} x \\ y \end{bmatrix}, K = \begin{bmatrix} 11 \\ -7 \end{bmatrix}$

$$\left[\begin{array}{cc|cc} 5 & -2 & 1 & 0 \\ 1 & -5 & 0 & 1 \end{array}\right]$$

$-5R_2 + R_1 \to R_2$

$$\left[\begin{array}{cc|cc} 5 & -2 & 1 & 0 \\ 0 & 23 & 1 & -5 \end{array}\right]$$

$\frac{1}{5}R_1 \to R_1$

$\frac{1}{23}R_2 \to R_2$

$$\left[\begin{array}{cc|cc} 1 & -\frac{2}{5} & \frac{1}{5} & 0 \\ 0 & 1 & \frac{1}{23} & -\frac{5}{23} \end{array}\right]$$

$\frac{2}{5}R_2 + R_1 \to R_1$

$$\left[\begin{array}{cc|cc} 1 & 0 & \frac{5}{23} & -\frac{2}{23} \\ 0 & 1 & \frac{1}{23} & -\frac{5}{23} \end{array}\right]$$

$X = A^{-1}K$

$$= \begin{bmatrix} \frac{5}{23} & -\frac{2}{23} \\ \frac{1}{23} & -\frac{5}{23} \end{bmatrix} \begin{bmatrix} 11 \\ -7 \end{bmatrix}$$

$$= \begin{bmatrix} \frac{5}{23}(11) + \left(-\frac{2}{23}\right)(-7) \\ \frac{1}{23}(11) + \left(-\frac{5}{23}\right)(-7) \end{bmatrix}$$

$$= \begin{bmatrix} 3 \\ 2 \end{bmatrix}$$

(3, 2)

13. $\begin{cases} -3x + y = -15 \\ 8x - 2y = 10 \end{cases}$

$A = \begin{bmatrix} -3 & 1 \\ 8 & -2 \end{bmatrix}, X = \begin{bmatrix} x \\ y \end{bmatrix}, K = \begin{bmatrix} -15 \\ 10 \end{bmatrix}$

$$\left[\begin{array}{cc|cc} -3 & 1 & 1 & 0 \\ 8 & -2 & 0 & 1 \end{array}\right]$$

$3R_1 - R_2 \to R_2$

$$\left[\begin{array}{cc|cc} 1 & -1 & -3 & -1 \\ 8 & -2 & 0 & 1 \end{array}\right]$$

$-8R_1 + R_2 \to R_2$

$$\left[\begin{array}{cc|cc} 1 & -1 & -3 & -1 \\ 0 & 6 & 24 & 9 \end{array}\right]$$

$\frac{1}{6}R_2 \to R_2$

$$\left[\begin{array}{cc|cc} 1 & -1 & -3 & -1 \\ 0 & 1 & 4 & \frac{3}{2} \end{array}\right]$$

$R_1 + R_2 \to R_1$

$$\left[\begin{array}{cc|cc} 1 & 0 & 1 & \frac{1}{2} \\ 0 & 1 & 4 & \frac{3}{2} \end{array}\right]$$

$X = A^{-1}K$

$$= \begin{bmatrix} 1 & \frac{1}{2} \\ 4 & \frac{3}{2} \end{bmatrix} \begin{bmatrix} -15 \\ 10 \end{bmatrix}$$

$$= \begin{bmatrix} 1(-15) + \left(\frac{1}{2}\right)(10) \\ 4(-15) + \left(\frac{3}{2}\right)(10) \end{bmatrix}$$

$$= \begin{bmatrix} -10 \\ -45 \end{bmatrix}$$

$(-10, -45)$

15. $A = \begin{bmatrix} \frac{3}{2} & \frac{1}{3} \\ 1 & \frac{1}{12} \end{bmatrix}, X = \begin{bmatrix} x \\ y \end{bmatrix}, K = \begin{bmatrix} 1 \\ \frac{1}{4} \end{bmatrix}$

$$\left[\begin{array}{cc|cc} \frac{3}{2} & \frac{1}{3} & 1 & 0 \\ 1 & \frac{1}{12} & 0 & 1 \end{array}\right]$$

$R_1 \leftrightarrow R_2$

$$\left[\begin{array}{cc|cc} 1 & \frac{1}{12} & 0 & 1 \\ \frac{3}{2} & \frac{1}{3} & 1 & 0 \end{array}\right]$$

$-\frac{3}{2}R_1 + R_2 \to R_2$

$$\left[\begin{array}{cc|cc} 1 & \frac{1}{12} & 0 & 1 \\ 0 & \frac{5}{24} & 1 & -\frac{3}{2} \end{array}\right]$$

$\frac{24}{5}R_2 \to R_2$

$$\left[\begin{array}{cc|cc} 1 & \frac{1}{12} & 0 & 1 \\ 0 & 1 & \frac{24}{5} & -\frac{36}{5} \end{array}\right]$$

$-\frac{1}{12}R_2 + R_1 \to R_1$

$$\left[\begin{array}{cc|cc} 1 & 0 & -\frac{2}{5} & \frac{8}{5} \\ 0 & 1 & \frac{24}{5} & -\frac{36}{5} \end{array}\right]$$

$X = A^{-1}K$

$$= \begin{bmatrix} -\frac{2}{5} & \frac{8}{5} \\ \frac{24}{5} & -\frac{36}{5} \end{bmatrix} \begin{bmatrix} 1 \\ \frac{1}{4} \end{bmatrix}$$

$$= \begin{bmatrix} \left(-\frac{2}{5}\right)(1) + \left(\frac{8}{5}\right)\left(\frac{1}{4}\right) \\ \left(\frac{24}{5}\right)(1) + \left(-\frac{36}{5}\right)\left(\frac{1}{4}\right) \end{bmatrix}$$

$$= \begin{bmatrix} 0 \\ 3 \end{bmatrix}$$

$(0, 3)$

17. $A = \begin{bmatrix} 1 & 1 & 1 \\ 3 & 1 & -1 \\ 2 & 1 & -1 \end{bmatrix}, X = \begin{bmatrix} x \\ y \\ z \end{bmatrix}, K = \begin{bmatrix} 6 \\ 6 \\ 4 \end{bmatrix}$

$$\left[\begin{array}{ccc|ccc} 1 & 1 & 1 & 1 & 0 & 0 \\ 3 & 1 & -2 & 0 & 1 & 0 \\ 2 & 1 & -1 & 0 & 0 & 1 \end{array}\right]$$

$-3R_1 + R_2 \to R_2$
$-2R_1 + R_3 \to R_3$

$$\left[\begin{array}{ccc|ccc} 1 & 1 & 1 & 1 & 0 & 0 \\ 0 & -2 & -4 & -3 & 1 & 0 \\ 0 & -1 & -3 & -2 & 0 & 1 \end{array}\right]$$

$-R_3 \leftrightarrow R_2$

$$\left[\begin{array}{ccc|ccc} 1 & 1 & 1 & 1 & 0 & 0 \\ 0 & 1 & 3 & 2 & 0 & -1 \\ 0 & -2 & -4 & -3 & 1 & 0 \end{array}\right]$$

$R_1 - R_2 \to R_1$
$2R_2 + R_3 \to R_3$

$$\left[\begin{array}{ccc|ccc} 1 & 0 & -2 & -1 & 0 & 1 \\ 0 & 1 & 3 & 2 & 0 & -1 \\ 0 & 0 & 2 & 1 & 1 & -2 \end{array}\right]$$

$R_1 + R_3 \to R_1$
$\frac{1}{2}R_3 \to R_3$

$$\left[\begin{array}{ccc|ccc} 1 & 0 & 0 & 0 & 1 & -1 \\ 0 & 1 & 3 & 2 & 0 & -1 \\ 0 & 0 & 1 & \frac{1}{2} & \frac{1}{2} & -1 \end{array}\right]$$

$-3R_3 + R_2 \to R_2$

$$\left[\begin{array}{ccc|ccc} 1 & 0 & 0 & 0 & 1 & -1 \\ 0 & 1 & 0 & \frac{1}{2} & -\frac{3}{2} & 2 \\ 0 & 0 & 1 & \frac{1}{2} & \frac{1}{2} & -1 \end{array}\right]$$

$X = A^{-1}K$

$$= \begin{bmatrix} 0 & 1 & -1 \\ \frac{1}{2} & -\frac{3}{2} & 2 \\ \frac{1}{2} & \frac{1}{2} & -1 \end{bmatrix} \begin{bmatrix} 6 \\ 6 \\ 4 \end{bmatrix}$$

$$= \begin{bmatrix} 0(6) + 1(6) + (-1)(4) \\ \left(\frac{1}{2}\right)(6) + \left(-\frac{3}{2}\right)(6) + 2(4) \\ \left(\frac{1}{2}\right)(6) + \left(\frac{1}{2}\right)(6) + (-1)(4) \end{bmatrix}$$

$$= \begin{bmatrix} 2 \\ 2 \\ 2 \end{bmatrix}$$

(2, 2, 2)

19. $A = \begin{bmatrix} 1 & -3 & 1 \\ 3 & -1 & 1 \\ 1 & 2 & -1 \end{bmatrix}, X = \begin{bmatrix} x \\ y \\ z \end{bmatrix}, K = \begin{bmatrix} 0 \\ 2 \\ 2 \end{bmatrix}$

$$\left[\begin{array}{ccc|ccc} 1 & -3 & 1 & 1 & 0 & 0 \\ 3 & -1 & 1 & 0 & 1 & 0 \\ 1 & 2 & -1 & 0 & 0 & 1 \end{array}\right]$$

$-3R_1 + R_2 \to R_2$
$R_1 - R_3 \to R_3$

$$\left[\begin{array}{ccc|ccc} 1 & -3 & 1 & 1 & 0 & 0 \\ 0 & 8 & -2 & -3 & 1 & 0 \\ 0 & -5 & 2 & 1 & 0 & -1 \end{array}\right]$$

$2R_2 + 3R_3 \to R_2$

$$\left[\begin{array}{ccc|ccc} 1 & -3 & 1 & 1 & 0 & 0 \\ 0 & 1 & 2 & -3 & 2 & -3 \\ 0 & -5 & 2 & 1 & 0 & -1 \end{array}\right]$$

$3R_2 + R_1 \to R_1$
$5R_2 + R_3 \to R_3$

$$\left[\begin{array}{ccc|ccc} 1 & 0 & 7 & -8 & 6 & -9 \\ 0 & 1 & 2 & -3 & 2 & -3 \\ 0 & 0 & 12 & -14 & 10 & -16 \end{array}\right]$$

$\frac{1}{12}R_3 \to R_3$

$$\left[\begin{array}{ccc|ccc} 1 & 0 & 7 & -8 & 6 & -9 \\ 0 & 1 & 2 & -3 & 2 & -3 \\ 0 & 0 & 1 & -\frac{7}{6} & \frac{5}{6} & -\frac{4}{3} \end{array}\right]$$

$-7R_3 + R_1 \to R_1$

$-2R_3 + R_2 \to R_2$

$$\left[\begin{array}{ccc|ccc} 1 & 0 & 0 & \frac{1}{6} & \frac{1}{6} & \frac{1}{3} \\ 0 & 1 & 0 & -\frac{2}{3} & \frac{1}{3} & -\frac{1}{3} \\ 0 & 0 & 1 & -\frac{7}{6} & \frac{5}{6} & -\frac{4}{3} \end{array}\right]$$

$X = A^{-1}K$

$$= \begin{bmatrix} \frac{1}{6} & \frac{1}{6} & \frac{1}{3} \\ -\frac{2}{3} & \frac{1}{3} & -\frac{1}{3} \\ -\frac{7}{6} & \frac{5}{6} & -\frac{4}{3} \end{bmatrix} \begin{bmatrix} 0 \\ 2 \\ 2 \end{bmatrix}$$

$$= \begin{bmatrix} \left(\frac{1}{6}\right)(0) + \left(\frac{1}{6}\right)(2) + \left(\frac{1}{3}\right)(2) \\ \left(-\frac{2}{3}\right)(0) + \left(\frac{1}{3}\right)(2) + \left(-\frac{1}{3}\right)(2) \\ \left(-\frac{7}{6}\right)(0) + \left(\frac{5}{6}\right)(2) + \left(-\frac{4}{3}\right)(2) \end{bmatrix}$$

$$= \begin{bmatrix} 1 \\ 0 \\ -1 \end{bmatrix}$$

(1, 0 −1)

21. $A = \begin{bmatrix} 1 & 1 & 1 \\ 3 & 1 & -1 \\ 2 & 2 & 1 \end{bmatrix}$, $X = \begin{bmatrix} x \\ y \\ z \end{bmatrix}$, $K = \begin{bmatrix} 6 \\ 4 \\ 4 \end{bmatrix}$

$$\left[\begin{array}{ccc|ccc} 1 & 1 & 1 & 1 & 0 & 0 \\ 3 & 1 & -1 & 0 & 1 & 0 \\ 2 & 2 & 1 & 0 & 0 & 1 \end{array}\right]$$

$-3R_1 + R_2 \to R_2$
$-2R_1 + R_3 \to R_3$

$$\left[\begin{array}{ccc|ccc} 1 & 1 & 1 & 1 & 0 & 0 \\ 0 & -2 & -4 & -3 & 1 & 0 \\ 0 & 0 & -1 & -2 & 0 & 1 \end{array}\right]$$

$-\frac{1}{2}R_2 \to R_2$
$-R_3 \to R_3$

$$\left[\begin{array}{ccc|ccc} 1 & 1 & 1 & 1 & 0 & 0 \\ 0 & 1 & 2 & \frac{3}{2} & -\frac{1}{2} & 0 \\ 0 & 0 & 1 & 2 & 0 & -1 \end{array}\right]$$

$R_1 - R_2 \to R_1$
$-2R_3 + R_2 \to R_2$

$$\left[\begin{array}{ccc|ccc} 1 & 0 & -1 & -\frac{1}{2} & \frac{1}{2} & 0 \\ 0 & 1 & 0 & -\frac{5}{2} & -\frac{1}{2} & 2 \\ 0 & 0 & 1 & 2 & 0 & -1 \end{array}\right]$$

$R_1 + R_3 \to R_1$

$$\left[\begin{array}{ccc|ccc} 1 & 0 & 0 & \frac{3}{2} & \frac{1}{2} & -1 \\ 0 & 1 & 0 & -\frac{5}{2} & -\frac{1}{2} & 2 \\ 0 & 0 & 1 & 2 & 0 & -1 \end{array}\right]$$

$$X = A^{-1}K$$

$$= \begin{bmatrix} \frac{3}{2} & \frac{1}{2} & -1 \\ -\frac{5}{2} & -\frac{1}{2} & 2 \\ 2 & 0 & -1 \end{bmatrix} \begin{bmatrix} 6 \\ 4 \\ 4 \end{bmatrix}$$

$$= \begin{bmatrix} \left(\frac{3}{2}\right)(6) + \left(\frac{1}{2}\right)(4) + (-1)(4) \\ \left(-\frac{5}{2}\right)(6) + \left(-\frac{1}{2}\right)(4) + (2)(4) \\ (2)(6) + (0)(4) + (-1)(4) \end{bmatrix}$$

$$= \begin{bmatrix} 7 \\ -9 \\ 8 \end{bmatrix}$$

(7, -9, 8)

23. Multiply the 1st by 5 and the 2nd equation by 10.

$$\begin{cases} 5x + y + 5z = 0.1 \\ 3x - 10y + 2z = 0 \\ x + y + z = 0.1 \end{cases}$$

$$A = \begin{bmatrix} 5 & 1 & 5 \\ 3 & -10 & 2 \\ 1 & 1 & 1 \end{bmatrix}, X = \begin{bmatrix} x \\ y \\ z \end{bmatrix}, K = \begin{bmatrix} 0.1 \\ 0 \\ 0.1 \end{bmatrix}$$

$$\left[\begin{array}{ccc|ccc} 5 & 1 & 5 & 1 & 0 & 0 \\ 3 & -10 & 2 & 0 & 1 & 0 \\ 1 & 1 & 1 & 0 & 0 & 1 \end{array}\right]$$

$R_1 \leftrightarrow R_3$

$$\left[\begin{array}{ccc|ccc} 1 & 1 & 1 & 0 & 0 & 1 \\ 3 & -10 & 2 & 0 & 1 & 0 \\ 5 & 1 & 5 & 1 & 0 & 0 \end{array}\right]$$

$-3R_1 + R_2 \to R_2$
$-5R_1 + R_3 \to R_3$

$$\left[\begin{array}{ccc|ccc} 1 & 1 & 1 & 0 & 0 & 1 \\ 0 & -13 & -1 & 0 & 1 & -3 \\ 0 & -4 & 0 & 1 & 0 & -5 \end{array}\right]$$

$-\frac{1}{4}R_3 \leftrightarrow R_2$

$$\left[\begin{array}{ccc|ccc} 1 & 1 & 1 & 0 & 0 & 1 \\ 0 & 1 & 0 & -\frac{1}{4} & 0 & \frac{5}{4} \\ 0 & -13 & -1 & 0 & 1 & -3 \end{array}\right]$$

$R_1 - R_2 \to R_1$
$13R_2 + R_3 \to R_3$

$$\left[\begin{array}{ccc|ccc} 1 & 0 & 1 & \frac{1}{4} & 0 & -\frac{1}{4} \\ 0 & 1 & 0 & -\frac{1}{4} & 0 & \frac{5}{4} \\ 0 & 0 & -1 & -\frac{13}{4} & 1 & \frac{53}{4} \end{array}\right]$$

$R_1 + R_3 \to R_1$
$-R_3 \to R_3$

$$\left[\begin{array}{ccc|ccc} 1 & 0 & 0 & -3 & 1 & 13 \\ 0 & 1 & 0 & -\frac{1}{4} & 0 & \frac{5}{4} \\ 0 & 0 & 1 & \frac{13}{4} & -1 & -\frac{53}{4} \end{array}\right]$$

$$X = A^{-1}K$$

$$= \begin{bmatrix} -3 & 1 & 13 \\ -\frac{1}{4} & 0 & \frac{5}{4} \\ \frac{13}{4} & -1 & -\frac{53}{4} \end{bmatrix} \begin{bmatrix} 0.1 \\ 0 \\ 0.1 \end{bmatrix}$$

$$= \begin{bmatrix} (-3)(0.1) + (1)(0) + (13)(0.1) \\ \left(-\frac{1}{4}\right)(0.1) + (0)(0) + \left(\frac{5}{4}\right)(0.1) \\ \left(\frac{13}{4}\right)(0.1) + (-1)(0) + \left(-\frac{53}{4}\right)(0.1) \end{bmatrix}$$

$$= \begin{bmatrix} 1 \\ 0.1 \\ -1 \end{bmatrix}$$

(1, 0.1, -1)

25. Let x = amount invested at 6%
y = amount invested at 18%

(a) $\begin{cases} x + y = 20{,}000 \\ 0.06x + 0.18y = 0.08(20{,}000) = 1600 \end{cases}$

$\begin{cases} x + y = 20{,}000 \\ 6x + 18y = 160{,}000 \end{cases}$

$$\left[\begin{array}{cc|cc} 1 & 1 & 1 & 0 \\ 6 & 18 & 0 & 1 \end{array}\right]$$

$-6R_1 + R_2 \to R_2$

$$\left[\begin{array}{cc|cc} 1 & 1 & 1 & 0 \\ 0 & 12 & -6 & 1 \end{array}\right]$$

$\frac{1}{12}R_2 \to R_2$

$$\left[\begin{array}{cc|cc} 1 & 1 & 1 & 0 \\ 0 & 1 & -\frac{1}{2} & \frac{1}{12} \end{array}\right]$$

$R_1 - R_2 \to R_1$

$$\left[\begin{array}{cc|cc} 1 & 0 & \frac{3}{2} & -\frac{1}{12} \\ 0 & 1 & -\frac{1}{2} & \frac{1}{12} \end{array}\right]$$

$$X = A^{-1}K$$

$$= \begin{bmatrix} \frac{3}{2} & -\frac{1}{12} \\ -\frac{1}{2} & \frac{1}{12} \end{bmatrix} \begin{bmatrix} 20{,}000 \\ 160{,}000 \end{bmatrix}$$

$$= \begin{bmatrix} \left(\frac{3}{2}\right)(20{,}000) + \left(-\frac{1}{12}\right)(160{,}000) \\ \left(-\frac{1}{2}\right)(20{,}000) + \left(\frac{1}{12}\right)(160{,}000) \end{bmatrix}$$

$$= \begin{bmatrix} 16666.67 \\ 3333.33 \end{bmatrix}$$

\$16,666.67 at 6%; \$3333.33 at 18%

(b) $\begin{cases} x + y = 20{,}000 \\ 0.06x + 0.18y = 0.10(20{,}000) = 2000 \end{cases}$

$\begin{cases} x + y = 20{,}000 \\ 6x + 18y = 200{,}000 \end{cases}$

$$X = A^{-1}K$$

$$= \begin{bmatrix} \frac{3}{2} & -\frac{1}{12} \\ -\frac{1}{2} & \frac{1}{12} \end{bmatrix} \begin{bmatrix} 20{,}000 \\ 200{,}000 \end{bmatrix}$$

$$= \begin{bmatrix} \left(\frac{3}{2}\right)(20{,}000) + \left(-\frac{1}{12}\right)(200{,}000) \\ \left(-\frac{1}{2}\right)(20{,}000) + \left(\frac{1}{12}\right)(200{,}000) \end{bmatrix}$$

$$= \begin{bmatrix} 13333.33 \\ 6666.67 \end{bmatrix}$$

\$13,333.33 at 6; \$6666.67 at 18%

(c) $\begin{cases} x + y = 20000 \\ 0.06x + 0.18y = 0.12(20000) = 2400 \end{cases}$

$\begin{cases} x + y = 20000 \\ 6x + 18y = 240000 \end{cases}$

$$X = A^{-1}K$$

$$= \begin{bmatrix} \frac{3}{2} & -\frac{1}{12} \\ -\frac{1}{2} & \frac{1}{12} \end{bmatrix} \begin{bmatrix} 20{,}000 \\ 240{,}000 \end{bmatrix}$$

$$= \begin{bmatrix} \left(\frac{3}{2}\right)(20{,}000) + \left(-\frac{1}{12}\right)(240{,}000) \\ \left(-\frac{1}{2}\right)(20{,}000) + \left(\frac{1}{12}\right)(240{,}000) \end{bmatrix}$$

$$= \begin{bmatrix} 10{,}000 \\ 10{,}000 \end{bmatrix}$$

\$10,000 at 6%; \$10,000 at 18%

27. Let x = amount at 4%
y = amount at 6%
z = amount at 20%

(a) $x + y + z = 12{,}000$
$x = 2z$
$0.04x + 0.06y + 0.20z = 0.08(12000)$

$$\begin{aligned} x + y + z &= 12000 \\ x \quad - 2z &= 0 \\ 4x + 6y + 20z &= 96000 \end{aligned}$$

$$\left[\begin{array}{ccc|ccc} 1 & 1 & 1 & 1 & 0 & 0 \\ 1 & 0 & -2 & 0 & 1 & 0 \\ 4 & 6 & 20 & 0 & 0 & 1 \end{array}\right]$$

$R_1 - R_2 \to R_2$
$4R_1 - R_3 \to R_3$

$$\left[\begin{array}{ccc|ccc} 1 & 1 & 1 & 1 & 0 & 0 \\ 0 & 1 & 3 & 1 & -1 & 0 \\ 0 & -2 & -16 & 4 & 0 & -1 \end{array}\right]$$

$R_1 - R_2 \to R_1$
$2R_2 + R_3 \to R_3$

$$\left[\begin{array}{ccc|ccc} 1 & 0 & -2 & 0 & 1 & 0 \\ 0 & 1 & 3 & 1 & -1 & 0 \\ 0 & 0 & -10 & 6 & -2 & -1 \end{array}\right]$$

$-\frac{1}{10} R_3 \to R_3$

$$\left[\begin{array}{ccc|ccc} 1 & 0 & -2 & 0 & 1 & 0 \\ 0 & 1 & 3 & 1 & -1 & 0 \\ 0 & 0 & 1 & -\frac{3}{5} & \frac{1}{5} & \frac{1}{10} \end{array}\right]$$

$R_1 + 2R_3 \to R_1$
$-3R_3 + R_2 \to R_2$

$$\left[\begin{array}{ccc|ccc} 1 & 0 & 0 & -\frac{6}{5} & \frac{7}{5} & \frac{1}{5} \\ 0 & 1 & 0 & \frac{14}{5} & -\frac{8}{5} & -\frac{3}{10} \\ 0 & 0 & 1 & -\frac{3}{5} & \frac{1}{5} & \frac{1}{10} \end{array}\right]$$

$$\begin{bmatrix} x \\ y \\ z \end{bmatrix} = \begin{bmatrix} -\frac{6}{5} & \frac{7}{5} & \frac{1}{5} \\ \frac{14}{5} & -\frac{8}{5} & -\frac{3}{10} \\ -\frac{3}{5} & \frac{1}{5} & \frac{1}{10} \end{bmatrix} \begin{bmatrix} 12000 \\ 0 \\ 96000 \end{bmatrix}$$

$$= \begin{bmatrix} 4800 \\ 4800 \\ 2400 \end{bmatrix}$$

He should invest \$4800 at 4%, \$4800 at 6% and \$2400 at 20%.

(b) $x + y + z = 12000$
$x = 2z$
$0.04x + 0.06y + 0.20z = 0.06(12000)$

$$\begin{aligned} x + y + z &= 12000 \\ x \quad -2z &= 0 \\ 4x + 6y + 20z &= 72000 \end{aligned}$$

$$\begin{bmatrix} x \\ y \\ z \end{bmatrix} = \begin{bmatrix} -\frac{6}{5} & \frac{7}{5} & \frac{1}{5} \\ \frac{14}{5} & -\frac{8}{5} & -\frac{3}{10} \\ -\frac{3}{5} & \frac{1}{5} & \frac{1}{10} \end{bmatrix} \begin{bmatrix} 12000 \\ 0 \\ 72000 \end{bmatrix}$$

$$= \begin{bmatrix} 0 \\ 12000 \\ 0 \end{bmatrix}$$

He should invest all \$12,000 at 6%.

(c) $x + y + z = 12000$
$x = 2z$
$0.04x + 0.06y + 0.20z = 0.12(12000)$

$$\begin{aligned} x + y + z &= 12000 \\ x \quad - 2z &= 0 \\ 4x + 6y + 20z &= 144000 \end{aligned}$$

$$\begin{bmatrix} x \\ y \\ z \end{bmatrix} = \begin{bmatrix} -\frac{6}{5} & \frac{7}{5} & \frac{1}{5} \\ \frac{14}{5} & -\frac{8}{5} & -\frac{3}{10} \\ -\frac{3}{5} & \frac{1}{5} & \frac{1}{10} \end{bmatrix} \begin{bmatrix} 12000 \\ 0 \\ 144000 \end{bmatrix}$$

$$= \begin{bmatrix} 14400 \\ -9600 \\ 7200 \end{bmatrix}$$

Cannot be done.

29. Let x = number of shirts
y = number of blouses
z = number of skirts

$$\begin{aligned} 0.1x + 0.2y + 0.2z &= 400 \\ 0.3x + 0.4y + 0.5z &= 980 \\ 0.08x + 0.06y + 0.05z &= 160 \end{aligned}$$

$$\begin{aligned} x + 2y + 2z &= 4000 \\ 3x + 4y + 5z &= 9800 \\ 8x + 6y + 5z &= 16000 \end{aligned}$$

$$\left[\begin{array}{ccc|ccc} 1 & 2 & 2 & 1 & 0 & 0 \\ 3 & 4 & 5 & 0 & 1 & 0 \\ 8 & 6 & 5 & 0 & 0 & 1 \end{array}\right]$$

$-3R_1 + R_2 \to R_2$
$-8R_1 + R_3 \to R_3$

$$\left[\begin{array}{ccc|ccc} 1 & 2 & 2 & 1 & 0 & 0 \\ 0 & -2 & -1 & -3 & 1 & 0 \\ 0 & -10 & -11 & -8 & 0 & 1 \end{array}\right]$$

$R_1 + R_2 \to R_1$
$-\frac{1}{2}R_2 \to R_2$
$-5R_2 + R_3 \to R_3$

$$\left[\begin{array}{ccc|ccc} 1 & 0 & 1 & -2 & 1 & 0 \\ 0 & 1 & \frac{1}{2} & \frac{3}{2} & -\frac{1}{2} & 0 \\ 0 & 0 & -6 & 7 & -5 & 1 \end{array}\right]$$

$-\frac{1}{6}R_3 \to R_3$

$$\left[\begin{array}{ccc|ccc} 1 & 0 & 1 & -2 & 1 & 0 \\ 0 & 1 & \frac{1}{2} & \frac{3}{2} & -\frac{1}{2} & 0 \\ 0 & 0 & 1 & -\frac{7}{6} & \frac{5}{6} & -\frac{1}{6} \end{array}\right]$$

$R_1 - R_3 \to R_1$
$-\frac{1}{2}R_3 + R_2 \to R_2$

$$\left[\begin{array}{ccc|ccc} 1 & 0 & 0 & -\frac{5}{6} & \frac{1}{6} & \frac{1}{6} \\ 0 & 1 & 0 & \frac{25}{12} & -\frac{11}{12} & \frac{1}{12} \\ 0 & 0 & 1 & -\frac{7}{6} & \frac{5}{6} & -\frac{1}{6} \end{array}\right]$$

$$\begin{bmatrix} x \\ y \\ z \end{bmatrix} = \begin{bmatrix} -\frac{5}{6} & \frac{1}{6} & \frac{1}{6} \\ \frac{25}{12} & -\frac{11}{12} & \frac{1}{12} \\ -\frac{7}{6} & \frac{5}{6} & -\frac{1}{6} \end{bmatrix} \begin{bmatrix} 4000 \\ 9800 \\ 16000 \end{bmatrix}$$

$$= \begin{bmatrix} 966.7 \\ 683.3 \\ 833.3 \end{bmatrix}$$

They should produce 967 shirts, 683 blouses and 833 skirts.

31. (a)
$$\begin{aligned} X + Y + Z &= 10 \\ 0.5X + 0.2Y + 0.4Z &= 3.5 \\ 0.1X + 0.06Y + 0.08Z &= 0.77 \end{aligned}$$

$$\begin{aligned} X + Y + Z &= 10 \\ 5X + 2Y + 4Z &= 35 \\ 10X + 6Y + 8Z &= 77 \end{aligned}$$

$$\left[\begin{array}{ccc|ccc} 1 & 1 & 1 & 1 & 0 & 0 \\ 5 & 2 & 4 & 0 & 1 & 0 \\ 10 & 6 & 8 & 0 & 0 & 1 \end{array}\right]$$

$-5R_1 + R_2 \to R_2$
$-2R_2 + R_3 \to R_3$

$$\left[\begin{array}{ccc|ccc} 1 & 1 & 1 & 1 & 0 & 0 \\ 0 & -3 & -1 & -5 & 1 & 0 \\ 0 & 2 & 0 & 0 & -2 & 1 \end{array}\right]$$

$-R_2 - R_3 \rightarrow R_2$

$$\left[\begin{array}{ccc|ccc} 1 & 1 & 1 & 1 & 0 & 0 \\ 0 & 1 & 1 & 5 & 1 & -1 \\ 0 & 2 & 0 & 0 & -2 & 1 \end{array}\right]$$

$R_1 - R_2 \rightarrow R_1$
$-2R_2 + R_3 \rightarrow R_3$

$$\left[\begin{array}{ccc|ccc} 1 & 0 & 0 & -4 & -1 & 1 \\ 0 & 1 & 1 & 5 & 1 & -1 \\ 0 & 0 & -2 & -10 & -4 & 3 \end{array}\right]$$

$-\frac{1}{2}R_3 \rightarrow R_3$

$$\left[\begin{array}{ccc|ccc} 1 & 0 & 0 & -4 & -1 & 1 \\ 0 & 1 & 1 & 5 & 1 & -1 \\ 0 & 0 & 1 & 5 & 2 & -\frac{3}{2} \end{array}\right]$$

$R_2 - R_3 \rightarrow R_2$

$$\left[\begin{array}{ccc|ccc} 1 & 0 & 0 & -4 & -1 & 1 \\ 0 & 1 & 0 & 0 & -1 & \frac{1}{2} \\ 0 & 0 & 1 & 5 & 2 & -\frac{3}{2} \end{array}\right]$$

$$\begin{bmatrix} X \\ Y \\ Z \end{bmatrix} = \begin{bmatrix} -4 & -1 & 1 \\ 0 & -1 & \frac{1}{2} \\ 5 & 2 & -\frac{3}{2} \end{bmatrix} \begin{bmatrix} 10 \\ 35 \\ 77 \end{bmatrix}$$

$$= \begin{bmatrix} 2 \\ 3.5 \\ 4.5 \end{bmatrix}$$

They should use $2g$ of X, $3.5g$ of Y, and $4.5g$ of Z.

(b) $$\begin{bmatrix} X \\ Y \\ Z \end{bmatrix} = \begin{bmatrix} -4 & -1 & 1 \\ 0 & -1 & \frac{1}{2} \\ 5 & 2 & -\frac{3}{2} \end{bmatrix} \begin{bmatrix} 10 \\ 35 \\ 78 \end{bmatrix}$$

$$= \begin{bmatrix} 3 \\ 4 \\ 3 \end{bmatrix}$$

They should use $3g$ of X, $4g$ of Y, and $3g$ of Z.

(c) $$\begin{bmatrix} X \\ Y \\ Z \end{bmatrix} = \begin{bmatrix} -4 & -1 & 1 \\ 0 & -1 & \frac{1}{2} \\ 5 & 2 & -\frac{3}{2} \end{bmatrix} \begin{bmatrix} 10 \\ 35 \\ 79 \end{bmatrix}$$

$$= \begin{bmatrix} 4 \\ 4.5 \\ 1.5 \end{bmatrix}$$

They should use $4g$ of X, $4.5g$ of Y, and $1.5g$ of Z.

35. $$y = \frac{kx^2}{\sqrt{z}}$$

$$10 = \frac{k(3)^2}{\sqrt{4}}$$

$$20 = 9k$$

$$\frac{20}{9} = k$$

$$y = \frac{20x^2}{9\sqrt{z}}$$

$$y = \frac{20(2)^2}{9\sqrt{9}} = \frac{80}{27}$$

37. $$\frac{x}{x+1} = \frac{x}{x+2}$$

$$x(x+2) = x(x+1)$$
$$x^2 + 2x = x^2 + x$$
$$2x = x$$
$$x = 0$$

Exercises 11.5

1. $\begin{vmatrix} 1 & 2 \\ 3 & 4 \end{vmatrix} = (1)(4) - (3)(2) = -2$

3. $\begin{vmatrix} 5 & 1 \\ -2 & 4 \end{vmatrix} = (5)(4) - (-2)(1) = 22$

5. $\begin{vmatrix} -3 & -1 \\ 2 & -2 \end{vmatrix} = (-3)(-2) - (2)(-1) = 8$

7. $\begin{vmatrix} 1 & 0 \\ 5 & 4 \end{vmatrix} = (1)(4) - (5)(0) = 4$

9. $\begin{vmatrix} 4 & 6 \\ 6 & 9 \end{vmatrix} = (4)(9) - (6)(6) = 0$

11. $\begin{vmatrix} 1 & 2 & -2 \\ 3 & -3 & 1 \\ -4 & 2 & -1 \end{vmatrix}$

$= 1\begin{vmatrix} -3 & 1 \\ 2 & -1 \end{vmatrix} - 2\begin{vmatrix} 3 & 1 \\ -4 & -1 \end{vmatrix} + (-2)\begin{vmatrix} 3 & -3 \\ -4 & 2 \end{vmatrix}$

$= (-3)(-1) - (2)(1) - 2[(3)(-1) - (-4)(1)]$
$\quad - 2[(3)(2) - (-4)(-3)]$
$= 1 - 2(1) - 2(-6)$
$= 11$

13. $\begin{vmatrix} 1 & 2 & 3 \\ 2 & 4 & 6 \\ 1 & 1 & 1 \end{vmatrix}$

$= 1\begin{vmatrix} 2 & 3 \\ 4 & 6 \end{vmatrix} - 1\begin{vmatrix} 1 & 3 \\ 2 & 6 \end{vmatrix} + 1\begin{vmatrix} 1 & 2 \\ 2 & 4 \end{vmatrix}$

$= [(2)(6) - (4)(3)] - [(1)(6) - (2)(3)]$
$\quad + [(1)(4) - (2)(2)]$
$= 0 - 0 + 0$
$= 0$

15. $\begin{vmatrix} -3 & 0 & 4 \\ 5 & 2 & -3 \\ 7 & 0 & 6 \end{vmatrix}$

$= 0\begin{vmatrix} 5 & -3 \\ 7 & 6 \end{vmatrix} + 2\begin{vmatrix} -3 & 4 \\ 7 & 6 \end{vmatrix} - 0\begin{vmatrix} -3 & 4 \\ 5 & -3 \end{vmatrix}$

$= 2[(-3)(6) - (7)(4)]$
$= -92$

17. $\begin{vmatrix} 1 & 0 & 0 \\ 0 & 1 & 0 \\ 0 & 0 & 1 \end{vmatrix}$

$= 1\begin{vmatrix} 1 & 0 \\ 0 & 1 \end{vmatrix} - 0\begin{vmatrix} 0 & 0 \\ 0 & 1 \end{vmatrix} + 0\begin{vmatrix} 0 & 1 \\ 0 & 0 \end{vmatrix}$

$= 1[(1)(1) - (0)(0)]$
$= 1$

19. $\begin{vmatrix} 2x & 4 \\ 3 & 5 \end{vmatrix} = 18$

$(2x)(5) - (3)(4) = 18$
$10x - 12 = 18$
$10x = 30$
$x = 3$

21. $\begin{vmatrix} x^2 & 5 \\ x & 1 \end{vmatrix} = 14$

$(x^2)(1) - 5(x) = 14$
$x^2 - 5x - 14 = 0$
$(x - 7)(x + 2) = 0$
$x - 7 = 0$ or $x + 2 = 0$
$x = 7$ or $x = -2$

23. $\begin{vmatrix} x & 3 \\ 2 & x + 2 \end{vmatrix} = 9$

$x(x + 2) - 2(3) = 9$
$x^2 + 2x - 6 = 9$
$x^2 + 2x - 15 = 0$
$(x + 5)(x - 3) = 0$
$x + 5 = 0$ or $x - 3 = 0$
$x = -5$ or $x = 3$

25. $D = \begin{vmatrix} 9 & 2 \\ 5 & 1 \end{vmatrix} = (9)(1) - (5)(2) = -1$

$D_x = \begin{vmatrix} 1 & 2 \\ 0 & 1 \end{vmatrix} = (1)(1) - (0)(2) = 1$

$D_y = \begin{vmatrix} 9 & 1 \\ 5 & 0 \end{vmatrix} = (9)(0) - (5)(1) = -5$

$x = \dfrac{D_x}{D} = \dfrac{1}{-1} = -1$

$y = \dfrac{D_y}{D} = \dfrac{-5}{-1} = 5$

27. $D = \begin{vmatrix} 2 & -4 \\ -3 & 6 \end{vmatrix} = (2)(6) - (-3)(-4) = 0$

No unique solution

29. $\begin{cases} 3x - 5y = 2 \\ 4x - 3y = 10 \end{cases}$

$D = \begin{vmatrix} 3 & -5 \\ 4 & -3 \end{vmatrix} = (3)(-3) - (4)(-5) = 11$

$D_x = \begin{vmatrix} 2 & -5 \\ 10 & -3 \end{vmatrix} = (2)(-3) - (10)(-5) = 44$

$D_y = \begin{vmatrix} 3 & 2 \\ 4 & 10 \end{vmatrix} = (3)(10) - (4)(2) = 22$

$x = \dfrac{D_x}{D} = \dfrac{44}{11} = 4$

$y = \dfrac{D_y}{D} = \dfrac{22}{11} = 2$

31. $\begin{cases} 2x - 7y = -4 \\ -4x + 3y = 8 \end{cases}$

$D = \begin{vmatrix} 2 & -7 \\ -4 & 3 \end{vmatrix} = (2)(3) - (-4)(-7) = -22$

$D_x = \begin{vmatrix} -4 & -7 \\ 8 & 3 \end{vmatrix} = (-4)(3) - (8)(-7) = 44$

$D_y = \begin{vmatrix} 2 & -4 \\ -4 & 8 \end{vmatrix} = (2)(8) - (-4)(-4) = 0$

$x = \dfrac{D_x}{D} = \dfrac{44}{-22} = -2$

$y = \dfrac{D_y}{D} = \dfrac{0}{-22} = 0$

33. $\begin{cases} 6x - 5y = -7 \\ -3x + 4y = -5 \end{cases}$

$D = \begin{vmatrix} 6 & -5 \\ -3 & 4 \end{vmatrix} = (6)(4) - (-3)(-5) = 9$

$D_x = \begin{vmatrix} -7 & -5 \\ -5 & 4 \end{vmatrix} = (-7)(4) - (-5)(-5) = -53$

$D_y = \begin{vmatrix} 6 & -7 \\ -3 & -5 \end{vmatrix} = (6)(-5) - (-3)(-7) = -51$

$x = \dfrac{D_x}{D} = \dfrac{-53}{9}$

$y = \dfrac{D_y}{D} = \dfrac{-51}{9} = \dfrac{-17}{3}$

35. $D = \begin{vmatrix} 2 & -9 \\ 3 & 5 \end{vmatrix} = (2)(5) - (3)(-9) = 37$

$D_s = \begin{vmatrix} 4 & -9 \\ 6 & 5 \end{vmatrix} = (4)(5) - (6)(-9) = 74$

$D_t = \begin{vmatrix} 2 & 4 \\ 3 & 6 \end{vmatrix} = (2)(6) - (3)(4) = 0$

$s = \dfrac{D_s}{D} = \dfrac{74}{37} = 2$

$t = \dfrac{D_t}{D} = \dfrac{0}{37} = 0$

37. $D = \begin{vmatrix} 6 & 7 \\ 5 & 8 \end{vmatrix} = (6)(8) - (5)(7) = 13$

$D_u = \begin{vmatrix} 3 & 7 \\ 9 & 8 \end{vmatrix} = (3)(8) - (9)(7) = -39$

$D_v = \begin{vmatrix} 6 & 3 \\ 5 & 9 \end{vmatrix} = (6)(9) - (5)(3) = 39$

$u = \dfrac{D_u}{D} = \dfrac{-39}{13} = -3$

$v = \dfrac{D_v}{D} = \dfrac{39}{13} = 3$

39. $x - 24y = 12$
$x - 24y = 18$

$D = \begin{vmatrix} 1 & -24 \\ 1 & -24 \end{vmatrix} = (1)(-24) - (1)(-24) = 0$

No unique solution

41. $D = \begin{vmatrix} 1 & 1 & 1 \\ 1 & -1 & 1 \\ -1 & 1 & 1 \end{vmatrix} = -4$

$D_x = \begin{vmatrix} 3 & 1 & 1 \\ 2 & -1 & 1 \\ 4 & 1 & 1 \end{vmatrix} = 2$

$D_y = \begin{vmatrix} 1 & 3 & 1 \\ 1 & 2 & 1 \\ -1 & 4 & 1 \end{vmatrix} = -2$

$D_z = \begin{vmatrix} 1 & 1 & 3 \\ 1 & -1 & 2 \\ -1 & 1 & 4 \end{vmatrix} = -12$

$x = \dfrac{D_x}{D} = \dfrac{2}{-4} = -\dfrac{1}{2}$

$y = \dfrac{D_y}{D} = \dfrac{-2}{-4} = \dfrac{1}{2}$

$z = \dfrac{D_z}{D} = \dfrac{-12}{-4} = 3$

43. $D = \begin{vmatrix} 3 & 4 & 2 \\ 2 & 3 & 1 \\ 6 & 1 & 5 \end{vmatrix} = -6$

$D_x = \begin{vmatrix} 1 & 4 & 2 \\ 1 & 3 & 1 \\ 1 & 1 & 5 \end{vmatrix} = -6$

$D_y = \begin{vmatrix} 3 & 1 & 2 \\ 2 & 1 & 1 \\ 6 & 1 & 5 \end{vmatrix} = 0$

$D_z = \begin{vmatrix} 3 & 4 & 1 \\ 2 & 3 & 1 \\ 6 & 1 & 1 \end{vmatrix} = 6$

$x = \dfrac{D_x}{D} = \dfrac{-6}{-6} = 1$

$y = \dfrac{D_y}{D} = \dfrac{0}{-6} = 0$

$z = \dfrac{D_z}{D} = \dfrac{6}{-6} = -1$

45. $D = \begin{vmatrix} 1 & 2 & 3 \\ 6 & 5 & 4 \\ 7 & 8 & 9 \end{vmatrix} = 0$

No unique solution

47. $D = \begin{vmatrix} 3 & -1 & 0 \\ 0 & 2 & 1 \\ 3 & 0 & 4 \end{vmatrix} = 21$

$D_x = \begin{vmatrix} 4 & -1 & 0 \\ 6 & 2 & 1 \\ 14 & 0 & 4 \end{vmatrix} = 42$

$D_y = \begin{vmatrix} 3 & 4 & 0 \\ 0 & 6 & 1 \\ 3 & 14 & 4 \end{vmatrix} = 42$

$D_z = \begin{vmatrix} 3 & -1 & 4 \\ 0 & 2 & 6 \\ 3 & 0 & 14 \end{vmatrix} = 42$

$$x = \frac{D_x}{D} = \frac{42}{21} = 2$$

$$y = \frac{D_y}{D} = \frac{42}{21} = 2$$

$$z = \frac{D_z}{D} = \frac{42}{21} = 2$$

49. $D = \begin{vmatrix} 6 & -2 & 4 \\ 12 & -4 & 8 \\ -3 & 1 & -2 \end{vmatrix} = 0$

No unique solution

51. $\begin{cases} x - y - z = 2 \\ -x + y + z = 3 \\ -x - y + z = 4 \end{cases}$

$$D = \begin{vmatrix} 1 & -1 & -1 \\ -1 & 1 & 1 \\ -1 & -1 & 1 \end{vmatrix} = 0$$

No unique solution

53. $D = \begin{vmatrix} 2 & 1 & 0 \\ -3 & 0 & 1 \\ 0 & 2 & 5 \end{vmatrix} = 11$

$$D_a = \begin{vmatrix} 8 & 1 & 0 \\ -13 & 0 & 1 \\ 0 & 2 & 5 \end{vmatrix} = 49$$

$$D_b = \begin{vmatrix} 2 & 8 & 0 \\ -3 & -13 & 1 \\ 0 & 0 & 5 \end{vmatrix} = -10$$

$$D_c = \begin{vmatrix} 2 & 1 & 8 \\ -3 & 0 & -13 \\ 0 & 2 & 0 \end{vmatrix} = 4$$

$$a = \frac{D_a}{D} = \frac{49}{11}$$

$$b = \frac{D_b}{D} = \frac{-10}{11}$$

$$c = \frac{D_c}{D} = \frac{4}{11}$$

55. $D = \begin{vmatrix} 68 & 85 \\ 920 & 743 \end{vmatrix} = (68)(743) - (920)(85) = -27676$

57. $D = \begin{vmatrix} 5 & -3 \\ 0.2 & 4 \end{vmatrix} = 20.6$

$$D_x = \begin{vmatrix} -0.181 & -3 \\ 0.2482 & 4 \end{vmatrix} = 0.0206$$

$$D_y = \begin{vmatrix} 5 & -0.181 \\ 0.2 & 0.2482 \end{vmatrix} = 1.2772$$

$$x = \frac{D_x}{D} = \frac{0.0206}{20.6} = 0.001$$

$$y = \frac{D_y}{D} = \frac{1.2772}{20.6} = 0.062$$

59. $D = \begin{vmatrix} 0.02 & 0.04 & -0.06 \\ 0.5 & 0.6 & -0.3 \\ 0.1 & -0.02 & 1 \end{vmatrix} = -0.00512$

$$D_x = \begin{vmatrix} 12 & 0.04 & -0.06 \\ 16 & 0.6 & -0.3 \\ 15 & -0.02 & 1 \end{vmatrix} = 6.8672$$

$$D_y = \begin{vmatrix} 0.02 & 12 & -0.06 \\ 0.5 & 16 & -0.3 \\ 0.1 & 15 & 1 \end{vmatrix} = -6.304$$

$$D_z = \begin{vmatrix} 0.02 & 0.04 & 12 \\ 0.5 & 0.6 & 16 \\ 0.1 & -0.02 & 15 \end{vmatrix} = -0.8896$$

$$x = \frac{D_x}{D} = \frac{6.8672}{-0.00512} = 1341.25$$

$$y = \frac{D_y}{D} = \frac{-6.304}{-0.00512} = 1231.25$$

$$z = \frac{D_z}{D} = \frac{-0.8896}{-0.00512} = 173.75$$

63. $\log_3 \frac{1}{81} = -4$

65. $\log_4 60 \approx 2.9534$
$\log_3 40 \approx 3.3578$
$\log_3 40$ is greater.

Exercises 11.6

1. $\begin{cases} x + y \le 6 & \text{(solid line)} \\ x + 2y \ge 3 & \text{(solid line)} \end{cases}$

Test point: (3, 1)

$3 + 1 \le 6$ $3 + 2(1) \ge 3$
$4 \le 6$ $5 \ge 3$
True   True

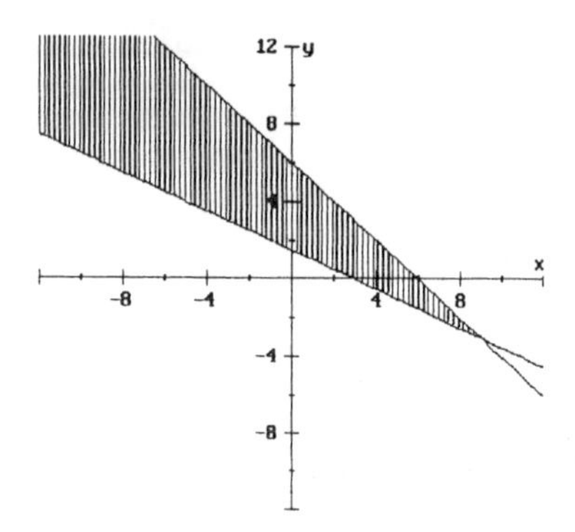

Test point: (3, 1)

$3 + 1 \le 6$ $3 + 2(1) \ge 3$
$4 \le 6$ $5 \ge 3$
True   True

3. $\begin{cases} 2y + x \ge 6 & \text{(solid line)} \\ 3y + x \ge 9 & \text{(solid line)} \end{cases}$

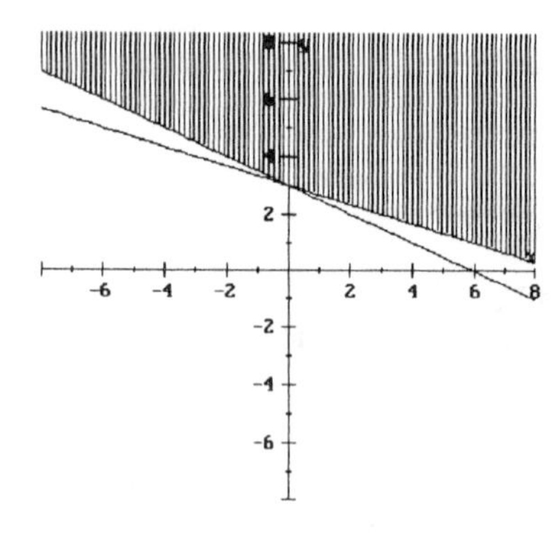

Test point: (1, 3)

$2(3) + 1 \ge 6$ $3(3) + 1 \ge 9$
$7 \ge 6$ $10 \ge 9$
True   True

5. $\begin{cases} x - y \ge 2 & \text{(solid line)} \\ y - x > -1 & \text{(dashed line)} \end{cases}$

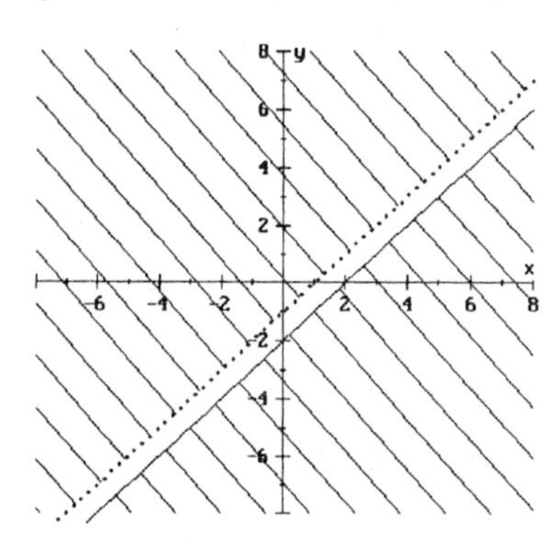

No solution, since the shadings do not overlap

7. $\begin{cases} x + y \le 5 & \text{(solid line)} \\ 2x + y \le 8 & \text{(solid line)} \end{cases}$

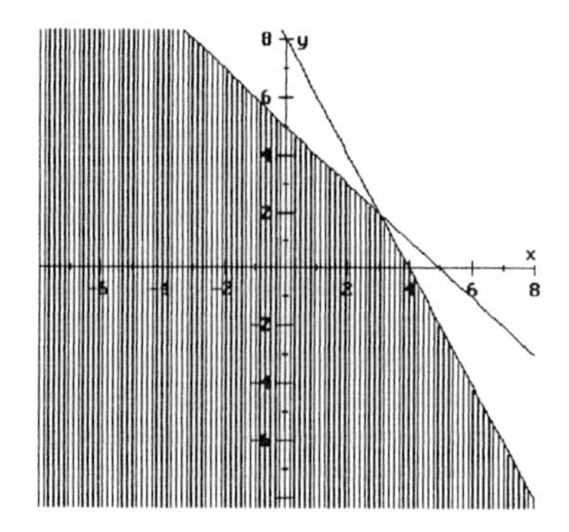

Test point: (0, 0)

$0 + 0 \le 5$ $2(0) + 0 \le 8$
$0 \le 5$ $0 \le 8$
True   True

9. $\begin{cases} 3x + 2y \le 12 & \text{(solid line)} \\ y \le x & \text{(solid line)} \end{cases}$

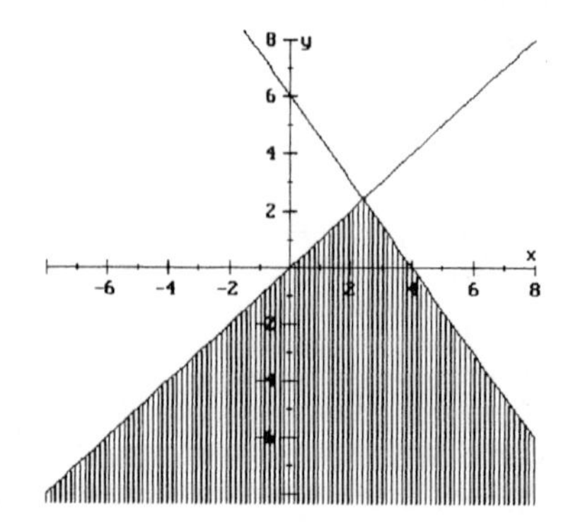

Test point: (0, −1)

$3(0) + 2(-1) \le 12$ $\quad -1 \le 0$
$-2 < 12$ $\quad$ True
True

11. $\begin{cases} 3x + 2y \ge 12 & \text{(solid line)} \\ y \ge x & \text{(solid line)} \end{cases}$

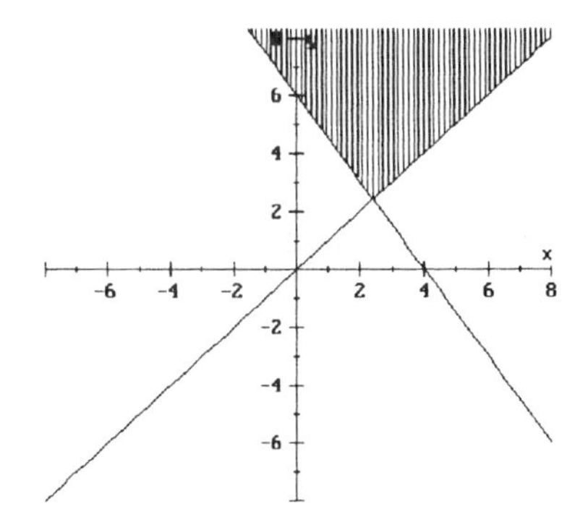

Test point: (0, 7)

$3(0) + 2(7) \ge 12$ $\quad 7 \ge 0$
$14 \ge 12$ $\quad$ True
True

13. $\begin{cases} 5x - 3y \le 15 & \text{(solid line)} \\ x < 3 & \text{(dashed line)} \end{cases}$

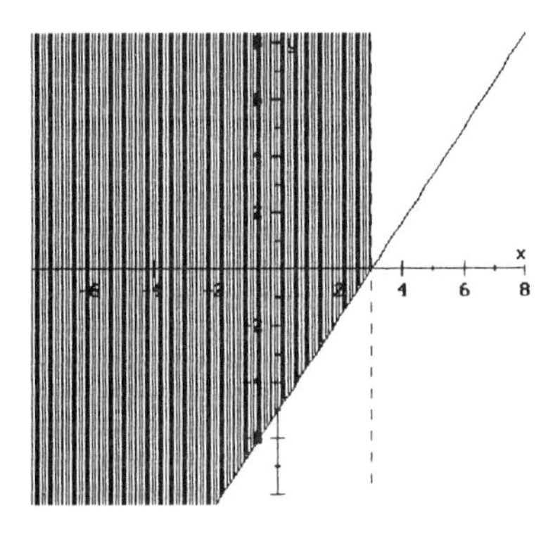

Test point: (0, 0)

$5(0) - 3(0) \le 15$ $\quad 0 < 3$
$0 \le 15$ $\quad$ True
True

15. $\begin{cases} 5x - 3y \ge 15 & \text{(solid line)} \\ x > 3 & \text{(dashed line)} \end{cases}$

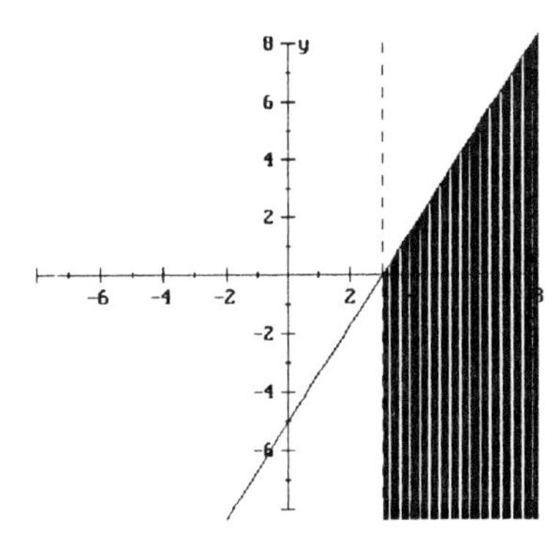

Test point: (4, 0)

$5(4) - 3(0) \ge 15$ $\quad 4 > 3$
$20 \ge 15$ $\quad$ True
True

17. $\begin{cases} x - 2y < 10 & \text{(dashed line)} \\ x \ge 2 & \text{(solid line)} \\ y \le 2 & \text{(solid line)} \end{cases}$

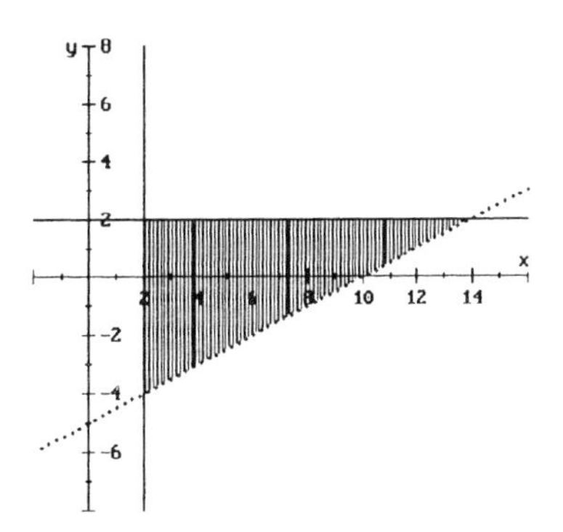

Test point: (3, 0)

$3 - 2(0) < 10$
$3 < 10$
True

$3 \ge 2$
True

$0 \le 2$
True

19. $\begin{cases} 4x + 3y \le 12 & \text{(solid line)} \\ x \ge 0 & \text{(solid line)} \\ y \ge 0 & \text{(solid line)} \end{cases}$

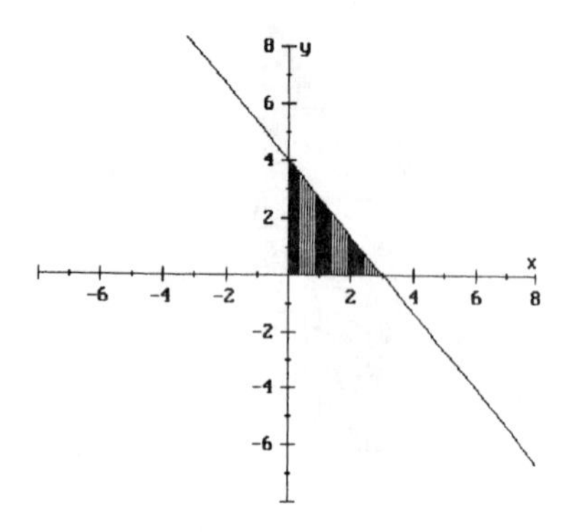

Test point: (1, 1)

$4(1) + 3(1) \le 12$
$7 \le 12$
True

$1 \ge 0$
True

$1 \ge 0$
True

21. $\begin{cases} x + 3y \ge 6 & \text{(solid line)} \\ 3x + 2y \le 18 & \text{(solid line)} \\ y \le 3 & \text{(solid line)} \\ x \ge 0 & \text{(solid line)} \end{cases}$

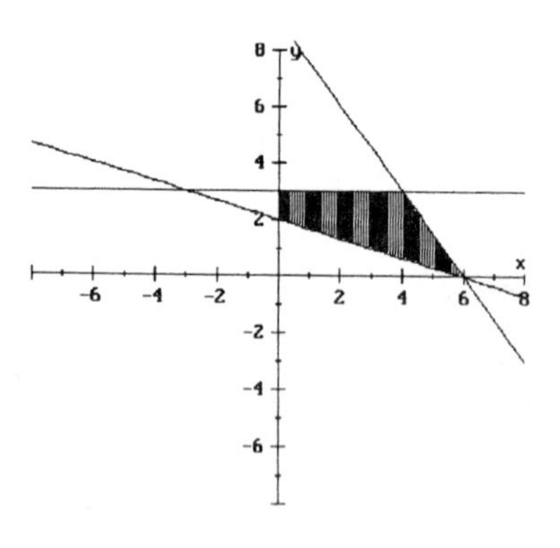

Test point: (1, 2)

$1 + 3(2) \ge 6$ $\quad 3(1) + 2(2) \le 18$
$7 \ge 6$ $\quad 7 \le 18$
True $\quad$ True

$2 \le 3$ $\quad 1 \ge 0$
True $\quad$ True

23. $\begin{cases} x + 2y \ge 4 & \text{(solid line)} \\ 2x - 4y \le -8 & \text{(solid line)} \\ x \ge 0 & \text{(solid line)} \\ y \ge 0 & \text{(solid line)} \end{cases}$

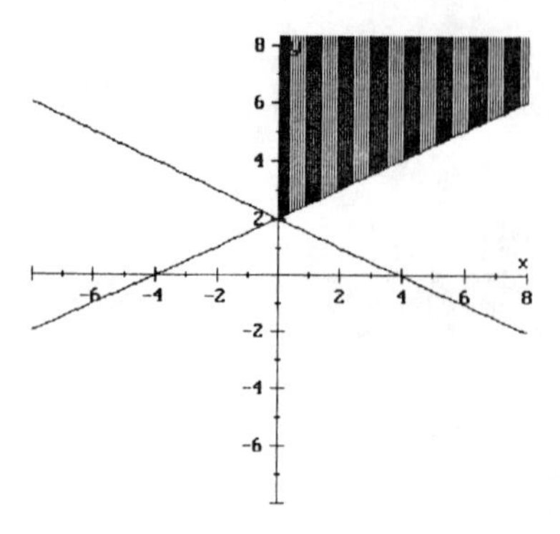

Test point: (1, 3)

$1 + 2(3) \ge 4$ $\quad 2(1) - 4(3) \le -8$
$7 \ge 4$ $\quad -10 \le -8$
True $\quad$ True

$1 \ge 0$ $\quad 3 \ge 0$
True $\quad$ True

25. x = number of single mattresses
y = number of double mattresses

$\begin{cases} 80x + 120y \le 8000 \\ 16x + 36y \le 2000 \\ x \ge 0 \\ y \ge 0 \end{cases}$

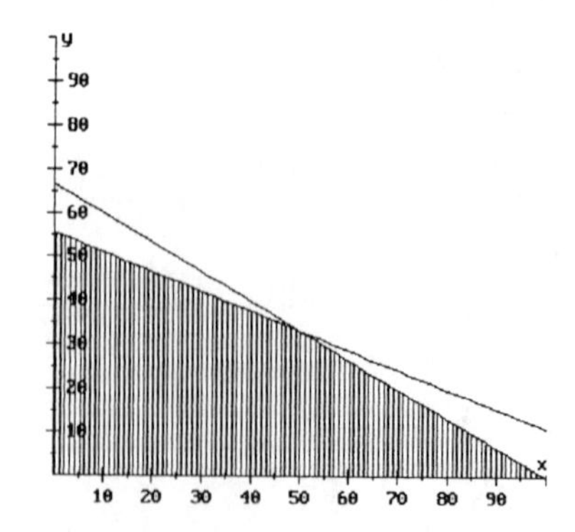

27. X = number of ounces of brand X
Y = number of ounces of brand Y

$$\begin{cases} 10X + 13Y \geq 21 \\ 0.33X + 0.37Y \leq 1 \\ X \geq 0 \\ Y \geq 0 \end{cases}$$

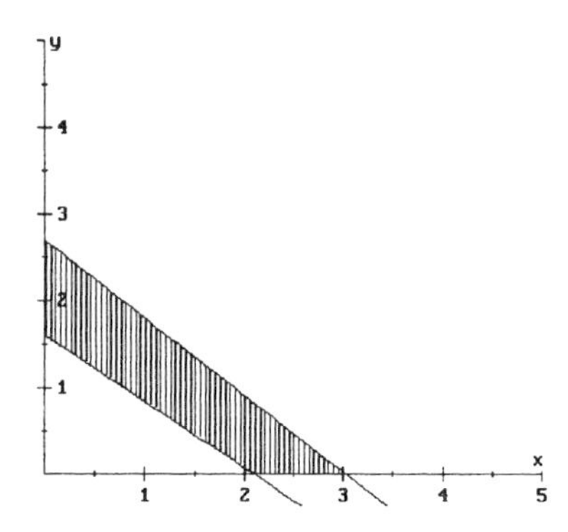

29. x = number of men's shoes
y = number of women's shoes

$$\begin{cases} x + \frac{3}{4}y \leq 500 \\ \frac{1}{3}x + \frac{1}{4}y \leq 150 \\ x \geq 0 \\ y \geq 0 \end{cases}$$

$$\begin{cases} 4x + 3y \leq 2000 \\ 4x + 3y \leq 1800 \\ x \geq 0 \\ y \geq 0 \end{cases}$$

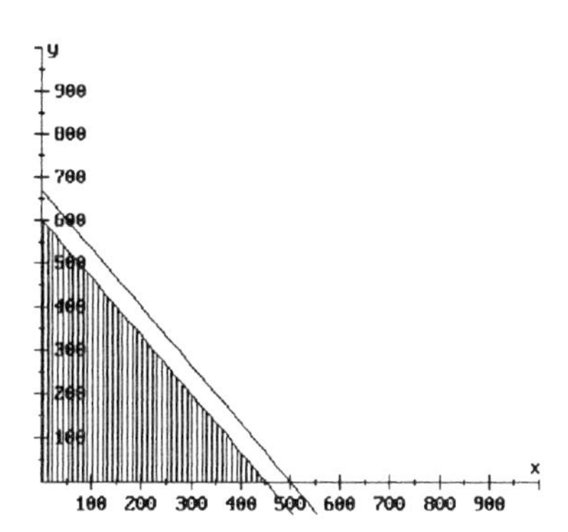

31. $\log_b \dfrac{\sqrt{x^3y^4}}{5b}$

$= \log_b \dfrac{(x^3y^4)^{1/2}}{5b}$

$= \log_b (x^3y^4)^{1/2} - \log_b (5b)$

$= \dfrac{1}{2} \log_b (x^3y^4) - \log_b (5b)$

$= \dfrac{1}{2}(\log_b x^3y^4) - [\log_b 5 + \log_b b]$

$= \dfrac{1}{2}(3 \log_b x + 4 \log_b y) - (\log_b 5 + 1)$

$= \dfrac{3}{2} \log_b x + 2 \log_b y - \log_b 5 - 1$

33. $100 \text{ wk} = \dfrac{100}{52} \text{ yr}$

$= \dfrac{25}{13} \text{ yr}$

$A = P\left(1 + \dfrac{r}{n}\right)^{nt}$

$= 6000\left(1 + \dfrac{0.072}{52}\right)^{(52)\left(\frac{25}{13}\right)}$

$= \$6890.37$

Exercises 11.7

1. $\begin{cases} x^2 + y^2 = 10 \\ x^2 - y^2 = 8 \end{cases}$

$$\begin{aligned} x^2 + y^2 &= 10 \\ x^2 - y^2 &= 8 \\ \hline 2x^2 &= 18 \\ x^2 &= 9 \\ x &= \pm 3 \end{aligned}$$

$x = 3$:
$(3)^2 + y^2 = 10$
$y^2 = 1$
$y = \pm 1$
$(3, 1), \ (3, -1)$

$x = -3$:
$$(-3)^2 + y^2 = 10$$
$$y^2 = 1$$
$$y = \pm 1$$
$(-3, 1),\ (-3, -1)$

$(3, 1),\ (3, -1),\ (-3, 1),\ (-3, -1)$

3. $\begin{cases} x^2 + y^2 = 25 \\ x - y = 5 \end{cases}$

$$x = 5 + y$$
$$(5 + y)^2 + y^2 = 25$$
$$25 + 10y + y^2 + y^2 = 25$$
$$2y^2 + 10y = 0$$
$$2y(y + 5) = 0$$
$2y = 0$ or $y + 5 = 0$
$y = 0$ or $y = -5$
$x = 5 + 0 = 5$ $\quad x = 5 + (-5) = 0$

$(5, 0),\ (0, -5)$

5. $\begin{cases} 16x^2 - 4y^2 = 64 \\ x^2 + y^2 = 9 \end{cases}$

$$\begin{array}{r} 4x^2 - y^2 = 16 \\ \underline{x^2 + y^2 = 9} \\ 5x^2 = 25 \end{array}$$
$$x^2 = 5$$
$$x = \pm\sqrt{5}$$

$x = \sqrt{5}$:
$$(\sqrt{5})^2 + y^2 = 9$$
$$y^2 = 4$$
$$y = \pm 2$$
$(\sqrt{5}, 2),\ (\sqrt{5}, -2)$

$x = -\sqrt{5}$:
$$(-\sqrt{5})^2 + y^2 = 9$$
$$y^2 = 4$$
$$y = \pm 2$$
$(-\sqrt{5}, 2),\ (-\sqrt{5}, -2)$

$(\sqrt{5}, 2),\ (\sqrt{5}, -2),\ (-\sqrt{5}, 2),\ (-\sqrt{5}, -2)$

7. $\begin{cases} x^2 + y^2 = 9 \\ 2x - y = 3 \end{cases}$

$$2x - y = 3$$
$$y = 2x - 3$$
$$x^2 + (2x - 3)^2 = 9$$
$$x^2 + 4x^2 - 12x + 9 = 9$$
$$5x^2 - 12x = 0$$
$$x(5x - 12) = 0$$
$x = 0$ or $5x - 12 = 0$
$y = 2(0) - 3 = -3$ $\quad x = \dfrac{12}{5}$

$$y = 2\left(\frac{12}{5}\right) - 3 = \frac{9}{5}$$

$(0, -3),\ \left(\dfrac{12}{5}, \dfrac{9}{5}\right)$

9. $\begin{cases} x^2 + y^2 = 13 \\ 3x^2 - y^2 = 3 \end{cases}$

$$\begin{array}{r} x^2 + y^2 = 13 \\ \underline{3x^2 - y^2 = 3} \\ 4x^2 = 16 \end{array}$$
$$x^2 = 4$$
$$x = \pm 2$$

$x = 2$:
$$2^2 + y^2 = 13$$
$$y^2 = 9$$
$$y = \pm 3$$
$(2, 3),\ (2, -3)$

$x = -2$:
$$(-2)^2 + y^2 = 13$$
$$y^2 = 9$$
$$y = \pm 3$$
$(-2, 3),\ (-2, -3)$

$(2, 3),\ (2, -3),\ (-2, 3),\ (-2, -3)$

11. $\begin{cases} x^2 - y = 0 \\ x^2 - 2x + y = 6 \end{cases}$

$$\begin{array}{r} x^2 \quad\quad - y = 0 \\ \underline{x^2 - 2x + y = 6} \\ 2x^2 - 2x \quad\quad = 6 \end{array}$$
$$2x^2 - 2x - 6 = 0$$
$$x^2 - x - 3 = 0$$

$$x = \frac{-(-1) \pm \sqrt{(-1)^2 - 4(1)(-3)}}{2(1)}$$

$$= \frac{1 \pm \sqrt{13}}{2}$$

$$x^2 - y = 0$$
$$x^2 = y$$

$x = \frac{1 + \sqrt{13}}{2}$:

$$y = \left(\frac{1 + \sqrt{13}}{2}\right)^2$$

$$= \frac{1 + 2\sqrt{13} + 13}{4}$$

$$= \frac{14 + 2\sqrt{13}}{4}$$

$$= \frac{7 + \sqrt{13}}{2}$$

$x = \frac{1 - \sqrt{13}}{2}$:

$$y = \left(\frac{1 - \sqrt{13}}{2}\right)^2$$

$$= \frac{1 - 2\sqrt{13} + 13}{4}$$

$$= \frac{14 - 2\sqrt{13}}{4}$$

$$= \frac{7 - \sqrt{13}}{2}$$

$$\left(\frac{1 + \sqrt{13}}{2}, \frac{7 + \sqrt{13}}{2}\right), \left(\frac{1 - \sqrt{13}}{2}, \frac{7 - \sqrt{13}}{2}\right)$$

13. $\begin{cases} y = 1 - x^2 \\ x + y = 2 \end{cases}$

$$x + (1 - x^2) = 2$$
$$0 = x^2 - x + 1$$
$$x = \frac{-(-1) \pm \sqrt{(-1)^2 - 4(1)(1)}}{2(1)}$$
$$= \frac{1 \pm \sqrt{-3}}{2}$$

No real solution

15. $\begin{cases} y = x^2 - 6x \\ y = x - 12 \end{cases}$

$$x^2 - 6x = x - 12$$
$$x^2 - 7x + 12 = 0$$
$$(x - 4)(x - 3) = 0$$

$x - 4 = 0$ or $x - 3 = 0$

$x = 4$ $\qquad$ $x = 3$

$y = 4 - 12 = -8$ $\qquad$ $y = 3 - 12 = -9$

$(4, -8)$, $(3, -9)$

17. $\begin{cases} x^2 + y^2 = 16 \\ x = y^2 - 16 \end{cases}$

$$x = y^2 - 16$$
$$x + 16 = y^2$$

$$x^2 + (x + 16) = 16$$
$$x^2 + x = 0$$
$$x(x + 1) = 0$$

$x = 0$ or $x + 1 = 0$

$x = -1$

$x = 0$:

$$0 + 16 = y^2$$
$$16 = y^2$$
$$\pm 4 = y$$

$x = -1$:

$$-1 + 16 = y^2$$
$$15 = y^2$$
$$\pm\sqrt{15} = y$$

$(0, 4)$, $(0, -4)$, $\left(-1, \sqrt{15}\right)$, $\left(-1, -\sqrt{15}\right)$

19. $\begin{cases} x^2 + y^2 - 25 = 0 \\ x + y - 7 = 0 \end{cases}$

$$x + y - 7 = 0$$
$$x = 7 - y$$

$$(7 - y)^2 + y^2 - 25 = 0$$
$$49 - 14y + y^2 + y^2 - 25 = 0$$
$$2y^2 - 14y + 24 = 0$$
$$y^2 - 7y + 12 = 0$$
$$(y - 4)(y - 3) = 0$$

$y - 4 = 0$ or $y - 3 = 0$

$y = 4$ $\qquad$ $y = 3$

$x = 7 - 4 = 3$ $\qquad$ $x = 7 - 3 = 4$

$(3, 4)$, $(4, 3)$

21. $\begin{cases} x^2 - y^2 = 4 \\ 2x^2 + y^2 = 16 \end{cases}$

$$y^2 = 16 - 2x^2$$
$$x^2 - (16 - 2x^2) = 4$$
$$x^2 - 16 + 2x^2 = 4$$
$$3x^2 = 20$$
$$x^2 = \frac{20}{3}$$
$$x = \pm\sqrt{\frac{20}{3}} = \pm\frac{2\sqrt{15}}{3}$$

$$x = \frac{2\sqrt{15}}{3}$$
$$y^2 = 16 - 2\left(\frac{2\sqrt{15}}{3}\right)^2$$
$$y^2 = \frac{8}{3}$$
$$y = \pm\sqrt{\frac{8}{3}} = \pm\frac{2\sqrt{6}}{3}$$
$$x = \frac{-2\sqrt{15}}{3}$$
$$y^2 = 16 - 2\left(-\frac{2\sqrt{15}}{3}\right)^2$$
$$y^2 = \frac{8}{3}$$
$$y = \pm\sqrt{\frac{8}{3}} = \pm\frac{2\sqrt{6}}{3}$$

$$\left(\frac{2\sqrt{15}}{3}, \frac{2\sqrt{6}}{3}\right), \left(\frac{2\sqrt{15}}{3}, -\frac{2\sqrt{6}}{3}\right),$$
$$\left(-\frac{2\sqrt{15}}{3}, \frac{2\sqrt{6}}{3}\right), \left(-\frac{2\sqrt{15}}{3}, -\frac{2\sqrt{6}}{3}\right)$$

23. $\begin{cases} 9x^2 + y^2 = 9 \\ 3x + y = 3 \end{cases}$

$$y = 3 - 3x$$
$$9x^2 + (3 - 3x)^2 = 9$$
$$9x^2 + (3 - 3x)^2 = 9$$
$$9x^2 + 9 - 18x + 9x^2 = 9$$
$$18x^2 - 18x = 0$$
$$18x(x - 1) = 0$$

$18x = 0$ or $x - 1 = 0$

$x = 0$ or $x = 1$

$y = 3 - 3(0) = 3$ $\quad$ $y = 3 - 3(1) = 0$

$(0, 3)$, $(1, 0)$

25. $\begin{cases} x^2 + y = 9 \\ 3x + 2y = 16 \end{cases}$

$$y = 9 - x^2$$
$$3x + 2(9 - x^2) = 16$$
$$3x + 18 - 2x^2 = 16$$
$$0 = 2x^2 - 3x - 2$$
$$0 = (2x + 1)(x - 2)$$

$2x + 1 = 0$ or $x - 2 = 0$

$x = -\frac{1}{2}$ $\quad$ $x = 2$

$y = 9 - \left(-\frac{1}{2}\right)^2 = \frac{35}{4}$ $\quad$ $y = 9 - 2^2 = 5$

$\left(-\frac{1}{2}, \frac{35}{4}\right)$, $(2, 5)$

27. $\begin{cases} x = y^2 - 3 \\ x + y = 9 \end{cases}$

$$(y^2 - 3) + y = 9$$
$$y^2 + y - 12 = 0$$
$$(y + 4)(y - 3) = 0$$

$y + 4 = 0$ or $y - 3 = 0$

$y = -4$ $\quad$ $y = 3$

$x = (-4)^2 - 3 = 13$ $\quad$ $x = 3^2 - 3 = 6$

$(13, -4)$, $(6, 3)$

29. $\begin{cases} x^2 + 4x + y^2 - 6y = 7 \\ 2x + y = 5 \end{cases}$

$$y = 5 - 2x$$
$$x^2 + 4x + (5-2x)^2 - 6(5-2x) = 7$$
$$x^2 + 4x + 25 - 20x + 4x^2 - 30 + 12x = 7$$
$$5x^2 - 4x - 12 = 0$$
$$(5x + 6)(x - 2) = 0$$

$5x + 6 = 0$ or $x - 2 = 0$

$x = -\frac{6}{5}$ $\qquad$ $x = 2$

$y = 5 - 2\left(-\frac{6}{5}\right) = \frac{37}{5}$ $\qquad$ $y = 5 - 2(2) = 1$

$\left(-\frac{6}{5}, \frac{37}{5}\right)$, $(2, 1)$

31. $\begin{cases} x^2 + y^2 = 10 \\ xy = 4 \end{cases}$

$$x = \frac{4}{y}$$
$$\left(\frac{4}{y}\right)^2 + y^2 = 10$$
$$\frac{16}{y^2} + y^2 = 10$$
$$16 + y^4 = 10y^2$$
$$y^4 - 10y^2 + 16 = 0$$
$$(y^2 - 8)(y^2 - 2) = 0$$

$y^2 - 8 = 0$ or $y^2 - 2 = 0$

$y^2 = 8$ $\qquad$ $y^2 = 2$

$y = \pm\sqrt{8} = \pm 2\sqrt{2}$ $\qquad$ $y = \pm\sqrt{2}$

$y = 2\sqrt{2}$ $\qquad$ $y = \sqrt{2}$

$x = \frac{4}{2\sqrt{2}} = \sqrt{2}$ $\qquad$ $x = \frac{4}{\sqrt{2}} = 2\sqrt{2}$

$y = -2\sqrt{2}$ $\qquad$ $y = -\sqrt{2}$

$x = \frac{4}{-2\sqrt{2}} = -\sqrt{2}$ $\qquad$ $x = \frac{4}{-\sqrt{2}} = -2\sqrt{2}$

$(\sqrt{2}, 2\sqrt{2})$, $(-\sqrt{2}, -2\sqrt{2})$, $(2\sqrt{2}, \sqrt{2})$, $(-2\sqrt{2}, -\sqrt{2})$

33. $\begin{cases} x^2 - y^2 = 4 \\ xy = 2\sqrt{3} \end{cases}$

$$x = \frac{2\sqrt{3}}{y}$$
$$\left(\frac{2\sqrt{3}}{y}\right)^2 - y^2 = 4$$
$$\frac{12}{y^2} - y^2 = 4$$
$$12 - y^4 = 4y^2$$
$$0 = y^4 + 4y^2 - 12$$
$$0 = (y^2 + 6)(y^2 - 2)$$

$y^2 + 6 = 0$ $\qquad$ $y^2 - 2 = 0$

$y^2 = -6$ $\qquad$ $y^2 = 2$

No Solution $\qquad$ $y = \pm\sqrt{2}$

$y = \sqrt{2}$ $\qquad$ $y = -\sqrt{2}$

$x = \frac{2\sqrt{3}}{\sqrt{2}} = \sqrt{6}$ $\qquad$ $x = \frac{2\sqrt{3}}{-\sqrt{2}} = -\sqrt{6}$

$(\sqrt{6}, \sqrt{2})$, $(-\sqrt{6}, -\sqrt{2})$

35. $\begin{cases} x^2 + y^2 = 25 \\ x^2 - xy + y^2 = 13 \end{cases}$

$$\begin{array}{r} x^2 \qquad + y^2 = 25 \\ \underline{-x^2 + xy - y^2 = -13} \\ xy \qquad = 12 \end{array}$$

$$x = \frac{12}{y}$$
$$\left(\frac{12}{y}\right)^2 + y^2 = 25$$
$$\frac{144}{y^2} + y^2 = 25$$
$$144 + y^4 = 25y^2$$
$$y^4 - 25y^2 + 144 = 0$$
$$(y^2 - 16)(y^2 - 9) = 0$$

$$y^2 - 16 = 0 \qquad y^2 - 9 = 0$$
$$y^2 = 16 \qquad y^2 = 9$$
$$y = \pm 4 \qquad y = \pm 3$$

$$y = 4 \qquad y = 3$$
$$x = \frac{12}{4} \qquad x = \frac{12}{3} = 4$$

$$y = -4 \qquad y = -3$$
$$x = \frac{12}{-4} = -3 \qquad x = \frac{12}{-3} = -4$$

(3, 4), (−3, −4), (4, 3), (−4, −3)

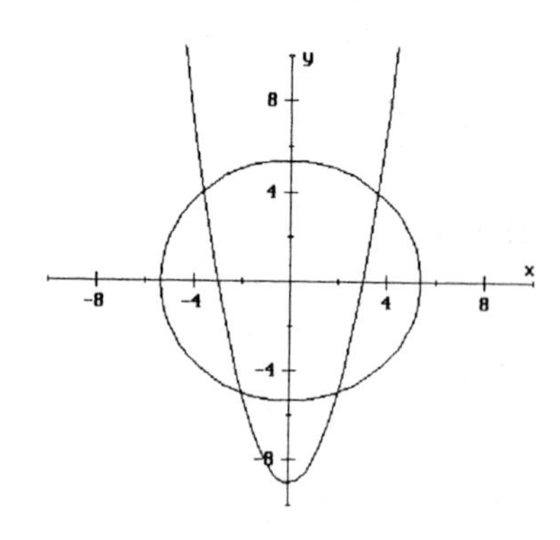

37. $\begin{cases} x^2 + y^2 = 29 \\ x - y = 3 \end{cases}$

$$x = 3 + y$$
$$(3 + y)^2 + y^2 = 29$$
$$9 + 6y + y^2 + y^2 = 29$$
$$2y^2 + 6y - 20 = 0$$
$$y^2 + 3y - 10 = 0$$
$$(y + 5)(y - 2) = 0$$
$$y + 5 = 0 \quad \text{or} \quad y - 2 = 0$$
$$y = -5 \qquad y = 2$$
$$x = 3 + (-5) = -2 \qquad x = 3 + 2 = 5$$

(−2, −5), (5, 2)

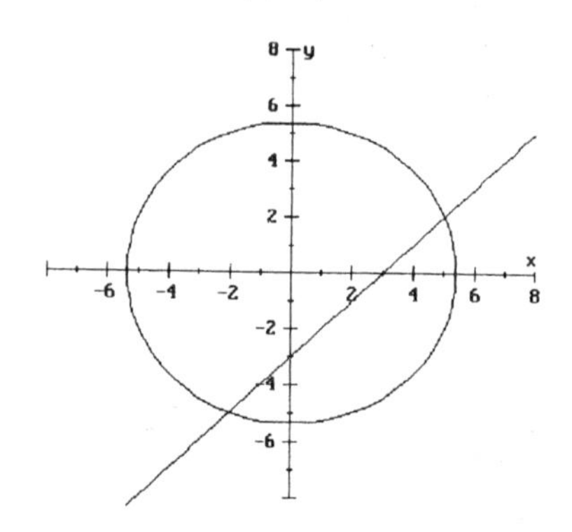

39. $\begin{cases} x^2 + y^2 = 29 \\ y = x^2 - 9 \end{cases}$

$$y = x^2 - 9$$
$$y + 9 = x^2$$

$$y + 9 + y^2 = 29$$
$$y^2 + y - 20 = 0$$
$$(y + 5)(y - 4) = 0$$
$$y + 5 = 0 \quad \text{or} \quad y - 4 = 0$$
$$y = -5 \qquad y = 4$$
$$x^2 = -5 + 9 \qquad x^2 = 4 + 9$$
$$x^2 = 4 \qquad x^2 = 13$$
$$x = \pm 2 \qquad x = \pm\sqrt{13}$$

(2, −5), (−2, −5), $(\sqrt{13}, 4)$, $(-\sqrt{13}, 4)$

41. $\begin{cases} x - y = 0 \\ 3y + x^2 = 4 \end{cases}$

$$x - y = 0$$
$$x = y$$

$$3(x) + x^2 = 4$$
$$x^2 + 3x - 4 = 0$$
$$(x + 4)(x - 1) = 0$$
$$x + 4 = 0 \quad \text{or} \quad x - 1 = 0$$
$$x = -4 \qquad x = 1$$
$$y = -4 \qquad y = 1$$

(−4, −4), (1, 1)

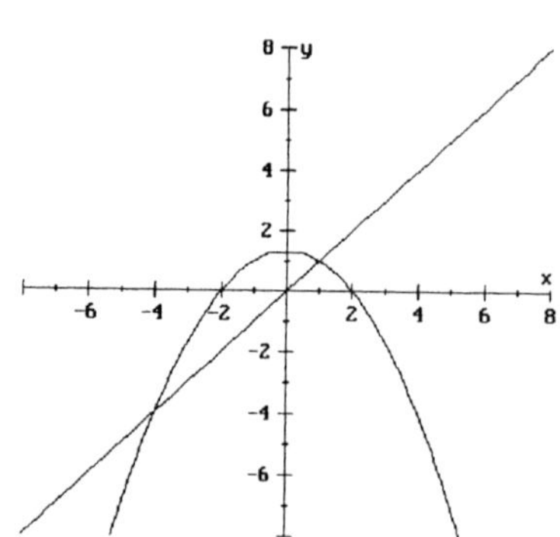

43. $\begin{cases} x^2 + y^2 = 5 \\ y = 8x^2 - 6 \end{cases}$

$$y = 8x^2 - 6$$
$$y + 6 = 8x^2$$
$$\frac{y + 6}{8} = x^2$$

$$\frac{y + 6}{8} + y^2 = 5$$

$$y + 6 + 8y^2 = 40$$
$$8y^2 + y - 34 = 0$$
$$(8y + 17)(y - 2) = 0$$

$$8y + 17 = 0 \quad \text{or} \quad y - 2 = 0$$
$$y = -\frac{17}{8} \qquad y = 2$$

$$-\frac{17}{8} + 6 = 8x^2 \qquad 2 + 6 = 8x^2$$

$$\frac{31}{8} = 8x^2 \qquad 8 = 8x^2$$

$$\frac{31}{64} = x^2 \qquad \pm 1 = x$$

$$\pm\frac{\sqrt{31}}{8} = x$$

$$\left(\frac{\sqrt{31}}{8}, -\frac{17}{8}\right), \left(-\frac{\sqrt{31}}{8}, -\frac{17}{8}\right),$$

$(1, 2)$, $(-1, 2)$

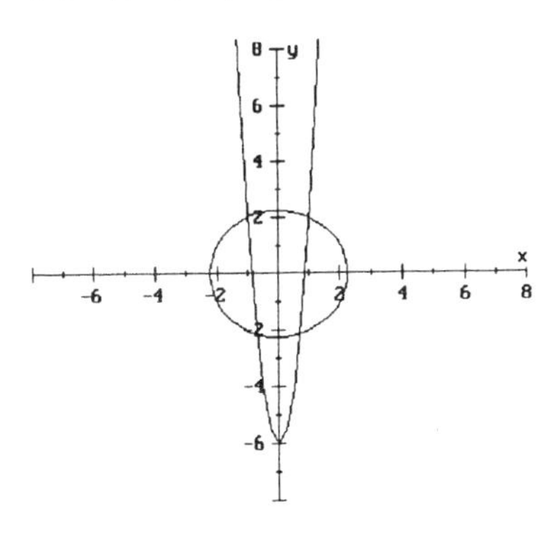

45. $\begin{cases} x^2 + y^2 = 1 \\ x^2 + 2x + y^2 = 0 \end{cases}$

$$\begin{aligned} -x^2 \qquad - y^2 &= -1 \\ x^2 + 2x + y^2 &= 0 \\ \hline 2x \qquad &= -1 \\ x &= -\frac{1}{2} \end{aligned}$$

$$\left(-\frac{1}{2}\right)^2 + y^2 = 1$$

$$y^2 = \frac{3}{4}$$

$$y = \pm\frac{\sqrt{3}}{2}$$

$$\left(-\frac{1}{2}, \frac{\sqrt{3}}{2}\right), \left(-\frac{1}{2}, -\frac{\sqrt{3}}{2}\right)$$

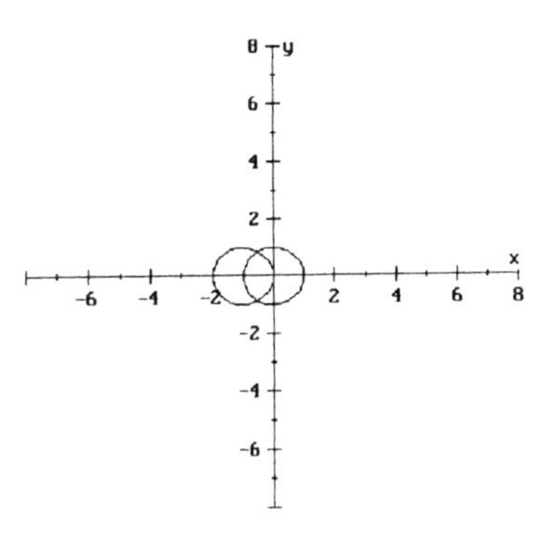

47. Let x = width
y = length

$$\begin{cases} xy = 210 \\ 2x + 2y = 59 \end{cases}$$

$$x = \frac{210}{y}$$

$$2\left(\frac{210}{y}\right) + 2y = 59$$

$$\frac{420}{y} + 2y = 59$$

$$420 + 2y^2 = 59y$$
$$2y^2 - 59y + 420 = 0$$
$$(2y - 35)(y - 12) = 0$$

$$2y - 35 = 0 \quad \text{or} \quad y - 12 = 0$$

$$y = \frac{35}{2} \qquad y = 12$$

$$x = \frac{210}{\frac{35}{2}} \qquad x = \frac{210}{12}$$

$$x = 12 \qquad x = 17.5$$

The dimensions are 12″ by 17.5″.

49. Let x = rate going,
then $x + 2$ = rate returning

Let y = time going,
then $y - \frac{1}{4}$ = time returning

$$\begin{cases} xy = 60 \\ (x + 12)\left(y - \frac{1}{4}\right) = 60 \end{cases}$$

$$x = \frac{60}{y}$$

$$\left(\frac{60}{y} + 12\right)\left(y - \frac{1}{4}\right) = 60$$

$$60 - \frac{15}{y} + 12y - 3 = 60$$

$$-\frac{15}{y} + 12y - 3 = 0$$

$$-15 + 12y^2 - 3y = 0$$
$$12y^2 - 3y - 15 = 0$$
$$4y^2 - y - 5 = 0$$
$$(y + 1)(4y - 5) = 0$$
$$y + 1 = 0 \qquad 4y - 5 = 0$$
$$y = -1 \qquad y = \frac{5}{4}$$

Since y must be positive, $y = \frac{5}{4}$.

$$y = \frac{5}{4}$$

$$y - \frac{1}{4} = \frac{5}{4} - \frac{1}{4} = 1$$

$$x = \frac{60}{\frac{5}{4}} = 48$$

$$x + 12 = 48 + 12 = 60$$

Going: rate = 48 mph; time = $\frac{5}{4}$ hr

Returning: rate = 60 mph; time = 1 hr

53. (a) $A(3) = 2000e^{0.0316(3)}$
$= 2198.9$ bacteria

(b)
$$5000 = 2000e^{0.0316t}$$
$$2.5 = e^{0.0316t}$$
$$\ln 2.5 = \ln e^{0.0316t}$$
$$\ln 2.5 = 0.0316t$$

$$\frac{\ln 2.5}{0.0316} = t$$

$$t \approx 29 \text{ hr}$$

55. $y = \left(\frac{2}{3}\right)^x$

x	y
-2	9/4
-1	3/2
0	1
1	2/3
2	4/9

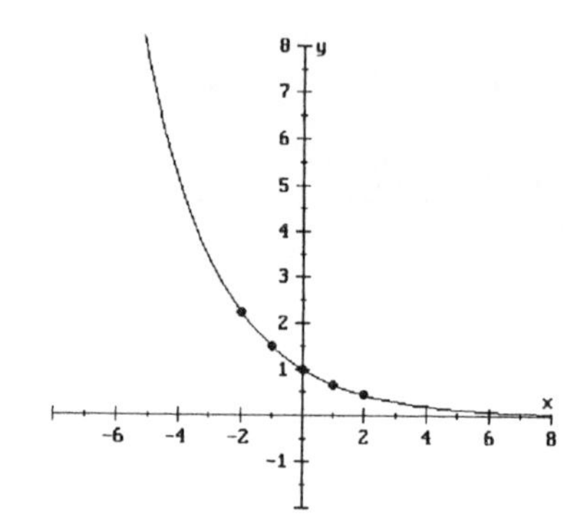

CHAPTER 11 REVIEW EXERCISES

1. $\begin{cases} x + 2y - z = 2 \\ 2x + 3y + 4z = 9 \\ 3x + y - 2z = 2 \end{cases}$

$$4x + 8y - 4z = 8$$
$$\underline{2x + 3y + 4z = 9}$$
$$6x + 11y = 17$$

$$2x + 3y + 4z = 9$$
$$\underline{6x + 2y - 4z = 4}$$
$$8x + 5y = 13$$

$$6x + 11y = 17$$
$$8x + 5y = 13$$

$$24x + 44y = 68$$
$$\underline{-24x - 15y = -39}$$
$$29y = 29$$
$$y = 1$$

$$6x + 11(1) = 17$$
$$6x = 6$$
$$x = 1$$

$$1 + 2(1) - z = 2$$
$$-z = -1$$
$$z = 1$$

$$x = 1, \quad y = 1, \quad z = 1$$

3. $\begin{cases} 2x - 3z = 7 \\ 3x + 4y = -11 \\ 3y - 2z = 0 \end{cases}$

$$\begin{array}{r} 4x - 6z = 14 \\ -9y + 6z = 0 \\ \hline 4x - 9y = 14 \end{array}$$

$$\begin{aligned} 4x - 9y &= 14 \\ 3x + 4y &= -11 \end{aligned}$$

$$\begin{array}{r} 16x - 36y = 56 \\ 27x + 36y = -99 \\ \hline 43x = -43 \\ x = -1 \end{array}$$

$$\begin{aligned} 2(-1) - 3z &= 7 \\ -3z &= 9 \\ z &= -3 \end{aligned}$$

$$\begin{aligned} 3y - 2(-3) &= 0 \\ 3y &= -6 \\ y &= -2 \end{aligned}$$

$x = -1, \quad y = -2, \quad z = -3$

5. $\begin{cases} 3x + y - z = 2 \\ 3x + y - z = 5 \\ x - y + z = 1 \end{cases}$

$$\begin{array}{r} -3x - y + z = -2 \\ 3x + y - z = 5 \\ \hline 0 = 3 \end{array}$$

No solution

7. $\left[\begin{array}{cc|c} 1 & 2 & 6 \\ 2 & 6 & 8 \end{array}\right]$

$-2R_1 + R_2 \to R_2$

$$\left[\begin{array}{cc|c} 1 & 2 & 6 \\ 0 & 2 & -4 \end{array}\right]$$

$\frac{1}{2}R_2 \to R_2$
$R_1 - R_2 \to R_1$

$$\left[\begin{array}{cc|c} 1 & 0 & 10 \\ 0 & 1 & -2 \end{array}\right]$$

9. $\left[\begin{array}{ccc|c} 2 & 2 & 6 & 4 \\ 2 & 5 & 9 & -2 \\ 1 & 2 & 3 & -1 \end{array}\right]$

$\frac{1}{2}R_1 \to R_1$
$R_1 - R_2 \to R_2$
$-2R_3 + R_1 \to R_3$

$$\left[\begin{array}{ccc|c} 1 & 1 & 3 & 2 \\ 0 & -3 & -3 & 6 \\ 0 & -2 & 0 & 6 \end{array}\right]$$

$(-1/3)R_2 \to R_2$

$$\left[\begin{array}{ccc|c} 1 & 1 & 3 & 2 \\ 0 & 1 & 1 & -2 \\ 0 & -2 & 0 & 6 \end{array}\right]$$

$2R_2 + R_3 \to R_3$
$R_1 - R_2 \to R_1$

$$\left[\begin{array}{ccc|c} 1 & 0 & 2 & 4 \\ 0 & 1 & 1 & -2 \\ 0 & 0 & 2 & 2 \end{array}\right]$$

$\frac{1}{2}R_3 \to R_3$

$$\left[\begin{array}{ccc|c} 1 & 0 & 2 & 4 \\ 0 & 1 & 1 & -2 \\ 0 & 0 & 1 & 1 \end{array}\right]$$

$-2R_3 + R_1 \to R_1$
$R_2 - R_3 \to R_2$

$$\left[\begin{array}{ccc|c} 1 & 0 & 0 & 2 \\ 0 & 1 & 0 & -3 \\ 0 & 0 & 1 & 1 \end{array}\right]$$

11. $\left[\begin{array}{cc|c} 1 & -2 & 1 \\ 2 & 3 & 9 \end{array}\right]$

$-2R_1 + R_2 \to R_2$

$$\left[\begin{array}{cc|c} 1 & -2 & 1 \\ 0 & 7 & 7 \end{array}\right]$$

$$\begin{cases} x - 2y = 1 \\ \quad 7y = 7 \end{cases}$$

$7y = 7$
$y = 1$

$x - 2(1) = 1$
$x = 3$

(3, 1)

13. $\left[\begin{array}{ccc|c} 1 & 2 & -1 & 5 \\ 1 & -1 & 1 & 0 \\ 1 & 1 & 2 & 1 \end{array}\right]$

$R_1 - R_2 \rightarrow R_2$
$R_1 - R_3 \rightarrow R_3$

$\left[\begin{array}{ccc|c} 1 & 2 & -1 & 5 \\ 0 & 3 & -2 & 5 \\ 0 & 1 & -3 & 4 \end{array}\right]$

$-3R_3 + R_2 \rightarrow R_3$

$\left[\begin{array}{ccc|c} 1 & 2 & -1 & 5 \\ 0 & 3 & -2 & 5 \\ 0 & 0 & 7 & -7 \end{array}\right]$

$$\begin{cases} x + 2y - z = 5 \\ \quad 3y - 2z = 5 \\ \quad\quad 7z = -7 \end{cases}$$

$7z = -7$
$z = -1$

$3y - 2(-1) = 5$
$3y = 3$
$y = 1$

$x + 2(1) - (-1) = 5$
$x = 2$

(2, 1, −1)

15. $\left[\begin{array}{cc|c} 1 & -3 & -1 \\ 2 & 1 & 5 \end{array}\right]$

$-2R_1 + R_2 \rightarrow R_2$

$\left[\begin{array}{cc|c} 1 & -3 & -1 \\ 0 & 7 & 7 \end{array}\right]$

$(1/7)R_2 \rightarrow R_2$

$\left[\begin{array}{cc|c} 1 & -3 & -1 \\ 0 & 1 & 1 \end{array}\right]$

$3R_2 + R_1 \rightarrow R_1$

$\left[\begin{array}{cc|c} 1 & 0 & 2 \\ 0 & 1 & 1 \end{array}\right]$

$x = 2, \quad y = 1$
(2, 1)

17. $5C = 5\begin{bmatrix} 2 & -1 \\ 0 & 5 \end{bmatrix}$

$= \begin{bmatrix} 10 & -5 \\ 0 & 25 \end{bmatrix}$

19. $C - D$

$= \begin{bmatrix} 2 & -1 \\ 0 & 5 \end{bmatrix} - \begin{bmatrix} -3 & 0 \\ 2 & 1 \end{bmatrix}$

$= \begin{bmatrix} 2 - (-3) & -1 - 0 \\ 0 - 2 & 5 - 1 \end{bmatrix}$

$= \begin{bmatrix} 5 & -1 \\ -2 & 4 \end{bmatrix}$

21. Not defined

23. CE

$= \begin{bmatrix} 2 & -1 \\ 0 & 5 \end{bmatrix}\begin{bmatrix} 1 & 5 & -2 \\ -1 & 0 & 6 \end{bmatrix}$

$= \begin{bmatrix} (2)(1) + (-1)(-1) & (2)(5) + (-1)(0) & (2)(-2) + (-1)(6) \\ (0)(1) + (5)(-1) & (0)(5) + (5)(0) & (0)(-2) + (5)(6) \end{bmatrix}$

$= \begin{bmatrix} 3 & 10 & -10 \\ -5 & 0 & 30 \end{bmatrix}$

25. $C + D$

$$= \begin{bmatrix} 2 & -1 \\ 0 & 5 \end{bmatrix} + \begin{bmatrix} -3 & 0 \\ 2 & 1 \end{bmatrix}$$

$$= \begin{bmatrix} -1 & -1 \\ 2 & 6 \end{bmatrix}$$

$(C + D)E$

$$= \begin{bmatrix} -1 & -1 \\ 2 & 6 \end{bmatrix} \begin{bmatrix} 1 & 5 & -2 \\ -1 & 0 & 6 \end{bmatrix}$$

$$= \begin{bmatrix} 0 & -5 & -4 \\ -4 & 10 & 32 \end{bmatrix}$$

27. $\begin{bmatrix} 2 & 3 \\ 3 & \frac{1}{2} \end{bmatrix} \begin{bmatrix} x \\ y \end{bmatrix} = \begin{bmatrix} 14 \\ 1 \end{bmatrix}$

$AX = K$
$X = A^{-1}K$

$$= \begin{bmatrix} -\frac{1}{2} \\ 5 \end{bmatrix}$$

$x = -\frac{1}{2}, \quad y = 5$

29. $\begin{bmatrix} 2 & -3 & 4 \\ 5 & 4 & -3 \\ 3 & -2 & 1 \end{bmatrix} \begin{bmatrix} x \\ y \\ z \end{bmatrix} = \begin{bmatrix} 18 \\ -4 \\ 12 \end{bmatrix}$

$AX = K$
$X = A^{-1}K$

$$= \begin{bmatrix} 2 \\ -2 \\ 2 \end{bmatrix}$$

$x = 2, \quad y = -2, \quad z = 2$

31. $\begin{vmatrix} 2 & 3 \\ 4 & -1 \end{vmatrix} = (2)(-1) - (4)(3) = -14$

33. $\begin{vmatrix} 1 & 3 & -2 \\ 2 & 4 & 3 \\ 5 & -1 & -3 \end{vmatrix}$

$$= 1\begin{vmatrix} 4 & 3 \\ -1 & -3 \end{vmatrix} - 3\begin{vmatrix} 2 & 3 \\ 5 & -3 \end{vmatrix} + (-2)\begin{vmatrix} 2 & 4 \\ 5 & -1 \end{vmatrix}$$

$= [-12 - (-3)] - 3[-6 - 15] - 2[-2 - 20]$
$= -9 + 63 + 44$
$= 98$

35. $D = \begin{vmatrix} 3 & 7 \\ 5 & 4 \end{vmatrix} = -23$

$D_x = \begin{vmatrix} 4 & 7 \\ 2 & 4 \end{vmatrix} = 2$

$D_y = \begin{vmatrix} 3 & 4 \\ 5 & 2 \end{vmatrix} = -14$

$$x = \frac{D_x}{D} = \frac{2}{-23} = -\frac{2}{23}$$

$$y = \frac{D_y}{D} = \frac{-14}{-23} = \frac{14}{23}$$

37. $D = \begin{vmatrix} 4 & 2 \\ 6 & 3 \end{vmatrix} = 0$

No unique solution

39. $D = \begin{vmatrix} 3 & 4 & 1 \\ 5 & -3 & 6 \\ 4 & -5 & -5 \end{vmatrix} = 318$

$D_x = \begin{vmatrix} 3 & 4 & 1 \\ 2 & -3 & 6 \\ 1 & -5 & -5 \end{vmatrix} = 192$

$D_y = \begin{vmatrix} 3 & 3 & 1 \\ 5 & 2 & 6 \\ 4 & 1 & -5 \end{vmatrix} = 96$

$D_z = \begin{vmatrix} 3 & 4 & 3 \\ 5 & -3 & 2 \\ 4 & -5 & 1 \end{vmatrix} = -6$

$$x = \frac{D_x}{D} = \frac{192}{318} = \frac{32}{53}$$

$$y = \frac{D_y}{D} = \frac{96}{318} = \frac{16}{53}$$

$$z = \frac{D_z}{D} = \frac{-6}{318} = -\frac{1}{53}$$

41. $\begin{cases} x^2 + y^2 = 8 \\ x + y = 4 \end{cases}$

$$x = 4 - y$$

$$\begin{aligned} (4 - y)^2 + y^2 &= 8 \\ 16 - 8y + y^2 + y^2 &= 8 \\ 2y^2 - 8y + 8 &= 0 \\ y^2 - 4y + 4 &= 0 \\ (y - 2)^2 &= 0 \\ y - 2 &= 0 \\ y &= 2 \\ x &= 4 - 2 = 2 \end{aligned}$$

(2, 2)

43. $\begin{cases} 2x^2 - y^2 = 14 \\ 2x - 3y = -8 \end{cases}$

$$\begin{aligned} 2x - 3y &= -8 \\ 2x &= 3y - 8 \\ x &= \frac{3}{2}y - 4 \end{aligned}$$

$$\begin{aligned} 2\left(\frac{3}{2}y - 4\right)^2 - y^2 &= 14 \\ 2\left(\frac{9}{4}y^2 - 12y + 16\right) - y^2 &= 14 \\ \frac{9}{2}y^2 - 24y + 32 - y^2 &= 14 \end{aligned}$$

$$\begin{aligned} 9y^2 - 48y + 64 - 2y^2 &= 28 \\ 7y^2 - 48y + 36 &= 0 \\ (y - 6)(7y - 6) &= 0 \end{aligned}$$

$y - 6 = 0$ or $7y - 6 = 0$

$y = 6$ $\qquad y = \frac{6}{7}$

$x = \frac{3}{2}(6) - 4 = 5$ $\qquad x = \frac{3}{2}\left(\frac{6}{7}\right) - 4 = -\frac{19}{7}$

$(5, 6)$, $\left(-\frac{19}{7}, \frac{6}{7}\right)$

45. $\begin{aligned} x^2 + y^2 &= 4 \\ 2x^2 + 3y^2 &= 18 \end{aligned}$

$$\begin{aligned} -2x^2 - 2y^2 &= -8 \\ 2x^2 + 3y^2 &= 18 \\ \hline y^2 &= 10 \\ y &= \pm\sqrt{10} \end{aligned}$$

$y = 10$:

$$\begin{aligned} x^2 + (\sqrt{10})^2 &= 4 \\ x^2 + 10 &= 4 \\ x^2 &= -6 \end{aligned}$$

No real solution

47. $\begin{cases} x + y \le 2 & \text{(solid line)} \\ 2x - 3y \le 6 & \text{(solid line)} \end{cases}$

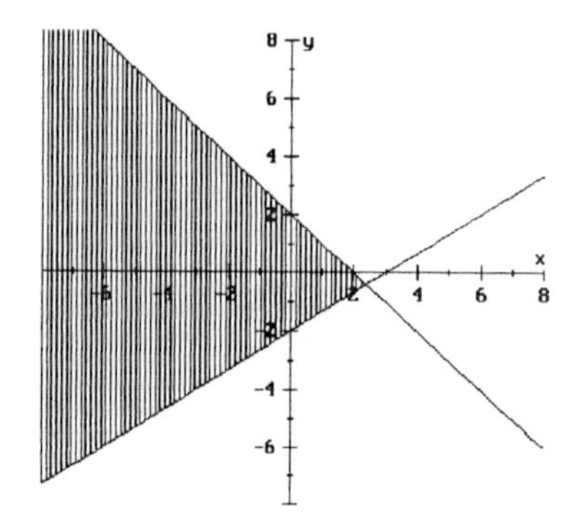

Test point: (0, 0)

$0 + 0 \le 2$ $\qquad 2(0) - 3(0) \le 6$

$0 \le 2$ $\qquad 0 \le 6$

True $\qquad$ True

49. $\begin{cases} 2x + 3y \le 12 & \text{(solid line)} \\ y < x & \text{(dashed line)} \\ x \ge 0 & \text{(solid line)} \\ y \ge 0 & \text{(solid line)} \end{cases}$

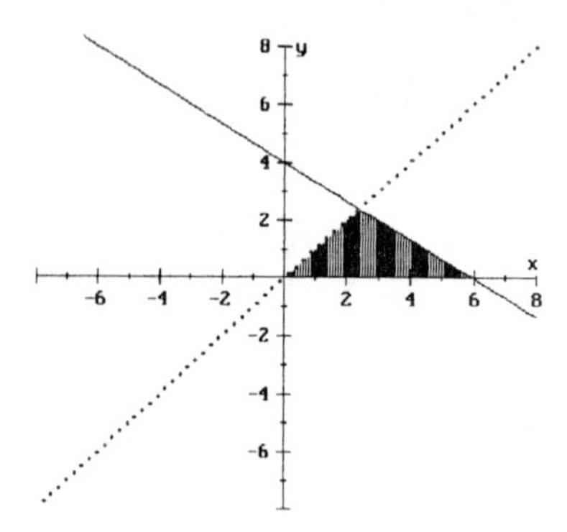

Test point: (2, 1)

$2(2) + 3(1) \le 12$ $\quad 1 < 2$

$7 \le 12$ $\quad$ True

True

$2 \ge 0$ $\quad 1 \ge 0$

True $\quad$ True

51. x = amount at 7.7%
y = amount at 8.6%
z = amount at 9.8%

$$\begin{cases} x + y + z = 20000 \\ y + z = 3000 + x \\ 0.077x + 0.086y + 0.098z = 1719.10 \end{cases}$$

$$\begin{cases} x + y + z = 20000 \\ -x + y + z = 3000 \\ 77x + 86y + 98z = 1719100 \end{cases}$$

$$\begin{bmatrix} 1 & 1 & 1 \\ -1 & 1 & 1 \\ 77 & 86 & 98 \end{bmatrix} \begin{bmatrix} x \\ y \\ z \end{bmatrix} = \begin{bmatrix} 20000 \\ 3000 \\ 1719100 \end{bmatrix}$$

$$AX = K$$
$$X = A^{-1}K$$
$$= \begin{bmatrix} 8500 \\ 5200 \\ 6300 \end{bmatrix}$$

\$8500 is invested at 7.7%, \$5200 at 8.6% and \$6300 at 9.8%.

53. x = width
y = length

$$\begin{cases} 2x + 2y = 41 \\ xy = 100 \end{cases}$$

$$x = \frac{100}{y}$$

$$2\left(\frac{100}{y}\right) + 2y = 41$$

$$\frac{200}{y} + 2y = 41$$

$$200 + 2y^2 = 41y$$
$$2y^2 - 41y + 200 = 0$$
$$(2y - 25)(y - 8) = 0$$

$2y - 25 = 0$ $\quad$ or $\quad y - 8 = 0$

$y = \frac{25}{2} = 12.5$ $\quad y = 8$

$x = \frac{100}{12.5} = 8$ $\quad x = \frac{100}{8} = 1$

The dimensions are 8 ft by 12.5 ft.

CHAPTER 11 PRACTICE TEST

1. $$\begin{cases} 2x - 3y + 4z = 2 \\ 3x + 2y - z = 10 \\ 2x - 4y + 3z = 3 \end{cases}$$

$$\begin{array}{r} 4x - 6y + 8z = 4 \\ 9x + 6y - 3z = 30 \\ \hline 13x \quad + 5z = 34 \end{array}$$

$$\begin{array}{r} 6x + 4y - 2z = 20 \\ 2x - 4y + 3z = 3 \\ \hline 8x \quad + z = 23 \end{array}$$

$$13x + 5z = 34$$
$$8x + z = 23$$

$$\begin{array}{r} 13x + 5z = 34 \\ -40x - 5z = -115 \\ \hline -27x = -81 \\ x = 3 \end{array}$$

$$8(3) + z = 23$$
$$z = -1$$

$$2(3) - 3y + 4(-1) = 2$$
$$-3y + 2 = 2$$
$$-3y = 0$$
$$y = 0$$

$x = 3,\ y = 0,\ z = -1$

3. (a) $$3A = \begin{bmatrix} 3 & 15 & -6 \\ -3 & 0 & 18 \\ 6 & 6 & 0 \end{bmatrix}$$

$$4B = \begin{bmatrix} 8 & -12 & 4 \\ 0 & 4 & -4 \\ 16 & 0 & 4 \end{bmatrix}$$

$3A - 4B$

$$= \begin{bmatrix} 3 & 15 & -6 \\ -3 & 0 & 18 \\ 6 & 6 & 0 \end{bmatrix} - \begin{bmatrix} 8 & -12 & 4 \\ 0 & 4 & -4 \\ 16 & 0 & 4 \end{bmatrix}$$

$$= \begin{bmatrix} -5 & 27 & -10 \\ -3 & -4 & 22 \\ -10 & 6 & -4 \end{bmatrix}$$

(b) AB

$$= \begin{bmatrix} 1 & 5 & -2 \\ -1 & 0 & 6 \\ 2 & 2 & 0 \end{bmatrix} \begin{bmatrix} 2 & -3 & 1 \\ 0 & 1 & -1 \\ 4 & 0 & 1 \end{bmatrix}$$

$$= \begin{bmatrix} (1)(2) + (5)(0) + (-2)(4) & (1)(-3) + (5)(1) + (-2)(0) & (1)(1) + (5)(-1) + (-2)(1) \\ (-1)(2) + (0)(0) + (6)(4) & (-1)(-3) + (0)(1) + (6)(0) & (-1)(1) + (0)(-1) + (6)(1) \\ (2)(2) + (2)(0) + (0)(4) & (2)(-3) + (2)(1) + (0)(0) & (2)(1) + (2)(-1) + (0)(1) \end{bmatrix}$$

$$= \begin{bmatrix} -6 & 2 & -6 \\ 22 & 3 & 5 \\ 4 & -4 & 0 \end{bmatrix}$$

5. (a) $\begin{vmatrix} 2 & 3 \\ -1 & 4 \end{vmatrix} = (2)(4) - (-1)(3) = 11$

(b) $\begin{vmatrix} 5 & 0 & 2 \\ 2 & 3 & 1 \\ 1 & 1 & 2 \end{vmatrix}$

$$=0\begin{vmatrix} 2 & 1 \\ 1 & 2 \end{vmatrix} + 3\begin{vmatrix} 5 & 2 \\ 1 & 2 \end{vmatrix} - 1\begin{vmatrix} 5 & 2 \\ 2 & 1 \end{vmatrix}$$

$= 0 + 3(10 - 2) - 1(5 - 4)$
$= 23$

7. x = number of \$5-bills
y = number of \$10-bills
z = number of \$20-bills

$$\begin{cases} 5x + 10y + 20z = 500 \\ y = 2z \\ x + y + z = 40 \end{cases}$$

$5x + 10(2z) + 20z = 500$
$x + 2z + z = 40$

$5x + 40z = 500$
$\underline{-5x - 15z = -200}$
$25z = 300$
$z = 12$

$x + 3(12) = 40$
$x = 4$
$y = 2(12) = 24$
There are 4 \$5-bills, 24 \$10-bills and 12 \$20-bills.

9. $\begin{cases} y < 4 & \text{(dashed line)} \\ x \le 2 & \text{(solid line)} \\ x + y > 3 & \text{(dashed line)} \end{cases}$

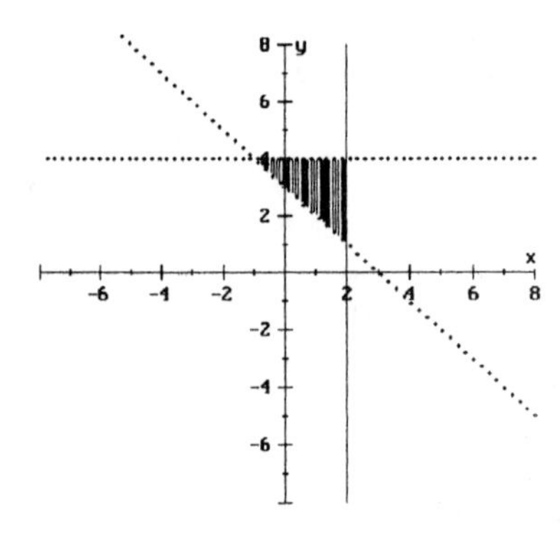

Test point: (1, 3)
$3 < 4$
True
$1 \le 2$
True
$1 + 3 > 3$
$4 > 3$ True

CHAPTERS 10 - 11 CUMULATIVE REVIEW

1. $\begin{cases} x + y + z = 6 \\ 2x + y - 2z = 6 \\ 3x - y + 3z = 10 \end{cases}$

$$\begin{array}{r} x + y + z = 6 \\ \underline{3x - y + 3z = 10} \\ 4x \quad + 4z = 16 \\ x \quad + z = 4 \end{array}$$

$$\begin{array}{r} 2x + y - 2z = 6 \\ \underline{3x - y + 3z = 10} \\ 5x \quad + z = 16 \end{array}$$

$$\begin{array}{r} x + z = 4 \\ 5x + z = 16 \end{array}$$

$$\begin{array}{r} -x - z = -4 \\ \underline{5x + z = 16} \\ 4x \quad = 12 \\ x = 3 \end{array}$$

$$\begin{array}{r} 3 + z = 4 \\ z = 1 \end{array}$$

$$\begin{array}{r} 3 + y + 1 = 6 \\ y = 2 \end{array}$$

$x = 3, \quad y = 2, \quad z = 1$

3. $\begin{cases} 3x - 4y + 5z = 1 \\ 2x - y + 3z = 2 \\ x - 2y + z = 3 \end{cases}$

$$\begin{array}{r} 3x - 4y + 5z = 1 \\ \underline{-8x + 4y - 12z = -8} \\ -5x \quad - 7z = -7 \end{array}$$

$$\begin{array}{r} -4x + 2y - 6z = -4 \\ \underline{x - 2y + z = 3} \\ -3x \quad - 5z = -1 \end{array}$$

$$\begin{array}{r} -5x - 7z = -7 \\ -3x - 5z = -1 \end{array}$$

$$\begin{array}{r} 15x + 21z = 21 \\ \underline{-15x - 25z = -5} \\ -4z = 16 \\ z = -4 \end{array}$$

$$\begin{array}{r} -3x - 5(-4) = -1 \\ -3x = -21 \\ x = 7 \end{array}$$

$$\begin{array}{r} 7 - 2y + (-4) = 3 \\ -2y = 0 \\ y = 0 \end{array}$$

$x = 7, \quad y = 0, \quad z = -4$

5. $\begin{cases} x - 2y + 3z = 4 \\ \frac{3}{2}x - 3y + \frac{9}{2}z = 6 \\ -3x + 6y - 9z = -12 \end{cases}$

$\begin{cases} x - 2y + 3z = 4 \\ 3x - 6y + 9z = 12 \\ -3x + 6y - 9z = -12 \end{cases}$

$$\begin{array}{r} 3x - 6y + 9z = 12 \\ \underline{-3x + 6y - 9z = -12} \\ 0 = 0 \end{array}$$

$\{(x, y, z) \mid x - 2y + 3z = 4\}$

7. $\begin{vmatrix} 3 & 2 \\ 4 & 5 \end{vmatrix} = (3)(5) - (4)(2) = 7$

9. $\begin{vmatrix} 1 & 2 \\ 2 & 4 \end{vmatrix} = (1)(4) - (2)(2) = 0$

11. $\begin{vmatrix} 4 & -1 & 2 \\ 2 & 1 & 0 \\ -1 & 2 & -3 \end{vmatrix}$

$= -2\begin{vmatrix} -1 & 2 \\ 2 & -3 \end{vmatrix} + 1\begin{vmatrix} 4 & 2 \\ -1 & -3 \end{vmatrix} - 0\begin{vmatrix} 4 & -1 \\ -1 & 2 \end{vmatrix}$

$= -2(3 - 4) + 1(-12 + 2) - 0$
$= -8$

13. $\begin{vmatrix} 5 & 4 & 3 \\ 2 & 0 & 0 \\ 3 & 1 & 1 \end{vmatrix}$

$= -2\begin{vmatrix} 4 & 3 \\ 1 & 1 \end{vmatrix} + 0\begin{vmatrix} 5 & 3 \\ 3 & 1 \end{vmatrix} - 0\begin{vmatrix} 5 & 4 \\ 3 & 1 \end{vmatrix}$

$= -2(4 - 3)$
$= -2$

15. $\begin{bmatrix} 1 & 2 \\ 2 & -3 \end{bmatrix} \begin{bmatrix} x \\ y \end{bmatrix} = \begin{bmatrix} 0 \\ 7 \end{bmatrix}$

$$AX = B$$
$$X = A^{-1}B$$
$$= \begin{bmatrix} 2 \\ -1 \end{bmatrix}$$

(2, -1)

17. $D = \begin{vmatrix} 3 & 7 \\ 10 & 5 \end{vmatrix} = -55$

$D_x = \begin{vmatrix} 2 & 7 \\ 11 & 5 \end{vmatrix} = -67$

$D_y = \begin{vmatrix} 3 & 2 \\ 10 & 11 \end{vmatrix} = 13$

$x = \frac{D_x}{D} = \frac{-67}{-55} = \frac{67}{55}$

$y = \frac{D_y}{D} = \frac{13}{-55} = -\frac{13}{55}$

19. $D = \begin{vmatrix} 2 & 3 & 4 \\ 5 & 2 & -3 \\ 3 & -7 & 5 \end{vmatrix} = -288$

$D_x = \begin{vmatrix} 3 & 3 & 4 \\ 2 & 2 & -3 \\ -7 & -7 & 5 \end{vmatrix} = 0$

$D_y = \begin{vmatrix} 2 & 3 & 4 \\ 5 & 2 & -3 \\ 3 & -7 & 5 \end{vmatrix} = -288$

$D_z = \begin{vmatrix} 2 & 3 & 3 \\ 5 & 2 & 2 \\ 3 & -7 & -7 \end{vmatrix} = 0$

$x = \frac{D_x}{D} = \frac{0}{-288} = 0$

$y = \frac{D_y}{D} = \frac{=288}{-288} = 1$

$z = \frac{D_z}{D} = \frac{0}{-288} = 0$

21. $\begin{cases} x + y = 1 \\ x^2 + y^2 = 5 \end{cases}$

$$x = 1 - y$$
$$(1 - y)^2 + y^2 = 5$$
$$1 - 2y + y^2 + y^2 = 5$$
$$2y^2 - 2y - 4 = 0$$
$$y^2 - y - 2 = 0$$
$$(y - 2)(y + 1) = 0$$

$y - 2 = 0$ or $y + 1 = 0$

$y = 2$ $\quad y = -1$

$x = 1 - 2 = -1$ $\quad x = 1 - (-1) = 2$

(-1, 2), (2, -1)

23. $\begin{cases} y - x^2 = 4 \\ x^2 + y = 1 \end{cases}$

$$-x^2 + y = 4$$
$$\underline{x^2 + y = 1}$$
$$2y = 5$$
$$y = \frac{5}{2}$$

$$x^2 + \frac{5}{2} = 1$$

$$x^2 = -\frac{3}{2}$$

No real solution

25. $\begin{cases} x^2 + y^2 = 10 \\ 3x^2 - 4y^2 = 23 \end{cases}$

$$4x^2 + 4y^2 = 40$$
$$\underline{3x^2 - 4y^2 = 23}$$
$$7x^2 = 63$$
$$x^2 = 9$$
$$x = \pm 3$$

$x = 3$:

$$3^2 + y^2 = 10$$
$$y^2 = 1$$
$$y = \pm 1$$

$x = -3$:

$$(-3)^2 + y^2 = 10$$
$$y^2 = 1$$
$$y = \pm 1$$

(3, 1), (3, -1), (-3, 1), (-3, -1)

27. $\begin{cases} x - y^2 = 3 \\ 3x - 2y = 9 \end{cases}$

$$x = 3 + y^2$$
$$3(3 + y^2) - 2y = 9$$
$$9 + 3y^2 - 2y = 9$$
$$3y^2 - 2y = 0$$
$$y(3y - 2) = 0$$

$y = 0$ or $3y - 2 = 0$

$x = 3 + 0^2 = 3$ $\quad y = \frac{2}{3}$

$$x = 3 + \left(\frac{2}{3}\right)^2 = \frac{31}{9}$$

$(3, 0)$, $\left(\frac{31}{9}, \frac{2}{3}\right)$

29. $\begin{cases} x + y \le 6 & \text{(solid line)} \\ 2x + y \ge 4 & \text{(solid line)} \end{cases}$

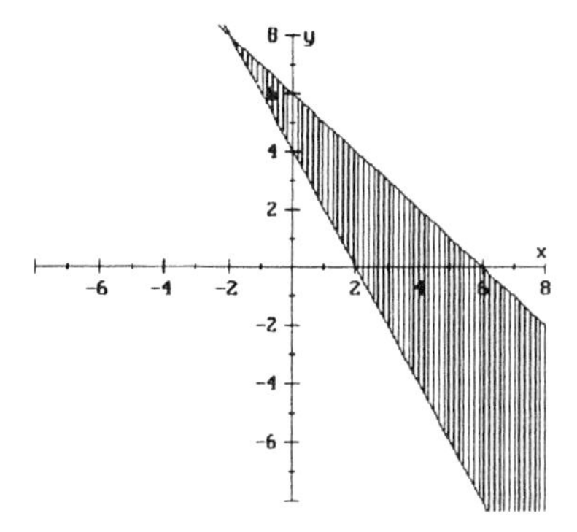

Test point: (3, 1)

$3 + 1 \le 6$ $\quad 2(3) + 1 \ge 4$

$4 \le 6$ $\quad 7 \ge 4$

True $\quad$ True

31. $\begin{cases} 2x + 3y < 12 & \text{(dashed line)} \\ x < y & \text{(dashed line)} \end{cases}$

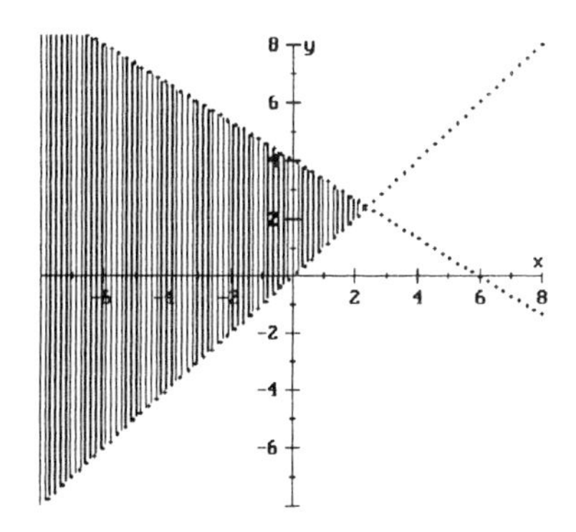

Test point: (0, 1)

$2(0) + 3(1) < 12$ $\quad 0 < 1$

$3 < 12$ $\quad$ True

True

33. $\begin{cases} x + y \le 4 & \text{(solid line)} \\ x - y \le 4 & \text{(solid line)} \\ x \ge 0 & \text{(solid line)} \end{cases}$

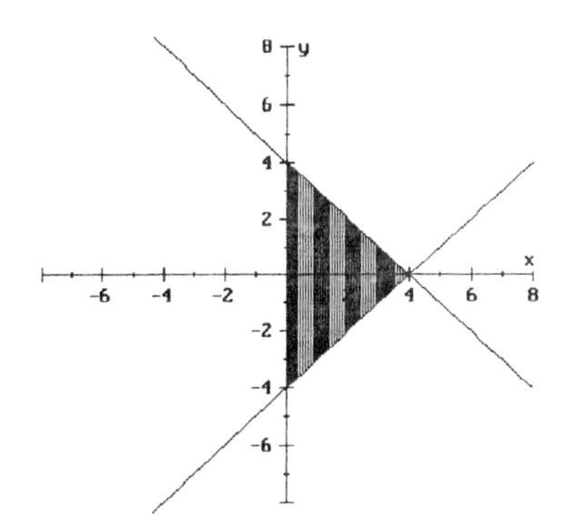

Test point: (1, 0)

$1 + 0 \le 4$

$1 \le 4$

True

$1 - 0 \le 4$

$1 \le 4$

True

$1 \ge 0$

True

35. $-3C = -3\begin{bmatrix} -1 & 2 \\ 5 & 0 \end{bmatrix}$

$$= \begin{bmatrix} 3 & -6 \\ -15 & 0 \end{bmatrix}$$

37. $3C + 5D$

$$= 3\begin{bmatrix} -1 & 2 \\ 5 & 0 \end{bmatrix} + 5\begin{bmatrix} -4 & 1 \\ 0 & 3 \end{bmatrix}$$

$$= \begin{bmatrix} -3 & 6 \\ 15 & 0 \end{bmatrix} + \begin{bmatrix} -20 & 5 \\ 0 & 15 \end{bmatrix}$$

$$= \begin{bmatrix} -23 & 11 \\ 15 & 15 \end{bmatrix}$$

39. $CD = \begin{bmatrix} -1 & 2 \\ 5 & 0 \end{bmatrix} \begin{bmatrix} -4 & 1 \\ 0 & 3 \end{bmatrix}$

$= \begin{bmatrix} (-1)(-4) + (2)(0) & (-1)(1) + (2)(3) \\ (5)(-4) + (0)(0) & (5)(1) + (0)(3) \end{bmatrix}$

$= \begin{bmatrix} 4 & 5 \\ -20 & 5 \end{bmatrix}$

41. $\begin{bmatrix} 3 & 2 \\ \frac{1}{2} & 3 \end{bmatrix} \begin{bmatrix} x \\ y \end{bmatrix} = \begin{bmatrix} 10 \\ 1 \end{bmatrix}$

$AX = K$
$X = A^{-1}K$

$= \begin{bmatrix} 3.5 \\ -0.25 \end{bmatrix}$

$x = 3.5, \quad y = -0.25$

43. $f(x) = 3^{x+1}$

x	y
-3	1/9
-2	1/3
-1	1
0	3
1	9

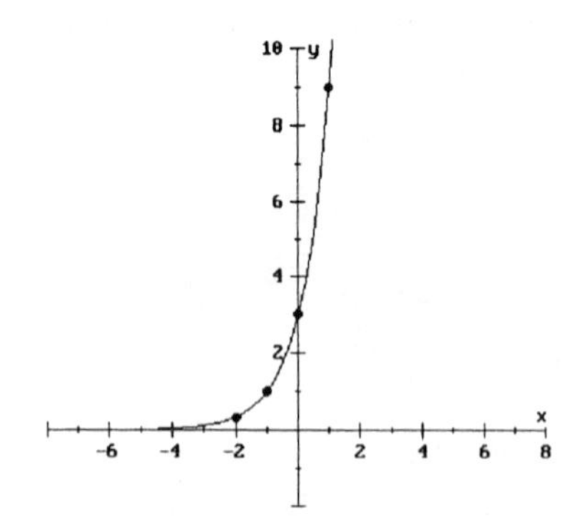

45. $f(x) = \log_5 x$
$y = \log_5 x$
$5^y = x$

x	y
1/25	-2
1/5	-1
1	0
5	1
25	2

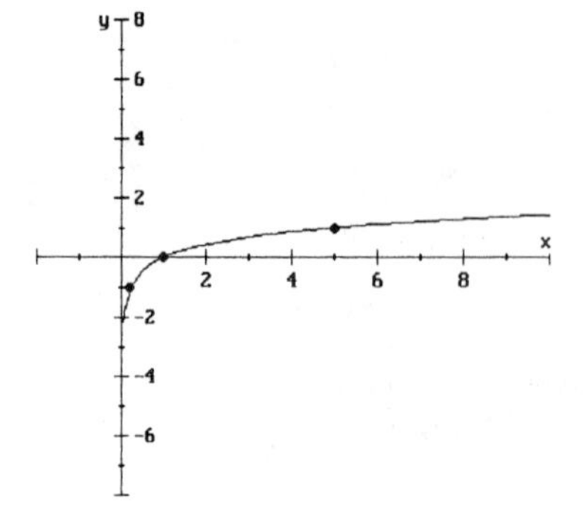

47. $\log_2 64 = 6$
$2^6 = 64$

49. $\sqrt[3]{125} = 5$
$125^{1/3} = 5$
$\log_{125} 5 = \frac{1}{3}$

51. $\log_{27} 81 = \frac{4}{3}$
$27^{4/3} = 81$

53. $\log_3 81 = 4$

55. $\log_4 \frac{1}{16} = x$
$4^x = \frac{1}{16}$
$4^x = 4^{-2}$
$x = -2$

57. $\log_9 \frac{1}{3} = x$
$9^x = \frac{1}{3}$
$(3^2)^x = 3^{-1}$
$3^{2x} = 3^{-1}$
$2x = -1$
$x = -\frac{1}{2}$

59. $\log_b 1 = x$

$$b^x = 1$$
$$b^x = b^0$$
$$x = 0$$

61. $\log_b \sqrt[3]{5xy}$

$$= \log_b (5xy)^{1/3}$$
$$= \frac{1}{3} \log_b (5xy)$$
$$= \frac{1}{3} (\log_b 5 + \log_b x + \log_b y)$$
$$= \frac{1}{3} \log_b 5 + \frac{1}{3} \log_b x + \frac{1}{3} \log_b y$$

63. $\log_3 \frac{x^2\sqrt{y}}{9wz}$

$$= \log_3 \frac{x^2 y^{1/2}}{9wz}$$
$$= \log_3 (x^2 y^{1/2}) - \log_3 (9wz)$$
$$= \log_3 x^2 + \log_3 y^{1/2} - (\log_3 9 + \log_3 w + \log_3 z)$$
$$= 2 \log_3 x + \frac{1}{2} \log_3 y - (2 + \log_3 w + \log_3 z)$$
$$= 2 \log_3 x + \frac{1}{2} \log_3 y - 2 - \log_3 w - \log_3 z$$

65. $\log 73{,}600 \approx 4.8669$

67. $\log x = 0.6085$

$$10^{0.6085} = x$$
$$x \approx 4.0598$$

69. $\log_9 384 = \frac{\log 384}{\log 9}$

$$\approx 2.7083$$

71. $\frac{1}{2} \log_8 x = \log_8 5$

$$\log_8 x^{1/2} = \log_8 5$$
$$x^{1/2} = 5$$
$$x = 5^2$$
$$x = 25$$

73. $5^x = \frac{1}{25}$

$$5^x = 5^{-2}$$
$$x = -2$$

75. $\log_6 x + \log_6 4 = 3$

$$\log_6 4x = 3$$
$$6^3 = 4x$$
$$216 = 4x$$
$$54 = x$$

77. $\frac{4^{x^2}}{2^x} = 64$

$$\frac{(2^2)^{x^2}}{2^x} = 2^6$$
$$\frac{2^{2x^2}}{2^x} = 2^6$$
$$2^{2x^2 - x} = 2^6$$
$$2x^2 - x = 6$$
$$2x^2 - x - 6 = 0$$
$$(2x + 3)(x - 2) = 0$$
$$2x + 3 = 0 \quad \text{or } x - 2 = 0$$
$$x = -\frac{3}{2} \quad \text{or} \quad x = 2$$

79. $9^x = 7^{x+3}$

$$\log 9^x = \log 7^{x+3}$$
$$x \log 9 = (x + 3) \log 7$$
$$x \log 9 = x \log 7 + 3 \log 7$$
$$x \log 9 - x \log 7 = 3 \log 7$$
$$x(\log 9 - \log 7) = 3 \log 7$$
$$x = \frac{3 \log 7}{\log 9 - \log 7}$$
$$x \approx 23.2288$$

81. $A = P\left(1 + \frac{r}{n}\right)^{nt}$

$$5000 = 3000\left(1 + \frac{0.08}{4}\right)^{4t}$$
$$\frac{5}{3} = 1.02^{4t}$$

$$\ln \frac{5}{3} = \ln 1.02^{4t}$$

$$\ln \frac{5}{3} = 4t \ln 1.02$$

$$\frac{\ln \frac{5}{3}}{4 \ln 1.02} = t$$

$$t \approx 6.449$$

It will take 6.449 years.

83.
$$\begin{aligned} A &= A_0 e^{rt} \\ 2 &= 20e^{-0.002t} \\ 0.1 &= e^{-0.002t} \\ \ln 0.1 &= \ln e^{-0.002t} \\ \ln 0.1 &= -0.002t \end{aligned}$$

$$\frac{\ln 0.1}{-0.002} = t$$

$$t \approx 1151$$

85.
$$\begin{aligned} pH &= -\log [H_3O^+] \\ 8.2 &= -\log [H_3O^+] \\ -8.2 &= \log [H_3O^+] \\ 10^{-8.2} &= [H_3O^+] \\ 6.31 \times 10^{-9} &= [H_3O^+] \end{aligned}$$

CHAPTERS 10 - 11 CUMULATIVE PRACTICE TEST

1. $\begin{cases} x + y + z = 6 \\ 3x + 2y - z = 11 \\ 2x - 4y - z = 12 \end{cases}$

$$\begin{aligned} x + y + z &= 6 \\ 3x + 2y - z &= 11 \\ \hline 4x + 3y &= 17 \end{aligned}$$

$$\begin{aligned} x + y + z &= 6 \\ 2x - 4y - z &= 12 \\ \hline 3x - 3y &= 18 \end{aligned}$$

$$\begin{aligned} 4x + 3y &= 17 \\ 3x - 3y &= 18 \\ \hline 7x &= 35 \\ x &= 5 \end{aligned}$$

$$\begin{aligned} 4(5) + 3y &= 17 \\ 3y &= -3 \\ y &= -1 \end{aligned}$$

$$\begin{aligned} 5 - 1 + z &= 6 \\ z &= 2 \end{aligned}$$

$x = 5, \quad y = -1, \quad z = 2$

3. (a) $D = \begin{vmatrix} 6 & 5 \\ 7 & 8 \end{vmatrix} = 13$

$$D_x = \begin{vmatrix} 13 & 5 \\ 26 & 8 \end{vmatrix} = -26$$

$$D_y = \begin{vmatrix} 6 & 13 \\ 7 & 26 \end{vmatrix} = 65$$

$$x = \frac{D_x}{D} = \frac{-26}{13} = -2$$

$$y = \frac{D_y}{D} = \frac{65}{13} = 5$$

(b) $D = \begin{vmatrix} 4 & -5 & 2 \\ 3 & 7 & -5 \\ 5 & -6 & 3 \end{vmatrix} = 28$

$$D_x = \begin{vmatrix} 17 & -5 & 2 \\ 2 & 7 & -5 \\ 21 & -6 & 3 \end{vmatrix} = 84$$

$$D_y = \begin{vmatrix} 4 & 17 & 2 \\ 3 & 2 & -5 \\ 5 & 21 & 3 \end{vmatrix} = -28$$

$$D_z = \begin{vmatrix} 4 & -5 & 17 \\ 3 & 7 & 2 \\ 5 & -6 & 21 \end{vmatrix} = 0$$

$$x = \frac{D_x}{D} = \frac{84}{28} = 3$$

$$y = \frac{D_y}{D} = \frac{-28}{28} = -1$$

$$z = \frac{D_z}{D} = \frac{0}{28} = 0$$

5. $\begin{cases} x - y \le 5 & \text{(solid line)} \\ x - 2y \le 0 & \text{(solid line)} \\ y \ge 0 & \text{(solid line)} \end{cases}$

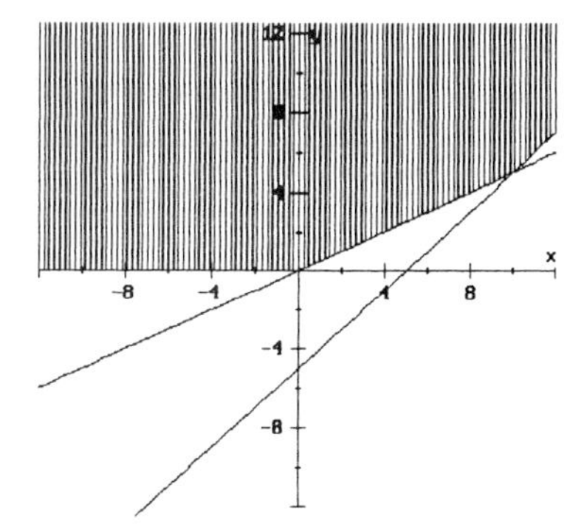

Test point: (0, 1)

$0 - 1 \le 5$

$-1 \le 5$

True

$0 - 2(1) \le 0$

$-2 \le 0$

True

$1 \ge 0$

True

7. $\left[\begin{array}{cc|cc} 2 & 4 & 1 & 0 \\ 3 & 2 & 0 & 1 \end{array}\right]$

$\frac{1}{2}R_1 \to R_1$

$\left[\begin{array}{cc|cc} 1 & 2 & \frac{1}{2} & 0 \\ 3 & 2 & 0 & 1 \end{array}\right]$

$-3R_1 + R_2 \to R_2$

$\left[\begin{array}{cc|cc} 1 & 2 & \frac{1}{2} & 0 \\ 0 & -4 & -\frac{3}{2} & 1 \end{array}\right]$

$-\frac{1}{4}R_2 \to R_2$

$\left[\begin{array}{cc|cc} 1 & 2 & \frac{1}{2} & 0 \\ 0 & 1 & \frac{3}{8} & -\frac{1}{4} \end{array}\right]$

$-2R_2 + R_1 \to R_1$

$\left[\begin{array}{cc|cc} 1 & 0 & -\frac{1}{4} & \frac{1}{2} \\ 0 & 1 & \frac{3}{8} & -\frac{1}{4} \end{array}\right]$

$$\begin{bmatrix} x \\ y \end{bmatrix} = \begin{bmatrix} -\frac{1}{4} & \frac{1}{2} \\ \frac{3}{8} & -\frac{1}{4} \end{bmatrix} \begin{bmatrix} -4 \\ -8 \end{bmatrix}$$

$$= \begin{bmatrix} -3 \\ \frac{1}{2} \end{bmatrix}$$

$x = -3, \ y = \frac{1}{2}$

9. $\log_8 \frac{1}{4} = -\frac{2}{3}$

$8^{-2/3} = \frac{1}{4}$

11. (a) $\log_b x\sqrt[3]{y}$

$= \log_b xy^{1/3}$

$= \log_b x + \log_b y^{1/3}$

$= \log_b x + \frac{1}{3}\log_b y$

(b) $\log_b \frac{x^3}{\sqrt{xy}}$

$$= \log_b \frac{x^3}{(xy)^{1/2}}$$

$$= \log_b x^3 - \log_b (xy)^{1/2}$$

$$= 3 \log_b x - \frac{1}{2} \log_b (xy)$$

$$= 3 \log_b x - \frac{1}{2}(\log_b x + \log_b y)$$

$$= 3 \log_b x - \frac{1}{2} \log_b x - \frac{1}{2} \log_b y$$

$$= \frac{5}{2} \log_b x - \frac{1}{2} \log_b y$$

13. $\log_8 985 = \frac{\log 985}{\log 8}$

$$\approx 3.3147$$

15.

$$A = P\left(1 + \frac{r}{n}\right)^{nt}$$

$$10000 = 5000\left(1 + \frac{0.10}{4}\right)^{4t}$$

$$2 = 1.025^{4t}$$

$$\log 2 = \log 1.025^{4t}$$

$$\log 2 = 4t \log 1.025$$

$$\frac{\log 2}{4 \log 1.025} = t$$

$$t \approx 7.018 \text{ yr}$$

APPENDIX A EXERCISES

1. True

3. False; $0 \not\subset W$

5. True

7. True

9. True

11. True

13. $\{1, 2, 3, 4, 5, 6, 7, 8, 9, 10, 11\}$

15. $\{7, 8, 9, 10, 11, 12\}$

17. $\{3, 4, 5, 6, 7, 8, 9, 10, 11, 12, 13, 14, 15, 16, 17\}$

19. $\emptyset$

21. $\{23, 29, 31, 37\}$

23. $\emptyset$

25. $\{0, 12, 24, 36, 48, 60, 72, \ldots\}$

27. $\{0, 5, 10, 15, 20, 25, 30, \ldots\}$

29. $\{1, 2, 3, 4, 6, 8, 12, 16, 24, 48\}$

31. $A = \{1, 2, 4, 5, 8\}$
    $B = \{4, 8, 12, 16, 20, 24, 28, 32\}$

    $A \cap B = \{4, 8\}$

33. $A \cup B = \{1, 2, 4, 5, 8, 12, 16, 20, 24, 28, 32\}$
    (see #31)

35. $D = \{6, 7, 8, 9\}$

    $D \cap C = \{7\}$

37. $A = \{1, 2, 4, 5, 8\}$

    $D = \{6, 7, 8, 9\}$

    $A \cap D = \{8\}$

APPENDIX B EXERCISES

1.
$$\begin{aligned} x^2 + y^2 - 4x + 12y - 18 &= 0 \\ (x^2 - 4x) + (y^2 + 12y) &= 18 \\ (x^2 - 4x + 4) + (y^2 + 12y + 36) &= 18 + 4 + 36 \\ (x - 2)^2 + (y + 6)^2 &= 58 \end{aligned}$$

$C = (2, -6)$
$r = \sqrt{58}$

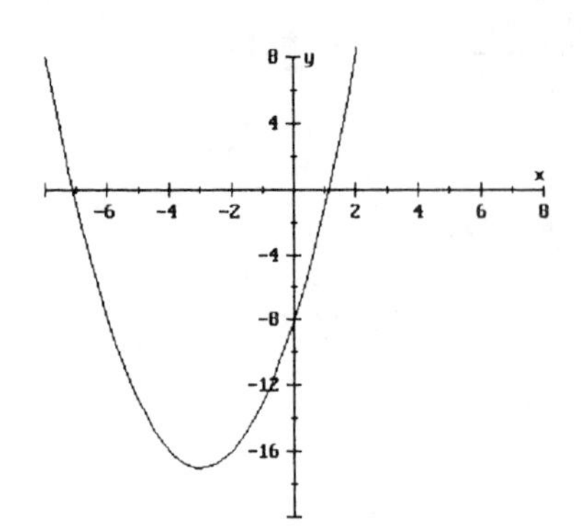

3.
$$\begin{aligned} -x^2 - y^2 - 8x + 12y + 15 &= 0 \\ x^2 + y^2 + 8x - 12y - 15 &= 0 \\ (x^2 + 8x) + (y^2 - 12y) &= 15 \\ (x^2 + 8x + 16) + (y^2 - 12y + 36) &= 15 + 16 + 36 \\ (x + 4)^2 + (y - 6)^2 &= 67 \end{aligned}$$

$C = (-4, 6)$
$r = \sqrt{67}$

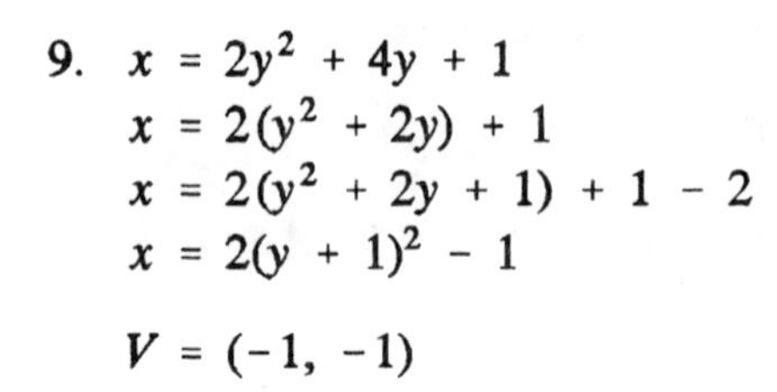

9.
$$\begin{aligned} x &= 2y^2 + 4y + 1 \\ x &= 2(y^2 + 2y) + 1 \\ x &= 2(y^2 + 2y + 1) + 1 - 2 \\ x &= 2(y + 1)^2 - 1 \end{aligned}$$

$V = (-1, -1)$

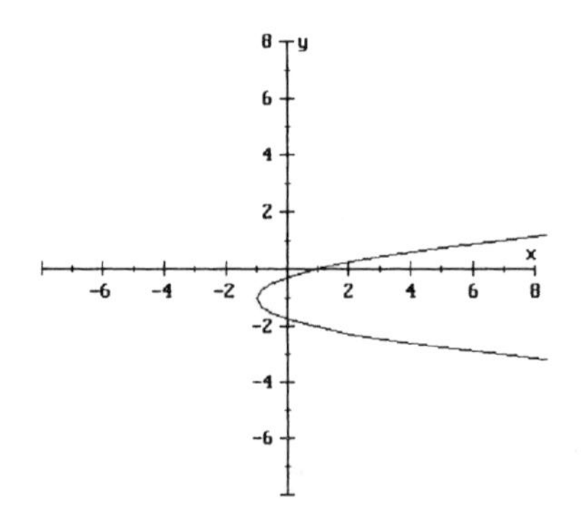

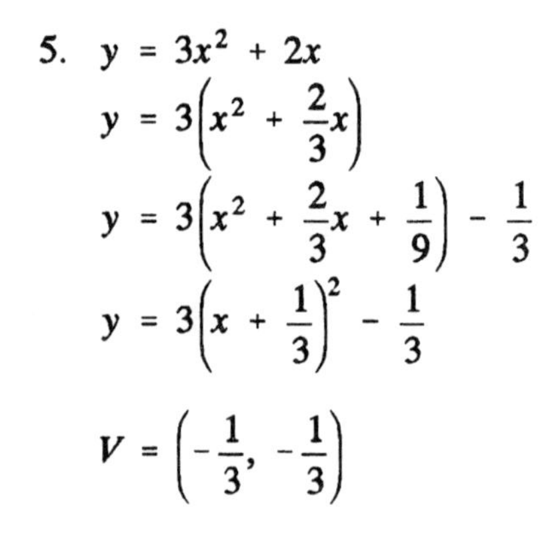

5.
$$\begin{aligned} y &= 3x^2 + 2x \\ y &= 3\left(x^2 + \frac{2}{3}x\right) \\ y &= 3\left(x^2 + \frac{2}{3}x + \frac{1}{9}\right) - \frac{1}{3} \\ y &= 3\left(x + \frac{1}{3}\right)^2 - \frac{1}{3} \end{aligned}$$

$V = \left(-\frac{1}{3}, -\frac{1}{3}\right)$

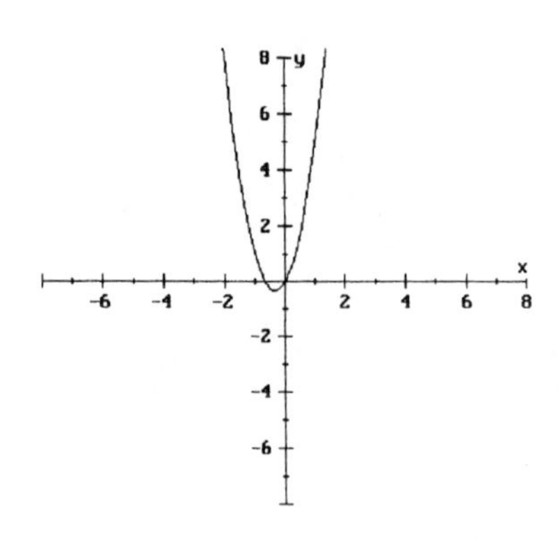

7.
$$\begin{aligned} y - x^2 &= 6x - 8 \\ y &= (x^2 + 6x) - 8 \\ y &= (x^2 + 6x + 9) - 8 - 9 \\ y &= (x + 3)^2 - 17 \end{aligned}$$

$V = (-3, -17)$

11.
$$\begin{aligned} x &= 2y^2 + y + 4 \\ x &= 2\left(y^2 + \frac{1}{2}y\right) + 4 \\ x &= 2\left(y^2 + \frac{1}{2}y + \frac{1}{16}\right) + 4 - \frac{1}{8} \\ x &= 2\left(y + \frac{1}{4}\right)^2 + \frac{31}{8} \end{aligned}$$

$V = \left(\frac{31}{8}, -\frac{1}{4}\right)$

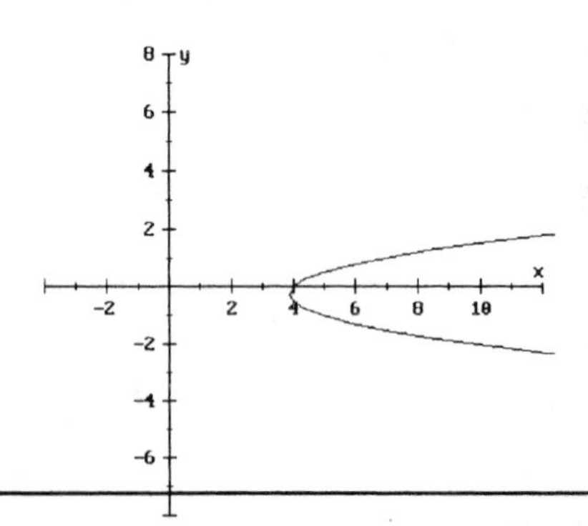

13. $\dfrac{x^2}{16} + \dfrac{y^2}{9} = 1$

$a^2 = 16 \qquad b^2 = 9$
$a = 4 \qquad b = 3$

$(4, 0), (-4, 0), (0, 3), (0, -3)$

15. $\dfrac{y^2}{36} + \dfrac{x^2}{25} = 1$

$a^2 = 25 \qquad b^2 = 36$
$a = 5 \qquad b = 6$

$(5, 0), (-5, 0), (0, 6), (0, -6)$

17. $\dfrac{x^2}{24} + \dfrac{y^2}{20} = 1$

$a^2 = 24 \qquad b^2 = 20$
$a = \sqrt{24} = 2\sqrt{6} \qquad b = \sqrt{20} = 2\sqrt{5}$

$(2\sqrt{6}, 0), (-2\sqrt{6}, 0), (0, 2\sqrt{5}), (0, -2\sqrt{5})$

19. $x^2 + \dfrac{y^2}{16} = 1$

$a^2 = 1 \qquad b^2 = 16$
$a = 1 \qquad b = 4$

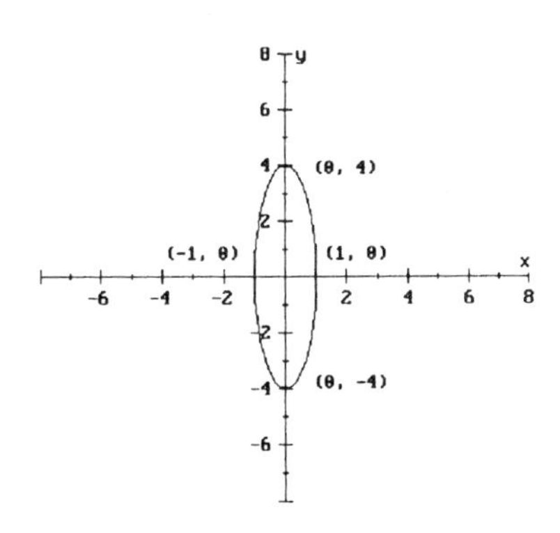

21. $4x^2 + 25y^2 = 100$

$\dfrac{x^2}{25} + \dfrac{y^2}{4} = 1$

$a^2 = 25 \qquad b^2 = 4$
$a = 5 \qquad b = 2$

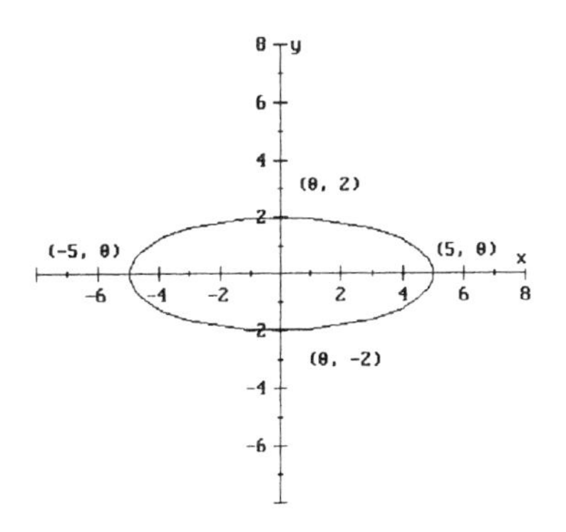

23. $8x^2 + 7y^2 = 56$

$\dfrac{x^2}{7} + \dfrac{y^2}{8} = 1$

$a^2 = 7 \qquad b^2 = 8$
$a = \sqrt{7} \qquad b = 2\sqrt{2}$

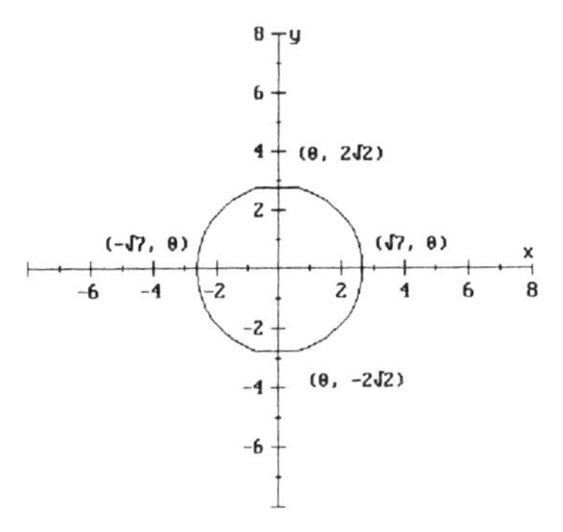

25. $25x^2 + 16y^2 = 1$

$\dfrac{x^2}{\frac{1}{25}} + \dfrac{y^2}{\frac{1}{16}} = 1$

$a^2 = \dfrac{1}{25} \qquad b^2 = \dfrac{1}{16}$
$a = \dfrac{1}{5} \qquad b = \dfrac{1}{4}$

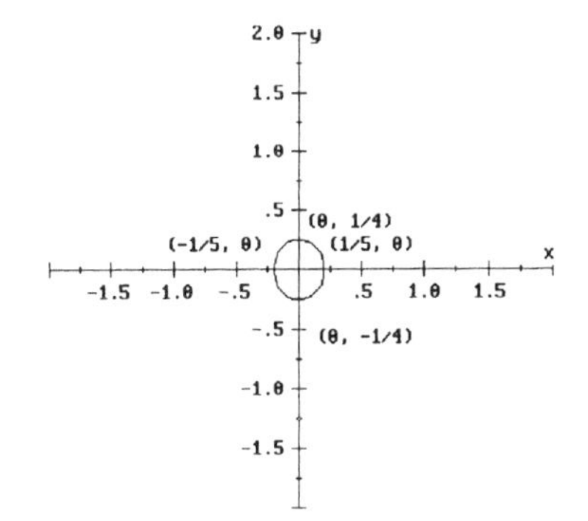

27. $a = \frac{1}{2}(50) = 25$

$b = 15$

$$\frac{x^2}{25^2} + \frac{y^2}{15^2} = 1$$

The truck is 11 ft wide. Find the height of the bridge for $x = 11$.

$$\frac{11^2}{25^2} + \frac{y^2}{15^2} = 1$$

$$\frac{121}{625} + \frac{y^2}{225} = 1$$

$$\frac{y^2}{225} = \frac{504}{625}$$

$$y^2 = \frac{4536}{25}$$

$$y = 13.47$$

The bridge is 13.47 ft high 11 ft right of the center line, so a 14 ft high truck will not be able to pass under it.

29. $\frac{x^2}{9} - \frac{y^2}{16} = 1$

$a^2 = 9 \qquad b^2 = 16$

$a = 3 \qquad b = 4$

Vertices: $(3, 0), (-3, 0)$

Asymptotes: $y = \pm\frac{b}{a}x$

$$y = \pm\frac{4}{3}x$$

31. $\frac{x^2}{36} - \frac{y^2}{25} = 1$

$a^2 = 36 \qquad b^2 = 25$

$a = 6 \qquad b = 5$

Vertices: $(6, 0), (-6, 0)$

Asymptotes: $y = \pm\frac{b}{a}x$

$$y = \pm\frac{5}{6}x$$

33. $\frac{y^2}{4} - \frac{x^2}{12} = 1$

$a^2 = 12 \qquad b^2 = 4$

$a = 2\sqrt{3} \qquad b = 2$

Vertices: $(0, 2), (0, -2)$

Asymptotes: $y = \pm\frac{b}{a}x$

$$y = \pm\frac{2}{2\sqrt{3}}x$$

$$y = \pm\frac{\sqrt{3}}{3}x$$

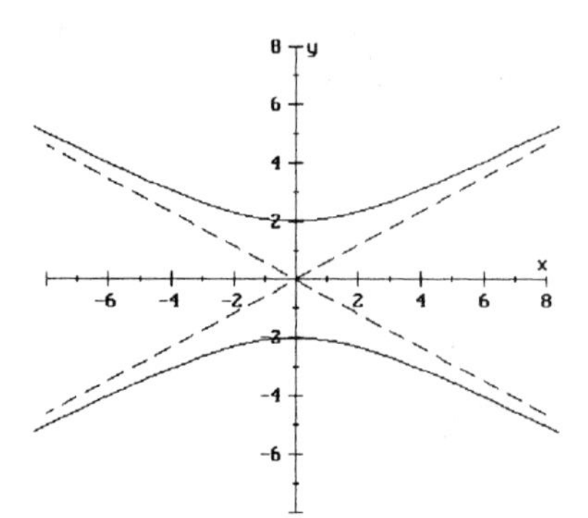

35. $2y^2 - x^2 = 4$

$$\frac{y^2}{2} - \frac{x^2}{4} = 1$$

$a^2 = 4 \qquad b^2 = 2$

$a = 2 \qquad b = \sqrt{2}$

Vertices: $(0, \sqrt{2}), (0, -\sqrt{2})$

Asymptotes: $y = \pm\frac{b}{a}x$

$$y = \pm\frac{\sqrt{2}}{2}x$$

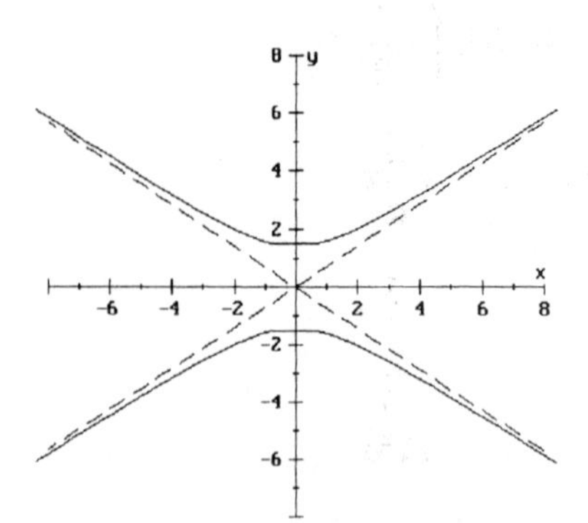

37. $12y^2 - 5x^2 = 60$

$$\frac{y^2}{5} - \frac{x^2}{12} = 1$$

$a^2 = 12 \qquad b^2 = 5$
$a = 2\sqrt{3} \qquad b = \sqrt{5}$

Vertices: $(0, \sqrt{5}), (0, -\sqrt{5})$

Asymptotes: $y = \pm\frac{b}{a}x$

$$y = \pm\frac{\sqrt{5}}{2\sqrt{3}}x$$

$$y = \pm\frac{\sqrt{15}}{6}x$$

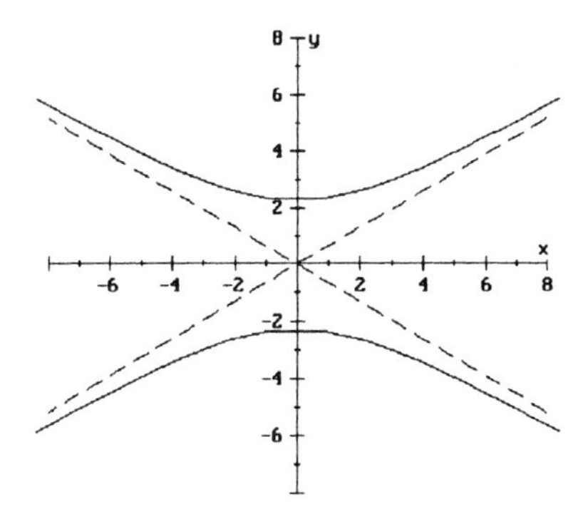

39. $225x^2 - y^2 = 25$

$$\frac{x^2}{\frac{1}{9}} - \frac{y^2}{25} = 1$$

$a^2 = \frac{1}{9} \qquad b^2 = 25$

$a = \frac{1}{3} \qquad b = 5$

Vertices: $\left(\frac{1}{3}, 0\right), \left(-\frac{1}{3}, 0\right)$

Asymptotes: $y = \pm\frac{b}{a}x$

$$y = \pm\frac{5}{\frac{1}{3}}x$$
$$y = \pm 15x$$

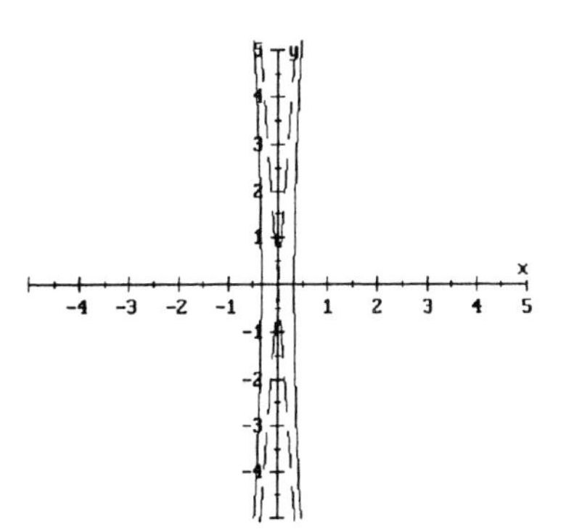

41. Circle

43. Ellipse

45. Parabola

47. Line

49. Hyperbola

51. Hyperbola

53. Circle

55. Circle

57. Hyperbola

59. $x^2 + y^2 = 16$
Circle
Center $= (0, 0)$
radius $= 4$

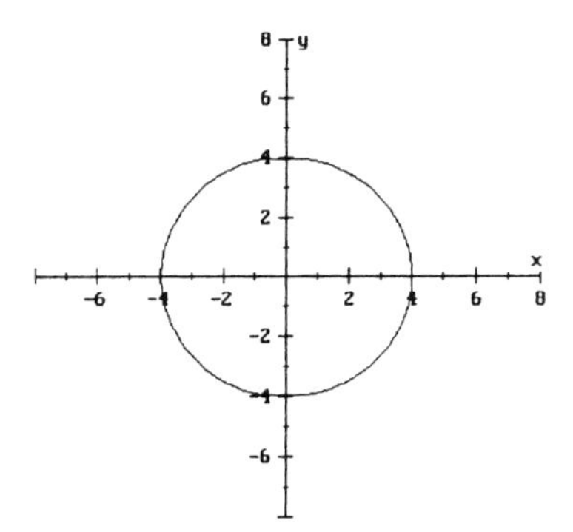

61. $x^2 + 2x - y = 9$

Parabola

$y = (x^2 + 2x) - 9$
$y = (x^2 + 2x + 1) - 9 - 1$
$y = (x + 1)^2 - 10$

Vertex $= (-1, -10)$

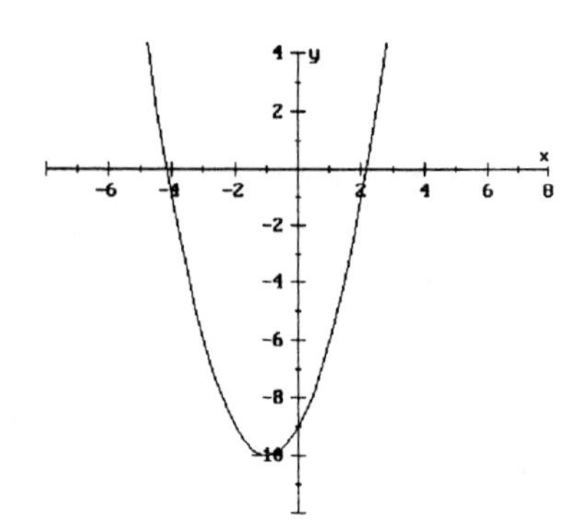

63. $x^2 - 100y^2 = 25$

$$\frac{x^2}{25} - \frac{y^2}{\frac{1}{4}} = 1$$

Hyperbola

Vertices: $(5, 0)$, $(-5, 0)$

Asymptotes: $y = \pm\frac{\frac{1}{2}}{5}x$

$$y = \pm\frac{1}{10}x$$

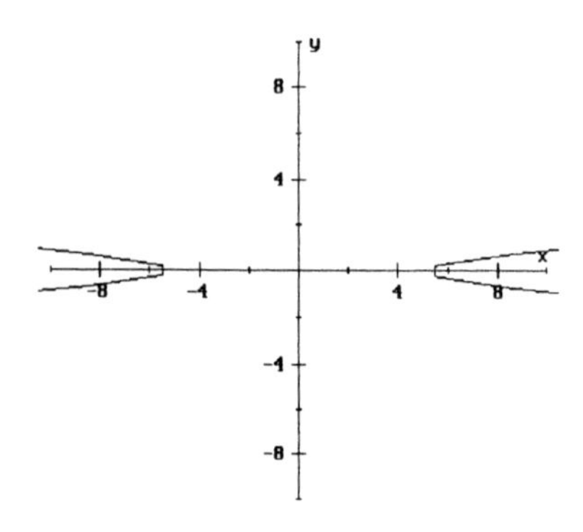

65. $$16x^2 - 9y^2 = 0$$
$$(4x - 3y)(4x + 3y) = 0$$

$4x - 3y = 0 \qquad 4x + 3y = 0$

$y = \frac{4}{3}x \qquad y = -\frac{4}{3}x$

pair of lines

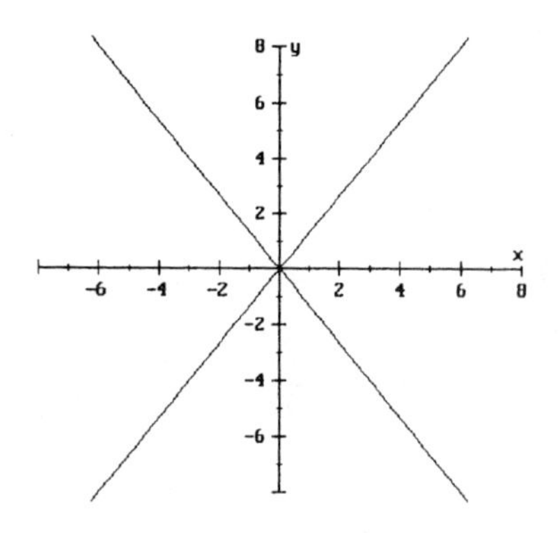

67. $3x^2 + 3y^2 = 24$
$x^2 + y^2 = 8$

Circle

Center $= (0, 0)$
radius $= 2\sqrt{2}$

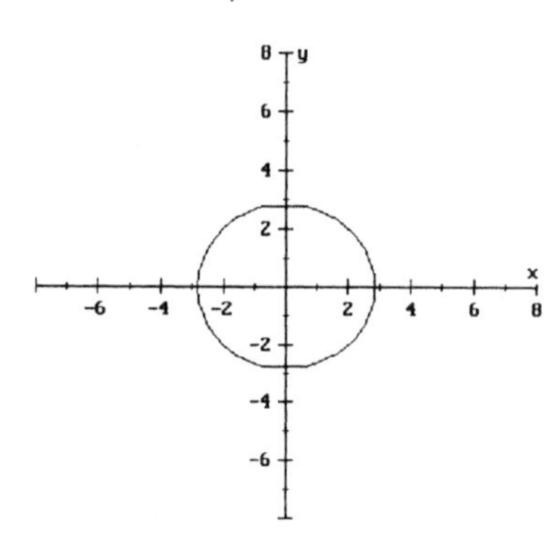

69. $2x^2 + 4x - y = -6$

Parabola

$y = 2x^2 + 4x + 6$
$y = 2(x^2 + 2x) + 6$
$y = 2(x^2 + 2x + 1) + 6 - 2$
$y = 2(x + 1)^2 + 4$

Vertex $= (-1, 4)$

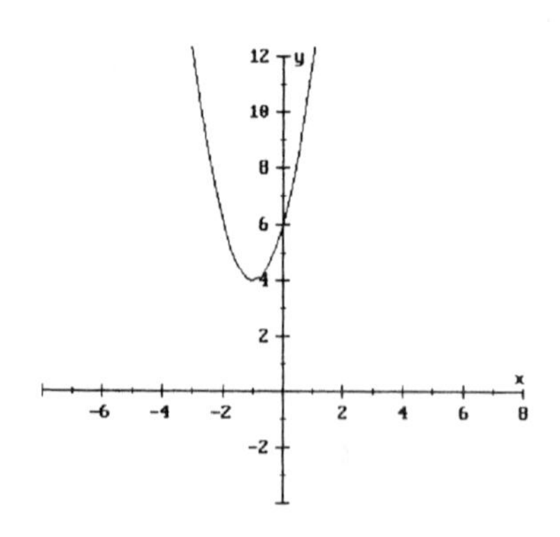

71. $2x^2 + y^2 = 8$

$$\frac{x^2}{4} + \frac{y^2}{8} = 1$$

Ellipse

$a^2 = 4 \qquad b^2 = 8$

$a = 2 \qquad b = 2\sqrt{2}$

Vertices: $(2, 0)$, $(-2, 0)$, $\left(0, 2\sqrt{2}\right)$, $\left(0, -2\sqrt{2}\right)$

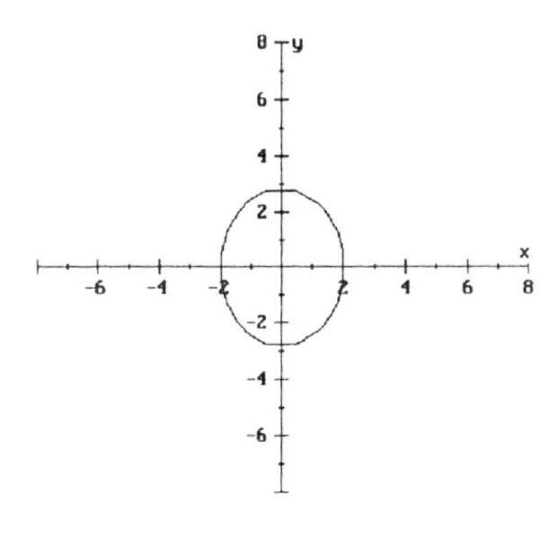

APPENDIX C

Exercises C.1

1. $a_n = 3n - 5$
   $a_1 = 3(1) - 5 = -2$
   $a_2 = 3(2) - 5 = 1$
   $a_3 = 3(3) - 5 = 4$
   $a_4 = 3(4) - 5 = 7$
   $a_{10} = 3(10) - 5 = 25$

3. $b_n = 4n + 3$
   $b_1 = 4(1) + 3 = 7$
   $b_2 = 4(2) + 3 = 11$
   $b_3 = 4(3) + 3 = 15$
   $b_4 = 4(4) + 3 = 19$
   $b_{10} = 4(10) + 3 = 43$

5. $c_i = 4^i$
   $c_1 = 4^1 = 4$
   $c_2 = 4^2 = 16$
   $c_3 = 4^3 = 64$
   $c_4 = 4^4 = 256$
   $c_{10} = 4^{10} = 1{,}048{,}576$

7. $x_j = \dfrac{j - 1}{j + 1}$

   $x_1 = \dfrac{1 - 1}{1 + 1} = 0$

   $x_2 = \dfrac{2 - 1}{2 + 1} = \dfrac{1}{3}$

   $x_3 = \dfrac{3 - 1}{3 + 1} = \dfrac{2}{4} = \dfrac{1}{2}$

   $x_4 = \dfrac{4 - 1}{4 + 1} = \dfrac{3}{5}$

   $x_{10} = \dfrac{10 - 1}{10 + 1} = \dfrac{9}{11}$

9. $x_n = \dfrac{(-1)^n}{n + 2}$

   $x_1 = \dfrac{(-1)^1}{1 + 2} = -\dfrac{1}{3}$

   $x_2 = \dfrac{(-1)^2}{2 + 2} = \dfrac{1}{4}$

   $x_3 = \dfrac{(-1)^3}{3 + 2} = -\dfrac{1}{5}$

   $x_4 = \dfrac{(-1)^4}{4 + 2} = \dfrac{1}{6}$

   $x_{10} = \dfrac{(-1)^{10}}{10 + 2} = \dfrac{1}{12}$

11. $a_n = \dfrac{(-1)^n n}{n + 1}$

   $a_1 = \dfrac{(-1)^1(1)}{1 + 1} = -\dfrac{1}{2}$

   $a_2 = \dfrac{(-1)^2(2)}{2 + 1} = \dfrac{2}{3}$

   $a_3 = \dfrac{(-1)^3(3)}{3 + 1} = -\dfrac{3}{4}$

   $a_4 = \dfrac{(-1)^4(4)}{4 + 1} = \dfrac{4}{5}$

   $a_{10} = \dfrac{(-1)^{10}(10)}{10 + 1} = \dfrac{10}{11}$

13. $a_n = 2^n + |n - 5|$
   $a_1 = 2^1 + |1 - 5|$
   $= 2 + 4$
   $= 6$

   $a_2 = 2^2 + |2 - 5|$
   $= 4 + 3$
   $= 7$

   $a_3 = 2^3 + |3 - 5|$
   $= 8 + 2$
   $= 10$

   $a_4 = 2^4 + |4 - 5|$
   $= 16 + 1$
   $= 17$

   $a_{10} = 2^{10} + |10 - 5|$
   $= 1024 + 5$
   $= 1029$

15. $b_n = 5 + (-0.1)^n$
$b_1 = 5 + (-0.1)^1 = 4.9$
$b_2 = 5 + (-0.1)^2 = 5.01$
$b_3 = 5 + (-0.1)^3 = 4.999$
$b_4 = 5 + (-0.1)^4 = 5.0001$
$b_{10} = 5 + (-0.1)^{10} = 5.0000000001$

17. $a_1 = 4 = 4 \cdot 1$
$a_2 = 8 = 4 \cdot 2$
$a_3 = 12 = 4 \cdot 3$
$a_4 = 16 = 4 \cdot 4$
$a_i = 4 \cdot i$
$a_i = 4i$

19. $a_1 = 1 = 2(1) - 1$
$a_2 = 3 = 2(2) - 1$
$a_3 = 5 = 2(3) - 1$
$a_4 = 7 = 2(4) - 1$
$a_5 = 9 = 2(5) - 1$
$a_i = 2(i) - 1$
$a_i = 2i - 1$

21. $a_1 = \frac{1}{2} = \frac{1}{1 + 1}$

$a_2 = \frac{2}{3} = \frac{2}{2 + 1}$

$a_3 = \frac{3}{4} = \frac{3}{3 + 1}$

$a_4 = \frac{4}{5} = \frac{4}{4 + 1}$

$a_i = \frac{i}{i + 1}$

23. $a_1 = \frac{1}{2} = \frac{(-1)^{1+1}}{1 + 1}$

$a_2 = -\frac{1}{3} = \frac{(-1)^{2+1}}{2 + 1}$

$a_3 = \frac{1}{4} = \frac{(-1)^{3+1}}{3 + 1}$

$a_4 = -\frac{1}{5} = \frac{(-1)^{4+1}}{4 + 1}$

$a_i = \frac{(-1)^{i+1}}{i + 1}$

25. $a_1 = -0.5 = 5(-0.1)^1$
$a_2 = 0.05 = 5(-0.1)^2$
$a_3 = -0.005 = 5(-0.1)^3$
$a_4 = 0.0005 = 5(-0.1)^4$
$a_i = 5(-0.1)^i$

27. 1st year: 26,000
2nd year: 26000 + 850 = 26,850
3rd year: 26000 + 2(850) = 27,700

$26,000, $26,850, $27,700

29. at end of 1st year = 20000 + 0.06(20000)
= 21200

at end of 2nd year = 21200 + 0.06(21200)
= 22472

at end of 3rd year = 22472 + 0.06(22472)
= 23820

21,200, 22,472, 23,820

31. at end of 1st year = 5000 + 0.06(5000)
= 5300

at end of 2nd year = 5300 + 0.06(5300)
= 5618

at end of 3rd year = 5618 + 0.06(5618)
= 5955.08

at end of 4th year = 5955.08 + 0.06(5955.08)
= 6312.38

$5300, $5618, $5955.08, $6312.38

33. at beginning of 1st year = 30,000

at beginning of 2nd year = 30,000 + 0.06(30,000)
= 31,800
at beginning of 3rd year = 31,800 + 0.06(31,800)
= 33,708

at beginning of 4th year = 33,708 + 0.06(33,708)
= 35,730.48

$30,000, $31,800, $33,708, $35,730.48

35. 1st bounce $= \frac{1}{2}(60) = 30$

2nd bounce $= \frac{1}{2}(30) = 15$

3rd bounce $= \frac{1}{2}(15) = 7.5$

4th bounce $= \frac{1}{2}(7.5) = 3.75$

5th bounce $= \frac{1}{2}(3.75) = 1.875$

30 ft, 15 ft, 7.5 ft, 3.75 ft, 1.875 ft

37. 1st month $= 200 + 0.01(5000 - 200)$
$= 248$

2nd month $= 200 + 0.01[5000 - 2(200)]$
$= 246$

3rd month $= 200 + 0.01[5000 - 3(200)]$
$= 244$

4th month $= 200 + 0.01[5000 - 4(200)]$
$= 242$

5th month $= 200 + 0.01[5000 - 5(200)]$
$= 240$

6th month $= 200 + 0.01[5000 - 6(200)]$
$= 238$

\$248, \$246, \$244, \$242, \$240, \$238

39. $a_1 = 4$
$a_2 = 5(a_1) = 5(4) = 20$
$a_3 = 5(a_2) = 5(20) = 100$
$a_4 = 5(a_3) = 5(100) = 500$
$a_5 = 5(a_4) = 5(500) = 2500$

41. $a_1 = 6$
$a_2 = 5(a_1) - 2$
$= 5(6) - 2$
$= 28$

$a_3 = 5(a_2) - 2$
$= 5(28) - 2$
$= 138$

$a_4 = 5(a_3) - 2$
$= 5(138) - 2$
$= 688$

$a_5 = 5(a_4) - 2$
$= 5(688) - 2$
$= 3438$

43. $a_1 = 4$
$a_2 = 5$
$a_3 = a_2 - a_1$
$= 5 - 4$
$= 1$

$a_4 = a_3 - a_2$
$= 1 - 5$
$= -4$

$a_5 = a_4 - a_3$
$= -4 - 1$
$= -5$

Exercises C.2

1. $S_5 = 2 + 5 + 8 + 11 + 14 = 40$

3. $S_6 = -2 + 0 + 2 + 4 + 6 + 8 = 18$

5. $S_3 = 7 + 4 + 1 = 12$

7. $a_1 = 3(1) + 1 = 4$
$a_2 = 3(2) + 1 = 7$

$S_2 = 4 + 7 = 11$

9. $b_1 = 5(1) - 1 = 4$
$b_2 = 5(2) - 1 = 9$
$b_3 = 5(3) - 1 = 14$

$S_3 = 4 + 9 + 14 = 27$

11. $x_1 = 2^1 = 2$
$x_2 = 2^2 = 4$
$x_3 = 2^3 = 8$
$x_4 = 2^4 = 16$
$x_5 = 2^5 = 32$

$S_5 = 2 + 4 + 8 + 16 + 32 = 62$

13. $y_1 = 2^1 + 1 = 3$
$y_2 = 2^2 + 1 = 5$
$y_3 = 2^3 + 1 = 9$
$y_4 = 2^4 + 1 = 17$
$y_5 = 2^5 + 1 = 33$
$y_6 = 2^6 + 1 = 65$

$S_6 = 3 + 5 + 9 + 17 + 33 + 65 = 132$

15. $a_1 = \dfrac{1}{1+1} = \dfrac{1}{2}$

$a_2 = \dfrac{2}{2+1} = \dfrac{2}{3}$

$a_3 = \dfrac{3}{3+1} = \dfrac{3}{4}$

$a_4 = \dfrac{4}{4+1} = \dfrac{4}{5}$

$S_4 = \dfrac{1}{2} + \dfrac{2}{3} + \dfrac{3}{4} + \dfrac{4}{5} = \dfrac{163}{60}$

17. $b_1 = \dfrac{(-1)^1}{1+1} = -\dfrac{1}{2}$

$b_2 = \dfrac{(-1)^2}{2+1} = \dfrac{1}{3}$

$b_3 = \dfrac{(-1)^3}{3+1} = -\dfrac{1}{4}$

$b_4 = \dfrac{(-1)^4}{4+1} = \dfrac{1}{5}$

$b_5 = \dfrac{(-1)^5}{5+1} = -\dfrac{1}{6}$

$S_5 = -\dfrac{1}{2} + \dfrac{1}{3} - \dfrac{1}{4} + \dfrac{1}{5} - \dfrac{1}{6} = -\dfrac{23}{60}$

19. $\sum_{i=1}^{5} i = 1 + 2 + 3 + 4 + 5 = 15$

21. $\sum_{i=1}^{7} 5i = 5(1) + 5(2) + 5(3) + 5(4) + 5(5)$
$+ 5(6) + 5(7)$
$= 140$

23. $\sum_{i=1}^{6} i^2 = 1^2 + 2^2 + 3^2 + 4^2 + 5^2 + 6^2 = 91$

25. $\sum_{m=2}^{5} 2m^3 = 2(2)^3 + 2(3)^3 + 2(4)^3 + 2(5)^3 = 448$

27. $\sum_{n=4}^{7} (2n+3) = [2(4)+3] + [2(5)+3]$
$+ [2(6)+3] + [2(7)+3]$
$= 11 + 13 + 15 + 17$
$= 56$

29. $\sum_{j=2}^{8} (4j+1) = [4(2)+1] + [4(3)+1] + [4(4)+1]$
$+ [4(5)+1] + [4(6)+1] + [4(7)+1]$
$+ [4(8)+1]$
$= 9 + 13 + 17 + 21 + 25 + 29 + 33$
$= 147$

31. $\sum_{i=2}^{6} (2i^2 - 1) = [2(2)^2 - 1] + [2(3)^2 - 1] + [2(4)^2 - 1]$
$+ [2(5)^2 - 1] + [2(6)^2 - 1]$
$= 7 + 17 + 31 + 49 + 71$
$= 175$

33. $\sum_{k=1}^{5} 2k$

35. $\sum_{k=1}^{7} k^2$

Exercises C.3

1. $a_1 = 3,\ d = 2$

$a_n = a_1 + (n-1)d$
$a_n = 3 + (n-1)2$

$a_2 = 3 + (2-1)2 = 5$
$a_3 = 3 + (3-1)2 = 7$
$a_{15} = 3 + (15-1)2 = 31$

3. $x_1 = 2,\ d = -4$

$$x_n = x_1 + (n - 1)d$$
$$x_n = 2 + (n - 1)(-4)$$

$$x_2 = 2 + (2 - 1)(-4) = -2$$
$$x_3 = 2 + (3 - 1)(-4) = -6$$
$$x_{15} = 2 + (15 - 1)(-4) = -54$$

5. $z_1 = -2,\ d = \frac{1}{2}$

$$z_n = z_1 + (n - 1)d$$
$$z_n = -2 + (n - 1)\frac{1}{2}$$

$$z_2 = -2 + (2 - 1)\frac{1}{2} = -\frac{3}{2}$$
$$z_3 = -2 + (3 - 1)\frac{1}{2} = -1$$
$$z_{15} = -2 + (15 - 1)\frac{1}{2} = 5$$

7. $d = 4 - 1 = 3$

$$a_5 = 10 + 3 = 13$$
$$a_6 = 13 + 3 = 16$$

$$a_n = a_1 + (n - 1)d$$
$$a_n = 1 + (n - 1)3$$
$$a_{15} = 1 + (15 - 1)3 = 43$$

9. $d = 0 - (-2) = 2$

$$a_5 = 4 + 2 = 6$$
$$a_6 = 6 + 2 = 8$$

$$a_n = a_1 + (n - 1)d$$
$$a_n = -2 + (n - 1)2$$
$$a_{15} = -2 + (15 - 1)2 = 26$$

11. $d = -3 - 2 = -5$

$$a_4 = -8 + (-5) = -13$$
$$a_5 = -13 + (-5) = -18$$

$$a_n = a_1 + (n - 1)d$$
$$a_n = 2 + (n - 1)(-5)$$
$$a_{15} = 2 + (15 - 1)(-5) = -68$$

13. $d = 16 - 8 = 8$

$$a_4 = 24 + 8 = 32$$
$$a_5 = 32 + 8 = 40$$

$$a_n = a_1 + (n - 1)d$$
$$a_n = 8 + (n - 1)8$$
$$a_{15} = 8 + (15 - 1)8 = 120$$

15. $d = \frac{5}{6} - \frac{1}{2} = \frac{1}{3}$

$$a_4 = \frac{7}{6} + \frac{1}{3} = \frac{3}{2}$$

$$a_5 = \frac{3}{2} + \frac{1}{3} = \frac{11}{6}$$

$$a_n = a_1 + (n - 1)d$$

$$a_n = \frac{1}{2} + (n - 1)\left(\frac{1}{3}\right)$$

$$a_{15} = \frac{1}{2} + (15 - 1)\left(\frac{1}{3}\right) = \frac{31}{6}$$

17. $d = -\frac{5}{6} - \left(-\frac{1}{3}\right) = -\frac{1}{2}$

$$a_3 = -\frac{5}{6} + \left(-\frac{1}{2}\right) = -\frac{4}{3}$$

$$a_4 = -\frac{4}{3} + \left(-\frac{1}{2}\right) = -\frac{11}{6}$$

$$a_n = a_1 + (n - 1)d$$

$$a_n = -\frac{1}{3} + (n - 1)\left(-\frac{1}{2}\right)$$
$$a_{15} = -\frac{1}{3} + (15 - 1)\left(-\frac{1}{2}\right) = -\frac{22}{3}$$

19. $a_k - a_j = (k - j)d$

$$85 - 25 = (15 - 5)d$$
$$60 = 10d$$
$$6 = d$$

$$a_i = a_1 + (n - 1)d$$
$$25 = a_1 + (5 - 1)(6)$$
$$25 = a_1 + 24$$
$$1 = a_1$$

21.
$$\begin{aligned} a_k - a_j &= (k - j)d \\ -41 - (-21) &= (12 - 7)d \\ -20 &= 5d \\ -4 &= d \end{aligned}$$

$$\begin{aligned} a_i &= a_1 + (n - 1)d \\ -21 &= a_1 + (7 - 1)(-4) \\ -21 &= a_1 - 24 \\ 3 &= a_1 \end{aligned}$$

$$\begin{aligned} a_i &= 3 + (n - 1)(-4) \\ a_5 &= 3 + (5 - 1)(-4) = -13 \end{aligned}$$

23. $d = 800$
3 years = 3(2) = 6 – 6 mo. periods
$a_1 = 26500$

$$\begin{aligned} a_n &= a_1 + (n - 1)d \\ a_n &= 26500 + (n - 1)800 \\ a_6 &= 26500 + (6 - 1)800 = 30{,}500 \end{aligned}$$

$30,500

25. $a_1 = 65$
$d = 5$

$$\begin{aligned} a_n &= a_1 + (n - 1)d \\ a_n &= 65 + (n - 1)5 \end{aligned}$$

20 weeks $= \dfrac{20}{2} = 10$ 2-week periods

$$a_{10} = 65 + (10 - 1)5 = 110$$

110 lb.

27. 16, 48, 80, ...
$d = 48 - 16 = 32$
$a_1 = 16$

$$\begin{aligned} a_n &= a_1 + (n - 1)d \\ a_n &= 16 + (n - 1)32 \\ a_6 &= 16 + (6 - 1)32 = 176 \end{aligned}$$

176 ft

29. $a_1 = 8,\ d = 2$

$$\begin{aligned} S_n &= \frac{n}{2}[2a_1 + (n - 1)d] \\ S_6 &= \frac{6}{2}[2(8) + (6 - 1)2] = 78 \end{aligned}$$

31. $a_1 = 5,\ d = -1$

$$\begin{aligned} S_n &= \frac{n}{2}[2a_1 + (n - 1)d] \\ S_5 &= \frac{5}{2}[2(5) + (5 - 1)(-1)] = 15 \end{aligned}$$

33. $a_1 = 4,\ d = \dfrac{1}{2}$

$$\begin{aligned} S_n &= \frac{n}{2}[2a_1 + (n - 1)d] \\ S_5 &= \frac{5}{2}\left[2(4) + (5 - 1)\frac{1}{2}\right] = 25 \end{aligned}$$

35. $d = -3 - 2 = -5$
$a_1 = 2$

$$S_n = \frac{n}{2}[2a_1 + (n - 1)d]$$

$$S_5 = \frac{5}{2}[2(2) + (5 - 1)(-5)] = -40$$

37. $d = 0 - (-2) = 2$
$a_1 = -2$

$$S_n = \frac{n}{2}[2a_1 + (n - 1)d]$$

$$S_6 = \frac{6}{2}[2(-2) + (6 - 1)2] = 18$$

39. $d = -\dfrac{5}{6} - \left(-\dfrac{1}{3}\right) = -\dfrac{1}{2}$

$$a_1 = -\frac{1}{3}$$

$$S_n = \frac{n}{2}[2a_1 + (n - 1)d]$$

$$S_4 = \frac{4}{2}\left[2\left(-\frac{1}{3}\right) + (4 - 1)\left(-\frac{1}{2}\right)\right] = -\frac{13}{3}$$

41. $a_1 = 3(1) = 3$
$a_6 = 3(6) = 18$

$$S_n = \frac{n}{2}(a_1 + a_n)$$

$$S_6 = \frac{6}{2}(3 + 18) = 63$$

43. $a_1 = 10(1) = 10$
$a_{10} = 10(10) = 100$

$$S_n = \frac{n}{2}(a_1 + a_n)$$

$$S_{10} = \frac{10}{2}(10 + 100) = 550$$

45. $a_1 = 4(1) + 3 = 7$
$a_5 = 4(5) + 3 = 23$

$$S_n = \frac{n}{2}(a_1 + a_n)$$

$$S_5 = \frac{5}{2}(7 + 23) = 75$$

47. $a_{30} = 30$
$a_{80} = 80$
$n = (80 - 30) + 1 = 51$

$$S_n = \frac{n}{2}(a_1 + a_n)$$

$$S_{51} = \frac{51}{2}(30 + 80) = 2805$$

49. $a_1 = 2,\ d = 2$

$$S_n = \frac{n}{2}[2a_1 + (n - 1)d]$$

$$S_{50} = \frac{50}{2}[2(2) + (50 - 1)2] = 2550$$

51. $a_1 = 8,\ d = 8$

$$S_n = \frac{n}{2}[2a_1 + (n - 1)d]$$

$$S_{10} = \frac{10}{2}[2(8) + (10 - 1)(8)] = 440$$

53. $a_k - a_j = (k - j)d$
$23 - 3 = (14 - 4)d$
$20 = 10d$
$2 = d$

$a_n = a_1 + (n - 1)d$
$3 = a_1 + (4 - 1)2$
$3 = a_1 + 6$
$-3 = a_1$

$$S_n = \frac{n}{2}[2a_1 + (n - 1)d]$$

$$S_{15} = \frac{15}{2}[2(-3) + (15 - 1)2] = 165$$

55. $a_k - a_j = (k - j)d$
$-58 - (-38) = (12 - 8)d$
$-20 = 4d$
$-5 = d$

$a_n = a_1 + (n - 1)d$
$-38 = a_1 + (8 - 1)(-5)$
$-38 = a_1 - 35$
$-3 = a_1$

$$S_n = \frac{n}{2}[2a_1 + (n - 1)d]$$

$$S_{15} = \frac{15}{2}[2(-3) + (15 - 1)(-5)] = -570$$

57. 16, 48, 80, . . .

$d = 48 - 16 = 32$
$a_1 = 16$

$$S_n = \frac{n}{2}[2a_1 + (n - 1)d]$$

$$S_6 = \frac{6}{2}[2(16) + (6 - 1)32] = 576$$

576 ft

59. Firm A: $a_1 = 30{,}000,\quad d = 2000$
$a_n = a_1 + (n - 1)d$
$a_n = 30{,}000 + (n - 1)2000$

Firm B: $a_1 = 32{,}000,\quad d = 1500$
$a_n = 32{,}000 + (n - 1)1500$

(a) Firm A:
$a_8 = 30{,}000 + (8 - 1)2000 = \$44{,}000$
Firm B:
$a_8 = 32000 + (8 - 1)1500 = \$42{,}500$

(b) Firm A:

$$S_n = \frac{n}{2}[2a_1 + (n - 1)d]$$

$$S_8 = \frac{8}{2}[2(30{,}000) + (8 - 1)2000]$$

$$= \$296{,}000$$

Firm B:

$$S_8 = \frac{8}{2}[2(32{,}000) + (8 - 1)1500]$$

$$= \$298{,}000$$

Exercises C.4

1. $a_1 = 2$

$$r = \frac{6}{2} = 3$$

$$a_n = a_1 r^{n-1}$$
$$a_6 = 2(3)^{6-1} = 486$$

3. $a_1 = 1$

$$r = \frac{6}{1} = 6$$

$$a_n = a_1 r^{n-1}$$
$$a_5 = 1(6)^{5-1} = 1296$$

5. $a_1 = 3$

$$r = \frac{1}{3}$$

$$a_n = a_1 r^{n-1}$$
$$a_8 = 3\left(\frac{1}{3}\right)^{8-1} = \frac{1}{729}$$

7. $a_1 = 5$

$$r = \frac{\frac{5}{6}}{5} = \frac{1}{6}$$

$$a_n = a_1 r^{n-1}$$
$$a_6 = 5\left(\frac{1}{6}\right)^{6-1} = \frac{5}{7776}$$

9. $a_1 = \frac{1}{2}$

$$r = \frac{\frac{1}{6}}{\frac{1}{2}} = \frac{1}{3}$$

$$a_n = a_1 r^{n-1}$$
$$a_5 = \frac{1}{2}\left(\frac{1}{3}\right)^{5-1} = \frac{1}{162}$$

11. $a_1 = 3,\ r = 2$

$$a_n = a_1 r^{n-1}$$
$$a_4 = 3(2)^{4-1} = 24$$

13. $b_1 = \frac{1}{2},\ r = \frac{1}{5}$

$$b_n = b_1 r^{n-1}$$
$$b_5 = \frac{1}{2}\left(\frac{1}{5}\right)^{5-1} = \frac{1}{1250}$$

15. $a_1 = 1,\ a_3 = 16$

$$a_n = a_1 r^{n-1}$$
$$16 = 1r^{3-1}$$
$$16 = r^2$$
$$r = \pm 4$$

$$a_5 = 1(\pm 4)^{5-1} = 256$$

17. $a_1 = 1,\ a_5 = \frac{1}{16}$

$$a_n = a_1 r^{n-1}$$

$$\frac{1}{16} = 1r^{5-1}$$

$$\frac{1}{16} = r^4$$

$$r = \pm\frac{1}{2}$$

$$a_7 = 1\left(\pm\frac{1}{2}\right)^{7-1} = \frac{1}{64}$$

19. $P = 16{,}000, \quad r = 1 - \frac{1}{6} = \frac{5}{6}$

$$A_i = 16{,}000\left(\frac{5}{6}\right)^i$$

$$A_4 = 16000\left(\frac{5}{6}\right) = 7716.05$$

\$7716.05

21. $P = 24{,}000, \quad r = 1 - \frac{1}{5} = \frac{4}{5}$

$$A_i = 24{,}000\left(\frac{4}{5}\right)^i$$

$$A_3 = 24{,}000\left(\frac{4}{5}\right)^3 = 12{,}288$$

\$12,228

23. $P = 100{,}000, \quad r = 1 + 0.05 = 1.05$

$$A_i = 100{,}000(1.05)^i$$
$$A_4 = 100{,}000(1.05)^4 = 121{,}551$$

The population wiil be 121,551.

25. $P = 10{,}000, \quad r = 1 + 0.08 = 1.08$

$$A_i = 10{,}000(1.08)^i$$
$$A_5 = 10{,}000(1.08)^5 = 14{,}693.28$$

\$14,693.28

27. $P = 5000, \quad r = 1 + 0.10 = 1.10$

$$A_i = 5000(1.10)^i$$
$$A_5 = 5000(1.10)^5 = 8052.55$$

\$8052.55

29. 1st rebound $= \frac{1}{2}(20) = 10$

$$a_1 = 10$$
$$r = \frac{1}{2}$$

$$a_n = a_1 r^{n-1}$$
$$a_5 = 10\left(\frac{1}{2}\right)^{5-1} = \frac{5}{8}$$

$\frac{5}{8}$ ft

31. $r = \frac{9}{3} = 3$

$$a_1 = 3$$

$$S_n = \frac{a_1(1 - r^n)}{1 - r}$$

$$S_6 = \frac{3(1 - 3^6)}{1 - 3} = 1092$$

33. $r = \frac{10}{5} = 2$

$$a_1 = 5$$

$$S_n = \frac{a_1(1 - r^n)}{1 - r}$$

$$S_6 = \frac{5(1 - 2^6)}{1 - 2} = 315$$

35. $r = \dfrac{\frac{2}{3}}{\frac{1}{3}} = 2$

$$a_1 = \frac{1}{3}$$

$$S_n = \frac{a_1(1 - r^n)}{1 - r}$$

$$S_6 = \frac{\frac{1}{3}(1 - 2^6)}{1 - 2} = 21$$

37. $a_1 = 1$

$$r = \frac{\frac{1}{10}}{1} = \frac{1}{10}$$

$$S_n = \frac{a_1(1 - r^n)}{1 - r}$$

$$S_5 = \frac{1\left[1 - \left(\frac{1}{10}\right)^5\right]}{1 - \frac{1}{10}} = \frac{11{,}111}{10{,}000}$$

39. $a_4 = \frac{1}{1000}$, let this be a_1

$$r = \frac{\frac{1}{10}}{1} = \frac{1}{10}$$

a_4 to a_8 is 5 terms, $n = 5$

$$S_n = \frac{a_1(1 - r^n)}{1 - r}$$

$$\sum_{k=4}^{8} a_k = \frac{\frac{1}{1000}\left[1 - \left(\frac{1}{10}\right)^5\right]}{1 - \frac{1}{10}} = \frac{11,111}{10,000,000}$$

41. $a_1 = \frac{1}{3}(30) = 10$

$r = \frac{1}{3}$

$$S_n = \frac{a_1(1 - r^n)}{1 - r}$$

$$S_5 = \frac{10\left[1 - \left(\frac{1}{3}\right)^5\right]}{1 - \frac{1}{3}} = 14\frac{76}{81}$$

$14\frac{76}{81}$ ft

43. Firm A: $a_1 = 30,000$

$r = 1 + 0.06 = 1.06$

Firm B: $a_1 = 34,000$

$r = 1 + 0.04 = 1.04$

(a) $a_n = a_1 r^{n-1}$

Firm A: $a_8 = 30,000(1.06)^{8-1} = \$45,108.91$

Firm B: $a_8 = 34,000(1.04)^{8-1} = \$44,741.68$

(b) $S_n = \frac{a_1(1 - r^n)}{1 - r}$

Firm A: $S_8 = \frac{30,000[1 - (1.06)^8]}{1 - 1.06} = \$296,924.04$

Firm B: $S_8 = \frac{34,000[1 - (1.04)^8]}{1 - 1.04} = \$313,283.69$

45. $a_1 = 2$

$r = \frac{1}{2}$

$$S = \frac{a_1}{1 - r}$$

$$S = \frac{2}{1 - \frac{1}{2}} = 4$$

47. $a_1 = \frac{1}{5}$

$$r = \frac{\frac{1}{10}}{\frac{1}{5}} = \frac{1}{2}$$

$$S = \frac{a_1}{1 - r}$$

$$S = \frac{\frac{1}{5}}{1 - \frac{1}{2}} = \frac{2}{5}$$

49. $a_1 = 5$

$$r = \frac{\frac{5}{3}}{5} = \frac{1}{3}$$

$$S = \frac{a_1}{1 - r}$$

$$S = \frac{5}{1 - \frac{1}{3}} = \frac{15}{2}$$

51. $a_1 = \frac{1}{5}$

$$r = \frac{\frac{2}{5}}{\frac{1}{5}} = 2$$

Since $|r| > 1$, the sum does not exist.

53. $a_1 = 0.35$
$r = 0.01$

$$S = \frac{a_1}{1 - r}$$

$$S = \frac{0.35}{1 - 0.01} = \frac{35}{99}$$

55. $a_1 = 0.0245$
$r = 0.0001$

$$S = \frac{a_1}{1 - r}$$

$$S = \frac{0.0245}{1 - 0.0001} = \frac{245}{9999}$$

57. dropped: 60 ft
1st rebound: $\frac{2}{3}(60) = 40$ ft
drops: 40 ft
2nd rebound: $\frac{2}{3}(40) = \frac{80}{3}$ ft
drops: $\frac{80}{3}$ ft
etc.

Total distance $= 60 + 40 + 40 + \frac{80}{3} + \frac{80}{3} + \ldots$

Consider the geometric series: $40 + \frac{80}{3} + \ldots$

The total distance is 60 plus this sum doubled.

$$\text{Total distance} = 60 + 2\left[\frac{40}{1 - \frac{2}{3}}\right]$$

$$= 60 + 240$$
$$= 300$$

300 ft

59. $a_1 = 10$
$r = \frac{2}{3}$

$$S = \frac{a_1}{1 - r}$$

$$S = \frac{10}{1 - \frac{2}{3}} = 30$$

30 ft

Exercises C.5

1. $6! = 6 \cdot 5 \cdot 4 \cdot 3 \cdot 2 \cdot 1 = 720$

3. $4! = 4 \cdot 3 \cdot 2 \cdot 1 = 24$

5. $0! = 1$

7. $(7 - 2)! = 5!$
$= 5 \cdot 4 \cdot 3 \cdot 2 \cdot 1$
$= 120$

9. $7! - 2 = 7 \cdot 6 \cdot 5 \cdot 4 \cdot 3 \cdot 2 \cdot 1 - 2$
$= 5040 - 2$
$= 5038$

11. $(2 \cdot 3)! = 6!$
$= 6 \cdot 5 \cdot 4 \cdot 3 \cdot 2 \cdot 1$
$= 720$

13. $2 \cdot 3! = 2 \cdot 3 \cdot 2 \cdot 1$
$= 12$

15. $\frac{5!}{1!4!} = \frac{5 \cdot 4 \cdot 3 \cdot 2 \cdot 1}{1 \cdot 4 \cdot 3 \cdot 2 \cdot 1}$

$= 5$

17. $\frac{8!}{5!3!} = \frac{8 \cdot 7 \cdot 6 \cdot 5 \cdot 4 \cdot 3 \cdot 2 \cdot 1}{5 \cdot 4 \cdot 3 \cdot 2 \cdot 1 \cdot 3 \cdot 2 \cdot 1}$

$= 56$

19. $\dfrac{12!}{3!4!5!} = \dfrac{12 \cdot 11 \cdot \overset{5}{10} \cdot 9 \cdot 8 \cdot 7 \cdot 6 \cdot 5 \cdot 4 \cdot 3 \cdot 2 \cdot 1}{3 \cdot 2 \cdot 1 \cdot 4 \cdot 3 \cdot 2 \cdot 1 \cdot 5 \cdot 4 \cdot 3 \cdot 2 \cdot 1}$

$= 27{,}720$

21. $(a + b)^6 = a^6 + \dfrac{6!}{5!1!}a^5b + \dfrac{6!}{4!2!}a^4b^2 + \dfrac{6!}{3!3!}a^3b^3 + \dfrac{6!}{2!4!}a^2b^4 + \dfrac{6!}{1!5!}ab^5 + b^6$

$= a^6 + 6a^5b + 15a^4b^2 + 20a^3b^3 + 15a^2b^4 + 6ab^5 + b^6$

23. $(a - b)^6 = [a + (-b)]^6$

$= a^6 + \dfrac{6!}{5!1!}a^5(-b) + \dfrac{6!}{4!2!}a^4(-b)^2 + \dfrac{6!}{3!3!}a^3(-b)^3 + \dfrac{6!}{2!4!}a^2(-b)^4 + \dfrac{6!}{1!5!}a(-b)^5 + (-b)^6$

$= a^6 - 6a^5b + 15a^4b^2 - 20a^3b^3 + 15a^2b^4 - 6ab^5 + b^6$

25. $(2a + 1)^5 = (2a)^5 + \dfrac{5!}{4!1!}(2a)^4 1 + \dfrac{5!}{3!2!}(2a)^3(1)^2 + \dfrac{5!}{2!3!}(2a)^2(1)^3 + \dfrac{5!}{1!4!}(2a)^1(1)^4 + (1)^5$

$= 32a^5 + 80a^4 + 80a^3 + 40a^2 + 10a + 1$

27. $(2a - 1)^5 = [2a + (-1)]^5$

$= (2a)^5 + \dfrac{5!}{4!1!}(2a)^4(-1) + \dfrac{5!}{3!2!}(2a)^3(-1)^2 + \dfrac{5!}{2!3!}(2a)^2(-1)^3 + \dfrac{5!}{1!4!}(2a)(-1)^4 + (-1)^5$

$= 32a^5 - 80a^4 + 80a^3 - 40a^2 + 10a - 1$

29. $(1 - 2a)^4 = [1 + (-2a)]^4$

$= 1^4 + \dfrac{4!}{3!1!}(1)^3(-2a) + \dfrac{4!}{2!2!}(1)^2(-2a)^2 + \dfrac{4!}{1!3!}(1)(-2a)^3 + (-2a)^4$

$= 1 - 8a + 24a^2 - 32a^3 + 16a^4$

31. $(a^2 + 2b)^5 = (a^2)^5 + \dfrac{5!}{4!1!}(a^2)^4(2b) + \dfrac{5!}{3!2!}(a^2)^3(2b)^2 + \dfrac{5!}{2!3!}(a^2)^2(2b)^3 + \dfrac{5!}{1!4!}(a^2)(2b)^4 + (2b)^5$

$= a^{10} + 10a^8b + 40a^6b^2 + 80a^4b^3 + 80a^2b^4 + 32b^5$

33. $(2x^2 - 3y^2)^4 = [2x^2 + (-3y^2)]^4$

$= (2x^2)^4 + \dfrac{4!}{3!1!}(2x^2)^3(-3y^2) + \dfrac{4!}{2!2!}(2x^2)^2(-3y^2)^2 + \dfrac{4!}{1!3!}(2x^2)(-3y^2)^3 + (-3y^2)^4$

$= 16x^8 - 96x^6y^2 + 216x^4y^4 - 216x^2y^6 + 81y^8$

35. $\left(\frac{x}{3} + 2\right)^4 = \left(\frac{x}{3}\right)^4 + \frac{4!}{3!1!}\left(\frac{x}{3}\right)^3(2) + \frac{4!}{2!2!}\left(\frac{x}{3}\right)^2(2)^2 + \frac{4!}{1!3!}\left(\frac{x}{3}\right)(2)^3 + 2^4$

$= \frac{x^4}{81} + \frac{8x^3}{27} + \frac{8x^2}{3} + \frac{32x}{3} + 16$

37. $(a^2)^6 + \frac{6!}{5!1!}(a^2)^5(-b^4) + \frac{6!}{4!2!}(a^2)^4(-b^4)^2 + \frac{6!}{3!3!}(a^2)^3(-b^4)^3$

$= a^{12} - 6a^{10}b^4 + 15a^8b^8 - 20a^6b^{12}$

39. $(2a)^8 + \frac{8!}{7!1!}(2a)^7(3b) + \frac{8!}{6!2!}(2a)^6(3b)^2 + \frac{8!}{5!3!}(2a)^5(3b)^3$

$= 256a^8 + 3072a^7b + 16128a^6b^2 + 48384a^5b^3$

41. $(3a^2)^6 + \frac{6!}{5!1!}(3a^2)^5(-2) + \frac{6!}{4!2!}(3a^2)^4(-2)^2 + \frac{6!}{3!3!}(3a^2)^3(-2)^3$

$= 729a^{12} - 2916a^{10} + 4860a^8 - 4320a^6$

43. 3rd term: $r = 3$
$k = r - 1 = 3 - 1 = 2$
$n = 8$

$\frac{n!}{(n-k)!k!}x^{n-k}y^k$

$= \frac{8!}{(8-2)!2!}x^{8-2}y^2$

$= \frac{8!}{6!2!}x^6y^2$

$= 28x^6y^2$

45. 4th term: $r = 4$
$k = r - 1 = 4 - 1 = 3$
$n = 6$

$\frac{n!}{(n-k)!k!}x^{n-k}y^k$

$= \frac{6!}{(6-3)!3!}a^{6-3}(-2b)^3$

$= \frac{6!}{3!3!}a^3(-8b^3)$

$= -160a^3b^3$

47. 3rd term: $r = 3$
$k = r - 1 = 3 - 1 = 2$
$n = 7$

$\frac{n!}{(n-k)!k!}x^{n-k}y^k$

$= \frac{7!}{(7-2)!2!}(3a^2)^{7-2}(-2)^2$

$= \frac{7!}{5!2!}(3a^2)^5(4)$

$= 20{,}412a^{10}$

49. 6th term: $r = 6$
$k = r - 1 = 6 - 1 = 5$
$n = 7$

$\frac{n!}{(n-k)!k!}x^{n-k}y^k$

$= \frac{7!}{(7-5)!5!}(3a^2)^{7-5}(-2)^5$

$= \frac{7!}{2!5!}(3a^2)^2(-32)$

$= -6048a^4$

51. 9th term: $r = 9$
$k = r - 1 = 9 - 1 = 8$
$n = 8$

$$\frac{n!}{(n-k)!k!}x^{n-k}y^k$$

$$= \frac{8!}{(8-8)!8!}(3a^2)^{8-8}(-2)^8$$

$$= \frac{8!}{0!8!}(3a^2)^0(256)$$

$$= 256$$

APPENDIX C REVIEW EXERCISES

1. $a_n = 2n - 5$

$$a_1 = 2(1) - 5 = -3$$
$$a_2 = 2(2) - 5 = -1$$
$$a_3 = 2(3) - 5 = 1$$
$$a_4 = 2(4) - 5 = 3$$
$$a_{12} = 2(12) - 5 = 19$$

3. $x_n = 2n^2$

$$x_1 = 2(1)^2 = 2$$
$$x_2 = 2(2)^2 = 8$$
$$x_3 = 2(3)^2 = 18$$
$$x_4 = 2(4)^2 = 32$$
$$x_{12} = 2(12)^2 = 288$$

5. $a_n = \dfrac{(-1)^n}{n+1}$

$$a_1 = \frac{(-1)^1}{1+1} = -\frac{1}{2}$$

$$a_2 = \frac{(-1)^2}{2+1} = \frac{1}{3}$$

$$a_3 = \frac{(-1)^3}{3+1} = -\frac{1}{4}$$

$$a_4 = \frac{(-1)^4}{4+1} = \frac{1}{5}$$

$$a_{12} = \frac{(-1)^{12}}{12+1} = \frac{1}{13}$$

7. $x_n = 3 + (-1)^n$

$$x_1 = 3 + (-1)^1 = 2$$
$$x_2 = 3 + (-1)^2 = 4$$
$$x_3 = 3 + (-1)^3 = 2$$
$$x_4 = 3 + (-1)^4 = 4$$
$$x_{12} = 3 + (-1)^{12} = 4$$

9. $3 = 5(1) - 2$
$8 = 5(2) - 2$
$13 = 5(3) - 2$
$18 = 5(4) - 2$

$$a_i = 5i - 2$$

11. $\dfrac{1}{2} = \dfrac{(-1)^{1+1}}{2(1)}$

$$-\frac{1}{4} = \frac{(-1)^{2+1}}{2(2)}$$

$$\frac{1}{6} = \frac{(-1)^{3+1}}{2(3)}$$

$$-\frac{1}{8} = \frac{(-1)^{4+1}}{2(4)}$$

$$a_i = \frac{(-1)^{i+1}}{2i}$$

13. At:
Start: \$23,000
6 mo: 23,000 + 700 = \$23,700
1st yr: 23,700 + 700 = \$24,400
1½ yr: 24,400 + 700 = \$25,100
2nd yr: 25,100 + 700 = \$25,800
2½ yr: 25,800 + 700 = \$26,500

15. $P = 500, \quad r = 1 + 0.12 = 1.12$

$$A_i = 500(1.12)^i$$
$$A_5 = 500(1.12)^5 = 881.17$$

\$881.17

17. $a_1 = 3, \quad d = 6 - 3 = 3$

$$S_n = \frac{n}{2}[2a_1 + (n - 1)d]$$

$$S_5 = \frac{5}{2}[2(3) + (5 - 1)3] = 45$$

19. $a_1 = 2, \quad r = \frac{4}{2} = 2$

$$S_n = \frac{a_1(1 - r^n)}{1 - r}$$

$$S_6 = \frac{2(1 - 2^6)}{1 - 2} = 126$$

21. $a_i = 3i - 1$

$$a_1 = 3(1) - 1 = 2$$
$$a_2 = 3(2) - 1 = 5$$
$$a_3 = 3(3) - 1 = 8$$

etc.

$a_1 = 2, \quad d = 5 - 2 = 3$

$$S_n = \frac{n}{2}[2a_1 + (n - 1)d]$$

$$S_6 = \frac{6}{2}[2(2) + (6 - 1)3] = 57$$

23. $\sum_{i=1}^{6} i = 1 + 2 + 3 + 4 + 5 + 6 = 21$

25. $\sum_{i=1}^{5} 2i = 2(1) + 2(2) + 2(3) + 2(4) + 2(5)$

$= 2 + 4 + 6 + 8 + 10$
$= 30$

27. $\sum_{n=1}^{6} 3n^2 = 3(1)^2 + 3(2)^2 + 3(3)^2 + 3(4)^2 + 3(5)^2$

$+ 3(6)^2$
$= 3 + 12 + 27 + 48 + 75 + 108$
$= 273$

29. $a_1 = 2$

$$a_2 = 2 + 3 = 5$$
$$a_3 = 5 + 3 = 8$$
$$a_n = a_1 + (n - 1)d$$
$$a_{10} = 2 + (10 - 1)3 = 29$$

31. $a_1 = 3, \quad d = 8 - 3 = 5$

$$a_5 = 18 + 5 = 23$$
$$a_6 = 23 + 5 = 28$$

$$a_n = a_1 + (n - 1)d$$
$$a_{10} = 3 + (10 - 1)5 = 48$$

33. $a_1 = 1, \quad d = -3 - 1 = -4$

$$a_5 = -11 + (-4) = -15$$
$$a_6 = -15 + (-4) = -19$$

$$a_n = a_1 + (n - 1)d$$
$$a_{10} = 1 + (10 - 1)(-4) = -35$$

35. $x_k - x_j = (k - j)d$

$$22 - 10 = (8 - 4)d$$
$$12 = 4d$$
$$3 = d$$

$$x_n = x_1 + (n - 1)d$$
$$10 = x_1 + (4 - 1)3$$
$$10 = x_1 + 9$$
$$1 = x_1$$

37. 16, 48, . . .

$a_1 = 16, \quad d = 48 - 16 = 32$

$$a_n = a_1 + (n - 1)d$$
$$a_8 = 16 + (8 - 1)32 = 240$$

240 ft

39. $S_n = \frac{n}{2}[2a_1 + (n - 1)d]$

$$S_{10} = \frac{10}{2}[2(5) + (10 - 1)3] = 185$$

41. $a_1 = 3, \quad d = 10 - 3 = 7$

$$S_n = \frac{n}{2}[2a_1 + (n - 1)d]$$

$$S_8 = \frac{8}{2}[2(3) + (8 - 1)7] = 220$$

43. $a_1 = 2(1) = 2, \quad a_{30} = 2(30) = 60$

$$S_n = \frac{n}{2}(a_1 + a_n)$$

$$S_{30} = \frac{30}{2}(2 + 60) = 930$$

45. $a_1 = 1, \quad d = 2$

$$S_n = \frac{n}{2}[2a_1 + (n - 1)d]$$

$$S_{30} = \frac{30}{2}[2(1) + (30 - 1)2] = 900$$

47. $a_k - a_j = (k - j)d$

$$8 - 5 = (8 - 40)d$$
$$3 = 4d$$
$$\frac{3}{4} = d$$

$$a_n = a_1 + (n - 1)d$$

$$5 = a_1 + (4 - 1)\frac{3}{4}$$

$$5 = a_1 + \frac{9}{4}$$

$$\frac{11}{4} = a_1$$

$$S_n = \frac{n}{2}[2a_1 + (n - 1)d]$$

$$S_{10} = \frac{10}{2}\left[2\left(\frac{11}{4}\right) + (10 - 1)\frac{3}{4}\right] = \frac{245}{4}$$

49. $a_1 = 2, \quad r = \frac{6}{2} = 3$

$$a_n = a_1 r^{n-1}$$
$$a_5 = 2(3)^{5-1} = 162$$

51. $a_1 = 1, \quad r = \frac{\frac{1}{3}}{1} = \frac{1}{3}$

$$a_n = a_1 r^{n-1}$$
$$a_6 = 1\left(\frac{1}{3}\right)^{6-1} = \frac{1}{243}$$

53. $a_n = a_1 r^{n-1}$

$$a_5 = 5(2)^{5-1} = 80$$

55. $a_1 = 5, \quad r = 2$

$$a_n = a_1 r^{n-1}$$
$$a_{12} = 5(2)^{12-1} = 10{,}240$$

$10,240

57. $a_1 = 3, \quad r = \frac{1}{3}$

$$S = \frac{a_1}{1 - r}$$

$$S = \frac{3}{1 - \frac{1}{3}} = \frac{9}{2}$$

59. $a_1 = \frac{1}{5}, \quad r = \frac{\frac{2}{5}}{\frac{1}{5}} = 2$

$|r| > 1$, sum does not exist

61. $a_1 = 0.64, \; r = 0.01$

$$S = \frac{a_1}{1 - r}$$

$$S = \frac{0.64}{1 - 0.01} = \frac{64}{99}$$

63. $8! = 8 \cdot 7 \cdot 6 \cdot 5 \cdot 4 \cdot 3 \cdot 2 \cdot 1 = 40{,}320$

65. $9! - 6! = 9 \cdot 8 \cdot 7 \cdot 6 \cdot 5 \cdot 4 \cdot 3 \cdot 2 \cdot 1$

$$- 6 \cdot 5 \cdot 4 \cdot 3 \cdot 2 \cdot 1$$
$$= 362{,}880 - 720$$
$$= 362{,}160$$

67. $(x - y)^5 = x^5 + \frac{5!}{4!1!}x^4(-y) + \frac{5!}{3!2!}x^3(-y)^2$

$+ \frac{5!}{2!3!}x^2(-y)^3 + \frac{5!}{1!4!}x(-y)^4 + (-y)^5$

$= x^5 - 5x^4y + 10x^3y^2 - 10x^2y^3 + 5xy^4 - y^5$

69. $(2a - 3b^2)^4 = (2a)^4 + \frac{4!}{3!1!}(2a)^3(-3b^2) + \frac{4!}{2!2!}(2a)^2(-3b^2)^2$

$+ \frac{4!}{1!3!}(2a)(-3b^2)^3 + (-3b^2)^4$

$= 16a^4 - 96a^3b^2 + 216a^2b^4 - 216ab^6 + 81b^8$

71. $(3a)^5 + \frac{5!}{4!1!}(3a)^4(-2b) + \frac{5!}{3!2!}(3a)^3(-2b)^2 + \frac{5!}{2!3!}(3a)^2(-2b)^3$

$= 243a^5 - 810a^4b + 1080a^3b^2 - 720a^2b^3$

73. 4th term: $r = 4$
$k = r - 1 = 4 - 1 = 3$
$n = 6$

$\frac{n!}{(n-k)!k!}x^{n-k}y^k = \frac{6!}{(6-3)!3!}(2a^2)^{6-3}(-3)^3$

$= \frac{6!}{3!3!}(2a^2)^3(-3)^3$

$= -4320a^6$

APPENDIX C PRACTICE TEST

1. (a) $x_1 = 3(1) + 1 = 4$
$x_2 = 3(2) + 1 = 7$
$x_3 = 3(3) + 1 = 10$
$x_9 = 3(9) + 1 = 28$

(b) $y_1 = 2(1)^3 - 1 = 1$
$y_2 = 2(2)^3 - 1 = 15$
$y_3 = 2(3)^3 - 1 = 53$
$y_9 = 2(9)^3 - 1 = 1457$

3. $\sum_{i=3}^{6}(2i^2 + 1) = [2(3)^2 + 1] + [2(4)^2 + 1] + [2(5)^2 + 1] + [2(6)^2 + 1]$
$= 19 + 33 + 51 + 73$
$= 176$

5. $a_n = a_1 r^{n-1}$

$A_5 = 4(2)^{5-1} = 64$

7. $a_1 = 2, \quad r = \frac{10}{2} = 5$

$$S_n = \frac{a_1(1 - r^n)}{1 - r}$$

$$S_4 = \frac{2(1 - 5^4)}{1 - 5} = 312$$

9. $a_1 = 12, \quad r = \frac{3}{5}$

$$S = \frac{a_1}{1 - r}$$

$$S = \frac{12}{1 - \frac{3}{5}} = 30$$

30 ft

11. 4th term: $r = 4$
$k = r - 1 = 4 - 1 = 3$
$n = 6$

$$\frac{n!}{(n - k)!k!}x^{n-k}y^k$$

$$= \frac{6!}{(6 - 3)!3!}(3a^2)^{6-3}(-1)^3$$

$$= \frac{6!}{3!3!}(3a^2)^3(-1)^3$$

$$= -540a^6$$